Universitext

Universitext

Universitext is a series of textbooks that presents material from a wide variety of mathematical disciplines at master's level and beyond. The books, often well class-tested by their author, may have an informal, personal even experimental approach to their subject matter. Some of the most successful and established books in the series have evolved through several editions, always following the evolution of teaching curricula, to very polished texts.

Thus as research topics trickle down into graduate-level teaching, first textbooks written for new, cutting-edge courses may make their way into *Universitext*.

More information about this series at http://www.springer.com/series/223

Douglas Farenick

Fundamentals of Functional Analysis

 Springer

Douglas Farenick
Department of Mathematics and Statistics
University of Regina
Regina, Saskatchewan, Canada

ISSN 0172-5939   ISSN 2191-6675 (electronic)
Universitext
ISBN 978-3-319-45631-7   ISBN 978-3-319-45633-1 (eBook)
DOI 10.1007/978-3-319-45633-1

Library of Congress Control Number: 2016954644

Mathematics Subject Classification (2010): 28-01, 46-01, 47-0, 54-01

Printed on acid-free paper

This Springer imprint is published by Springer Nature
The registered company is Springer International Publishing AG
The registered company address is: Gewerbestrasse 11, 6330 Cham, Switzerland

To Katherine

Preface

Functional analysis tightly interweaves concepts from analysis, topology, geometry, and algebra. This book explores the fabric of functional analysis by providing a careful presentation of the fundamental results that are essential for the contemporary mathematician. Organised into four thematic sections of (i) topology, (ii) measure theory and integration, (iii) Banach spaces, and (iv) operator theory, the treatment is entirely self-contained with the exception of a brief introductory section in Chapter 1 that draws attention to, but does not develop, the foundations upon which functional analysis rests (e.g., the axiom of choice and the completeness of the real number system).

The reader is assumed to have knowledge of basic real and complex analysis at the level of the classic monographs of Rudin [49] and Brown and Churchill [8] and to possess an understanding of algebra and linear algebra at the level of, say, Herstein's well-known book [33]. However, as it is virtually impossible to study the fundamental results of functional analysis without a working facility with topology and measure theory, these two subjects form the book's starting point. Both topics are introduced on the assumption that the reader has no prior knowledge of them. The reader who is familiar with these subjects can select a later entry point to the book, knowing that he or she may refer back to the topological and measure-theoretic results as needed.

The topology treated herein is essentially of the point-set variety, the first chapter of which covers the generic features in topology (subspaces, products, continuity, quotients) while examining some specific and important examples in detail (such as the Cantor set and the Cantor ternary function). The second half of the study of topology concentrates on those structures and results that are required by functional analysis: compactness and local compactness, Urysohn's Lemma, second countability, and the Stone-Čech compactification, for example. A novel feature of Chapter 2 is the inclusion of a purely set-theoretic proof of the Tietze extension theorem.

The primary examples of Banach spaces are drawn from vector spaces of integrable functions, and among the most important of such spaces are the ones in which the underlying measure space is a topological space. Therefore, the second thematic section of the book is devoted to the basics of measure theory, with particular emphasis on regular Borel measures and Lebesgue measure. Because spaces of complex measures also form a Banach space of interest, these are examined as well. The Lebesgue integral is also developed and the essential inequalities of analysis (Jensen, Hölder, Minkowski) are established.

The third part of the book is devoted to Banach and Hilbert spaces, duality, and convexity. Equipped with topology and measure theory, the important examples in functional analysis are studied, with L^p spaces and spaces of continuous complex-valued functions on a compact Hausdorff space being among these. The Riesz representation theorem for the dual of $C(X)$, where X is compact Hausdorff, is proved only in the case where the topology on X is metrisable. As it turns out, this is the only case that is required for subsequent results in the book, including Choquet's integral representations for elements of compact convex sets.

The final part of the book concerns operator theory, beginning with a study of bounded linear operators acting on Banach spaces and the spectral features of operators. Because spectral theory is complex-analytic in nature, it seems appropriate at this stage of the book to introduce and study spectral theory in the context of Banach algebras. Although the treatment is relatively brief, the main results about abelian Banach algebras are established, and a few applications to classical analysis are made. The final two chapters deal with operators acting on Hilbert spaces and self-adjoint algebras of such operators. The first of these two chapters provides the basic theory of Hilbert space operators and includes the spectral theory of compact normal operators, as well as a detailed analysis of the Banach space of trace-class operators. Matrices of operators and some of the most widely used operator inequalities are developed. The study of von Neumann algebras and C^*-algebras in the final chapter is approached as a natural continuation of basic Hilbert space operator theory and brings the book to a close by touching upon some of its topological and measure-theoretic beginnings.

What is not in this book? I did not touch upon any of the multivariable aspects of functional analysis, which explains the absence of product measures, Fubini's theorem, and tensor products. And while the core topics of functional analysis do come very close to those of harmonic analysis, I felt that the treatment of Haar measure and the Fourier transform, for example, would be better suited to books devoted to harmonic analysis. Some of the material that one would need for applications of functional analysis to partial differential equations or physics, such as unbounded operators and distributions, is also omitted. Although complex analysis has a major role in the theory of Banach algebras and non-self-adjoint operator algebras, my approach here has been to view the study of functional analysis as a continuation of real analysis, which accounts somewhat for the inclusion of measure theory and self-adjoint algebras of Hilbert space operators at the cost of excluding Banach spaces of analytic functions and, for example, holomorphic functional calculus.

This book can be read for self-study or be used as a text for a course on one of the thematic sections or for a sequence of courses. It may also be used as a reference work, keeping in mind the omission of certain significant topics as mentioned above.

I have discussed many aspects of functional analysis with numerous colleagues, friends, and students over the years, and I am grateful for all that I have learned from them. I wish to acknowledge, in particular, Angshuman Bhattacharya and Sarah Plosker, who patiently worked through some of the material on convexity with me, and Martín Argerami, who used early versions of the measure theory and integration chapters in his courses on real analysis.

In writing this book, I am especially indebted to Gwyneth Bergman for her editorial work on initial drafts of the manuscript and to anonymous reviewers who provided thoughtful criticism and suggestions. Lastly, I wish to thank Donna Chernyk and Mahmood Alaghmandan for getting me started on this project and for their gentle, regular reminders to get it completed.

Regina, Saskatchewan, Canada   Douglas Farenick
June 2016

Contents

Part I
Topology

Chapter 1
Topological Spaces

At its most intuitive level, topology is the study of geometric objects and their continuous deformations. Such objects might be concrete and easy to visualise, such as a disc or an annulus in the plane $\mathbb{R}^2$, or a sphere or torus in $\mathbb{R}^3$, while other such geometric objects might be a great deal more abstract in nature, as is the case with higher-dimensional spheres and tori, groups of unitary matrices, and the unit balls of the dual spaces of normed vector spaces. Fortunately, the concepts and methods of topology apply at a very general level, thereby providing us with rather deep insight into the nature of a variety of interesting mathematical structures.

This chapter is devoted to the introduction and study of generic features and constructions of spaces in topology. The study of topological spaces that have certain special properties is taken up in the second chapter. A topological space is defined by set-theoretic axioms; for this reason, we begin with a very brief word on set theory.

1.1 Sets and Partial Orderings

It is difficult to think of a branch of mathematics that does not depend crucially on the formalism and language of set theory. Elementary uses of set theory are so common as to be unnoticed: set union and intersection, subsets, Cartesian products, relations and functions, and so forth. However, certain subtle aspects of set theory (the Axiom of Choice, for example) lead to powerful tools (such as Zorn's Lemma) that are fundamental features of many areas of modern mathematics.

If one is not precise about what is meant by the term "set", then it is possible to introduce undesired paradoxes. In a non-rigorous discussion, it is common to refer to a set as a collection of objects in which elements of the collection are determined or

© Springer International Publishing Switzerland 2016
D. Farenick, *Fundamentals of Functional Analysis*, Universitext,
DOI 10.1007/978-3-319-45633-1_1

described in some definitive manner. The set H of all hydrogen atoms, for example, contains each and every hydrogen atom and no other elements; that is, $x \in H$ if and only if x is a hydrogen atom.

A paradox put forward by the philosopher Bertrand Russell shows that a more rigorous definition of set is required. Specifically, consider the "set"

$$R = \{x \mid x \text{ is a set and } x \notin x\}.$$

By definition, $R \in R$ if and only if $R \notin R$, and therein lies the paradox.

To avoid contradictory assertions such as Russell's paradox, various axioms for set theory have been developed to formally describe what sets are and what operations on sets may be carried out. The most widely used (and accepted) of these axiomatic systems are the Zermelo-Fraenkel Axioms. The ZF Axioms assert, for example, the existence of the empty set and the existence of an infinite set. These axioms also define precisely the concept of subset and they confirm that the familiar operations of set union and set intersection as valid operations in set theory.

If X and Y are sets, then the *Cartesian product* of X and Y is denoted by $X \times Y$, and consists of all ordered pairs (x, y) in which $x \in X$ and $y \in Y$. More generally, if $X_1, \ldots, X_n$ are sets, then $X_1 \times \cdots \times X_n$ is the set of all n-tuples $(x_1, \ldots, x_n)$ such each $x_j \in X_j$ for each $j = 1, \ldots, n$.

If Y is a subset of a set X, then the notation Y^c or $X \setminus Y$ will be used to denote the *complement of Y in X*: namely, the set of all $x \in X$ for which $x \notin Y$.

Recall the following basic fact regarding unions and intersections.

Proposition 1.1. *If $\{Y_\alpha\}_{\alpha \in \Lambda}$ and $\{Z_\beta\}_{\beta \in \Omega}$ are families of subsets of a set X, then*

$$\left(\bigcup_{\alpha \in \Lambda} Y_\alpha\right) \cap \left(\bigcup_{\beta \in \Omega} Z_\beta\right) = \bigcup_{(\alpha,\beta) \in \Lambda \times \Omega} Y_\alpha \cap Z_\beta, \text{ and}$$

$$\left(\bigcap_{\alpha \in \Lambda} Y_\alpha\right) \cup \left(\bigcap_{\beta \in \Omega} Z_\beta\right) = \bigcap_{(\alpha,\beta) \in \Lambda \times \Omega} Y_\alpha \cup Z_\beta.$$

With respect to Cartesian products, union and intersection satisfy

$$\left(\bigcup_{\alpha \in \Lambda} Y_\alpha\right) \times \left(\bigcup_{\beta \in \Omega} Z_\beta\right) = \bigcup_{(\alpha,\beta) \in \Lambda \times \Omega} Y_\alpha \times Z_\beta, \text{ and}$$

$$\left(\bigcap_{\alpha \in \Lambda} Y_\alpha\right) \times \left(\bigcap_{\beta \in \Omega} Z_\beta\right) = \bigcap_{(\alpha,\beta) \in \Lambda \times \Omega} Y_\alpha \times Z_\beta.$$

With respect to complements, DeMogan's Laws hold:

$$\left(\bigcup_{\alpha \in \Lambda} X_\alpha\right)^c = \bigcap_{\alpha \in \Lambda} X_\alpha^c \text{ and } \left(\bigcap_{\alpha \in \Lambda} X_\alpha\right)^c = \bigcup_{\alpha \in \Lambda} X_\alpha^c.$$

A more extensive review of the ZF Axioms will not be undertaken here because it is unnecessary for the subjects addressed in this book. However, it is worth noting that quite recent research in functional analysis has drawn substantially upon sophisticated techniques and results in abstract set theory, showing that set theory has a role in analysis which reaches far beyond the foundational role that we are focusing upon here.

In addition to the ZF Axioms, it is essential in functional analysis to use one additional axiom: the Axiom of Choice. In informal language, the Axiom of Choice asserts that if X is a set of pairwise disjoint nonempty sets, then there exists a set C that contains exactly one element from each set in X. (Note, here, that X is a set of sets.) Conceptually, C is constructed by "choosing" one element from each set in X. The Zermelo-Fraenkel Axioms together with the Axiom of Choice are referred to as the ZFC Axioms.

The existence of the set $\mathbb{N} = \{1, 2, 3, \ldots\}$ of natural numbers results from an axiomatic system developed by G. Peano. The "$\ldots$" in $\{1, 2, 3, \ldots\}$ is not, of course, rigorously defined; making formal sense of this unspecified portion of $\mathbb{N}$ is exactly what the Peano axioms were created to do. Without pursuing these axioms in more detail, let us accept as given that the Peano axioms establish $\mathbb{N}$ as a set in accordance with the ZF axioms. (See, for example, [21, Chapter 1] for a detailed treatment of these facts.)

In contrast to the subtle nature of the axioms that define a set, the definitions of relations and functions are simple. A relation R from a set X to a set Y is a subset of $X \times Y$, whereas a function is a relation F from X to Y such that, for any $x \in X$, if $y_1, y_2 \in Y$ satisfy $(x, y_1) \in F$ and $(x, y_2) \in F$, then necessarily $y_1 = y_2$. It is cumbersome to express functions as subsets of Cartesian products, and we normally write $y = f(x)$ if $(x, y) \in F$ (and so the symbol f depends on the set F).

Recall that if X and Y are sets, then a function $f : X \to Y$ is injective if $f(x_1) = f(x_2)$ holds only for $x_1 = x_2$, and is surjective if for every $y \in Y$ there is at least one $x \in X$ such that $y = f(x)$. A function that is both injective and surjective is said to be bijective.

Definition 1.2. Two sets, X and Y, are in *bijective correspondence* if there exists a bijection $f : X \to Y$.

The following theorem is a useful criterion for bijective correspondence.

Theorem 1.3 (Schroeder-Bernstein). *If, for sets X and Y, there exist injective functions $g : X \to Y$ and $h : Y \to X$, then X and Y are in bijective correspondence.*

In comparing sets X to the set $\mathbb{N}$ of natural numbers, the Schroeder-Bernstein Theorem has the following simpler form.

Theorem 1.4. *The following conditions are equivalent for an infinite set X:*

1. *there exists an injection $g : X \to \mathbb{N}$;*
2. *there exists a surjection $h : \mathbb{N} \to X$;*
3. *there exists a bijection $f : X \to \mathbb{N}$.*

Recall the concept of countability, which is an essential idea in topology and analysis.

Definition 1.5. A set X is *countable* if X is in bijective correspondence with some subset of $\mathbb{N}$.

Theorem 1.4 allows one to deduce, for example, that subsets of countable sets are countable and that the union of a countable collection of countable sets is a countable set.

The formal notion of partial order allows for a qualitative comparison of certain elements in a set, and it also leads to the important concept of a maximal element.

Definition 1.6. A *partial order* on a set $\mathfrak{S}$ is a relation denoted by $\preceq$ that has the following properties:

1. (reflexivity) for all $\alpha \in \mathfrak{S}, \alpha \preceq \alpha$;
2. (antisymmetry) for all $\alpha, \beta \in \mathfrak{S}, \alpha \preceq \beta$ and $\beta \preceq \alpha$ implies that $\beta = \alpha$;
3. (transitivity) for all $\alpha, \beta, \gamma \in \mathfrak{S}, \alpha \preceq b$ and $\beta \preceq \gamma$ implies that $\alpha \preceq \gamma$.

If, in addition, a partial order $\preceq$ satisfies property (iv) below, then $\preceq$ is called a *linear order*.

(iv) (comparability) for all $\alpha, \beta \in \mathfrak{S}$, either $\alpha \preceq \beta$ or $\beta \preceq \alpha$.

A partial order $\preceq$ on a set $\mathfrak{S}$ can be used to assert that β is in some sense larger than α, if it happens to be true that $\alpha \preceq \beta$. This manner of thinking leads to a natural notion of largest element.

Definition 1.7. In a partially ordered set $\mathfrak{S}$, an element $\alpha \in \mathfrak{S}$ is said to be a *maximal element* if for $\gamma \in \mathfrak{S}$ the relation $\alpha \preceq \gamma$ implies that $\gamma = \alpha$. If $\mathfrak{E} \subseteq \mathfrak{S}$, then an element $\alpha \in \mathfrak{S}$ is an *upper bound* for $\mathfrak{E}$ if $\beta \preceq \alpha$ for every $\beta \in \mathfrak{E}$.

Because a partially ordered set need not be linearly ordered, if a partially ordered set $\mathfrak{S}$ has a maximal element, then this maximal element may or may not be unique. The existence of a maximal element is established by Zorn's Lemma.

Theorem 1.8 (Zorn's Lemma). *If $\mathfrak{S}$ is a nonempty partially ordered set such that every linearly ordered subset $\mathfrak{E} \subseteq \mathfrak{S}$ (where the linear order on $\mathfrak{E}$ is inherited from the partial order on $\mathfrak{S}$) has an upper bound in $\mathfrak{S}$, then $\mathfrak{S}$ has a maximal element.*

A detailed discussion of the facts reviewed above can be found in [21, 41, 49].

1.2 Completeness of the Real Number System

In the foundations of arithmetic, one begins with the Peano axioms that define the set $\mathbb{N}$ of natural numbers and then proceeds to construct the set of $\mathbb{Z}$ of integers by considering formal differences $n - m$ of natural numbers $n, m \in \mathbb{N}$. The set $\mathbb{Z}$ of integers is more than just a set—it is an arithmetic system as well. Specifically, in the language of modern algebra, $\mathbb{Z}$ is a commutative unital ring and, therefore, it has a field of fractions, which is the countable set $\mathbb{Q}$ of rational numbers. Constructing the real numbers system $\mathbb{R}$ from the rational field $\mathbb{Q}$ is a fairly elaborate process due to R. Dedekind; see [49] for the details of the construction. The end result, namely the uncountable set $\mathbb{R}$ of real numbers, is once again a field, but it is no ordinary field, as $\mathbb{R}$ is distinguished by its completeness as both a metric space and an ordered space.

By "metric" is meant the usual distance between real numbers; that is, if $x, y \in \mathbb{R}$, then the distance between x and y is the absolute value $|x - y|$ of $x - y$. By "order" is meant the usual order on the real numbers whereby, if $x, y \in \mathbb{R}$, then exactly one of the following three statements holds true: (i) $x < y$, (ii) $y < x$, or (iii) $y = x$.

Definition 1.9. If S is a nonempty set of real numbers, then

1. S is *bounded above* if there exists a real number z such that $s \le z$ for every $s \in S$, and
2. S is *bounded below* if there exists a real number y such that $y \le s$ for every $s \in S$.

A real number z that satisfies condition (1) in Definition 1.9 is called an *upper bound* for S, while the real number y that satisfies condition (2) is called a *lower bound* for S.

Definition 1.10. A real number z is the *supremum* (or least upper bound) of a nonempty set S of real numbers if

1. z is an upper bound for S, and
2. $z \le g$ for every upper bound g of S.

The definition of *infimum* (or greatest lower bound) of S is similarly defined.

The completeness of $\mathbb{R}$ as an ordered space (which is to say that suprema exist) is one of the most important and extraordinary features of the real number system.

Theorem 1.11 (Least Upper Bound Principle). *If $S \subseteq \mathbb{R}$ is a nonempty subset that is bounded above, then S has a supremum in $\mathbb{R}$.*

To discuss the completeness of $\mathbb{R}$ as a metric space, recall the formal definition of sequence.

Definition 1.12. An *infinite sequence* in a set X is a function $f : \mathbb{N} \to X$.

Definition 1.12 makes no claims about whether the function $f : \mathbb{N} \to X$ is injective or surjective, as neither need hold. It is typical to denote the element $f(n) \in X$ by x_n, for each $n \in \mathbb{N}$, and to write the infinite sequence f as $\{x_n\}_{n \in \mathbb{N}}$.

Throughout this book, we shall drop the adjective "infinite" and refer to infinite sequences as *sequences*.

Definition 1.13. A sequence $\{x_n\}_{n\in\mathbb{N}}$ is:

1. *bounded above* if there is a real number z such that $x_n \leq z$ for every $n \in \mathbb{N}$;
2. *bounded below* if there is a real number y such that $y \leq x_n$ for every $n \in \mathbb{N}$;
3. *bounded* if it is bounded above and below;
4. *convergent* if there is a real number x with the property that for each $\varepsilon > 0$ there exists $N \in \mathbb{N}$ such that $|x - x_n| < \varepsilon$ for all $n \geq N$; and
5. a *Cauchy sequence* for each $\varepsilon > 0$ there exists $N \in \mathbb{N}$ such that $|x_m - x_n| < \varepsilon$ for all $m, n \geq N$.

The following theorem establishes the completeness of $\mathbb{R}$ as a metric space.

Theorem 1.14. *Every Cauchy sequence of real numbers is convergent.*

Theorems 1.11 and 1.14, the proofs of which can be found in [49], underly all of real and functional analysis.

Definition 1.15. A set X has the *cardinality of the continuum* if X is in bijective correspondence with $\mathbb{R}$.

The cardinal number c shall henceforth denote the cardinality of the continuum.

1.3 Topological Spaces

Although the notion of a "topology" is abstract, the definition itself is motivated by our experience with the real numbers system. For example, an intersection of a finite number of open intervals results in the empty set or another open interval; and while the union of a family of open intervals need not be an open interval, it is clear that such a set retains features exhibited by open intervals. Such basic observations lead to the following definition.

Definition 1.16. A *topology* on a set X is a collection $\mathscr{T}$ of subsets of X such that:

1. $\mathscr{T}$ contains both the empty set $\emptyset$ and the set X itself;
2. $\mathscr{T}$ is closed under arbitrary unions:

$$\bigcup_{\alpha\in\Lambda} U_\alpha \in \mathscr{T},$$

for every family $\{U_\alpha\}_{\alpha\in\Lambda}$ of sets $U_\alpha \in \mathscr{T}$;
3. $\mathscr{T}$ is closed under finite intersections:

$$\bigcap_{k=1}^{n} U_k \in \mathscr{T},$$

for every finite collections $\{U_k\}_{k=1}^n$ of sets $U_k \in \mathcal{T}$.

The pair $(X, \mathcal{T})$, but more often just X itself, is called a *topological space*. Elements $U \in \mathcal{T}$ are called *open sets*.

Example 1.17 (Examples of Topologies). *The following collections of sets form a topology on a nonempty set X:*

1. *the* indiscrete topology $\mathcal{T}_{\text{indiscr}} = \{\emptyset, X\}$;
2. *the* discrete topology $\mathcal{T}_{\text{discr}} = \{U \mid U \subseteq X\}$, *which is the power set* $\mathcal{P}(X)$ *of X*;
3. *the* co-finite topology $\mathcal{T}_{\text{co-fin}} = \{\emptyset\} \cup \{F^c \mid F \subseteq X, F \text{ is finite}\}$, *which consists of the empty set and the complements of all finite subsets of X; and*
4. *the* co-countable topology $\mathcal{T}_{\text{co-cntbl}} = \{\emptyset\} \cup \{F^c \mid F \subseteq X, F \text{ is countable}\}$, *consisting of the empty set and the complements of all countable subsets of X.*

Proof. The indiscrete and discrete topologies trivially satisfy the axioms for a topology on X.

To verify that $\mathcal{T}_{\text{co-cntbl}}$ is indeed a topology on X, note that $\emptyset \in \mathcal{T}_{\text{co-cntbl}}$ by construction and that $X = \emptyset^c \in \mathcal{T}_{\text{co-cntbl}}$ because the empty set $\mathcal{T}_{\text{co-cntbl}}$ is countable. If we now suppose that $\{U_\alpha\}_\alpha$ is a family of sets $U_\alpha \in \mathcal{T}_{\text{co-cntbl}}$, then for each α there is a countable subset $F_\alpha \subseteq X$ such that $U_\alpha = F_\alpha^c$. Therefore,

$$\bigcup_\alpha U_\alpha = \bigcup_\alpha F_\alpha^c = \left(\bigcap_\alpha F_\alpha\right)^c \in \mathcal{T}_{\text{co-cntbl}},$$

because $\bigcap_\alpha F_\alpha$ is a subset of each of the sets F_α and is, therefore, countable. This shows that $\mathcal{T}_{\text{co-cntbl}}$ is closed under arbitrary unions. Finally, if $U_1, \dots, U_n \in \mathcal{T}_{\text{co-cntbl}}$, and if $F_1, \dots, F_n \subseteq X$ are countable subsets for which each $U_j = F_j^c$, then

$$\bigcap_{j=1}^n U_j = \bigcap_{j=1}^n F_j^c = \left(\bigcup_{j=1}^n F_j\right)^c \in \mathcal{T}_{\text{co-cntbl}}$$

because a finite union of countable sets is countable. Hence, $\mathcal{T}_{\text{co-cntbl}}$ is closed under finite unions.

The proof that $\mathcal{T}_{\text{co-fin}}$ is a topology on X is similar and simpler. $\square$

Definition 1.18. If a set X is endowed with two topologies, say $\mathcal{T}$ and $\mathcal{T}'$, and if $\mathcal{T} \subseteq \mathcal{T}'$, then we say that $\mathcal{T}$ is *coarser than* $\mathcal{T}'$, and that $\mathcal{T}'$ is *finer than* $\mathcal{T}$.

Using the terminology above, the indiscrete topology is the coarsest topology that a set X admits, whereas the discrete topology is the finest topology on X.

To arrive at further examples of topological spaces, it is useful to have tools that offer simple ways to prescribe what is meant by an open set. Two such tools are to be found in the notions of basis and subbasis for a topology.

Definition 1.19. A collection $\mathscr{B}$ of subsets of a set X is called a *basis of subsets of* X if:

1. for every $x \in X$, there is a $B \in \mathscr{B}$ with $x \in B$; and
2. for all $B_1, B_2 \in \mathscr{B}$, and every $x \in B_1 \cap B_2$, there is a $B_3 \in \mathscr{B}$ such that $x \in B_3 \subseteq B_1 \cap B_2$.

The relevance of the definition of basis is revealed by the following proposition.

Proposition 1.20. *If $\mathscr{B}$ is a basis of subsets of X and if $\mathscr{T}$ is the collection of all subsets $U \subseteq X$ with the property that for each $x \in U$ there is a $B \in \mathscr{B}$ with $x \in B \subseteq U$, then $\mathscr{T}$ is a topology on X.*

Proof. It is clear that both the empty set $\emptyset$ and X satisfy the requirements for membership in $\mathscr{T}$.

Suppose that $\{U_\alpha\}_{\alpha \in \Lambda}$ is a family of sets $U_\alpha \in \mathscr{T}$ and let $U = \bigcup_\alpha U_\alpha$. Choose any $x \in U$; thus, $x \in U_\alpha$ for some $\alpha \in \Lambda$. Because $U_\alpha \in \mathscr{T}$, there is a $B \in \mathscr{B}$ such that $x \in B \subseteq U_\alpha$. As $U_\alpha \subseteq U$, we conclude that $U \in \mathscr{T}$ and, therefore, that $\mathscr{T}$ is closed under arbitrary unions.

To show that $\mathscr{T}$ is closed under finite intersections, it is enough to show, for all $U_1, U_2 \in \mathscr{T}$, that $U_1 \cap U_2 \in \mathscr{T}$, and to then proceed inductively. If $U_1, U_2 \in \mathscr{T}$ satisfy $U_1 \cap U_2 = \emptyset$, then $U_1 \cap U_2 \in \mathscr{T}$ trivially. Therefore, suppose that there exists an $x \in U_1 \cap U_2$. Because $\mathscr{B}$ is a basis, there are $B_1, B_2 \in \mathscr{B}$ such that $x \in B_k \subseteq U_k$, for $k = 1, 2$, and there is a $B_3 \in \mathscr{B}$ such that $x \in B_3 \subseteq B_1 \cap B_2$. Hence, $x \in B_3 \subseteq B_1 \cap B_2 \subseteq U_1 \cap U_2$, which proves that $U_1 \cap U_2 \in \mathscr{T}$. Proceeding by induction, we conclude that $\mathscr{T}$ is closed under finite intersections. $\square$

Proposition 1.20 illustrates that to specify a topology on a set X, it is sufficient to specify a basis $\mathscr{B}$ of subsets of X for the topology.

Definition 1.21. If $\mathscr{B}$ is a basis of subsets of X, and if $\mathscr{T}$ is the collection of all subsets $U \subseteq X$ for which each $x \in U$ is contained in at least one $B \in \mathscr{B}$ satisfying $B \subseteq U$, then $\mathscr{T}$ is called the *topology generated by* $\mathscr{B}$.

To this point we have demonstrated that a basis of subsets determine a topology. Conversely, every topology $\mathscr{T}$ on X admits a basis $\mathscr{B}$ of subsets of X such that $\mathscr{T}$ is the topology generated by $\mathscr{B}$.

Proposition 1.22. *If $\mathscr{T}$ is a topology on X, and if $\mathscr{B}$ is the collection of subsets with the property that, for every $U \in \mathscr{T}$ and each $x \in U$, there is a $B \in \mathscr{B}$ with $x \in B \subseteq U$, then $\mathscr{B}$ is a basis of subsets and $\mathscr{T}$ is the topology generated by $\mathscr{B}$.*

Proof. Exercise 1.103. $\square$

In light of Propositions 1.20 and 1.22, the following definition is made.

Definition 1.23. If $\mathscr{T}$ is a topology on X, then a *basis for the topology $\mathscr{T}$* is a collection $\mathscr{B}$ of subsets of X such that (i) $\mathscr{B}$ is a basis of subsets of X (in the sense of Definition 1.19) and (ii) $\mathscr{T}$ is the topology generated by $\mathscr{B}$ (in the sense of Definition 1.22). The elements of a basis $\mathscr{B}$ for a topology $\mathscr{T}$ on X are called *basic open sets*.

Proposition 1.24. *If $\mathcal{T}$ is a topology on X, and if $\mathcal{B}$ is a basis for the topology $\mathcal{T}$, then the following statements are equivalent for a subset $U \subseteq X$:*

1. *$U \in \mathcal{T}$;*
2. *there is a family $\{B_\alpha\}_\alpha$ of subsets $B_\alpha \in \mathcal{B}$ such that $U = \bigcup_\alpha B_\alpha$.*

Proof. Exercise 1.104. □

Proposition 1.20 and our knowledge of the real number system lead us to endow the set $\mathbb{R}$ of real numbers with a topology.

Definition 1.25. Assume that $a, b \in \mathbb{R}$ with $a < b$.

1. The subset of $\mathbb{R}$ denoted by (a, b), and defined by $(a, b) = \{x \in \mathbb{R} \mid a < x < b\}$, is called an *open interval*.
2. The subset of $\mathbb{R}$ denoted by $[a, b]$, and defined by $[a, b] = \{x \in \mathbb{R} \mid a \leq x \leq b\}$, is called a *closed interval*.
3. The subsets of $\mathbb{R}$ denoted by $[a, b)$ and $(a, b]$, and defined by $[a, b) = \{x \in \mathbb{R} \mid a \leq x < b\}$ and $(a, b] = \{x \in \mathbb{R} \mid a < x \leq b\}$, are called *half-open intervals*.

The term "open" in the definition above is the traditional terminology of calculus and real analysis and does not, *a priori*, refer to an open set in the sense of topology. However, in an appropriate topology on $\mathbb{R}$, these open intervals will in fact be open sets.

Proposition 1.26. *Let $\mathcal{B}$ be the set of all finite open intervals with rational end points. That is, assume that*

$$\mathcal{B} = \{(p, q) \subset \mathbb{R} \mid p, q \in \mathbb{Q}, \, p < q\}.$$

Then $\mathcal{B}$ is a basis of subsets of $\mathbb{R}$.

Proof. Exercise 1.105. □

Definition 1.27. The topology on the real number system $\mathbb{R}$ generated by the basis

$$\mathcal{B} = \{(p, q) \mid p, q \in \mathbb{Q}, \, p < q\}$$

of subsets of $\mathbb{R}$ is called the *standard topology of $\mathbb{R}$.*

In elementary real analysis, open sets in $\mathbb{R}$ are defined in a different fashion. The next proposition reconciles these two definitions, showing that the standard topology on $\mathbb{R}$ yields open sets that are familiar from calculus and real analysis.

Proposition 1.28. *If $\mathbb{R}$ is endowed with the standard topology, then the following statements are equivalent for a subset $U \subseteq \mathbb{R}$:*

1. *U is a open set;*
2. *for every $x \in U$ there is a $\varepsilon > 0$ such that*

$$(x - \varepsilon, x + \varepsilon) \subseteq U.$$

Proof. Suppose that assertion (1) holds. Suppose that $U \subseteq \mathbb{R}$ is an open set, and let $\mathscr{B}$ denote by $\mathscr{B}$ the basis of Proposition 1.26 for the standard topology $\mathscr{T}$ on $\mathbb{R}$. Choose $x \in U$. As $\mathscr{B}$ is a basis for $\mathscr{T}$, there exists $B \in \mathscr{B}$ such that $x \in B \subseteq U$. By definition, there are $p, q \in \mathbb{Q}$ such that $B = (p,q)$. If $\varepsilon = \min\{q - x, x - p\}$, then $(x - \varepsilon, x + \varepsilon) \subseteq (p,q)$. Hence,

$$(x - \varepsilon, x + \varepsilon) \subseteq (p, q) \subseteq U.$$

Conversely, suppose that statement (2) holds, and that $U \subseteq \mathbb{R}$ satisfies hypothesis (2). Thus, if $x \in U$, then there is a $\varepsilon > 0$ such that $(x - \varepsilon, x + \varepsilon) \subseteq U$. Between any two real numbers there is a rational number, and so let $p_x, q_x \in \mathbb{Q}$ be such that $x - \varepsilon < p_x < x$ and $x < q_x < x + \varepsilon$. Then $(p_x, q_x) \in \mathscr{B}$ and $x \in (p_x, q_x) \subset (x - \varepsilon, x + \varepsilon) \subseteq U$. Continuing this procedure for each $x \in U$ leads to:

$$U \subseteq \bigcup_{x \in U}(p_x, q_x) \subseteq U.$$

The inclusions above show that U is a union of the family $\{(p_x, q_x)\}_{x \in U}$ of basic open sets; hence, U is an open set in the standard topology of $\mathbb{R}$.   □

Corollary 1.29. *Open intervals are open sets in the standard topology of $\mathbb{R}$.*

The order completeness of the real number system allows for a description of open sets in $\mathbb{R}$ in terms of pairwise disjoint open intervals.

Proposition 1.30 (Cantor's Lemma). *If $U \subseteq \mathbb{R}$ is open in the standard topology of $\mathbb{R}$, then there is a countable family $\{J_n\}_{n \in \mathbb{N}}$ such that:*

1. *J_n is an open interval, for all n;*
2. *$J_n \cap J_m = \emptyset$, if $m \neq n$; and*
3. *$\bigcup_n J_n = U$.*

Proof. For each $x \in U$ there exists $\varepsilon > 0$ such that $(x - \varepsilon, x + \varepsilon) \subseteq U$, by Proposition 1.28. Therefore, if $x \in U$, then

$$a_x = \inf\{u \in \mathbb{R} \,|\, (u, x) \subseteq U\} \text{ and } b_x = \sup\{z \in \mathbb{R} \,|\, (x, z) \subseteq U\}$$

are well defined. It may be that $a_x = -\infty$ or $b_x = +\infty$, or both. If b_x is finite, then $b_x \notin U$; likewise, if a_x is finite, then $a_x \notin U$. Hence, each $x \in U$ determines an open interval $(a_x, b_x) \subseteq U$, which shall be denoted by $J(x)$. Observe that

$$U = \bigcup_{x \in U} J(x)$$

and that $J(x) = J(y)$, for all $y \in J(x)$. Indeed, if $x_1, x_2 \in U$, then either $J(x_1) = J(x_2)$ or $J(x_1) \cap J(x_2) = \emptyset$. Thus, U is the union of a family of pairwise disjoint open intervals. In each open interval there is a rational number, and so each of these intervals $J(x)$ can be labeled by a rational number. Because there are countably many rational numbers, there are also countably many distinct intervals of the type $J(x)$. □

While the use of a basis to specify a topology on X is convenient, it is often even more convenient to define a topology by way of a much smaller collection of sets.

Definition 1.31. A collection $\mathscr{S}$ of subsets of X is a *subbasis of subsets of X* if $\bigcup_{S \in \mathscr{S}} S = X$.

The concept of subbasis is important in topology because of the next result.

Proposition 1.32. *If $\mathscr{S}$ is a subbasis of subsets of X, then*

$$\mathscr{B} = \left\{ \bigcap_{j=1}^{n} S_j \,|\, n \in \mathbb{N}, S_1, \ldots, S_n \in \mathscr{S} \right\}$$

is a basis of subsets of X.

Proof. Exercise 1.108. □

1.4 Metric Topologies

Metrics provide a measure of the distance between any two points in a set; as a consequence, metrics can be used to specify a topology on the set.

Definition 1.33. A *metric* on a set X is a function $d : X \times X \to [0, \infty)$ such that, for all $x, y, z \in X$,

1. $d(x, x) = 0$,
2. $d(x, y) = 0$ only if $y = x$,
3. $d(x, y) = d(y, x)$, and
4. $d(x, z) \leq d(x, y) + d(y, z)$.

The pair (X, d) is called a *metric space*.

The inequality $d(x, z) \leq d(x, y) + d(y, z)$ in Definition 1.33 is called the *triangle inequality*.

Less formally, one can refer to X itself as a metric space, rather than the pair (X, d), if it is understood that d is the underlying metric on X.

The Euclidean metric on $\mathbb{R}^2$ is derived from the Pythagorean theorem, and it admits a formulation in higher dimensions as well.

Definition 1.34. The *Euclidean metric* on $\mathbb{R}^n$ is the function $d_2 : \mathbb{R}^n \times \mathbb{R}^n \to [0, \infty)$ defined by

$$d_2(x,y) = \sqrt{\sum_{j=1}^{n}(x_j - y_j)^2}, \quad \forall x, y \in \mathbb{R}^n. \tag{1.1}$$

Proposition 1.35. *The Euclidean metric on $\mathbb{R}^n$ is a metric in the sense of* Definition 1.33.

Proof. The function d_2 in (1.1) plainly satisfies the first three conditions of Definition 1.33. All that remains is to prove the triangle inequality. To this end, an intermediate inequality is required.

Suppose that $u, v \in \mathbb{R}^n$ and $u \neq 0$. Let $f_{u,v}$ denote the quadratic polynomial

$$f_{u,v}(t) = \sum_{j=1}^{n}(tu_j + w_j)^2 = t^2 \sum_{j=1}^{n} u_j^2 + 2t \sum_{j=1}^{n} u_j w_j + \sum_{j=1}^{n} w_j^2.$$

Thus, $f_{u,v}(t) \geq 0$ for all $t \in \mathbb{R}$ and $f_{u,v}(t_0) = 0$ if and only if $t_0 u_j = -w_j$ for every $j = 1, \ldots, n$. Therefore, because $u \neq 0$, there is at most one $t_0 \in \mathbb{R}$ that satisfies $f_{u,v}(t_0) = 0$. On the other hand, the quadratic formula yields

$$t_0 = \frac{-2\sum_j u_j w_j \pm \sqrt{(2\sum_j u_j w_j)^2 - 4(\sum_j u_j^2)(\sum_j w_j^2)}}{2\sum_j u_j^2},$$

and so the discriminant $(2\sum_j u_j w_j)^2 - 4(\sum_j u_j^2)(\sum_j w_j^2)$ is a nonpositive real number. Hence,

$$(\sum_{j=1}^{n} u_j w_j)^2 \leq (\sum_{j=1}^{n} u_j^2)(\sum_{j=1}^{n} w_j^2),$$

implying that

$$\left| \sum_{j=1}^{n} u_j v_j \right| \leq \sqrt{\left(\sum_{j=1}^{n} u_j^2\right)\left(\sum_{j=1}^{n} v_j^2\right)}.$$

If $x, y, z \in \mathbb{R}^n$, then

$$d_2(x,z)^2 = \sum_{i=1}^{n}(x_i - z_i)^2 \sum_{i=1}^{n}([x_i - y_i] + [y_i - z_i])^2$$

$$= \sum_{i=1}^{n}(x_i - y_i)^2 + 2\sum_{i=1}^{n}(x_i - y_i)(y_i - z_i) + \sum_{i=1}^{n}(y_i - z_i)^2$$

$$\leq \sum_{i=1}^{n}(x_i - y_i)^2 + 2\sqrt{\sum_{i=1}^{n}(x_i - y_i)^2 \sum_{i=1}^{n}(y_i - z_i)^2} + \sum_{i=1}^{n}(y_i - z_i)^2$$

$$= \left(\sqrt{\sum_{i=1}^{n}(x_i - y_i)^2} + \sqrt{\sum_{i=1}^{n}(y_i - z_i)^2} \right)^2$$

$$= (d_2(x,y) + d_2(y,z))^2.$$

Hence, the d_2 is indeed a metric in the sense of Definition 1.33. $\qquad\square$

Corollary 1.36. *The formula $d(x,y) = |x - y|$, for $x, y \in \mathbb{R}$, defines a metric on $\mathbb{R}$.*

Proof. The function d is simply the Euclidean metric d_2 on $\mathbb{R}^n$ in the case where $n = 1$. $\qquad\square$

Definition 1.37. The inequality

$$\left| \sum_{j=1}^{n} u_j v_j \right| \leq \sqrt{\left(\sum_{j=1}^{n} u_j^2 \right) \left(\sum_{j=1}^{n} v_j^2 \right)}, \qquad (1.2)$$

for real numbers $u_1, \ldots, u_n, v_1, \ldots, v_n$ is called the *Cauchy-Schwarz* inequality.

The Euclidean metric is not the only metric of interest.

Proposition 1.38. *The following functions d_1 and d_∞ are metrics on $\mathbb{R}^n$:*

$$d_1(x,y) = \sum_{j=1}^{n} |x_j - y_j| \qquad (1.3)$$

$$d_\infty(x,y) = \max_{1 \leq j \leq n} |x_j - y_j|. \qquad (1.4)$$

Proof. Exercise 1.111. $\qquad\square$

The reason for labelling these metrics as d_1, d_2, and d_∞ will be apparent in subsequent chapters, particularly in connection with Banach space theory.

Proposition 1.39. *If d is a metric on a set X and if*

$$\mathscr{B}_d = \{B_r(x) \mid x \in X, r \in \mathbb{R}, r > 0\}, \tag{1.5}$$

where $B_r(x) = \{y \in X \mid d(x,y) < r\}$, then $\mathscr{B}$ is a basis.

Proof. We are to verify that $\mathscr{B}_d$ has the following properties:

1. for each $x \in X$ there is a $B \in \mathscr{B}_d$ such that $x \in B$;
2. for every $B_1, B_2 \in \mathscr{B}_d$ and each $x \in B_1 \cap B_2$, there is a set $B_3 \in \mathscr{B}_d$ such that $x \in B_3 \subseteq B_1 \cap B_2$.

The first item is obvious, as $x \in B_r(x)$ for every $x \in X$ and every $r > 0$. Assume now that $B_1, B_2 \in \mathscr{B}_d$ and $x \in B_1 \cap B_2$. There are $x_j \in X$ and $r_j \in \mathbb{R}$ such that $B_j = B_{r_j}(x_j)$. Thus, $d(x,x_j) < r_j$ for each j, and so $0 < s_j = r_j - d(x,x_j)$. Let $r = \min\{s_1, s_2\}$. Then $x \in B_r(x) \subseteq B_{r_1}(x_1) \cap B_{r_2}(x_2)$. $\qquad\square$

Definition 1.40. The topology $\mathscr{T}_d$ on X induced by the basis $\mathscr{B}_d$ of Proposition 1.39 is called a *metric topology*.

Thus, a metric space (X,d) is a topological space $(X, \mathscr{T}_d)$.

The proof of Proposition 1.28 is easily adapted to the case of metric spaces to yield that following characterisation of open sets.

Proposition 1.41. *In a metric space (X,d), a subset $U \subseteq X$ is an open set if and only if for each $x \in U$ there is an $\varepsilon > 0$ such that $B_\varepsilon(x) \subseteq U$.*

Corollary 1.42. *The metric topology on $\mathbb{R}$ induced by the metric $d(x,y) = |x-y|$, for $x, y \in \mathbb{R}$, coincides with the standard topology on $\mathbb{R}$.*

Proof. Propositions 1.28 and 1.41 indicate that the metric topology and the standard topology have exactly the same open sets. $\qquad\square$

It is important to note that different metrics on a set X can induce the same topology.

Example 1.43. *The metrics d_1 and d_2 on $\mathbb{R}^n$ induce the same topology; that is, $\mathscr{T}_{d_1} = \mathscr{T}_{d_2}$.*

Proof. First notice that if $\alpha_1, \ldots, \alpha_n \in \mathbb{R}$ are nonnegative, then by a binomial expansion we see that

$$(\alpha_1 + \cdots + \alpha_n)^2 \geq \alpha_1^2 + \ldots \alpha_n^2.$$

Taking square roots yields

$$\sum_{j=1}^{n} \alpha_j \geq \left(\sum_{j=1}^{n} \alpha_j^2 \right)^{1/2},$$

which implies that $d_2(x,y) \leq d_1(x,y)$ for all $x, y \in \mathbb{R}^n$.

On the other hand, the Cauchy-Schwarz inequality (1.2) yields

$$\sum_{j=1}^{n} \alpha_j = \sum_{j=1}^{n} \alpha_j(1) \leq \left(\sum_{j=1}^{n} \alpha_j^2\right)^{1/2} \left(\sum_{j=1}^{n} 1^2\right)^{1/2} = \sqrt{n} \left(\sum_{j=1}^{n} \alpha_j^2\right)^{1/2}.$$

Hence, $d_1(x,y) \leq \sqrt{n} d_2(x,y)$, for all $x, y \in \mathbb{R}^n$.

Let $B_r^{d_j}(x)$ denote a basic open set in $\mathscr{T}_{d_j}$. To show that $\mathscr{T}_{d_1} = \mathscr{T}_{d_2}$, it is enough, by Proposition 1.24, to show that $B_r^{d_1}(x) \in \mathscr{T}_{d_2}$ and $B_r^{d_2}(x) \in \mathscr{T}_{d_1}$ for all $x \in X$ and $r > 0$. To this end, if $y \in B_r^{d_1}(x)$, then $d_2(x,y) \leq d_1(x,y) < r$ implies that $y \in B_r^{d_2}(x)$; thus, $B_r^{d_1}(x) \subseteq B_r^{d_2}(x) \in \mathscr{T}_{d_2}$. Similarly, if $y \in B_r^{d_2}(x)$, then $d_1(x,y) \leq \sqrt{n} d_2(x,y) < \sqrt{n} r$ implies that $y \in B_{\sqrt{n}r}^{d_1}(x)$; that is, $B_r^{d_2}(x) \subseteq B_{\sqrt{n}r}^{d_1}(x) \in \mathscr{T}_{d_1}$. □

One can also alter a given metric d on a space X, without changing the topology, in such a way that the distances between points measured by the new metric are no greater than 1.

Proposition 1.44. *If (X,d) is a metric space, then the function $d_b : X \times X \to \mathbb{R}$, defined by*

$$d_b(x,y) = \min\{d(x,y), 1\}, \tag{1.6}$$

is a metric X. Furthermore, the metric topologies $\mathscr{T}_d$ and $\mathscr{T}_{d^b}$ on X coincide.

Proof. Exercise 1.113. □

The standard topology on $\mathbb{R}$ was first introduced here by specifying a certain basis for the topology, and it was noted subsequently that the topology is actually induced by a metric on $\mathbb{R}$. This represents one example of a metrisable space.

Definition 1.45. A topological space $(X, \mathscr{T})$ is *metrisable* if there exists a metric d on X such that $\mathscr{T} = \mathscr{T}_d$, where $\mathscr{T}_d$ is the metric topology induced by the metric d.

There are several instances throughout this book where we will need to determine whether a certain topological space is a metrisable space.

1.5 Subspaces and Product Spaces

Definition 1.46. The *subspace topology* $\mathscr{T}_Y$ on a subset $Y \subseteq X$ in a topological space $(X, \mathscr{T})$ is the collection

$$\mathscr{T}_Y = \{U \cap Y \mid U \in \mathscr{T}\}$$

of subsets of Y.

It is straightforward to verify that $\mathscr{T}_Y$ satisfies the axioms for a topology. Hence, $(Y, \mathscr{T}_Y)$ is indeed a topological space. A subset $W \subseteq Y$ is said to be *open in Y* if $W \in \mathscr{T}_Y$. It need not be true that an open set of Y is an open set of X. For example, with respect to the standard topology of $\mathbb{R}$, the set $W = \{x \in \mathbb{R} \,|\, 0 \leq x < 1/2\}$ is open in $Y = \{x \in \mathbb{R} \,|\, 0 \leq x < 1\}$, but not in $\mathbb{R}$.

Proposition 1.47. *If Y is an open set in a topological space $(X, \mathscr{T})$, then $W \subseteq Y$ is open in Y if and only if W is open in X.*

Proof. If $W \in \mathscr{T}_Y$, then there is a $U \in \mathscr{T}$ such that $W = Y \cap U$. As $Y \in \mathscr{T}$ also and $\mathscr{T}$ is closed under the intersection of two of its elements, $W \in \mathscr{T}$. Conversely, if $W \in \mathscr{T}$, then $W = W \cap Y$ implies that $W \in \mathscr{T}_Y$.   □

Proposition 1.48. *If $\mathscr{B}$ is a basis for a topology $\mathscr{T}$ on X and if $Y \subseteq X$, then the collection $\mathscr{B}_Y$ defined by $\mathscr{B}_Y = \{Y \cap B \,|\, B \in \mathscr{B}\}$ is a basis for the subspace topology $\mathscr{T}_Y$ on Y.*

Proof. Exercise 1.115.   □

The space $\mathbb{R}^n$ is a metric space via the Euclidean metric. However, $\mathbb{R}^n$ is also the Cartesian product of n copies of $\mathbb{R}$. Given that $\mathbb{R}$ itself is a topological space, one expects that there is a way in which these identical copies of $\mathbb{R}$ induce a topology on $\mathbb{R}^n$. This is indeed the case, and the resulting topology is called the *product topology*. Throughout this book we will make use of Cartesian products of both finite and infinite numbers of spaces, and so we consider the most general case here.

Let Λ be a set, and suppose that, for each $\alpha \in \Lambda$, $(X_\alpha, \mathscr{T}_\alpha)$ is a topological space. The Cartesian product of the family $\{X_\alpha\}_{\alpha \in \Lambda}$ is defined by

$$\prod_{\alpha \in \Lambda} X_\alpha = \{(x_\alpha)_{\alpha \in \Lambda} \,|\, x_\alpha \in X_\alpha\},$$

where $(x_\alpha)_{\alpha \in \Lambda}$ denotes the function $\Lambda \to \bigcup_{\alpha \in \Lambda} X_\alpha$, whose value at α is x_α. More informally, we consider $(x_\alpha)_{\alpha \in \Lambda}$ to be a tuple of elements, one from each X_α. If $Y_\alpha \subseteq X_\alpha$ for every $\alpha \in \Lambda$, then $\prod_\alpha Y_\alpha$ is considered as a subset of $\prod_\alpha X_\alpha$ in the natural way.

Proposition 1.49. *Assume that $\{(X_\alpha, \mathscr{T}_\alpha)\}_{\alpha \in \Lambda}$ is a family of topological spaces and define the following subcollections of subsets of $\prod_\alpha X_\alpha$:*

$$\mathscr{B}^\square = \{\textstyle\prod_\alpha U_\alpha \,|\, U_\alpha \in \mathscr{T}_\alpha, \, \forall \alpha \in \Lambda\}$$
$$\mathscr{B}^\times = \{\textstyle\prod_\alpha U_\alpha \,|\, U_\alpha \in \mathscr{T}_\alpha, \, \forall \alpha \in \Lambda, \, and$$
$$U_\alpha = X_\alpha \text{ for all but at most a finite number of } \alpha\}.$$

Then $\mathscr{B}^\square$ and $\mathscr{B}^\times$ are bases of subsets of $\prod_\alpha X_\alpha$.

The proof of Proposition 1.49 is a direct application of the definition. If Λ is a finite set, then $\mathscr{B}^{\square} = \mathscr{B}^{\times}$; however, this is no longer the case when Λ is an infinite set.

Definition 1.50. If $\{(X_\alpha, \mathscr{T}_\alpha)\}_{\alpha \in \Lambda}$ is a family of topological spaces, then

1. the *box topology* on $\prod_\alpha X_\alpha$ is the topology $\mathscr{T}^{\square}$ induced by the basis $\mathscr{B}^{\square}$, and
2. the *product topology* on $\prod_\alpha X_\alpha$ is the topology $\mathscr{T}^{\times}$ induced by the basis $\mathscr{B}^{\times}$.

Because $\mathscr{B}^{\times} \subseteq \mathscr{B}^{\square}$, the box topology is finer than the product topology; but if Λ is a finite set, then these two topologies coincide.

The following technical result is another consequence of the definitions.

Proposition 1.51. *Suppose that* $\{(X_\alpha, \mathscr{T}_\alpha)\}_{\alpha \in \Lambda}$ *is a family of topological spaces, and, for each* $\alpha \in \Lambda$*, that* $\mathscr{B}_\alpha$ *is a basis for* $\mathscr{T}_\alpha$*,* Y_α *is a subspace of* X_α*, and* $\mathscr{T}_{Y,\alpha}$ *is the subspace topology of* Y_α*.*

1. *If* $\mathscr{B}^{\square} = \{\prod_\alpha B_\alpha \mid B_\alpha \in \mathscr{B}_\alpha\}$*, then* $\mathscr{B}^{\square}$ *is a basis for the box topology of* $\prod_\alpha X_\alpha$*. Moreover, the box topology of the product space* $\prod_\alpha Y_\alpha$ *coincides with the subspace topology of* $\prod_\alpha Y_\alpha$ *inherited from the box topology of* $\prod_\alpha X_\alpha$*.*
2. *If* $\mathscr{B}^{\times} = \{\prod_\alpha B_\alpha \mid B_\alpha \in \mathscr{B}_\alpha \text{ and } B_\alpha = X_\alpha \text{ for all but finitely many } \alpha\}$*, then* $\mathscr{B}^{\times}$ *is a basis for the product topology of* $\prod_\alpha X_\alpha$*. Moreover, the product topology of the product space* $\prod_\alpha Y_\alpha$ *coincides with the subspace topology of* $\prod_\alpha Y_\alpha$ *inherited from the product topology of* $\prod_\alpha X_\alpha$*.*

Another way to induce a topology on a Cartesian product of topological spaces is to take advantage of special features that the family might have, as in the case of products of metric spaces.

Proposition 1.52. *If* $\{(X, d_\alpha)\}_{\alpha \in \Lambda}$ *is a family of metric spaces, and if* $d_{\alpha,b}$ *denotes the metric on* X_α *as given by equation (1.6) of Proposition 1.44, then the formula*

$$d((x_\alpha)_\alpha, (y_\alpha)_\alpha) = \sup_\alpha (d_{\alpha,b}(x_\alpha, y_\alpha)) \tag{1.7}$$

defines a metric on $\prod_\alpha X_\alpha$*.*

Proof. Exercise 1.114. □

Definition 1.53. The metric d of Proposition 1.52 is called the *uniform metric* on $\prod_\alpha X_\alpha$ and the resulting topology $\mathscr{T}_d$ is called the *uniform metric topology*.

Proposition 1.54. *If* $\{(X, d_\alpha)\}_{\alpha \in \Lambda}$ *is a family of metric spaces, then the uniform metric topology on* $\prod_\alpha X_\alpha$ *is finer than the product topology on* $\prod_\alpha X_\alpha$*.*

Proof. Let $U = \prod_\alpha U_\alpha \subseteq \prod_\alpha X_\alpha$ be a basic open set in the product topology. Thus, there is a finite set $F = \{\alpha_1, \ldots, \alpha_n\} \subset \Lambda$ such that $U_\alpha = X_\alpha$ for all $\alpha \notin F$. Choose $x = (x_\alpha)_\alpha \in U$. For each $j = 1, \ldots, n$ there is a $\varepsilon_j > 0$ such that $B_{\varepsilon_j}(x_{\alpha_j}) \subseteq U_{\alpha_j}$. Let ε_x be the minimum of these ε_j and consider the basic open set $B_{\varepsilon_x}(x)$ in the uniform metric topology of $\prod_\alpha X_\alpha$. If $y \in B_{\varepsilon_x}(x)$, then $y_\alpha \in X_\alpha = U_\alpha$ for $\alpha \notin F$; and, for each $j = 1, \ldots, n$, $d_{\alpha_j,b}(x_{\alpha_j}, y_{\alpha_j}) < \varepsilon_x \leq \varepsilon_j$, which implies that $y_{\alpha_j} \in U_{\alpha_j}$. Hence, $B_{\varepsilon_x}(x) \subseteq U$.

That is,

$$U = \bigcup_{x \in U} B_{\varepsilon_x}(x) \in \mathscr{T}^\times.$$

Hence, the uniform metric topology $\mathscr{T}_d$ is finer than the product topology $\mathscr{T}^\times$. $\square$

Definition 1.55. If $\{(X_\alpha, \mathscr{T}_\alpha)\}_{\alpha \in \Lambda}$ is a family of topological spaces such that each $X_\alpha = X$ and $\mathscr{T}_\alpha = \mathscr{T}$ for some topological space $(X, \mathscr{T})$, then the Cartesian product of the family $\{(X_\alpha, \mathscr{T}_\alpha)\}_{\alpha \in \Lambda}$ is denoted by X^Λ and the $\mathscr{T}^\square$ and $\mathscr{T}^\times$ denote the box and product topologies on X^Λ, respectively. For $\Lambda = \{1, \ldots, n\} \subset \mathbb{N}$, the notation X^n is used for X^Λ.

Of special interest are the product spaces $\mathbb{R}^n$ and $\mathbb{R}^{\mathbb{N}}$.

Proposition 1.56. *The metric topology on $\mathbb{R}^n$ induced by the Euclidean metric coincides with the product topology of $\mathbb{R}^n$.*

Proof. Let $\mathscr{T}_{d_2}$ and $\mathscr{T}^\times$ denote the metric and product topologies on $\mathbb{R}^n$, and let basic open sets of $\mathscr{T}_{d_2}$ be denoted by $B_r(x)$.

Suppose that $U \subseteq \mathbb{R}^n$ is an open set in the product topology of $\mathbb{R}^n$, and let $x = (x_1, \ldots, x_n) \in U$. Thus, there is a basic open set $B_x \in \mathscr{T}^\times$ such that $x \in B_x \subset \mathbb{R}^n$. Such a basic open set B_x has the form $B_x = \prod_{j=1}^m (p_j^x, q_j^x)$ for some $p_j^x, q_j^x \in \mathbb{Q}, j = 1, \ldots, n$. Now if $r_x = \min_{1 \le j \le n} \{|x_j - p_j^x|, |x_j - q_j^x|\}$, then for every $y = (y_1, \ldots, y_n) \in B_{r_x}(x)$ we have that

$$|x_i - y_i| \le \left(\sum_{j=1}^n |x_j - y_j|^2 \right)^{1/2} < r = \min_{1 \le j \le n} \{|x_j - p_j^x|, |x_j - q_j^x|\}$$

for each $i = 1, \ldots, n$. Hence, $y_i \in (p_i^x, q_i^x)$ for every i, implying that $y \in B_x$. Thus,

$$U = \bigcup_{x \in U} B_{r_x}(x) \in \mathscr{T}_{d_2}.$$

Conversely, if U is an open set in the metric topology of $\mathbb{R}^n$ and if $x = (x_1, \ldots, x_n) \in U$, then there exists $\varepsilon_x > 0$ such that $B_{\varepsilon_x}(x) \subseteq U$ (Proposition 1.41). For each $j = 1, \ldots, n$, choose $p_j^x, q_j^x \in \mathbb{Q}$ such that $p_j^x \in (x_j - \frac{\varepsilon_x}{\sqrt{n}}, x_j)$ and $q_j^x \in (x_j, x_j + \frac{\varepsilon_x}{\sqrt{n}})$. Thus, if $y = (y_1, \ldots, y_n) \in \prod_j (p_j^x, q_j^x)$, then

$$d_2(x, y) = \left(\sum_{j=1}^n |x_j - y_j|^2 \right)^{1/2} < \left(\sum_{j=1}^n \left(\frac{\varepsilon_x}{\sqrt{n}} \right)^2 \right)^{1/2} = \varepsilon_x.$$

Hence,

$$U = \bigcup_{x \in U} \left(\prod_j (p_j^x, q_j^x) \right) = \bigcup_{x \in U} B_{\varepsilon_x}(x) \in \mathscr{T}^{\times}.$$

Thus, the product topology and metric topology of $\mathbb{R}^n$ coincide. □

Proposition 1.56 asserts that the product topology on $\mathbb{R}^n$ is metrisable. Because different metrics on a given space can yield the same topologies, it can be difficult to tell from topological properties alone whether the topology on a given topological space $(X, \mathscr{T})$ is induced by some metric. Nevertheless, we have the following information about $\mathbb{R}^{\mathbb{N}}$ in its product topology.

Proposition 1.57. *The topological space* $(\mathbb{R}^{\mathbb{N}}, \mathscr{T}^{\times})$ *is metrisable.*

Proof. By Propositions 1.28 and 1.44, the standard topology on $\mathbb{R}$ is induced by the metric $d(x, y) = \min(|x - y|, 1)$. Basic open sets in this metric topology will be denoted by $B_{\varepsilon}^d(x)$.

Define $\rho : \mathbb{R}^{\mathbb{N}} \times \mathbb{R}^{\mathbb{N}} \to \mathbb{R}$ by

$$\rho((x_n)_n, (y_n)_n) = \sup_n \left(\min \left(\frac{|x_n - y_n|}{n}, \frac{1}{n} \right) \right) = \sup_n \left(\frac{1}{n} d(x_n, y_n) \right),$$

and note that ρ is a metric on $\mathbb{R}^{\mathbb{N}}$. Basic open sets in this metric topology will be denoted by $B_{\varepsilon}^{\rho}(x)$.

Let $U = \prod_n U_n$ be an open set in the product topology of $\mathbb{R}^{\mathbb{N}}$. Thus, there is a finite subset $F = \{n_1, \ldots, n_k\} \subset \mathbb{N}$ such that $n_1 < \cdots < n_k$ and $U_n = \mathbb{R}$ for all $n \notin F$. Select $x = (x_n)_n \in U$ and let $\varepsilon_j > 0$ be such that $B_{\varepsilon_j}^d(x_{n_j}) \subseteq U_{n_j}$, for $j = 1, \ldots, k$. Now let ε_x be the minimum of $\frac{\varepsilon_j}{n_j}$, where $j = 1, \ldots, k$, and note that if $y \in B_{\varepsilon_x}^{\rho}(x)$, then $y \in U$. Hence, U is a union of sets of the form $B_{\varepsilon_x}^{\rho}(x)$, for $x \in U$, which implies that U is open in the metric topology $\mathscr{T}_{\rho}$ of $\mathbb{R}^{\mathbb{N}}$.

Conversely, suppose that $B_r^{\rho}(x)$, where $x = (x_n)_n$, is a basic open set in the metric topology $\mathscr{T}_{\rho}$ of $\mathbb{R}^{\mathbb{N}}$. Choose $k \in \mathbb{N}$ such that $\frac{1}{k} < \frac{r}{2}$, and for each $n = 1, \ldots, k$ define $U_n = B_{r/2}^d(x_n)$, which is an open set in the standard topology of $\mathbb{R}$. Let $U_n = \mathbb{R}$ for every $n > k$. Consider $U_x = \prod_n U_n$, which is a basic open set in the product topology of $\mathbb{R}^{\mathbb{N}}$ and which contains x. If $y = (y_n)_n \in U_x$, then $d(x_n, y_n) < \frac{r}{2}$ if $1 \leq n \leq k$, and $d(x_n, y_n) \leq 1$ for all $n > k$. Thus, $\rho(x, y) = \sup_n \frac{1}{n} d(x_n, y_n) < \frac{r}{2}$, which implies that $U_x \subseteq B_r^{\rho}(x)$. Hence,

$$B_r^{\rho}(x) = \bigcup_{x \in U} U_x$$

implies that $B_r^{\rho}(x)$ is open in the product topology $\mathscr{T}^{\times}$ of $\mathbb{R}^{\mathbb{N}}$. □

1.6   Closures, Interiors, and Limit Points

Definition 1.58. A subset $F \subseteq X$ in a topological space $(X, \mathcal{T})$ is a *closed set* if there is an open set $U \in \mathcal{T}$ such that $F = U^c$.

Because the complements of closed sets are open sets, $\emptyset$ and X are both open and closed. Furthermore, arbitrary intersections of closed sets are closed, and finite unions of closed sets are closed.

Example 1.59. *In the standard topology of* $\mathbb{R}$, *closed intervals are closed sets.*

Proof. If $a, b \in \mathbb{R}$, with $a < b$, then $[a, b]$ is the complement of the union of the open sets $\bigcup_{x<a}(x, a)$ and $\bigcup_{b<y}(b, y)$, proving that $[a, b]$ is a closed set. $\square$

The next class of closed sets is used extensively in the study of topological spaces.

Example 1.60. *In the Euclidean metric space* $(\mathbb{R}^{n+1}, d_2)$, *where* $n \geq 1$, *the* n-*sphere*

$$S^n = \{x \in \mathbb{R}^{n+1} \mid d_2(x, 0) = 1\} = \left\{ x \in \mathbb{R}^{n+1} \mid \sum_{j=1}^{n} x_j^2 = 1 \right\}$$

is closed.

Proof. The set

$$U = B_1(0) \cup \bigcup_{x \in \mathbb{R}^{n+1}, \, d_2(x,0)>1} B_{d_2(x,0)-1}(x)$$

is open in $\mathbb{R}^{n+1}$, and S^n is its complement. $\square$

Proposition 1.61. *If* Y *is a subspace of a topological space* X, *then* $F \subseteq Y$ *is closed in* Y *if and only if there is a closed set* $C \subseteq X$ *in* X *such that* $F = Y \cap C$.

Proof. If $F \subseteq Y$ is closed in Y, then $F^c \cap Y$ is open in Y, and thus $F^c \cap Y = Y \cap U$ for some open set U in X. Let $C = U^c$, which is closed in X, and note that $F = Y \cap C$.

Conversely, if there is a closed set $C \subseteq X$ in X such that $F = Y \cap C$, then $U = C^c$ is open in X, and so $Y \cap U$ is open in Y. Thus, $F^c \cap Y = Y \cap (Y \cap U^c) = Y \cap U$, which is open in Y; hence, F is closed in Y. $\square$

Definition 1.62. Assume that $Y \subseteq X$ is a subset of a topological space X.

1. The *closure* of Y is the subset $\overline{Y}$ of X defined by

$$\overline{Y} = \bigcap_{c \supseteq Y \text{ and } c \text{ is closed}} C.$$

2. The *interior* of Y is the subset int Y of X defined by

$$\text{int } Y = \bigcup_{U \subseteq Y \text{ and } U \text{ is open}} U.$$

3. If $\overline{Y} = X$, then Y is said to be *dense* in X.

The closure operation has the following properties.

Proposition 1.63 (Closures of Unions and Intersections). *Assume that Y and Z are subsets of X, and that $\{Y_\alpha\}_{\alpha \in \Lambda}$ is a collection of subsets of X. The following statements hold:*

1. *if $Y \subseteq Z$, then $\overline{Y} \subseteq \overline{Z}$;*
2. $\overline{Y \cup Z} = \overline{Y} \cup \overline{Z}$;
3. $\overline{\bigcup_\alpha Y_\alpha} \subseteq \bigcup_\alpha \overline{Y_\alpha}$, *and equality need not be attained;*
4. $\overline{\bigcap_\alpha Y_\alpha} \subseteq \bigcap_\alpha \overline{Y_\alpha}$, *and equality need not be attained;*
5. *if $Y \subseteq Z$, then $\overline{Z} \setminus \overline{Y} \subseteq \overline{Z \setminus Y}$, and equality need not be attained.*

Proof. For (1), the definition of closure yields $A \subseteq \overline{A}$ for every subset A of X. Thus, $Y \subseteq Z \subseteq \overline{Z}$ implies that $\overline{Z}$ is a closed set containing Y. Hence, $\overline{Y} \subseteq \overline{Z}$, by definition.

To prove (2), note that $Y \cup Z \subseteq \overline{Y} \cup \overline{Z}$, which is closed. Hence, $\overline{Y \cup Z} \subseteq \overline{Y} \cup \overline{Z}$, by definition of closure. Conversely, $Y \subseteq (Y \cup Z)$ implies, by (1), that $\overline{Y} \subseteq \overline{Y \cup Z}$. Likewise, $\overline{Z} \subseteq \overline{Y \cup Z}$, and so $\overline{Y} \cup \overline{Z} \subseteq \overline{Y \cup Z}$.

The proofs of the remaining assertions are left to the reader (Exercise 1.116). $\quad\square$

If one has a nested inclusion $Z \subseteq Y \subseteq X$, then it is possible to consider the closure of Z relative to the topology of X or to the subspace topology of Y. The following proposition indicates the relationship between these closures.

Proposition 1.64. *Assume that $Z \subseteq Y \subseteq X$, and denote the closure in Y of a subset $A \subseteq Y$ by $\overline{A}^Y$ and the closure in X of a subset $B \subseteq X$ by $\overline{B}^X$. Then $\overline{Z}^Y = Y \cap \overline{Z}^X$.*

Proof. By Proposition 1.61, $Y \cap \overline{Z}^X$ is closed in Y; as this set also contains Z, it must contain the closure $\overline{Z}^Y$ of Z in Y. Furthermore, $\overline{Z}^Y$ has the form $Y \cap C$, for some closed set C in X, again by Proposition 1.61. Therefore, $\overline{Z}^X \subseteq C$, as C is closed in X and contains Z. Thus, $Y \cap \overline{Z}^X \subseteq Y \cap C = \overline{Z}^Y$ and, hence, $\overline{Z}^Y = Y \cap \overline{Z}^X$. $\quad\square$

The main proposition relating closures and interiors is as follows.

Proposition 1.65. *If $Y \subseteq X$ is a subset of a topological space X, then*

1. $\overline{Y}^c = \text{int}\,(Y^c)$ *and*
2. $(\text{int } Y)^c = \overline{Y^c}$.

Proof. Exercise 1.117. $\quad\square$

To obtain a clearer understanding of what is added to a set Y when passing from Y to its closure $\overline{Y}$, the notion of (topological) limit point is introduced.

Definition 1.66. A *neighbourhood* of a point x in a topological space X is an open subset U of X such that U contains x.

Proposition 1.67. *The following statements are equivalent for a subset Y of a topological space X:*

1. $x \in \overline{Y}$;
2. $U \cap Y \neq \emptyset$ *for every neighbourhood U of x.*

Proof. Assume that $x \in \overline{Y}$, and that U is a neighbourhood of x. If it were true that $U \cap Y = \emptyset$, then U would be a subset of Y^c and, thus, would be contained in the interior of Y^c. However, $\text{int}(Y^c) = \overline{Y}^c$ (Proposition 1.65); therefore, $x \in U \subseteq \overline{Y}^c$, which is in contradiction to $x \in \overline{Y}$.

Conversely, assume that $x \in X$, and that $U \cap Y \neq \emptyset$ for every neighbourhood U of x. Also assume, contrary to what we aim to prove, that $x \notin \overline{Y}$. Thus, $x \in \overline{Y}^c = \text{int}(Y^c)$, which implies that $\text{int}(Y^c)$ is a neighbourhood of x. Therefore, by hypothesis, $\text{int}(Y^c) \cap Y \neq \emptyset$. This is, however, a contradiction of the fact that $\text{int}(Y^c) \cap Y \subseteq Y^c \cap Y = \emptyset$. Hence, it must be that $x \in \overline{Y}$. $\qquad\square$

The passage from Y to $\overline{Y}$ is a matter of adding the limit points of Y.

Definition 1.68. An element $x \in X$ is a *limit point* of a subset $Y \subseteq X$ if, for every neighbourhood U of x, there is an element $y \in Y$ such that $y \in U$ and $y \neq x$. The set of limit points of Y is denoted by $L(Y)$.

Proposition 1.69. $\overline{Y} = Y \cup L(Y)$.

Proof. If $x \in L(Y)$, then $U \cap Y \neq \emptyset$ for every neighbourhood U of x. Thus, $L(Y) \subseteq \overline{Y}$, and therefore $Y \cup L(Y) \subseteq \overline{Y}$.

Conversely, suppose that $x \in \overline{Y}$ and $x \notin Y$. By virtue of $x \in \overline{Y}$, $U \cap Y \neq \emptyset$ for every neighbourhood U of x. Because $x \notin Y$, for each neighbourhood U there must be some $y \in Y$ with $y \in U$. Hence, $y \in L(Y)$, which proves that $\overline{Y} \subseteq Y \cup L(Y)$. $\qquad\square$

The topological boundary is another important closed set associated with an arbitrary set.

Definition 1.70. The *boundary* of a subset $Y \subseteq X$ in a topological space X is the set ∂Y defined by

$$\partial Y = \overline{Y} \cap \overline{Y^c}.$$

Proposition 1.71. *If $Y \subset X$, then:*

1. $\partial Y = \overline{Y} \setminus \text{int} Y$;
2. $\overline{Y} = \text{int} Y \cup \partial Y$;
3. $\text{int} Y = \overline{Y} \setminus \partial Y$;
4. Y *is closed if and only if $\partial Y \subseteq Y$; and*
5. Y *is open if and only if $Y \cap \partial Y = \emptyset$.*

Proof. Exercise 1.123. $\qquad\square$

1.7 Continuous Functions

The interesting functions between topological spaces are the continuous ones. (We shall use the terms "function" and "map" interchangeably.) To define what is meant by a continuous function, we first recall some notation.

If X and Y are sets, if $U \subseteq X$ and $V \subseteq Y$ are subsets, and if $f : X \to Y$ is a function, then $f(U)$ and $f^{-1}(V)$ denote the subsets of Y and X, respectively, defined by

$$f(U) = \{f(x) \mid x \in U\}$$
$$f^{-1}(V) = \{x \in X \mid f(x) \in V\}.$$

Observe that $f\left(f^{-1}(V)\right) \subseteq V$ and $U \subseteq f^{-1}\left(f(U)\right)$.

Definition 1.72. If X and Y are topological spaces, then a function $f : X \to Y$ is *continuous* if $f^{-1}(V)$ is an open set in X for every open subset V of Y.

The definition of continuity is global, but it is convenient to be able to discuss continuity as a local property.

Definition 1.73. A function $f : X \to Y$ between topological spaces X and Y is *continuous at a point* $x \in X$ if, for every neighbourhood $V \subseteq Y$ of $f(x)$, there is a neighbourhood $U \subseteq X$ of x such that $f(U) \subseteq V$.

Global continuity is the same as local continuity at each point of the space.

Proposition 1.74. *The following statements are equivalent for a function* $f : X \to Y$ *between topological spaces X and Y:*

1. f is continuous;
2. f is continuous at every point $x \in X$.

Proof. Suppose that f is continuous, that $x \in X$, and that $V \subseteq Y$ is any neighbourhood of $f(x)$. The continuity of f implies that $U = f^{-1}(V)$ is an open neighbourhood of x. Further, $f(U) = f\left(f^{-1}(V)\right) \subseteq V$, and so f is continuous at x.

Conversely, suppose that f is continuous at each $x \in X$. Choose any open set $V \subseteq Y$, and consider $U = f^{-1}(V)$. For each $x \in U$, the set V is a neighbourhood of x. Because f is continuous at each x, there is a neighbourhood U_x of x such that $f(U_x) \subseteq V$. Thus, $U_x \subseteq f^{-1}(V) = U$, and so

$$U = \bigcup_{x \in U} \{x\} \subseteq \bigcup_{x \in U} U_x \subseteq U.$$

Hence, U is the union of a family of open sets and is, therefore, open. This proves that f is a continuous function. $\square$

In the context of metric spaces, the continuity of a function is given by the following familiar criterion:

Proposition 1.75. *If X and Y are metric spaces, with metrics d_X and d_Y, respectively, then a function $f : X \to Y$ is continuous if and only if for each $x_0 \in X$ and for every $\varepsilon > 0$ there is a $\delta > 0$ such that*

$$d_Y\left(f(x_0), f(x)\right) < \varepsilon \text{ for all } x \in X \text{ that satisfy } d_X(x, x_0) < \delta.$$

Proof. Exercise 1.125.   □

For each fixed $y \in X$ in a metric space (X, d), one can ask whether the function $d_y : X \to \mathbb{R}$ defined by $d_y(x) = d(x, y)$, for $x \in X$, is continuous. In other words, is the metric d continuous in each of its variables? This is indeed true, but a more general result, Proposition 1.77 below, will be necessary for our study of operator theory.

Definition 1.76. If S is a nonempty subset of a metric space (X, d) and if $x \in X$, then the *distance from x to S* is the real number denoted by $\text{dist}(x, S)$ and defined by

$$\text{dist}(x, S) = \inf\{d(x, s) \,|\, s \in S\}. \tag{1.8}$$

Proposition 1.77. *If S is a nonempty subset of a metric space (X, d), then the function $d_S : X \to \mathbb{R}$ defined by $d_S(x) = \text{dist}(x, S)$ is continuous.*

Proof. By definition, if $x \in X$, then $\text{dist}(x, S) \leq d(x, s)$ for every $s \in S$. Thus, by the triangle inequality, $d_S(x) \leq d(x, s) \leq d(x, y) + d(y, s)$ for all $s \in S$ and $y \in X$. That is, by varying s through S,

$$\text{dist}(x, S) - d(x, y) \leq \inf_{s \in S} d(y, S) = \text{dist}(y, S).$$

Hence, $d_S(x) - d_S(y) \leq d(x, y)$. By interchanging the roles of x and y we obtain $d_S(y) - d_S(x) \leq d(x, y)$, and so $|d_S(x) - d_S(y)| \leq d(x, y)$ for all $x, y \in X$. An application of Proposition 1.75 now yields the continuity of d_S.   □

Another useful criterion for continuity is given by the following proposition, which involves closed sets and closures of sets.

Proposition 1.78. *The following statements are equivalent for a map $f : X \to Y$ of topological spaces X and Y:*

1. f is continuous;
2. $f(\overline{A}) \subseteq \overline{f(A)}$ for every subset $A \subseteq X$; and
3. $f^{-1}(C)$ is closed in X for every closed set C in Y.

Proof. (1) $\Rightarrow$ (2). Assume that f is continuous and that $A \subseteq X$. If $x \in \overline{A}$ and if V is a neighbourhood of $f(x)$, then $f^{-1}(V)$ is a neighbourhood of x, and so $A \cap f^{-1}(V)$ is nonempty. Now, if $z \in A \cap f^{-1}(V)$, then $f(z) \in V \cap f(A)$, which is to say that the neighbourhood V of $f(x)$ has nonempty intersection with $f(A)$. Hence, $f(x) \in \overline{f(A)}$.

(2) $\Rightarrow$ (3). Suppose that $f(\overline{A}) \subseteq \overline{f(A)}$ for every subset $A \subseteq X$. Select a closed set $C \subseteq Y$ and let $A = f^{-1}(C)$. Thus, $f(A) \subseteq C$ and, therefore, $\overline{f(A)} \subseteq C$. By hypothesis, $f(\overline{A}) \subseteq \overline{f(A)}$ and so $f(\overline{A}) \subseteq C$. Hence, $\overline{A} \subseteq f^{-1}\left(f(\overline{A})\right) \subseteq f^{-1}(C) = A \subseteq \overline{A}$ implies that $\overline{A} = A$. That is, $f^{-1}(C)$ is closed.

(3) $\Rightarrow$ (1). If $V \subseteq Y$ is open, then $f^{-1}(V^c)$ is closed, by hypothesis. Because $f^{-1}(V^c) = f^{-1}(V)^c$, we deduce that $f^{-1}(V)$ is open, and so f is continuous. $\square$

Continuous maps possess the following basic properties, each of which is readily verified.

Proposition 1.79. *The following functions are continuous:*

1. *every constant map, which is to say that if X and Y are topological spaces, and if $y_0 \in Y$, then the function $f : X \to Y$ given by $f(x) = y_0$, for all $x \in X$, is continuous;*
2. *the composition of continuous functions;*
3. *restrictions of continuous functions to subspaces; and*
4. *the inclusion map $\iota_Y : Y \to X$, $\iota_Y(x) = x$ for all $x \in Y$, for every subspace $Y \subseteq X$.*

1.8   The Cantor Ternary Set and Ternary Function

The "middle third" of the interval $[0, 1]$ is the open interval $(1/3, 2/3)$. If one removes the middle third from $[0, 1]$, then the closed set

$$\mathscr{C}_1 = [0, 1/3] \cup [2/3, 1]$$

remains. Note that $\mathscr{C}_1$ is a union of two closed subintervals, each of length $1/3$.

The middle thirds of $\mathscr{C}_1$ are the open intervals $(1/9, 2/9)$, which is the middle third of $[0, 1/3]$, and $(7/9, 8/9)$, which is the middle third of $[2/3, 1]$. If these middle thirds are removed from $\mathscr{C}_1$, then one is left with

$$\mathscr{C}_2 = [0, 1/9] \cup [2/9, 1/3] \cup [2/3, 7/9] \cup [8/9, 1],$$

which is a union of four subintervals, each of length $1/(3^2)$.

Proceed by induction. Once $\mathscr{C}_{n-1}$ has been constructed as a union of 2^{n-1} closed subintervals F_k, remove the middle third from each F_k to obtain $\mathscr{C}_n$, a union of 2^n closed subintervals, each of length $1/(3^n)$.

Definition 1.80. The *Cantor ternary set* is the subset $\mathscr{C} \subset [0, 1]$ defined by

$$\mathscr{C} = \bigcap_{n \in \mathbb{N}} \mathscr{C}_n.$$

Proposition 1.81. *The Cantor set $\mathscr{C}$ is a nonempty closed set with no interior.*

Proof. Each $\mathscr{C}_n$ is closed, and therefore so is $\mathscr{C}$. Moreover, $\mathscr{C}$ contains the endpoints of each subinterval in each $\mathscr{C}_n$; thus, $\mathscr{C}$ is nonempty.

Suppose that U is an open subset of $\mathscr{C}$. Then U contains an open interval J, and J is a subset of each $\mathscr{C}_n$. For fixed n, J must lie in one of the closed subintervals that form $\mathscr{C}_n$, and such intervals have length $(1/3)^n$. Thus, the length of J is at most $(1/3)^n$. But this length of J holds for all n, which implies that J is length zero; that is, $J = \emptyset$. Hence, $U = \emptyset$. $\square$

Another way to characterise $\mathscr{C}$ is by considering the ternary expansion of each $\zeta \in [0,1]$. Recall that every real number admits a decimal representation. A very similar argument show that every $\zeta \in [0,1]$ can be expressed in ternary form as

$$\zeta = \sum_{k=1}^{\infty} \frac{c_k}{3^k}, \quad \text{where } c_k \in \{0,1,2\} \quad \forall k \in \mathbb{N}. \tag{1.9}$$

The representation in (1.9) is not unique, because the real number $1/3$, for example, can be expressed in two ways: (i) $c_1 = 1$ and $c_k = 0$ for all $k \geq 2$; and (ii) $c_1 = 0$ and $c_k = 2$ for all $k \geq 2$. This is the only kind of ambiguity that can arise, and this ambiguity does not have any bearing on the description of $\mathscr{C}$ that follows.

It is also convenient to adopt base-3 notation for $\zeta \in [0,1]$:

$$\zeta = (0.c_1c_2c_3\cdots)_3 \quad \text{means} \quad \zeta = \sum_{k=1}^{\infty} \frac{c_k}{3^k}.$$

In this notation, for example,

$$\tfrac{1}{3} = (0.10000\cdots)_3 \quad \tfrac{1}{3} = (0.02222\cdots)_3$$

$$\tfrac{2}{9} = (0.02000\cdots)_3 \quad \tfrac{2}{9} = (0.01222\cdots)_3$$

$$\tfrac{7}{9} = (0.21000\cdots)_3 \quad \tfrac{8}{9} = (0.22000\cdots)_3$$

and so forth.

Proposition 1.82. $\zeta \in \mathscr{C}$ *if and only if* $\zeta = (0.c_1c_2c_3\cdots)_3$, *where* $c_k \in \{0,2\}$ *for every* $k \in \mathbb{N}$.

Proof. The left and right end points of any one of the closed intervals that make up $\mathscr{C}_n$ will have ternary form

$$(0.g_1g_2g_3\cdots g_{n-1}1000\cdots)_3 \quad \text{and} \quad (0.g_1g_2g_3\cdots g_{n-1}2000\cdots)_3$$

respectively, where $g_j \in \{0,2\}$ for all $1 \leq j \leq (n-1)$. Thus,

$$\zeta = (0.c_1c_2c_3\cdots)_3$$

is in this closed interval if and only if $g_j = c_j$ for all $1 \leq j \leq (n-1)$. $\square$

Proposition 1.83. $\mathscr{C}$ *and* $\mathbb{R}$ *have the same cardinality.*

Proof. On the one hand, the map that sends a binary sequence $b = (b_1b_2\ldots)$ to the real number $\displaystyle\sum_{k=1}^{\infty} \frac{2b_k}{3^k}$ is an injection of the set $\mathscr{B}$ of all binary sequences into the

Cantor set $\mathscr{C}$. On the other hand, the cardinality of $\mathscr{B}$ is that of the power set $\mathscr{P}(\mathbb{N})$ of $\mathbb{N}$, which in turn has the cardinality c of the continuum $\mathbb{R}$. Hence, there is an injection of $\mathbb{R}$ into the Cantor set $\mathscr{C}$, and so $\mathscr{C}$ and $\mathbb{R}$ are in bijective correspondence by the Schroeder-Bernstein Theorem (Theorem 1.3). □

Another application of Proposition 1.82 leads to a very interesting continuous function Φ on $[0,1]$ that has quite remarkable features. Let $\phi : \mathscr{C} \to [0,1]$ be the function that sends the ternary form of $\zeta \in \mathscr{C}$ to a number in binary form: namely,

$$\phi\left((0.c_1c_2\cdots)_3\right) = \sum_{k=1}^{\infty} \frac{c_k/2}{2^k}. \tag{1.10}$$

Definition 1.84. The function $\Phi : [0,1] \to [0,1]$ defined by

$$\Phi(x) = \sup\{\phi(y) \,|\, y \in \mathscr{C} \text{ and } y \le x\},$$

for $x \in [0,1]$, and where $\phi : \mathscr{C} \to [0,1]$ is the function defined by equation (1.10), is called the *Cantor ternary function*.

Definition 1.85. If $J \subseteq \mathbb{R}$ is an interval, then a function $f : J \to \mathbb{R}$ is *monotone increasing* if $f(x_1) \le f(x_2)$, for all $x_1, x_2 \in J$ such that $x_1 \le x_2$. If $f(x_1) < f(x_2)$, for all $x_1, x_2 \in J$ such that $x_1 < x_2$, then f is *strictly monotone increasing*.

Proposition 1.86. *The Cantor ternary function Φ is a monotone increasing continuous function and Φ maps the Cantor set $\mathscr{C}$ onto $[0,1]$.*

Proof. The map $\phi : \mathscr{C} \to [0,1]$ is clearly surjective and monotone increasing; therefore, the same is true of $\Phi : [0,1] \to [0,1]$.

To verify that Φ is continuous, select $y_0 \in [0,1]$ and let $\varepsilon > 0$. If $y_0 \notin \mathscr{C}$, then $y_0 \notin \mathscr{C}_n$ for some $n \in \mathbb{N}$. Thus, y_0 is contained in one of the open intervals that have been removed from $\mathscr{C}_{n-1}$; denote such an interval by (a, b). The endpoints a and b are the left and right endpoints of a closed subinterval of $\mathscr{C}_n$, and so $\Phi(x) = \phi(a)$ for all $x \in (a, b)$. This shows that Φ is constant on (a, b), implying that Φ is continuous at the point $y_0 \in (a, b)$.

However, if $y_0 \in \mathscr{C}$, then choose $m \in \mathbb{N}$ such that $(1/2)^m < \varepsilon$. Let $\delta = (1/3)^m$. Suppose that $x \in [0,1]$ is such that $|x - y_0| < \delta$. Without loss of generality, we may suppose that $x < y_0$; thus,

$$y_0 - x = \sum_{k=m+1}^{\infty} \frac{\omega_k}{3^k}, \quad \text{for some } \omega_k \in \{0,1,2\}.$$

That is, in any ternary expansion of x, the first m ternary digits of x coincide with those of y_0. Let $\alpha_1, \ldots, \alpha_m \in \{0,2\}$ be such that $y_0 = (0.\alpha_1 \cdots \alpha_m \cdots)_3$ and consider $z_0 = (0.\alpha_1 \cdots \alpha_m 00 \cdots)_3 \in \mathscr{C}$. Therefore,

$$z_0 \le x \le y_0 \;\Rightarrow\; \Phi(z_0) \le \Phi(x) \le \Phi(y_0).$$

The inequality

$$\Phi(y_0) - \Phi(z_0) = (0.\frac{\alpha_1}{2}, \cdots \frac{\alpha_m}{2} \cdots)_2 - (0.\frac{\alpha_1}{2}, \cdots \frac{\alpha_m}{2} 00 \cdots)_2 < (1/2)^m = \varepsilon$$

implies $|\Phi(y_0) - \Phi(x)| < \varepsilon$, proving the continuity of Φ at y_0.　　　□

The proof of Proposition 1.86 shows that the Cantor ternary function is constant on open intervals in the complement of $\mathscr{C}$. We shall have need of this fact later, and so this feature is recorded below for future reference.

Corollary 1.87. *If J is any open interval in $[0, 1] \backslash \mathscr{C}$, then Φ is constant on J.*

1.9　Weak Topologies and Continuous Maps of Product Spaces

Traditionally one begins with topological spaces and considers continuous maps between them. In many instances, especially in functional analysis, it is beneficial to reverse this process to allow functions to determine the topology.

Proposition 1.88. *If $\{(Y_\alpha, \mathscr{T}_\alpha)\}_{\alpha \in \Lambda}$ is a family of topological spaces, X is a set, and $g_\alpha : X \to Y_\alpha$ is a function, for each $\alpha \in \Lambda$, then*

1. *there is a coarsest topology on X in which each function $g_\alpha : X \to Y_\alpha$ is continuous, and*
2. *for every topological space Z, a map $f : Z \to X$ is continuous if and only if $g_\alpha \circ f : Z \to Y_\alpha$ is continuous for all α.*

Proof. For each $\beta \in \Lambda$ let $\mathscr{S}_\beta = \{g_\beta^{-1}(U_\beta) \mid U_\beta \in \mathscr{T}_\beta\}$ and let $\mathscr{S} = \bigcup_\beta \mathscr{S}_\beta$. The collection $\mathscr{S}$ is plainly a subbasis, and thus induces a topology on X which we denote by $\mathscr{T}_{\mathrm{wk}}$. Observe that each $g_\alpha : (X, \mathscr{T}_{\mathrm{wk}}) \to (Y_\alpha, \mathscr{T}_\alpha)$ is continuous. Furthermore, if $\mathscr{T}'$ is any other topology on X in which every g_α is continuous, then $\mathscr{T}'$ must contain every set of the form $g_\alpha^{-1}(U_\alpha)$, where $U_\alpha \in \mathscr{T}_\alpha$. Hence, $\mathscr{T}_{\mathrm{wk}} \subseteq \mathscr{T}'$.

Next, fix a topological space Z and assume that $f : Z \to X$ is a function. If f is continuous, then so is every $g_\alpha \circ f : Z \to Y_\alpha$, because the composition of continuous maps is continuous. Conversely, if it is assumed now that $g_\alpha \circ f : Z \to Y_\alpha$ is continuous for all α, then for every α and every open set $U_\alpha \subseteq Y_\alpha$, the set $f^{-1}(g^{-1}(U_\alpha))$ is open in Z. Thus, $f^{-1}(S)$ is open in Z for every S in the subbasis $\mathscr{S}$, and so $f^{-1}(B)$ is open in Z for every B in the basis $\mathscr{B}$ for $\mathscr{T}_{\mathrm{wk}}$ generated by $\mathscr{S}$. By Exercise 1.126, this implies that f continuous.　　　□

A result related to Proposition 1.88 is as follows.

Proposition 1.89. *If $\{(Y_\alpha, \mathscr{T}_\alpha)\}_{\alpha \in \Lambda}$ is a family of topological spaces, Y is a set, and $g_\alpha : X_\alpha \to Y$ is a function, for each $\alpha \in \Lambda$, then*

1. *there is a finest topology on Y in which each function $g_\alpha : X_\alpha \to Y$ is continuous, and*
2. *for every topological space Z, a map $f : Y \to Z$ is continuous if and only if $f \circ g_\alpha : X_\alpha \to Z$ is continuous for all α.*

Proof. Exercise 1.128. □

Definition 1.90. Each of the topologies introduced in Propositions 1.88 and 1.89 is called the *weak topology* induced by the family of functions $\{g_\alpha\}_{\alpha \in \Lambda}$.

Turning now to product spaces, we begin with a definition.

Definition 1.91. Assume that $\{(X_\alpha, \mathscr{T}_\alpha)\}_{\alpha \in \Lambda}$ is a family of topological spaces. For each $\beta \in \Lambda$, the map $p_\beta : \prod_{\alpha \in \Lambda} X_\alpha \to X_\beta$ defined by

$$p_\beta ((x_\alpha)_\alpha) = x_\beta,$$

is called a *projection map*.

Observe that p_β maps each $x \in \prod_{\alpha \in \Lambda} X_\alpha$ onto the "β-th coordinate" of x. Furthermore, if U_β is an open set in X_β, then $p_\beta^{-1}(U_\beta) = \prod_\alpha A_\alpha$, where $A_\alpha = X_\alpha$ for all $\alpha \neq \beta$, and $A_\beta = U_\beta$. Hence, the maps p_β are continuous with respect to both the product and the box topology of $\prod_{\alpha \in \Lambda} X_\alpha$.

Proposition 1.92. *In the product topology of $\prod_{\alpha \in \Lambda} X_\alpha$, a map $f : Z \to \prod_{\alpha \in \Lambda} X_\alpha$ is continuous if and only if $p_\alpha \circ f : Z \to X_\alpha$ is continuous for every $\alpha \in \Lambda$.*

Proof. If $f : Z \to \prod_{\alpha \in \Lambda} X_\alpha$ is continuous, then so is $p_\alpha \circ f$ because the composition of continuous maps is continuous.

Conversely, suppose that $p_\alpha \circ f : Z \to X_\alpha$ is continuous for every $\alpha \in \Lambda$. Fix $\beta \in \Lambda$ and suppose that $U_\beta \subseteq X_\beta$ is open. Let $W_\beta = p_\beta^{-1}(U_\beta)$, which is open in $\prod_{\alpha \in \Lambda} X_\alpha$; thus, $(p_\beta \circ f)^{-1}(U_\beta) = f^{-1}(W_\beta)$ is open in Z. Now if B is a basic open set in $\prod_{\alpha \in \Lambda} X_\alpha$, then there are $\beta_1, \ldots, \beta_n \in \Lambda$ and open sets $U_{\beta_j} \subseteq X_{\beta_j}$, for $j = 1, \ldots, n$, such that $B = \bigcap_{j=1}^n W_{\beta_j}$, where $W_{\beta_j} = p_{\beta_j}^{-1}(U_{\beta_j})$. Hence, $f^{-1}(B) = \bigcap_{j=1}^n f^{-1}(W_{\beta_j})$ is open in Z. By Exercise 1.126, this implies that f is continuous. □

One implication in Proposition 1.92 fails for the box topology.

Example 1.93. *If $\mathbb{R}^{\mathbb{N}}$ has the box topology, and if $p_n : \mathbb{R}^{\mathbb{N}} \to \mathbb{R}$ denotes the projection onto the n-th coordinate, then there exists a function $f : \mathbb{R} \to \mathbb{R}^{\mathbb{N}}$ such that $p_n \circ f : \mathbb{R} \to \mathbb{R}$ is continuous for all $n \in \mathbb{N}$, yet f itself is not continuous.*

Proof. Let $f : \mathbb{R} \to \mathbb{R}^{\mathbb{N}}$ be defined by $f(x) = (x, x, x, \ldots)$. Thus, $p_n \circ f(x) = x$, which is plainly continuous. However, if $V \subset \mathbb{R}^{\mathbb{N}}$ is the open set $\prod_n (\frac{-1}{n}, \frac{1}{n}) \in \mathscr{T}^\square$, then $f^{-1}(V) = \{x \in \mathbb{R} \mid x \in (\frac{-1}{n}, \frac{1}{n}) \; \forall n \in \mathbb{N}\} = \{0\}$, which is not open. Thus, f is not continuous. □

1.10   Quotient Spaces

Suppose that $\sim$ is an equivalence relation on a set X. For each $x \in X$, the equivalence class of x is denoted by $\dot{x}$; thus,

$$\dot{x} = \{y \in X \mid y \sim x\}.$$

If $X/\sim$ denotes the set $\{\dot{x} \mid x \in X\}$ of equivalence classes of X, and if $(X, \mathscr{T})$ is a topological space, then it is possible to endow $X/\sim$ with the structure of a topological space.

Proposition 1.94. *If $\sim$ is an equivalence relation on a topological space $(X, \mathscr{T})$, and if $q : X \to X/\sim$ is the canonical quotient map $q(x) = \dot{x}$, then*

$$\mathscr{T}_{\text{quo}} = \{V \subseteq (X/\sim) \mid q^{-1}(V) \in \mathscr{T}\}$$

is a topology on $X/\sim$ and the function $q : X \to X/\sim$ is continuous.

The proof of Proposition 1.94 is a matter of verifying the definitions.

Definition 1.95. The topology $\mathscr{T}_{\text{quo}}$ in Proposition 1.94 is called the *quotient topology* on $X/\sim$.

Two common quotient structures in algebra arise from subspaces of vector spaces and normal subgroups of groups. First, if $L \subseteq V$ is a linear subspace of a vector space V, then the relation $v \sim w$ if and only if $v - w \in L$ is an equivalence relation on V and the space of equivalence classes, which is denoted by V/L, has the structure of a vector space under the operations $\alpha \dot{v} = (\dot{\alpha v})$ and $\dot{v} + \dot{w} = (\dot{v + w})$. Thus, if it assumed that $\mathbb{R}$ has the standard topology, and if L is a subspace of $\mathbb{R}^n$, then the vector space $\mathbb{R}^n/L$ is, as a topological space, a quotient space.

A second familiar quotient structure occurs in group theory. If H is a normal subgroup of a group G with binary operation $*$, then the relation $a \sim b$ if and only if $b^{-1} * a \in$ H is an equivalence relation on G and the space of equivalence classes, which is denoted by G/H, has the structure of a group under the binary operation $\dot{*}$ given by $\dot{a} \dot{*} \dot{b} = \dot{a * b}$. Of particular interest is the additive abelian group $(\mathbb{R}, +)$ and its (normal) subgroup $(\mathbb{Z}, +)$, which results in the quotient group $\mathbb{R}/\mathbb{Z}$.

Example 1.96. *The function $f : \mathbb{R}/\mathbb{Z} \to \mathbb{R}^2$, defined by $f(\dot{t}) = (\cos 2\pi t, \sin 2\pi t)$, for $t \in \mathbb{R}$, is continuous.*

Proof. Let $F : \mathbb{R} \to \mathbb{R}^2$ be the function $F(t) = (\cos 2\pi t, \sin 2\pi t)$. If p_1 and p_2 denote the projections onto the first and second coordinates, respectively, then $p_j \circ F$ is a trigonometric function $\mathbb{R} \to \mathbb{R}$, and is therefore continuous. By Proposition 1.92, the continuity of each $p_j \circ F$ implies the continuity of F. Hence, $F^{-1}(V)$ is open in $\mathbb{R}$ for every open subset $V \subseteq \mathbb{R}^2$.

Observe that $F(t) = f(\dot{t})$ for every $t \in \mathbb{R}$; hence, if V is open in $\mathbb{R}^2$, then

$$q^{-1}\left(f^{-1}(V)\right) = q^{-1}(\{\dot{t} \mid f(\dot{t}) \in V\}) = \{t \mid F(t) \in V\} = F^{-1}(V),$$

which, by definition of the quotient topology, implies that $f^{-1}(V)$ is open in $\mathbb{R}/\mathbb{Z}$. Hence, f is a continuous map. □

A general method for analysing the continuity of functions defined on quotient spaces is given by the next result.

Proposition 1.97. *If X and Y are topological spaces, and if $\sim$ is an equivalence relation on the space X, then the following statements are equivalent for a function $g : (X/\sim) \to Y$:*

1. g is continuous, considered as a map $(X/\sim) \to Y$;
2. $g \circ q$ is continuous, considered as a map $X \to Y$.

Proof. By definition of quotient topology, the canonical projection $q : X \to X/\sim$ is continuous. Hence, if g is continuous, then so is $g \circ q$.

Conversely, assume that $g \circ q$ is continuous. Select an open set V in Y; thus, $U = (g \circ q)^{-1}(V)$ is open in X. Because

$$U = (g \circ q)^{-1}(V) = q^{-1}(\{\dot{x} \mid g(\dot{x}) \in V\}) = q^{-1}\left(g^{-1}(V)\right),$$

the set $g^{-1}(V)$ is open in $X/\sim$, by definition of the quotient topology. □

1.11 Topological Equivalence

Definition 1.98. Assume that X and Y are topological spaces.

1. A bijective function $f : X \to Y$ in which both f and f^{-1} are continuous is called a *homeomorphism*.
2. The topological spaces X and Y are said to be *homeomorphic* if there exists a homeomorphism $f : X \to Y$ (or $g : Y \to X$).

The notation $X \simeq Y$ is used to denote that X and Y are homeomorphic spaces.

In some sense, the goal of topology is to identify topological spaces up to homeomorphism. This objective, however, is hugely unrealistic, and therefore topologists have introduced other invariants of topological spaces that are less stringent than that of topological equivalence. Even a basic question such as "Are the topological spaces $\mathbb{R}^n$ and $\mathbb{R}^m$ homeomorphic if $n \neq m$?" requires sophisticated tools to resolve. (The answer to the question is no.)

Example 1.99. *If $a, b \in \mathbb{R}$ are such that $a < b$, then $(a, b) \simeq \mathbb{R}$.*

Proof. First note that $g : (-\frac{\pi}{2}, \frac{\pi}{2}) \to \mathbb{R}$, where $g(\theta) = \tan(\theta)$, is a continuous bijection with continuous inverse $g^{-1}(s) = \tan^{-1}(s)$. Hence, $(-\frac{\pi}{2}, \frac{\pi}{2}) \simeq \mathbb{R}$. Now consider the straight line L in $\mathbb{R}^2$ that passes through the points $(a, b) \in \mathbb{R}^2$ and $(-\frac{\pi}{2}, \frac{\pi}{2}) \in \mathbb{R}^2$, and let $F : \mathbb{R} \to \mathbb{R}$ be the equation of this line—that is, F is the function whose graph is L. The function F is a homeomorphism; therefore, the restriction f of F to the interval (a,b) is a homeomorphism of (a,b) and $(-\frac{\pi}{2}, \frac{\pi}{2})$. Hence, $(a,b) \simeq (-\frac{\pi}{2}, \frac{\pi}{2})$ and, by transitivity of topological equivalence (Exercise 1.130), we deduce that $(a,b) \simeq \mathbb{R}$. $\qquad\square$

Recall from Example 1.60 that the 1-sphere is the unit circle in $\mathbb{R}^2$, which we assume to have the subspace topology. The following example shows that the quotient space $\mathbb{R}/\mathbb{Z}$ is topologically equivalent to S^1.

Example 1.100. $\mathbb{R}/\mathbb{Z} \simeq S^1$.

Proof. Example 1.96 already shows that the map $f : \mathbb{R}/\mathbb{Z} \to S^1$ defined by $f(\dot{t}) = (\cos 2\pi t, \sin 2\pi t)$, for $t \in \mathbb{R}$, is continuous. As f is plainly bijective, all that remains is to show that f^{-1} is continuous. As in Example 1.96, let $F : \mathbb{R} \to \mathbb{R}^2$ denote the function $F(t) = (\cos 2\pi t, \sin 2\pi t)$.

Now, if $U \in \mathbb{R}/\mathbb{Z}$ is open, then $\left(f^{-1}\right)^{-1}(U) = f(U) = F\left(q^{-1}(U)\right)$; thus, the aim is to prove that $F(W)$ is open in $\mathbb{R}^2$, where $W \subseteq \mathbb{R}$ is the open set $W = q^{-1}(U)$. Select $t_0 \in W$ and let $\varepsilon > 0$ be such that $V = (t_0 - \varepsilon, t_0 + \varepsilon) \subseteq W$ and $\varepsilon < 1/2$. By Exercise 1.112, $d_2(F(t), F(t_0)) = 2|\sin(\pi(t - t_0))|$. Therefore, if $t \in V$, then $\pi|t - t_0| < \pi\epsilon < \pi/2$, and the fact that the sine function is strictly monotone increasing on $(0, 1/2)$, leads to $d_2(F(t), F(t_0)) < \sin(\pi\varepsilon)$. Hence, F maps V into the basic open set $B_{\sin(\pi\varepsilon)}(F(t_0))$. Conversely, if $(s, r) \in B_{\sin(\pi\varepsilon)}(F(t_0))$, then $\pi\varepsilon < \pi/2$ implies that the Euclidean distance between (s, r) and $F(t_0)$ is less than $\sin(\pi/2) = 1$. Hence, there is semicircular arc in S^1 that contains both (s, r) and $F(t_0)$, and so there is a $t \in \mathbb{R}$ with $|t - t_0| \leq \frac{1}{2}$ and $F(t) = (s, r)$. Therefore, $2 \sin \pi\varepsilon > (2|\sin(\pi(t - t_0))|) = 2\sin(\pi|t - t_0|)$ implies, by the monotonicity of the sine function on the interval $[0, \pi/2]$, that $\pi|t - t_0| < \pi\varepsilon$, and so $t \in V$. This proves that F maps V onto $S^1 \cap B_{\sin(\pi\varepsilon)}(F(t_0))$. Carrying this procedure out for every $t_0 \in W$ shows that $F(W)$ is open in S^1, which completes the proof of the continuity of f^{-1}. $\qquad\square$

Problems

1.101. Let X be the set of rational numbers q for which $0 < q < 1$ and Λ be the set of irrational numbers λ such that $0 < \lambda < 1$. For each $\lambda \in \Lambda$, let X_λ be the set of all sequences of elements in X with limit λ. Prove the following assertions.

1. Each X_λ is an infinite set.
2. $X_\lambda \cap X_{\lambda'}$ is a finite set, for every pair of distinct irrationals $\lambda, \lambda' \in \Lambda$.

1.102. Let $\mathcal{T}$ be the collection of all subsets $U \subseteq \mathbb{N}$ with the property that a natural number n belongs to U only if every divisor $k \in \mathbb{N}$ of n belongs to U.

1. Prove that $\mathcal{T}$ is a topology on $\mathbb{N}$.
2. Determine whether $\mathcal{T}$ coincides with the discrete topology on $\mathbb{N}$.

1.103. Prove that if $\mathscr{T}$ is a topology on X, and if $\mathscr{B}$ is the collection of subsets with the property that, for every $U \in \mathscr{T}$ and each $x \in U$, there is a $B \in \mathscr{B}$ with $x \in B \subseteq U$, then $\mathscr{B}$ is a basis of subsets and $\mathscr{T}$ is the topology generated by $\mathscr{B}$.

1.104. Prove that if $\mathscr{T}$ is a topology on X and if $\mathscr{B}$ is a basis for the topology $\mathscr{T}$, then the following statements are equivalent for a subset $U \subseteq X$:

1. $U \in \mathscr{T}$;
2. there is a family $\{B_\alpha\}_\alpha$ of subsets $B_\alpha \in \mathscr{B}$ such that $U = \bigcup_\alpha B_\alpha$.

1.105. Prove that $\mathscr{B} = \{(p,q) \subset \mathbb{R} \,|\, p,q \in \mathbb{Q},\ p < q\}$ is a basis of subsets of $\mathbb{R}$.

1.106. Consider $\mathbb{Z}$, the set of integers. Fix a prime number p. For every natural $k \in \mathbb{N}$ and integer $a \in \mathbb{Z}$, let

$$B_{k,a} = \left\{a + bp^k \,|\, b \in \mathbb{Z}\right\}.$$

Show that the collection $\mathscr{B} = \{B_{k,a}\}_{(k,a) \in \mathbb{N} \times \mathbb{Z}}$ is a basis for a topology on $\mathbb{Z}$.

1.107. Let

$$\mathscr{B} = \{[a,b) \,|\, a,b \in \mathbb{R},\ a \leq b\}.$$

1. Prove that $\mathscr{B}$ is a basis.
2. The topology on $\mathbb{R}$ induced by this basis is called the *lower-limit topology*. Prove that the lower-limit topology on $\mathbb{R}$ is strictly finer than the standard topology on $\mathbb{R}$.

1.108. Prove that if $\mathscr{S}$ is a subbasis of subsets of X, then

$$\mathscr{B} = \left\{\bigcap_{j=1}^{n} \,|\, n \in \mathbb{N},\ S_1, \ldots, S_n \in \mathscr{S}\right\}$$

is a basis of subsets of X.

1.109. Let (X,d) be a metric space. Prove that a subset $U \subseteq X$ is an open set if and only if, for each $x \in U$, there is an $\varepsilon > 0$ such that $B_\varepsilon(x) \subseteq U$.

1.110. Show that if $\left|\sum_{j=1}^{n} u_j v_j\right| = \sqrt{\left(\sum_{j=1}^{n} u_j^2\right)\left(\sum_{j=1}^{n} v_j^2\right)}$, for real numbers $u_1, \ldots, u_n$ and $v_1, \ldots, v_n$, then there is a $\lambda \in \mathbb{R}$ such that $v_j = \lambda u_j$ for every $j = 1, \ldots, n$.

1.111. Consider the functions $\mathbb{R}^n \times \mathbb{R}^n \to \mathbb{R}$ defined by

$$d_1(x,y) = \sum_{j=1}^n |x_j - y_j|$$
$$d_\infty(x,y) = \max_{1 \le j \le n} |x_j - y_j|,$$

for $x, y \in \mathbb{R}^n$.

1. Prove that d_1 and d_∞ are metrics on $\mathbb{R}^n$.
2. Prove that d_1 and d_∞ determine the same topology on $\mathbb{R}^n$

1.112. If $t, t_0 \in \mathbb{R}$, and if $u_t = (\cos 2\pi t, \sin 2\pi t)$ and $u_{t_0} = (\cos 2\pi t_0, \sin 2\pi t_0)$, then show that $d_2(u_t, u_{t_0}) = 2|\sin(\pi(t - t_0))|$ in the Euclidean metric space $(\mathbb{R}^2, d_2)$.

1.113. Assume that (X, d) is a metric space.

1. Prove that the formula $d_b(x, y) = \min\{d(x, y), 1\}$, for $x, y \in X$, defines a metric on X.
2. Prove that the metric topologies $\mathscr{T}_d$ and $\mathscr{T}_{d^b}$ on X coincide.

1.114. If $\{(X, d_\alpha)\}_{\alpha \in \Lambda}$ is a family of metric spaces, and if $d_{\alpha,b}$ denotes the metric on X_α as given by equation (1.6) of Proposition 1.44, then prove that the formula

$$d((x_\alpha)_\alpha, (y_\alpha)_\alpha) = \sup_\alpha (d_{\alpha,b}(x_\alpha, y_\alpha))$$

defines a metric on $\prod_\alpha X_\alpha$.

1.115. Prove that if $\mathscr{B}$ is a basis for a topology $\mathscr{T}$ on X and if $Y \subseteq X$, then the collection $\mathscr{B}_Y$ defined by $\mathscr{B}_Y = \{Y \cap B \mid B \in \mathscr{B}\}$ is a basis for the subspace topology $\mathscr{T}_Y$ on Y.

1.116. Assume that $\{Y_\alpha\}_{\alpha \in \Lambda}$ is a collection of subsets of a set X.

1. Prove that the following statements hold:

   a. $\bigcup_\alpha \overline{Y}_\alpha \subseteq \overline{\bigcup_\alpha Y_\alpha}$;

   b. $\overline{\bigcap_\alpha Y_\alpha} \subseteq \bigcap_\alpha \overline{Y}_\alpha$;

   c. if $Y \subseteq Z$, then $\overline{Z} \setminus \overline{Y} \subseteq \overline{Z \setminus Y}$.

2. For each of the statements above, find an example to show that equality in the inclusion is not achieved.

1.117. If $Y \subseteq X$ is a subset of a topological space X, then prove that $\overline{Y}^c = \operatorname{int}(Y^c)$ and $(\operatorname{int} Y)^c = \overline{Y^c}$.

1.118. Let $A \subset X$ and $B \subset Y$, and prove the following assertions for the product topology on $X \times Y$:

1. $\overline{A \times B} = \overline{A} \times \overline{B}$;
2. $\operatorname{int}(A \times B) = \operatorname{int}(A) \times \operatorname{int}(B)$.

1.119. Suppose that X is a topological space and that $U \subseteq X$. Prove that the following statements are equivalent:

1. U is an open set in X;
2. for every subset $Y \subseteq X$, $\overline{U \cap \overline{Y}} = \overline{U \cap Y}$.

1.120. Assume that $\mathbb{R}$ has the standard topology and consider the following subsets:

$$A = \left\{ \frac{1}{m} + \frac{1}{n} \,\middle|\, m, n \in \mathbb{N} \right\} \quad \text{and} \quad B = \{0\} \cup \left\{ \frac{1}{n} \,\middle|\, n \in \mathbb{N} \right\}.$$

1. Prove that $B = L(A)$, the set of limit points of A.
2. Determine $L(B)$.

1.121. Let $Y \subset \mathbb{R}^2$ be the set

$$Y = \left\{ (x, \sin(1/x)) \in \mathbb{R}^2 \,\middle|\, 0 < x \leq 1 \right\}.$$

Determine the closure of Y in the standard topology of $\mathbb{R}^2$.

1.122. If $Y \subseteq X$, then prove that $\operatorname{int} Y = Y \setminus L(Y^c)$, where $L(Y^c)$ is the set of limit points of Y^c.

1.123. Prove the following statements for a subset Y of a topological space X.

1. $\partial Y = \overline{Y} \setminus \operatorname{int} Y$.
2. $\overline{Y} = \operatorname{int} Y \cup \partial Y$.
3. $\operatorname{int} Y = \overline{Y} \setminus \partial Y$.
4. Y is closed if and only if $\partial Y \subseteq Y$.
5. Y is open if and only if $Y \cap \partial Y = \emptyset$.

1.124. Determine the closure and boundary of $B_r(x) = \{y \in X \,|\, d(x,y) < r\}$ in a metric space (X, d).

1.125. Assume that X and Y are metric spaces and denote their metrics by d_X and d_Y, respectively. Prove that a function $f : X \to Y$ is continuous at a point $x_0 \in X$ if and only if for every $\varepsilon > 0$ there is a $\delta > 0$ such that

$$d_Y\left(f(x_0), f(x)\right) < \varepsilon \quad \text{for all } x \in X \text{ that satisfy } d_X(x, x_0) < \delta.$$

1.126. Let X and Y be topological spaces and let $\mathscr{B}_Y$ be a basis for the topology on Y. Prove that a map $f : X \to Y$ is continuous if and only if $f^{-1}(B)$ is open in X for every $B \in \mathscr{B}_Y$.

1.127. Assume that A and B are closed subsets of a topological space X and that $A \cup B = X$. Prove that if $g : A \to Y$ and $h : B \to Y$ are continuous maps for which $g(x) = h(x)$ for every $x \in A \cap B$, then the map $f : X \to Y$ in which $f(x) = g(x)$ for $x \in A$ and $f(x) = h(x)$ for $x \in B$ is continuous.

1.128. Assume that $\{X_\alpha\}_{\alpha \in \Lambda}$ is a family of topological spaces and that Y is a set. Suppose that $g_\alpha : X_\alpha \to Y$ is a function, for each $\alpha \in \Lambda$. Prove that

1. there is a finest topology on Y in which each function $g_\alpha : X_\alpha \to Y$ is continuous, and
2. for every topological space Z, a map $f : Y \to Z$ is continuous if and only if $f \circ g_\alpha : X_\alpha \to Z$ is continuous for all α.

1.129. Prove that a quotient space $X/\sim$ is compact, if X is compact.

1.130. Prove that if $X \simeq Y$ and $Y \simeq Z$, then $X \simeq Z$.

1.131. Consider $\mathbb{R}^n$ as a metric space with respect to the Euclidean metric, and suppose that $x, y \in \mathbb{R}^n$ and that $r, s \in \mathbb{R}$ are positive. Prove that $B_r(x) \simeq B_s(y)$.

1.132. Suppose that $1 \leq k < n$ and that L is a k-dimensional vector space of $\mathbb{R}^n$. Prove that $\mathbb{R}^n/L \simeq \mathbb{R}^{n-k}$.

1.133. Assume that $\{0,1\}^{\mathbb{N}}$ has the product topology and consider the Cantor ternary set $\mathscr{C}$.

1. Prove that the topological spaces $\{0,1\}^{\mathbb{N}}$ and $\mathscr{C}$ are homeomorphic.
2. Prove that if $\mathscr{C}^{\mathbb{N}}$ has the product topology, then $\mathscr{C}$ and $\mathscr{C}^{\mathbb{N}}$ are homeomorphic.
3. Assuming that $[0,1]^{\mathbb{N}}$ has the product topology, prove that there exists a continuous surjection $f : \mathscr{C} \to [0,1]^{\mathbb{N}}$.

Chapter 2
Topological Spaces with Special Properties

Generic features of topological spaces and their continuous maps were considered in the previous chapter. This chapter investigates certain qualitative features of topological spaces: compactness (how small is a space?), normality (how separated can disjoint closed sets in a space be?), second countability (what is the smallest cardinality of the basis for the topology?), and connectedness (how disperse or disjoint is a space?). This chapter will also introduce the notion of a net, which is a natural extension of the concept of a sequence, and show how properties of nets capture some of the topological features of prime interest in abstract analysis.

2.1 Compact Spaces

Definition 2.1. Suppose that $\{U_\alpha\}_{\alpha \in \Lambda}$ is a family of open subsets of a topological space X and that $Y \subseteq X$ is a subset.

1. The family $\{U_\alpha\}_{\alpha \in \Lambda}$ is an *open cover* of Y if $\bigcup_\alpha U_\alpha \supseteq Y$.

2. If $\{U_\alpha\}_{\alpha \in \Lambda}$ is an *open cover* of Y and if $\Omega \subseteq \Lambda$, then the family $\{U_\omega\}_{\omega \in \Omega}$ is a *subcover* of $\{U_\alpha\}_{\alpha \in \Lambda}$ if $\{U_\omega\}_{\omega \in \Omega}$ is also cover of Y.

Definition 2.2. A subset K of a topological space X is a *compact set* if every open cover of K admits a finite subcover.

Thus, $K \subseteq X$ is a compact subset if, for every family $\{U_\alpha\}_{\alpha \in \Lambda}$ of open sets $U_\alpha \subseteq X$ for which $\bigcup_\alpha U_\alpha \supseteq K$, there is a finite set $F = \{\alpha_1, \ldots, \alpha_n\} \subseteq \Lambda$ such that $\bigcup_{j=1}^n U_{\alpha_j} \supseteq K$.

© Springer International Publishing Switzerland 2016
D. Farenick, *Fundamentals of Functional Analysis*, Universitext,
DOI 10.1007/978-3-319-45633-1_2

The compactness of a space K indicates that K is, in some sense, rather small. For example, the space $\mathbb{R}$, which may be viewed as a line of infinite length, is not compact because the open cover $\{B_\varepsilon(q)\}_{q \in \mathbb{Q}}$ of $\mathbb{R}$, for some fixed $\varepsilon > 0$, does not admit a finite subcover.

Example 2.3. *Every closed interval $[a,b]$ is a compact subset of $\mathbb{R}$.*

Proof. Suppose that $\{U_\alpha\}_{\alpha \in \Lambda}$ is an open cover of $[a,b]$ in $\mathbb{R}$, and let

$$K = \left\{ x \in (a,b] \mid \text{there is a finite subset } F \subseteq \Lambda \text{ such that } [a,x] \subseteq \bigcup_{\alpha \in F} U_\alpha \right\}.$$

Choose $\alpha_0 \in \Lambda$ such that $a \in U_{\alpha_0}$. Hence, there is a $\varepsilon_0 > 0$ such that $[a, a+\varepsilon_0) \subseteq U_{\alpha_0}$, and therefore $x \in K$ for every $x \in (a, a+\varepsilon_0)$. Because K is nonempty and is bounded above by $b \in \mathbb{R}$, the supremum of K, $c = \sup K$, exists.

Select $\alpha_1 \in \Lambda$ such that $c \in U_{\alpha_1}$. Because U_{α_1} is open, there is a $\varepsilon_1 > 0$ such that $(c - \varepsilon_1, c] \subset U_{\alpha_1}$. And because c is the least upper bound for K, there exists $z \in K$ such that $z \in (c - \varepsilon_1, c)$. Because $z \in K$, there is a finite subset $F \subset \Lambda$ for which $[a,z] \subseteq \bigcup_{\alpha \in F} U_\alpha$. Therefore,

$$[a,c] = [a,z] \bigcup (c - \varepsilon_1, c] \subseteq \left(\bigcup_{\alpha \in F} U_\alpha \right) \bigcup U_{\alpha_1} = \bigcup_{\beta \in F'} U_\beta,$$

where $F' = F \cup \{\alpha_1\}$, implying that $c \in K$.

Now if it were true that $c \neq b$, then we would have $(c, c+\varepsilon) \subseteq U_{\alpha_1}$ for some $\varepsilon > 0$. By selecting any $w \in (c, c+\varepsilon)$ we would obtain $[c,w] \subset U_{\alpha_1}$, and so $[a,w] = [a,c] \cup [c,w] \subseteq \bigcup_{\beta \in F'} U_\beta$, implying that $w \in K$. But $w \in K$ would be in contradiction to $c = \sup K$. Hence, it must be that $b \in K$, which proves that $[a,b]$ is compact. $\square$

A convenient characterisation of compactness is given by Proposition 2.5 below.

Definition 2.4. *A family $\{E_\alpha\}_{\alpha \in \Lambda}$ of subsets $E_\alpha \subseteq X$ has the finite intersection property if $\bigcap_{\alpha \in F} E_\alpha \neq \emptyset$ for every finite subset $F \subseteq \Lambda$.*

Proposition 2.5. *The following statements are equivalent for a topological space X:*

1. *X is compact;*
2. *$\bigcap_{\alpha \in \Lambda} F_\alpha \neq \emptyset$, for every family $\{F_\alpha\}_{\alpha \in \Lambda}$ of closed sets $F_\alpha \subseteq X$ with the finite intersection property.*

Proof. Exercise 2.81. $\square$

In a metric space (X,d), if $x_1, x_2 \in X$, and if $0 < \varepsilon \leq \frac{1}{2}d(x_1, x_2)$, then $B_\varepsilon(x_1) \cap B_\varepsilon(x_2) = \emptyset$. Thus, the neighbourhoods $B_\varepsilon(x_1)$ and $B_\varepsilon(x_2)$ of x_1 and x_2, respectively, separate the points x_1 and x_2. Topological spaces with this separation property are called T_2-spaces or Hausdorff spaces.

Definition 2.6. A topological space X is a *Hausdorff space* if, for every pair of distinct points $x, y \in X$, there are neighbourhoods U and V of x and y, respectively, such that $U \cap V = \emptyset$.

The definition of "topology" is so general that the axioms on their own are insufficient to settle the question of whether point sets (that is, sets of the form $\{x\}$, for $x \in X$) are closed. By adding the Hausdorff separation axiom, this fact about closedness can be deduced.

Proposition 2.7. *If Y is a finite set in a Hausdorff space X, then Y is closed.*

Proof. Exercise 2.83. □

The Hausdorff property also has a role in determining which subsets of a space are compact.

Proposition 2.8. *Assume that $K \subseteq X$.*

1. *If X is compact and K is closed, then K is compact.*
2. *If X is Hausdorff and K is compact, then K is closed.*

Proof. Suppose that X is compact and K is closed, and suppose that $\{F_\alpha\}_{\alpha \in \Lambda}$ is a family of closed sets $F_\alpha \subseteq K$ with the finite intersection property. Because K is closed, the family $\{F_\alpha\}_{\alpha \in \Lambda}$ is also closed in X (Proposition 1.61); and because X is compact, $\bigcap_{\alpha \in \Lambda} F_\alpha \neq \emptyset$, by Proposition 2.5. Hence, K is compact.

Next, suppose that X is Hausdorff and K is compact. Let $U = K^c$ and choose $x \in U$. Because X is Hausdorff, for every $y \in K$ there are disjoint neighbourhoods U_y and V_y of x and y, respectively. By the compactness of K, the open cover $\{V_y\}_{y \in K}$ admits a finite subcover $\{V_{y_j}\}_{j=1}^n$. The corresponding sets $U_{y_1}, \ldots, U_{y_n}$ are neighbourhoods of x, and so $W_x = \bigcap_{j=1}^n U_{y_j}$ is a neighbourhood of x disjoint from $\bigcup_{j=1}^n V_{y_j}$ and, hence, disjoint from K. Thus, $K^c = \bigcup_{x \in K^c} W_x$ is open, which implies that K is closed. □

Compactness is a topological property that is preserved under continuous maps of spaces.

Proposition 2.9. *Suppose that $f : X \to Y$ is a continuous maps of topological spaces.*

1. *If X is compact, then $f(X)$ is a compact subset of Y.*
2. *If X is compact, Y is Hausdorff, and if f is a bijection, then f^{-1} is continuous.*

Proof. Assuming that X is compact, let $\{V_\alpha\}_{\alpha \in \Lambda}$ be an open cover of $f(X)$. Thus, $\{f^{-1}(V_\alpha)\}_{\alpha \in \Lambda}$ is an open cover of X and therefore admits a finite subcover $\{f^{-1}(V_\alpha)\}_{\alpha \in F}$ for some finite subset $F \subseteq \Lambda$. The inclusion

$$f(X) \subseteq f\left(\bigcup_{\alpha \in F} f^{-1}(V_\alpha)\right) \subseteq \bigcup_{\alpha \in F}(V_\alpha),$$

implies that $\{(V_\alpha)\}_{\alpha \in F}$ is a finite subcover of $f(X)$, and so $f(X)$ is compact.

If, in addition to X being compact, Y is Hausdorff and f is bijective, then consider the function $g = f^{-1} : Y \to X$. If $K \subseteq X$ is closed, then K is compact; and, by what we just proved, $g^1(K) = f(K)$ is compact in Y. Because Y is Hausdorff, $f(K)$ is closed. Hence, $g^{-1}(K)$ is closed in Y for every closed set K in X, which proves that $g = f^{-1}$ is continuous.   □

The second assertion of Proposition 2.9 says that *if $f : X \to Y$ is a continuous bijection of a compact space X to a Hausdorff space Y, then f is a homeomorphism*, which simplifies considerably the proof in Example 1.100 that the quotient space $\mathbb{R}/\mathbb{Z}$ and the unit circle S^1 are homeomorphic.

Example 2.10. $\mathbb{R}/\mathbb{Z} \simeq S^1$.

Proof. Example 1.96 already shows that the map $f : \mathbb{R}/\mathbb{Z} \to S^1$ defined by $f(\dot{t}) = (\cos 2\pi t, \sin 2\pi t)$, for $t \in \mathbb{R}$, is continuous. As f is plainly bijective, and because S^1 is Hausdorff (as a subspace of the metric space $\mathbb{R}^2$), all that remains is to show that $\mathbb{R}/\mathbb{Z}$ is compact. To this end, let $\{V_\alpha\}_{\alpha \in \Lambda}$ be an open cover of $\mathbb{R}/\mathbb{Z}$. Thus, $\{q^{-1}(V_\alpha)\}_{\alpha \in \Lambda}$ is an open cover of $\mathbb{R}$ and, in particular, of the closed interval $[0, 1]$. By the compactness of $[0, 1]$, there exists a finite subset $F \subseteq \Lambda$ such that $\{q^{-1}(V_\alpha)\}_{\alpha \in F}$ is a cover of $[0, 1]$. The inclusion $\mathbb{R}/\mathbb{Z} = q([0, 1]) \subseteq \bigcup_{\alpha \in F}(V_\alpha)$ shows that $\mathbb{R}/\mathbb{Z}$ is compact. Hence, by Proposition 2.9, f is a homeomorphism.   □

Another application of Proposition 2.9 is Proposition 2.12 below, which is similar in essence to the familiar theorem from group theory that the range of a group homomorphism $\varphi : G \to H$ is isomorphic to the quotient group $G/\ker \varphi$.

Definition 2.11. A topological space Y is *a quotient* of a topological space X if there exists an equivalence relation on X such that Y and the quotient space $X/\sim$ are homeomorphic.

Proposition 2.12. *If X and Y are compact Hausdorff spaces, and if $f : X \to Y$ is a continuous surjection, then Y is a quotient of X.*

Proof. Define a relation $\sim$ on X by $x_1 \sim x_2$, if $f(x_1) = f(x_2)$. It is plain to see that $\sim$ is an equivalence relation. Define a map $g : (X/\sim) \to Y$ by $g(\dot{x}) = f(x)$. Note that g is well defined and that $f = g \circ q$, where $q : X \to X/\sim$ is the canonical quotient map. Hence, by Proposition 1.97, the continuity of f and q imply the continuity of g. The continuous map g is plainly a bijection. Therefore, in light of the fact that $X/\sim$ is compact (Exercise 1.129) and Y is Hausdorff, Proposition 2.9 implies that g is a homeomorphism.   □

The following proposition, which asserts that a continuous real-valued function on a compact space achieves both its maximum and minimum values, is one of the single-most important results concerning continuous functions on compact sets.

Proposition 2.13. *If $f : X \to \mathbb{R}$ is a continuous map of a compact space X, then there are $x_0, x_1 \in X$ such that $f(x_0) \le f(x) \le f(x_1)$ for all $x \in X$.*

Proof. Let $\varepsilon > 0$ be given, and consider the open cover $\{B_\varepsilon(f(x))\}_{x \in X}$ of $f(X)$. By Proposition 2.9, $f(X)$ is a compact subset of $\mathbb{R}$; thus, there are $x_1, \ldots, x_n \in X$ such that $\{B_\varepsilon(f(x_j))\}_{j=1}^n$ covers $f(X)$. Hence, $f(X)$ is contained within the union of a finite number of finite open intervals, which implies that $f(X)$ is bounded both below and above. Therefore, by the completeness property of the real numbers, $c = \inf f(X)$ and $d = \sup f(X)$ exist.

Suppose that $c \notin f(X)$; then $(c, c + \varepsilon) \cap f(X) \neq \emptyset$ for every $\varepsilon > 0$. Therefore, $\{(y, \infty)\}_{y \in f(X)}$ is an open cover of $f(X)$, and this cover admits a finite subcover $\{(y_j, \infty)\}_{j=1}^n$ for some $y_1, \ldots, y_n \in f(X)$. Let $y = \min\{y_1, \ldots, y_n\}$. However, $y \in f(X) \subset (y, \infty)$ and $y \notin (y, \infty)$ is a contradiction. Hence, it must be that $c \in f(X)$, and so $c = f(x_0)$, for some $x_0 \in X$. A similar argument yields $d = f(x_1)$, for some $x_1 \in X$. $\qquad\square$

The closed interval $[0, 1]$ is compact, and one might imagine that the same would be true of the closed unit square, $[0, 1] \times [0, 1]$, and, in higher dimensions, of the closed unit n-cube $[0, 1]^n$. Such is indeed the case as a consequence of the following theorem of Tychonoff, which is a powerfully general result.

Theorem 2.14 (Tychonoff). *If $\{X_\alpha\}_{\alpha \in \Lambda}$ is a family of compact spaces, then the product space $\prod_{\alpha \in \Lambda} X_\alpha$ is compact in the product topology.*

Proof. Let $X = \prod_{\alpha_\Lambda} X_\alpha$, and suppose that $\mathscr{G} = \{G_\gamma\}_{\gamma \in \Gamma}$ is a family of subsets of X with the finite intersection property. By Exercise 2.82, to prove that X is compact it is sufficient to prove that $\bigcap_{\gamma \in \Gamma} \overline{G}_\gamma \neq \emptyset$.

In what follows, $\mathscr{E}$ shall denote an arbitrary family of subsets of X. Consider

$$\mathfrak{S} = \{\mathscr{E} \subseteq \mathscr{P}(X) \mid \mathscr{E} \supseteq \mathscr{G} \text{ and } \mathscr{E} \text{ has the finite intersection property}\}.$$

Impose a partial order on $\mathscr{G}$ by inclusion: $\mathscr{E} \preceq \mathscr{E}'$ if $\mathscr{E} \subseteq \mathscr{E}'$.

Suppose that $\mathfrak{L} \subseteq \mathfrak{S}$ is an arbitrary totally ordered subset, and define $\mathscr{U} = \bigcup_{\mathscr{E} \in \mathfrak{L}} \mathscr{E}$.

Select $n \in \mathbb{N}$, and any $E_1, \ldots, E_n \in \mathscr{U}$. Because $\mathfrak{L}$ is totally ordered, there exists $\mathscr{E} \in \mathfrak{L}$ such that $E_j \in \mathscr{E}$ for $j = 1, \ldots, n$. The finite intersection property of the family $\mathscr{E}$ yields $\bigcap_{j=1}^n E_j \neq \emptyset$, which proves that $\mathscr{U}$ has the finite intersection property. Moreover, $\mathscr{U}$ plainly contains $\mathscr{G}$. Hence, $\mathscr{U} \in \mathfrak{S}$ and $\mathscr{E} \preceq \mathscr{U}$ for every $\mathscr{E} \in \mathfrak{L}$; that is, $\mathscr{U}$ is an upper bound in $\mathfrak{S}$ for the totally ordered subset $\mathfrak{L}$. Therefore, by Zorn's Lemma (Theorem 1.8), $\mathscr{S}$ contains a maximal element, which we denote by $\mathscr{M}$.

Two observations concerning $\mathscr{M}$ are:

(i) if $Y \subseteq X$ satisfies $Y \cap M \neq \emptyset$ for every $M \in \mathscr{M}$, then $Y \in \mathscr{M}$; and

(ii) if $M_1, \ldots, M_n \in \mathscr{M}$, then $\bigcap_{j=1}^n M_j \in \mathscr{M}$.

The first observation above is verified by noting that the hypothesis implies that $\{Y\} \cup \mathcal{M}$ is in $\mathfrak{S}$, and that $\mathcal{M} \preceq \{Y\} \cup \mathcal{M}$, which can occur only if $\mathcal{M} = \{Y\} \cup \mathcal{M}$. Therefore, $Y \in \mathcal{M}$. The second observation is a consequence of the first: let Y be the set $Y = \bigcap_{j=1}^{n} M_j$, and make use of the fact that $\mathcal{M}$ has the finite intersection property.

For each $\alpha \in \Lambda$, consider the family $\mathcal{E}_\alpha = \{p_\alpha(M)\}_{M \in \mathcal{M}}$ of subsets of X_α, where each $p_\alpha : X \to X_\alpha$ is the projection map. Fix α, and select finitely many sets in $\mathcal{E}_\alpha$, say $p_\alpha(M_1), \ldots, p_\alpha(M_n)$. Because $\bigcap_{j=1}^{n} M_j \in \mathcal{M}$ by observation (ii), it is also true that $\bigcap_{j=1}^{n} p_\alpha(M_j) \neq \emptyset$. Thus, the family $\mathcal{E}_\alpha$ has the finite intersection property in the compact space X_α, and so, by Exercise 2.82, $\bigcap_{M \in \mathcal{M}} \overline{p_\alpha(M)} \neq \emptyset$.

Suppose now that $x = (x_\alpha)_{\alpha \in \Lambda}$, where each $x_\alpha \in \bigcap_{M \in \mathcal{M}} \overline{p_\alpha(M)}$. If $\alpha \in \Lambda$ and V_α is a neighbourhood of x_α in X_α, then $U_\alpha = p_\alpha^{-1}(V_\alpha)$ is a basic open set in X containing x. Moreover, because $x_\alpha \in \overline{p_\alpha(M)}$ for every $M \in \mathcal{M}$, we deduce that $V_\alpha \cap p_\alpha(M) \neq \emptyset$ for every $M \in \mathcal{M}$. That is, for each $M \in \mathcal{M}$ there is a $y^M \in X$ such that $y^M \in p_\alpha^{-1}(V_\alpha) \cap M$. Hence, $U_\alpha \cap M \neq \emptyset$ for every $M \in \mathcal{M}$. By observation (i), we deduce that $U_\alpha \in \mathcal{M}$.

Lastly, suppose that $B \subseteq X$ is a basic open set containing x. Thus, there are $\alpha_1, \ldots, \alpha_n \in \Lambda$ such that $B = \bigcap_{j=1}^{n} p_{\alpha_1}^{-1}(V_{\alpha_j}) = \bigcap_{j=1}^{n} U_{\alpha_j}$, where V_{α_j} is a neighbourhood of x_{α_j} in X_{α_j}. Because each $U_{\alpha_j} \in \mathcal{M}$, observation (i) yields $B \in \mathcal{M}$. Hence, $x \in B$ and $B \cap M \neq \emptyset$ for every $M \in \mathcal{M}$ implying that $x \in \bigcap_{M \in \mathcal{M}} \overline{M}$. Now because $\mathcal{M} \supseteq \mathcal{G}$, it is also true that $\bigcap_{G \in \mathcal{G}} \overline{G} \neq \emptyset$, thereby proving that X is compact. □

Another use of compactness in analysis arises from the notion of convergence.

Definition 2.15. A sequence $\{x_n\}_{n \in \mathbb{N}}$ of elements x_n in a topological space X is *convergent* if there exists an element $x \in X$ with the property that for every neighbourhood U of x there is a positive integer $k_U \in \mathbb{N}$ such that $x_n \in U$ for every $n \geq k_U$. Such an element x is called a *limit* of the sequence.

In Hausdorff spaces, limits of convergent sequences are unique.

Proposition 2.16. *If x and x' are limits of a convergent sequence $\{x_n\}_{n \in \mathbb{N}}$ in a Hausdorff space X, then $x' = x$.*

Proof. Exercise 2.88. □

In metric spaces, compactness has numerous advantageous features, such as the uniform continuity of continuous functions (Proposition 2.18 below). A key lemma for the analysis of compact sets in metric spaces is the following result, the proof of which is outlined in Exercise 2.85.

Lemma 2.17. *If $\{U_\alpha\}_{\alpha \in \Lambda}$ is an open cover of a compact metric space (X,d), then there exists a $\delta > 0$ such that for each nonempty subset $S \subseteq X$ for which $\sup\{d(s_1,s_2) \,|\, s_1,s_2 \in S\} < \delta$ there is a $\alpha_0 \in \Lambda$ such that $S \subseteq U_{\alpha_0}$.*

Proof. Exercise 2.85.   □

Proposition 2.18 (Uniform Continuity). *Suppose that $f : X \to Y$ is a continuous function, where (X,d_X) and (Y,d_Y) are metric spaces. If X is compact, then for every $\varepsilon > 0$ there exists a $\delta > 0$ such that $d_Y(f(x_1),f(x_2)) < \varepsilon$ for all $x_1,x_2 \in X$ that satisfy $d_X(x_1,x_2) < \delta$.*

Proof. Let $\varepsilon > 0$ and consider the open covering $\{f^{-1}(B_\varepsilon(y))\}_{y \in Y}$ of X. By Lemma 2.17, there is a $\delta' > 0$ for which any nonempty subset $S \subseteq X$ that satisfies $\sup\{d(s_1,s_2) \,|\, s_1,s_2 \in S\} < \delta'$ also satisfies $S \subseteq f^{-1}(B_\varepsilon(y))$ for at least one $y \in Y$. Therefore, if $\delta = \delta'/2$ and if $x_1,x_2 \in X$ are such that $d_X(x_1,x_2) < \delta$, then taking $S = B_\delta(x_1)$ leads to $B_\delta(x_1) \subset f^{-1}(B_\varepsilon(y))$ for some $y \in Y$, implying that $d_Y(f(x_1),f(x_2)) < \varepsilon$.   □

Another major feature of compactness in metric spaces is the following characterisation of compactness in terms of convergent subsequences.

Theorem 2.19. *The following statements are equivalent for a metric space (X,d):*

1. *X is compact;*
2. *if $\{x_n\}_{n \in \mathbb{N}}$ is a sequence in X, then there is a subsequence $\{x_{n_j}\}_{j \in \mathbb{N}}$ of $\{x_n\}_{n \in \mathbb{N}}$ that is convergent to some $x \in X$.*

Proof. Suppose first that X is compact. If $\{x_n\}_{n \in \mathbb{N}}$ is a finite set, then some elements in the sequence are repeated infinitely often. The constant subsequence extracted from an element $x \in \{x_n\}_{n \in \mathbb{N}}$ that is repeated infinitely often is trivially convergent to x. Therefore, suppose that $Y = \{x_n\}_{n \in \mathbb{N}}$ is an infinite set of elements of X. Consider the set $L(Y)$ of limit points of Y, and recall that $x \in L(Y)$ if and only if $(U \setminus \{x\}) \cap Y \neq \emptyset$ for every neighbourhood U of x. Hence, if $x \in L(Y)$, then for each $j \in \mathbb{N}$ there is an element $x_{n_j} \in (B_{1/j}(x) \setminus \{x\}) \cap Y$, and so the subsequence $\{x_{n_j}\}_{j \in \mathbb{N}}$ of $\{x_n\}_{n \in \mathbb{N}}$ converges to x.

The only issue left to resolve is whether it is indeed true that $L(Y) \neq \emptyset$. This is settled by using the compactness and Hausdorff properties of X. If $L(Y)$ contained an element x, then $(U \setminus \{x\}) \cap Y \neq \emptyset$ for every neighbourhood U of x. Furthermore, more is true: namely, $U \cap Y$ is infinite for every neighbourhood U of x. To prove this assertion, suppose that U is a neighbourhood of x and that $U \cap Y$ is finite. Thus, $(U \setminus \{x\}) \cap Y$ is also a finite set, and therefore, by Proposition 2.7, $(U \setminus \{x\}) \cap Y$ is a closed set. The complement W of $(U \setminus \{x\}) \cap Y$ is open and, hence, so is $V = U \cap W$. Because V is a neighbourhood of $x \in L(Y)$, $(V \setminus \{x\}) \cap Y$ is nonempty. But $(V \setminus \{x\}) \cap Y \neq \emptyset$ is in contradiction to $(V \setminus \{x\}) \cap Y = W^c \cap W = \emptyset$. Therefore, it must be that $U \cap Y$ is infinite.

To complete the verification that $L(Y) \neq \emptyset$, suppose, on the contrary, that the set $L(Y) = \emptyset$. Thus, the equality of $\overline{Y}$ and $Y \cup L(Y)$ implies that Y is a closed set in a compact space X; therefore, Y is also compact. The fact that $x_n \notin L(Y)$, for each $n \in \mathbb{N}$, implies that there is a neighbourhood U_n of x_n such that $U_n \cap Y$ is a finite set (by the previous paragraph). From the open cover $\{U_n\}_{n \in \mathbb{N}}$ of the compact space Y, extract a finite subcover $\{U_{n_j}\}_{j=1}^{m}$. Thus, $Y \subseteq \bigcup_{j=1}^{m} U_{n_j}$. However, $Y \cap U_{n_j}$ if finite for each j, and so Y itself must be a finite set, which is a contradiction. Therefore, it must be that $L(Y) \neq \emptyset$.

Conversely, suppose that every sequence in X admits a convergent subsequence. Let $\{U_\alpha\}_{\alpha \in \Lambda}$ be an arbitrary open cover of X. We claim that the conclusion of Lemma 2.17 holds: namely, that there exists a $\delta > 0$ such that, for each nonempty subset $S \subseteq X$ for which $\sup\{d(s_1, s_2) \mid s_1, s_2 \in S\} < \delta$, there is a $\alpha_S \in \Lambda$ such that $S \subseteq U_{\alpha_S}$. If this were not true, then for each $n \in \mathbb{N}$ there would exist a subset $S_n \subseteq X$ such that $d(x, y) < \frac{1}{n}$ for all $x, y \in S_n$ and $S_n \not\subseteq U_\alpha$ for every $\alpha \in \Lambda$. Selecting an element x_n from each set S_n yields a sequence $\{x_n\}_{n \in \mathbb{N}}$ and, by hypothesis, a convergent subsequence $\{x_{n_j}\}_{j \in \mathbb{N}}$. If x is the limit of the convergent subsequence $\{x_{n_j}\}_{j \in \mathbb{N}}$, then $x \in U_{\alpha_0}$ for some $\alpha_0 \in \Lambda$ and there exists a $\varepsilon > 0$ such that $B_\varepsilon(x) \subseteq U_{\alpha_0}$. Furthermore, there exists a $N_1 \in \mathbb{N}$ such that $x_{n_j} \in B_{\varepsilon/2}(x)$ for all $n_j \geq N_1$. By the assumption on the sets S_n, there also exists $N_2 \geq N_1$ such that $S_{n_j} \subseteq B_{\varepsilon/2}(x_{n_j})$ if $n_j \geq N_2$. This would then imply that $S_{n_j} \subseteq U_{\alpha_0}$ for any n_j that satisfies $n_j \geq N_2$, which is in contradiction to the assumption.

As shown in the previous paragraph, the covering $\{U_\alpha\}_{\alpha \in \Lambda}$ of X yields a $\delta > 0$ such that for each nonempty subset $S \subseteq X$ for which $\sup\{d(s_1, s_2) \mid s_1, s_2 \in S\} < \delta$ there is a $\alpha_S \in \Lambda$ such that $S \subseteq U_{\alpha_S}$. Let $\varepsilon = \delta/3$ and consider the open cover of X given by $\{B_\varepsilon(x)\}_{x \in X}$. If $\{B_\varepsilon(x)\}_{x \in X}$ were not to admit a finite subcover of X, then for any $x_1 \in X$ there would exist an element $x_2 \in X$ such that $x_2 \notin B_\varepsilon(x_1)$; likewise, there would exist $x_3 \in X$ such that $x_3 \notin B_\varepsilon(x_2) \cup B_\varepsilon(x_1)$. Indeed, continuing by induction, if $x_1, \ldots, x_n \in X$ are chosen so that $d(x_i, x_j) \geq \varepsilon$ whenever $j \neq i$, then there would also exist $x_{n+1} \in X$ such that $x_{n+1} \notin \cup_{j=1}^{n} B_\varepsilon(x_j)$. However, this would yield a sequence $\{x_n\}_{n \in \mathbb{N}}$ for which $d(x_i, x_j) \geq \varepsilon$ for all $i, j \in \mathbb{N}$ with $j \neq i$, thereby making it impossible for $\{x_n\}_{n \in \mathbb{N}}$ to admit a convergent subsequence. Hence, it must be that $\{B_\varepsilon(x)\}_{x \in X}$ admits a finite subcover $\{B_\varepsilon(z_j)\}_{j=1}^{n}$ of X for some elements $z_1, \ldots, z_n \in X$. Because, for fixed j, $d(x, z_j) < \frac{2}{3}\delta < \delta$ for all $x \in B_\varepsilon(z_j)$, there exists $\alpha_j \in \Lambda$ with $B_\varepsilon(z_j) \subseteq U_{\alpha_j}$. Hence, the covering $\{U_\alpha\}_{\alpha \in \Lambda}$ of X yields a finite subcovering $\{U_{\alpha_j}\}_{j=1}^{n}$, which implies that X is compact. $\qquad \square$

Compactness is a global feature of a topological space. A variant of compactness, defined below, is a local feature.

Definition 2.20. A topological space X is *locally compact* if for each $x \in X$ there is a neighbourhood U of x and a compact subset $K \subseteq X$ such that $x \in U \subseteq K$.

By definition, every compact space is locally compact. The following example is a very familiar non-compact space that exhibits the local compactness property.

Example 2.21. $\mathbb{R}^n$ *is locally compact.*

Proof. Choose any $x = (x_1, \ldots, x_n) \in \mathbb{R}^n$. For each j let $(a_j, b_j) \subset \mathbb{R}$ be an open interval that contains x_j, and let $U = \prod_{j=1}^{n}(a_j, b_j)$ and $K = \prod_{j=1}^{n}[a_j, b_j]$. By definition of product topology, U is open; and by Tychonoff's Theorem, K is compact. As $x \in U \subset K$, this shows that $\mathbb{R}^n$ is locally compact. □

If one imagines the real line $\mathbb{R}$ as being anchored on the left by $-\infty$ and on the right by $+\infty$, then joining these two ends yields a circle, thereby embedding the noncompact space $\mathbb{R}$ into the compact space S^1. This conceptual idea is made rigorous in the following manner.

Theorem 2.22 (One-Point Compactification). *If X is a non-compact locally compact Hausdorff space, then there is a compact Hausdorff space $\tilde{X}$ such that*

1. $X \subset \tilde{X}$ and $\tilde{X} \setminus X$ is a singleton set, and
2. X is open and dense in $\tilde{X}$.

Proof. Let ∞ denote an element that is not in X and let $\tilde{X} = X \cup \{\infty\}$. Define the following collection $\mathscr{B}$ of subsets of $\tilde{X}$:

$$\mathscr{B} = \{U \subseteq X \,|\, U \text{ is open in } X\} \bigcup \{\tilde{X} \setminus K \,|\, K \text{ is compact in } X\}.$$

To prove that $\mathscr{B}$ is a basis, note that, if $x \in X$, then $x \in U$ for some open subset U of X. Also, $\infty \in \tilde{X} \setminus \{x\}$. Hence, for every $x \in \tilde{X}$ there is a $B \in \mathscr{B}$ such that $x \in B$. Suppose now that $B_1, B_2 \in \mathscr{B}$ and that $x \in B_1 \cap B_2$. If B_1 and B_2 are open sets in X, then let $B = B_1 \cap B_2 \in \mathscr{B}$ to obtain $x \in B \subseteq B_1 \cap B_2$. Next, if B_1 is open in X and if $B_2 = \tilde{X} \setminus K$ for some compact $K \subseteq X$, then necessarily $x \neq \infty$ and we may let $B = B_1 \cap (X \setminus K)$ to obtain a $B \in \mathscr{B}$ with $x \in B \subseteq B_1 \cap B_2$. Lastly, assume that $B_j = \tilde{X} \setminus K_j$ for some compact $K_j \subseteq X$, $j = 1, 2$. Let $K = K_1 \cup K_2$, which is compact in X, so that $\tilde{X} \setminus K = B_1 \cap B_2$, yielding an element $B \in \mathscr{B}$ with $x \in B \subseteq B_1 \cap B_2$.

Let $\mathscr{T}$ be the topology on $\tilde{X}$ with basis $\mathscr{B}$. Choose any neighbourhood V of ∞, and let $B \in \mathscr{B}$ be a basic open set with $\infty \in B \subseteq V$. Thus, $B = \tilde{X} \setminus K$ for some compact set K. Because $K \neq X$, there is an $x \in X$ such that $x \notin K$. Hence, $V \cap X \supseteq B \cap X \neq \emptyset$, which implies that $\infty \in \overline{X}$ (the closure of X in $(\tilde{X}, \mathscr{T})$). Hence, X is open, because $X \in \mathscr{B}$, and dense in $\tilde{X}$.

The proof that $\tilde{X}$ is Hausdorff is left as an exercise (Exercise 2.93). To show compactness, let $\{V_\alpha\}_{\alpha \in \Lambda}$ be an open cover of $\tilde{X}$. Thus, there is a $\beta \in \Lambda$ for which $\infty \in V_\beta$. Choose any $B \in \mathscr{B}$ for which $\infty \in B \subseteq V_\beta$; thus, $B = \tilde{X} \setminus K$ for some compact subset $K \subseteq X$. As $\{X \cap V_\alpha\}_{\alpha \in \Lambda}$ is an open cover of K, there are $\alpha_1, \ldots, \alpha_n$ such that $K \subseteq \bigcup_{j=1}^{n}(X \cap V_{\alpha_j})$. Hence $\{V_\beta\} \bigcup \{V_{\alpha_j}\}_{j=1}^{n}$ is a finite subcover of $\tilde{X}$. □

Definition 2.23. The compact Hausdorff space $\tilde{X}$ in Proposition 2.22 is called the *one-point compactification* of X.

2.2 Topological Properties Described by Nets

Formally, a sequence in a set X is a function $f : \mathbb{N} \to X$. One can of course replace $\mathbb{N}$ by some other set Λ. However, $\mathbb{N}$ is not simply a set, it also comes with a linear order $\leq$. Therefore, to extend the notion of sequence to more general contexts, it is useful to consider certain partially ordered sets $(\Lambda, \preceq)$.

Definition 2.24. If $\preceq$ is a partial order on a partially ordered set Λ, then Λ is a *directed set* if for each pair of $\alpha, \beta \in \Lambda$ there is a $\gamma \in \Lambda$ such that $\alpha \preceq \gamma$ and $\beta \preceq \gamma$.

Definition 2.25. A *net* in a set X is a function $\varphi : \Lambda \to X$ for some partially ordered directed set $(\Lambda, \preceq)$.

As with sequences, it is notationally economical to express a net $\varphi : \Lambda \to X$ by its values $x_\alpha = \varphi(\alpha)$, $\alpha \in \Lambda$. Thus, a net in X is a family $\{x_\alpha\}_{\alpha \in \Lambda}$ of elements $x_\alpha \in X$ for some directed set $(\Lambda, \preceq)$.

Definition 2.26. If X is a topological space, then a net $\{x_\alpha\}_{\alpha \in \Lambda}$ in X is *convergent* to $x \in X$ if for every open neighbourhood $U \subseteq X$ of x there is a $\alpha_0 \in \Lambda$ such that $x_\alpha \in U$ for all $\alpha \in \Lambda$ satisfying $\alpha_0 \preceq \alpha$.

The notation $x = \lim_\alpha x_\alpha$ will be used to signify that a net $\{x_\alpha\}_{\alpha \in \Lambda}$ in X is convergent to $x \in X$.

The following proposition shows the relationship between topological closure and the convergence of nets.

Proposition 2.27. *If Y is a nonempty subset of a topological space X, then the following statements are equivalent for an element $x \in A$:*

1. $x \in \overline{Y}$;
2. there exists a net $\{y_\alpha\}_{\alpha \in \Lambda}$ in Y such that $x = \lim_\alpha y_\alpha$.

Proof. The proof of (2) implies (1) is immediate from the definition of convergent net (Exercise 2.95).

Therefore, suppose that $x \in \overline{Y}$. Let Λ be the set of all open subsets $U \subseteq X$ for which $x \in U$, and let $U \preceq V$ denote $V \subseteq U$, for $U, V \in \Lambda$. Note that $\preceq$ is a partial order on Λ. Furthermore, if $U, V \in \Lambda$ and if $W = U \cap V$, then $W \in \Lambda$ and $U \preceq W$ and $V \preceq W$. Hence, Λ is a directed set.

For each $U \in \Lambda$, choose $y_U \in U \cap Y$ (such an element $y_U \in Y$ exists because $x \in \overline{Y}$), and define the map $\varphi : \Lambda \to Y$ by $\varphi(U) = y_U$. The net $\{y_U\}_{U \in \Lambda}$ has the property that for every open neighbourhood V of x there exists an element $V_0 \in \Lambda$ (namely, $V_0 = V$) such that $y_W \in U$ for all $W \in \Lambda$ for which $V_0 \preceq W$. That is, the net $\{y_U\}_{U \in \Lambda}$ in Y converges to x. $\square$

The use of nets allows for the following rephrasing of the property of continuity for functions on topological spaces.

Proposition 2.28. *If X and Y are topological spaces, then the following statements are equivalent for a function $f : X \to Y$:*

1. *f is continuous;*
2. *for every convergent net $\{x_\alpha\}_{\alpha \in \Lambda}$ in X with limit $x \in X$, $\{f(x_\alpha)\}_{\alpha \in \Lambda}$ is a convergent net in Y with limit $f(x)$.*

Proof. Exercise 2.96.   □

In analysis, the language of nets gives a very convenient method for certain proofs. Possibly the most important of these methods is provided by Proposition 2.31 below, which characterises compactness in terms of convergent subnets.

Definition 2.29. A *cofinal* subset of a directed set $(\Lambda, \preceq)$ is a subset $\Delta \subset \Lambda$ with the property that for each $\lambda \in \Lambda$ there exists $\delta \in \Delta$ such that $\lambda \preceq \delta$.

Note that if Δ is a cofinal subset of a directed set $(\Lambda, \preceq)$, and if $\delta_1, \delta_2 \in \Delta$, then in considering δ_1 and δ_2 as elements of Λ there necessarily exists $\lambda \in \Lambda$ with $\delta_1 \preceq \lambda$ and $\delta_2 \preceq \lambda$. Because Δ is cofinal, there in turn exists $\delta \in \Delta$ with $\lambda \preceq \delta$. Hence, $\delta_1 \preceq \delta$ and $\delta_2 \preceq \delta$, which shows that $(\Delta, \preceq)$ is itself a directed set.

Definition 2.30. Assume that $\varphi : \Lambda \to X$, for some directed set $(\Lambda, \preceq)$. A *subnet* of the net $\varphi : \Lambda \to X$ is a function $\vartheta : \Omega \to X$ of the form $\vartheta = \varphi \circ \psi$, where $\psi : \Omega \to \Lambda$ is a function on a directed set $(\Omega, \tilde{\preceq})$ such that

1. $\psi(\Omega)$ is a cofinal subset of Λ, and
2. $\psi(\omega_1) \preceq \psi(\omega_2)$, for all $\omega_1, \omega_2 \in \Omega$ with $\omega_1 \tilde{\preceq} \omega_2$.

For the purposes of simplified notation, if $\{x_\alpha\}_{\alpha \in \Lambda}$ denotes a net, then $\{x_\omega\}_{\omega \in \Omega}$ shall denote a subset of $\{x_\alpha\}_{\alpha \in \Lambda}$.

Proposition 2.31. *The following statements are equivalent for a topological space X:*

1. *X is compact;*
2. *every net in X admits a convergent subnet.*

Proof. We shall make use of the criterion of Proposition 2.5: namely, that X is compact if and only if $\bigcap_{\alpha \in \Lambda} F_\alpha \neq \emptyset$ for every family $\{F_\alpha\}_{\alpha \in \Lambda}$ of closed sets with the finite intersection property.

To begin, suppose that X is compact and let $\{x_\alpha\}_{\alpha \in \Lambda}$ be an arbitrary net in X. Formally, there is a function $\varphi : \Lambda \to X$ on a directed set $(\Lambda, \preceq)$ such that $\varphi(\alpha) = x_\alpha$ for every $\alpha \in \Lambda$. For each $\alpha \in \Lambda$, let $S_\alpha = \{x_\beta \in X \,|\, \alpha \preceq \beta\}$. Because Λ is a directed set, $S_{\alpha_1} \cap S_{\alpha_2} \neq \emptyset$ for all $\alpha_1, \alpha_2 \in \Lambda$. Hence, the collection $\{S_\alpha \,|\, \alpha \in \Lambda\}$ has the finite intersection property and so, by the compactness of X, there exists $x \in X$ such that $x \in \overline{S}_\alpha$ for every $\alpha \in \Lambda$.

Let $\mathcal{O} = \{U \subseteq X \,|\, U \text{ is an open set, and } x \in U\}$ and, for each $U \in \mathcal{O}$, define $\Lambda_U = \{\alpha \in \Lambda \,|\, x_\alpha \in U\}$. Each of the sets Λ_U is cofinal in Λ. To verify this assertion, choose $U \in \mathcal{O}$ and $\alpha \in \Lambda$. Because $x \in \overline{S}_\alpha$, the open set U has nonempty intersection with S_α. Thus, there is some $x_\beta \in S_\alpha \cap U$. Because the element β satisfies $\alpha \preceq \beta$ and is, by definition, an element of Λ_U, the subset Λ_U is therefore cofinal.

Consider the subset Ω of $\Lambda \times \mathscr{O}$ that consists of all (α, U) for which $\alpha \in \Lambda_U$, and define a partial order $\tilde{\preceq}$ on Ω by

$$(\alpha, U) \tilde{\preceq} (\beta, V), \text{ if } \alpha \preceq \beta \text{ and } V \subseteq U.$$

If $(\alpha, U), (\beta, V) \in \Omega$, then let $\gamma \in \Lambda$ be an element for which $\alpha \preceq \gamma$ and $\beta \preceq \gamma$. Let $W = U \cap V$. Because Λ_W is confinal in Λ, there is a $\delta \in \Lambda_W$ such that $\gamma \preceq \delta$. Hence, the ordered pair (δ, W) is an element of Ω and satisfies $(\alpha, U) \tilde{\preceq} (\delta, W)$ and $(\beta, V) \tilde{\preceq} (\delta, W)$, which proves that $(\Omega, \tilde{\preceq})$ is a directed set.

The function $\psi : \Omega \to \Lambda$ defined by $\psi(\alpha, U) = \alpha$ plainly gives rise to a subnet $\{x_{(\alpha, U)}\}_{(\alpha, U) \in \Omega}$ of $\{x_\alpha\}_{\alpha \in \Lambda}$. To show that this subnet converges to x, select any open set $U \subseteq X$ that contains x. Pick $\alpha_0 \in \Lambda$. Because Λ_U is cofinal, there is a $\alpha \in \Lambda_U$ such that $\alpha_0 \preceq \alpha$. Therefore, $(\alpha, U) \in \Omega$. If $(\beta, V) \in \Omega$ satisfies $(\alpha, U) \tilde{\preceq} (\beta, V)$, then $x_{(\beta, V)} \in V \subseteq U$. Hence, the subnet $\{x_{(\alpha, U)}\}_{(\alpha, U) \in \Omega}$ of $\{x_\alpha\}_{\alpha \in \Lambda}$ is convergent.

Conversely, suppose that every net in X admits a convergent subnet. Suppose that $\{F_\alpha\}_{\alpha \in \Lambda}$ is a family of closed sets that has the finite intersection property. Let $\mathscr{F} = \{F \subset \Lambda \mid F \text{ is finite}\}$ and define $\preceq$ on $\mathscr{F}$ by $F \preceq G$ if $G \subseteq F$. Because $\{F_\alpha\}_{\alpha \in \Lambda}$ has the finite intersection property, $(\mathscr{F}, \preceq)$ is a directed set. For each $F \in \mathscr{F}$, the set $\bigcap_{\alpha \in F} F_\alpha$ is nonempty; thus, select x_F in this intersection. Now define $\varphi : \mathscr{F} \to X$ by $\varphi(F) = x_F$; that is, consider the net $\{x_F\}_{F \in \mathscr{F}}$. By hypothesis, $\{x_F\}_{F \in \mathscr{F}}$ admits a convergent subset net $\{x_\omega\}_{\omega \in \Omega}$, with limit $x \in X$, where without loss of notational generality we may assume that Ω is a cofinal subset of $\mathscr{F}$.

Fix $\beta \in \Lambda$. We shall prove that $x \in F_\beta$, which will imply that $\bigcap_{\alpha \in \Lambda} F_\alpha$ is nonempty. To this end, let U be any open neighbourhood U of x. Thus, there is a $G_0 \in \Omega$ such that $x_G \in U$ for all $G \in \Omega$ for which $G_0 \preceq G$. Now because $\{\beta\} \in \mathscr{F}$, there is an $F \in \mathscr{F}$ with $\{\beta\} \preceq F$ and $G_0 \preceq F$. As Ω is cofinal in $\mathscr{F}$, we may in fact assume that $F \in \Omega$. Hence, $F \subseteq \{\beta\}$ implies that $x_F \in F_\beta$. Furthermore, $G_0 \preceq F$ implies that $x_F \in U$. Hence, $U \cap F_\beta$ is nonempty. Because U is an arbitrary open neighbourhood of x, we deduce that $x \in \overline{F_\beta} = F_\beta$. $\square$

Observe that the statement of Proposition 2.31 is almost superficially trivial, while all of the underlying topology embodied by the statement is buried within the proof. This is what makes the use of nets so compelling in analysis.

The Hausdorff property may also be characterised by a property of nets; however, the proof of Proposition 2.32 below is not nearly as subtle as the proof of Proposition 2.31.

Proposition 2.32. *The following statements are equivalent for a topological space X:*

1. X is Hausdorff;
2. every convergent net $\{x_\alpha\}_{\alpha \in \Lambda}$ in X has a unique limit point.

Proof. Exercise 2.97. $\square$

2.3 Normal Spaces

Definition 2.33. A topological space X is *normal* if

1. $\{x\}$ is a closed set for every $x \in X$, and
2. for every pairs of disjoint closed sets $C, F \subseteq X$ there exist disjoint open sets $U, V \subseteq X$ with $C \subseteq U$ and $F \subseteq V$.

The notion of normality is a separation property, not unlike the separation property that defines a Hausdorff space.

Proposition 2.34. *Every compact Hausdorff space is normal.*

Proof. Suppose that X is a compact Hausdorff space. By the Hausdorff property, $\{x\}$ is closed for every $x \in X$ (Proposition 2.7). Therefore, assume that $C, F \subseteq X$ are closed sets such that $C \cap F = \emptyset$. Because C and F are closed and X is compact, both C and F are compact.

Select $x \in F$. For each $y \in C$ there are neighbourhoods V_y of x and U_y of y such that $U_y \cap V_y = \emptyset$. The family $\{U_y\}_{y \in C}$ is an open cover of C and therefore admits a finite subcover $\{U_{y_j}\}_{j=1}^n$ for some $y_1, \dots, y_n \in C$. Let $V_x = \bigcap_{j=1}^n V_{y_j}$ and $U_x = \bigcup_{j=1}^n U_{y_j}$; thus, V_x is a neighbourhood of x, $U_x \supseteq C$, and $U_x \cap V_x = \emptyset$.

Carrying out the procedure above for every $x \in F$ leads to an open cover $\{V_x\}_{x \in F}$ of F. By the compactness of F, there is a finite subcover $\{V_{x_i}\}_{i=1}^m$ for some $x_1, \dots, y_m \in F$. Let $V = \bigcup_{i=1}^m V_{x_i}$ and $U = \bigcap_{i=1}^m U_{x_i}$, which are open sets for which $C \subseteq U$, $F \subseteq V$, and $U \cap V = \emptyset$. Hence, X is normal. $\qquad\square$

Another class of normal spaces is that of metric spaces.

Proposition 2.35. *Every metric space is normal.*

Proof. Suppose that X is a metric space. By the Hausdorff property of metric spaces, $\{x\}$ is closed for every $x \in X$.

Next, suppose that $C, F \subseteq X$ are closed sets such that $C \cap F = \emptyset$. Select $x \in C$. Then x is contained in the open set F^c, and so there is a $\varepsilon_x > 0$ such that $B_{\varepsilon_x}(x) \cap F = \emptyset$. Similarly, for each $y \in F$ there is a $\varepsilon_y > 0$ such that $B_{\varepsilon_y}(y) \cap C = \emptyset$. Let

$$U = \bigcup_{x \in C} B_{\varepsilon_x/2}(x) \quad \text{and} \quad V = \bigcup_{y \in F} B_{\varepsilon_y/2}(y).$$

Thus, U and V are open sets such that $C \subseteq U$ and $F \subseteq V$. If, for some $x \in C$ and $y \in F$, $B_{\varepsilon_x/2}(x) \cap B_{\varepsilon_y/2}(y)$ were nonempty, then via $z \in B_{\varepsilon_x/2}(x) \cap B_{\varepsilon_y/2}(y)$ we would obtain

$$d(x, y) \le d(x, z) + d(z, y) < \varepsilon_x/2 + \varepsilon_y/2 \le \max\{\varepsilon_x, \varepsilon_y\},$$

implying that $y \in B_{\varepsilon_x}(x)$, or $x \in B_{\varepsilon_y}(y)$, both of which are in contradiction of the fact that $B_{\varepsilon_y}(y) \cap C = B_{\varepsilon_x}(x) \cap F = \emptyset$. Hence, $B_{\varepsilon_x/2}(x) \cap B_{\varepsilon_y/2}(y) = \emptyset$ for all $x \in C$ and $y \in F$, and therefore $U \cap V = \emptyset$.   □

An alternate characterisation of normality is given by the following proposition.

Proposition 2.36. *The following statements are equivalent for a topological space X:*

1. *X is normal;*
2. *X has the properties that*

   a. *$\{x\}$ is a closed set, for all $x \in X$, and*
   b. *for every closed set F and open set U for which $F \subseteq U$, there is an open set V with $F \subseteq V \subseteq \overline{V} \subseteq U$.*

Proof. Exercise 2.98.   □

The following theorem captures the most important feature of normal spaces.

Theorem 2.37 (Tietze Extension Theorem). *If A is a closed subset of a normal topological space X, and if $f : A \rightarrow [0,1]$ is a continuous function, then there exists a continuous function $F : X \rightarrow [0,1]$ such that $F(x) = f(x)$ for every $x \in A$.*

The function F above is called an *extension* of f.

Proof. Select $p \in \mathbb{Q}$ and set $A_p = f^{-1}((-\infty, p])$ and $B_p = f^{-1}([p, \infty))$, which are closed subsets of A and hence of X (since A is closed); let $U_p = B_p^c$, which is an open subset of X. Observe that, if $p, q \in \mathbb{Q}$ satisfy $p \leq q$, then $A_p \subseteq A_q$ and $U_p \subseteq U_q$; and, if $p < q$, then $A_p \subseteq U_q$.

Let $\preceq$ be the partial order on $\mathbb{Q} \times \mathbb{Q}$ in which $(p,q) \preceq (p',q')$, if $p \leq p'$ and $q \leq q'$, and let $\mathscr{P} = \{(p,q) \in \mathbb{Q} \times \mathbb{Q} \mid 0 \leq p < q \leq 1\}$.

The set $\mathscr{P}$ is countable, and so there is an enumeration $\{(p_n, q_n)\}_{n \in \mathbb{N}}$ of its elements. With the first of these elements, (p_1, q_1), the inclusion $A_{p_1} \subseteq U_{q_1}$ and the normality of X imply that there is an open set $V_1 \subseteq X$ such that $A_{p_1} \subseteq V_1 \subseteq \overline{V}_1 \subseteq U_{q_1}$ (Proposition 2.36). Consider now the next element of $\mathscr{P}$, namely (p_2, q_2). The partial order $\preceq$ on $\mathbb{Q} \times \mathbb{Q}$ is not a total order; therefore, we must consider the following subcases.

(i) $(p_1, q_1) \preceq (p_2, q_2)$: The inclusions $A_{p_1} \subseteq A_{p_2} \subseteq U_{q_2}$ and $\overline{V}_1 \subseteq U_{q_1} \subseteq U_{q_2}$ imply, by Proposition 2.36, the existence of an open set $V_2 \subseteq X$ such that

$$\left(A_{p_2} \cup \overline{V}_1\right) \subseteq V_2 \subseteq \overline{V}_2 \subseteq U_{q_2}.$$

In particular, $\overline{V}_1 \subseteq V_2$.

(ii) $(p_2, q_2) \preceq (p_1, q_1)$: We have that $A_{p_2} \subseteq A_{p_1} \subseteq U_{q_1}$, because $p_2 \leq p_1 < q_1$, and we know already that $A_{p_1} \subseteq V_1$. Thus, $A_{p_2} \subseteq U_{q_1} \cap V_1$. Again, by Proposition 2.36, there is an open set $V_2 \subseteq X$ for which

$$A_{p_2} \subseteq V_2 \subseteq \overline{V}_2 \subseteq (U_{q_1} \cap V_1).$$

In particular, $\overline{V}_2 \subseteq V_1$.

(iii) Neither $(p_1, q_1) \preceq (p_2, q_2)$ nor $(p_2, q_2) \preceq (p_1, q_1)$: In this case, $p_2 < q_2$, and Proposition 2.36 implies that there is an open set $V_2 \subseteq X$ such that $A_{p_2} \subseteq V_2 \subseteq \overline{V}_2 \subseteq U_{q_2}$.

Therefore, from what has been argued thus far, we have that: (a) $A_{p_k} \subseteq V_k \subseteq \overline{V}_k \subseteq U_{q_k}$ for $k = 1, 2$; (b) $\overline{V}_1 \subseteq V_2$, if $(p_1, q_1) \preceq (p_2, q_2)$; (c) $\overline{V}_2 \subseteq V_1$, if $(p_2, q_2) \preceq (p_1, q_1)$.

Based on the arguments above, proceed by induction. Suppose that open sets $V_k \subseteq X$ have been constructed for $k = 1, \ldots, n-1$ with the properties

(a) $A_{p_k} \subseteq V_k \subseteq \overline{V}_k \subseteq U_{q_k}$ for $k = 1, \ldots, n-1$, and
(b) $\overline{V}_j \subseteq V_k$, if $(p_j, q_j) \preceq (p_k, q_k)$.

Define

$$\mathscr{J}_n = \{ j \mid 1 \leq j \leq n-1 \text{ and } (p_j, q_j) \preceq (p_n, q_n) \} \text{ and}$$
$$\mathscr{K}_n = \{ k \mid 1 \leq k \leq n-1 \text{ and } (p_n, q_n) \preceq (p_k, q_k) \}.$$

By Proposition 2.36, there exists an open set $V_n \subseteq X$ such that

$$A_{p_n} \cup \left(\bigcup_{j \in \mathscr{J}_n} \overline{V}_j \right) \subseteq V_n \subseteq \overline{V}_n \subseteq \left[U_{q_n} \cap \left(\bigcup_{k \in \mathscr{K}_n} U_{q_k} \right) \right].$$

Relabel the sets V_n as V_{pq}, where $p = p_n$ and $q = q_n$. Thus, by the principle of mathematical induction, there exists a family $\{V_{pq}\}_{(p,q) \in \mathscr{P}}$ of open sets $V_{pq} \subseteq X$ for which $A_p \subseteq V_{pq} \subseteq \overline{V}_{pq} \subseteq U_q$, for all $(p, q) \in \mathscr{P}$, and $\overline{V}_{pq} \subseteq V_{p'q'}$, if $(p, q) \preceq (p', q')$.

Let $X_p = \bigcap_{q > p} \overline{V}_{pq}$ for each $p \in \mathbb{Q}$, and observe that $X_p = \emptyset$ for every $p < 0$, and that $X_p = X$ for every $p \geq 1$. For each $(p, q) \in \mathscr{P}$, choose $t \in \mathbb{Q}$ such that $p < t < q$; in so doing, we have the inclusions:

$$X_p \subseteq \overline{V}_{pt} \subseteq V_{tq} \subseteq \bigcap_{s > q} \overline{V}_{qs} = X_q.$$

Note also that $A_p \subseteq \bigcap_{q > q} \overline{V}_{pq} = X_p$, for every $p \in \mathbb{Q} \cap [0, 1)$. Thus,

$$A_p = X_p \cap A = \left(\bigcap_{q > p} \overline{V}_{pq} \right) \cap A \subseteq \left(\bigcap_{q > p} U_q \cap A \right) = A_p.$$

Therefore, $\{X_p\}_{p \in \mathbb{Q}}$ is a family of closed sets such that $X_p \subseteq \operatorname{int} X_q$, for all $q > p$, and $X_p \cap A = A_p$, for all p.

The extension $F : X \to [0,1]$ of $f : A \to [0,1]$ is defined as follows: for each $x \in X$, let

$$F(x) = \inf \{p \in \mathbb{Q} \,|\, x \in X_p\}.$$

Because $X_p = \emptyset$ for every $p < 0$, and $X_p = X$ for every $p \geq 1$, for each $x \in X$ the set $\{p \in \mathbb{Q} \,|\, x \in X_p\}$ is nonempty and is bounded below by 0 and above by 1; hence, F is a well-defined function.

To show that F is indeed an extension of f, recall that $X_p \cap A = A_p$. Thus, if $x \in A$, then

$$F(x) = \inf \{p \in \mathbb{Q} \,|\, x \in X_p\} = \inf \{p \in \mathbb{Q} \,|\, x \in A_p\} = f(x),$$

by the continuity of f.

All that remains, therefore, is to verify that F is continuous. Observe that F has the following two properties:

1. if $x \in X_p$, then $F(x) \leq p$, and
2. if $x \notin \operatorname{int} X_q$ for some $q \in \mathbb{Q}$, then $q \leq F(x)$

The first of these properties above follows from the definition of F. For the second property, suppose that $x \notin \operatorname{int} X_q$ and $F(x) < q$. Because F is defined as an infimum, there must be a rational $p < q$ for which $x \in X_p$. But $p < q$ implies that $X_p \subseteq \operatorname{int} X_q$, in contradiction to the hypothesis. Therefore, it must be true that $q \leq F(x)$.

Fix $x \in X$, and let W be a neighbourhood in $\mathbb{R}$ of $F(x)$. Select $p, q \in \mathbb{Q}$ such that $p < q$ and $F(x) \in (p,q) \subseteq W$. Define $U = \operatorname{int} X_q \cap X_p^c$. Because $p < F(x)$, $x \notin X_p$; and because $F(x) < q$, $x \in \operatorname{int} X_q$. Thus, $x \in U$, and so U is a neighbourhood of x. Lastly, if $z \in U$, then $F(z) < q$, because $z \in \operatorname{int} X_q$, and $p < F(z)$, because $z \notin X_p$. Hence, $F(U) \subseteq W$, which proves that F is continuous at $x \in X$. Since the choice of x is arbitrary, we conclude that F is continuous at every $x \in X$. $\qquad\square$

Corollary 2.38. *If A is a closed subset of a normal topological space X, and if $f : A \to \mathbb{C}$ is a bounded continuous function, then there exists a continuous function $F : X \to \mathbb{C}$ such that $F(x) = f(x)$ for every $x \in A$ and $\sup_{x \in X} |F(x)| = \sup_{a \in A} |f(a)|$.*

Proof. Let $\alpha = \sup_{a \in A} |f(a)|$ so that the range of $f_1 = \frac{1}{\alpha} f$ is in the closed unit disc of $\mathbb{C}$. The range of the real and imaginary parts $\Re f_1$ and $\Im f_1$ of f_1, which themselves are continuous functions on A, lie in the closed interval $[-1,1]$. Therefore, the ranges of $g = \frac{1}{2}(\Re f_1 + 1)$ and $h = \frac{1}{2}(\Im f_1 + 1)$ are contained in the closed interval $[0,1]$. By the Tietze Extension Theorem, each of the continuous functions g and h admits continuous extensions $G : X \to [0,1]$ and $H : X \to [0,1]$, respectively, and so $F = \alpha((2G - 1) + i(2H - 1))$ is a continuous extension of f with the property that $\sup_{x \in X} |F(x)| = \sup_{a \in A} |f(a)|$. $\qquad\square$

Corollary 2.39 (Urysohn's Lemma). *If A and B are disjoint closed subsets of a normal space X, then there exists a continuous function $f : X \to [0,1]$ such that $f(A) = \{0\}$ and $f(B) = \{1\}$.*

Proof. Let $h : A \cup B \to [0,1]$ be defined by $h(x) = 0$, if $x \in A$, and $h(x) = 1$, if $x \in B$. Because $A \cap B = \emptyset$, there is no ambiguity in the definition of h. Moreover, as $h^{-1}(F)$ is closed in $A \cup B$ for every closed $F \subseteq [0,1]$, the map h is continuous. By the Tietze Extension Theorem, h admits a continuous extension $f : X \to [0,1]$.

The next result, Proposition 2.41 below, establishes is a widely used technical tool in analysis.

Definition 2.40. If X is a topological space, then the *support of a continuous function $f : X \to \mathbb{C}$* is the set $\operatorname{supp} f \subseteq X$ defined by

$$\operatorname{supp} f = \overline{\{x \in X \mid f(x) \neq 0\}}.$$

Note that the range $f(X)$ of f satisfies

$$f(\operatorname{supp} f) \subseteq f(X) \subseteq f(\operatorname{supp} f) \cup \{0\}.$$

Therefore, if f has compact support, then $f(\operatorname{supp} f)$ is compact, and so the range $f(X)$ of f is also compact.

Proposition 2.41 (Partitions of Unity). *If $\{U_j\}_{j=1}^n$ is an open cover of a normal space X, then there exist continuous functions $h_1, \ldots, h_n : X \to \mathbb{R}$ such that*

1. *$0 \leq h_j(x) \leq 1$, for every $x \in X$, and each $j = 1, \ldots, n$,*
2. *$\operatorname{supp} h_j \subseteq U_j$, for $j = 1, \ldots, n$, and*
3. *$\displaystyle\sum_{j=1}^n h_j(x) = 1$, for every $x \in X$.*

Proof. Let $F_1 = \left(\displaystyle\bigcup_{j=2}^n U_j \right)$, which is a closed subset of U_1. Because X is normal, there is an open set V_1 such that $F_1 \subseteq V_1 \subseteq \overline{V}_1 \subseteq U_1$. Note that $F_1 \subseteq V_1$ implies that $\{V_1\} \cup \{U_j\}_{j=2}^n$ is an open cover of X. By induction, if open sets $V_1, \ldots V_{k-1}$ have been constructed such that $\{V_\ell\}_{\ell=1}^{k-1} \cup \{U_j\}_{j=k}^n$ is an open cover of X such that $\overline{V}_\ell \subseteq U_\ell$ for all $\ell = 1, \ldots, k-1$, then the closed set $F_k = \left(\displaystyle\bigcup_{\ell=1}^{k-1} V_j \right)^c \cap \left(\displaystyle\bigcup_{j=k+1}^n U_j \right)^c$ will lie in some open set V_k such that $\overline{V}_k \subseteq U_k$ (by the normality of X), and therefore, $\{V_\ell\}_{\ell=1}^k \cup \{U_j\}_{j=k+1}^n$ is an open cover of X. Hence, after n steps, there exists an open cover $\{V_j\}_{j=1}^n$ of X with the property that $\overline{V}_j \subseteq U_j$, for each j.

Applying the argument of the previous paragraph to the open cover $\{V_j\}_{j=1}^n$ of X produces an open cover $\{W_j\}_{j=1}^n$ of X such that $\overline{W}_j \subseteq V_j$, for each j. Use Urysohn's

Lemma to select, for each j, a continuous function $g_j : X \to [0,1]$ with $g_j(\overline{W_j}) = \{1\}$ and $g_j(V_j^c) = \{0\}$. Hence, the support of g_j is contained in $\overline{V_j} \subseteq U_j$. Because $\{W_j\}_{j=1}^n$ covers X, $\sum_{j=1}^n g_j(x) \geq 1$ for every $x \in X$. Thus, for each j, the continuous function $h_j : X \to \mathbb{R}$ defined by

$$h_j(x) = \left(\sum_{\ell=1}^n g_\ell(x), \right)^{-1} g_j(x)$$

satisfies $\operatorname{supp} h_j \subseteq U_j$ and $0 \leq h_j(x) \leq 1$, for all $x \in X$. Moreover, if $x \in X$, then

$$\sum_{j=1}^n h_j(x) = 1.$$ □

Definition 2.42. The family $\{h_1, \ldots, h_n\}$ in Proposition 2.41 is called a *partition of unity of X subordinate to the open cover* $\{U_j\}_{j=1}^n$.

Locally compact Hausdorff spaces need not be compact; however, it is still possible to establish a Tietze Extension Theorem in this setting, which has important consequences in analysis (such as Theorem 5.43).

Theorem 2.43 (Tietze Extension Theorem, II). *If X is a locally compact Hausdorff space, if $K \subset X$ is compact, and if $K \subseteq U$ for some open set $U \subseteq X$, then every continuous function $f : K \to \mathbb{C}$ extends to a continuous bounded function $F : X \to \mathbb{C}$ such that*

1. *$\operatorname{supp} F$ is a compact subset of U,*
2. *$F(x) = 0$ for all $x \in U^c$, and*
3. *$\max_{x \in X} |F(x)| = \max_{y \in K} |f(y)|$.*

Proof. By Exercise 2.91, there exists an open subset $W \subseteq X$ such that $K \subseteq W \subseteq U$ and $\overline{W}$ is compact. Let $\tilde{X}$ be the one-point compactification of X. Because $\tilde{X}$ is compact and Hausdorff, $\tilde{X}$ is normal. Note that $X \setminus K$ a basic open set in the topology of $\tilde{X}$, and so K is a closed subset of $\tilde{X}$. Hence, K is compact subset of $\tilde{X}$. By Corollary 2.38, $f : K \to \mathbb{C}$ admits a continuous extension $\tilde{f} : \tilde{X} \to \mathbb{C}$ such that $\max_{x \in \tilde{X}} |\tilde{f}(x)| = \max_{y \in K} |f(y)|$. In addition, because $\tilde{X} \setminus W$ is closed in $\tilde{X}$ and is disjoint from K, Urysohn's Lemma yields a function $\tilde{h} : \tilde{X} \to [0,1]$ such that $\tilde{h}(K) = \{1\}$ and $\tilde{h}(\tilde{X} \setminus W) = \{0\}$. Hence, the continuous and bounded function $F : X \to \mathbb{C}$ given by $F = (\tilde{h} \cdot \tilde{f})_{|X}$ satisfies $F(x) = f(x)$ for all $x \in K$, $F(x) = 0$ for all $x \notin W$, and $\sup_{x \in X} |F(x)| = \max_{y \in K} |f(y)|$. Furthermore, because $\operatorname{supp} F \subseteq \overline{W}$ and $\overline{W}$ is compact, we deduce that F has compact support and, therefore, the supremum $\sup_{x \in X} |F(x)|$ is achieved at some point $x_0 \in X$.   □

The following consequence of Theorem 2.43 will be used extensively in subsequent chapters.

Corollary 2.44 (Urysohn's Lemma, II). *If K and U are nonempty subsets of a locally compact Hausdorff space X, and if K is compact, if U is open, and if $K \subseteq U$, then there exists a continuous function $f : X \to [0, 1]$ such that $f(K) = \{1\}$ and* supp f *is a compact subset of U.*

Proof. Let $f_0 : K \to \mathbb{R}$ be given by $f_0(x) = 1$, for all $x \in K$, and apply Theorem 2.43 to produce the desired extension $f : X \to \mathbb{R}$ with all the stated properties. That the range of f lies in the interval $[0, 1]$ is a consequence of the proof of Theorem 2.43 and the Tietze Extension Theorem. $\qquad\square$

2.4 Properties of Metric Spaces

Topological spaces that are very large topologically can behave poorly. A case in point is a product space with the box topology. One measure of topological smallness is compactness. Another smallness quality involves countability features of a space.

Definition 2.45. A topological space $(X, \mathscr{T})$ is:

1. *separable*, if there is a countable subset $Y \subseteq X$ such that Y is dense in X (that is, $\overline{Y} = X$);
2. *second countable*, if there is a countable basis $\mathscr{B}$ for the topology $\mathscr{T}$ of X.

Proposition 2.46. *Every second countable space is separable.*

Proof. Suppose that $\mathscr{B} = \{B_n\}_{n \in \mathbb{N}}$ is a countable basis for the topology $\mathscr{T}$ of a topological space $(X, \mathscr{T})$. For each n select $x_n \in B_n$ and let $Y = \{x_n\}_n$, which is a countable subset of X. Suppose now that $x \in X$ and U is a neighbourhood of x. Because $\mathscr{B}$ is a basis, there is a basic open set B_n for which $x \in B_n \subseteq U$. Now since $x_n \in B_n$, we have that $x_n \in U \cap Y$, implying that $U \cap Y \neq \emptyset$. Hence, $\overline{Y} = X$. $\qquad\square$

The converse to Proposition 2.46 is not true in general, but it is true in an important special case.

Proposition 2.47. *Every separable metric space is second countable.*

Proof. Exercise 2.99. $\qquad\square$

There are separable spaces that fail to be second countable (see Proposition 2.79, for example). The following important theorem determines precisely when a compact Hausdorff space is metrisable.

Theorem 2.48 (Compact Metrisable Spaces). *The following statements are equivalent for a compact Hausdorff space X:*

1. *X is metrisable;*
2. *X is second countable.*

Proof. Suppose that X is metrisable. Let d be a metric on X that induces the topology of X. For each $n \in \mathbb{N}$, the family $\{B_{1/n}(x)\}_{x \in X}$ is an open covering of X; thus, there is a finite subcovering $\{B_{1/n}(x_{n,j})\}_{j=1}^{n_k}$ of X. Let $\mathscr{B} = \bigcup_{n \in \mathbb{N}} \{B_{1/n}(x_{n,j})\}_{j=1}^{n_k}$, which is a countable collection. We claim that $\mathscr{B}$ is a basis for the topology on X. To this end, let $U \in X$ be open and consider $x \in U$. Because X is a metric space, there is an $n \in \mathbb{N}$ such that $B_{1/n}(x) \subseteq U$. Likewise, there is an $x' \in \{x_{3n,1}, \dots, x_{3n,(3n)_k}\}$ for which $x \in B_{1/3n}(x')$. Now since $B_{1/3n}(x') \subset B_{1/n}(x)$, there is a set $B \in \mathscr{B}$ with $x \in B \subseteq U$. Hence, $\mathscr{B}$ is a basis for X, which implies that X is second countable.

Conversely, suppose that X is second countable, and let $\mathscr{B} = \{B_n\}_{n \in \mathbb{N}}$ be a countable basis for the topology of X. Let

$$\mathscr{I} = \{(m,n) \in \mathbb{N} \times \mathbb{N} \mid \overline{B}_m \subset B_n\}.$$

By Proposition 2.34, every compact Hausdorff space is normal. Hence, by Urysohn's Lemma, for each $(m,n) \in \mathscr{I}$, there is a continuous map $f_{(m,n)} : X \to [0,1]$ for which $f_{(m,n)}(\overline{B}_m) = \{1\}$ and $f_{(m,n)}(B_n^c) = \{0\}$.

Select $x \in X$ and a neighbourhood U of x. Thus, there is a basic open set $B_n \in \mathscr{B}$ with $x \in B_n \subseteq U$. Because $\{x\}$ is a closed set and X is normal, Proposition 2.36 asserts that there exists an open set V with $x \in V \subseteq \overline{V} \subseteq B_n$. Since V is a neighbourhood of x, there is some $B_m \in \mathscr{B}$ with $x \in B_m \subseteq V$. Hence, $\overline{B}_m \subseteq \overline{V} \subseteq B_n$; therefore, $(m,n) \in \mathscr{I}$, $f_{(m,n)}(x) = 1$, and $f_{(m,n)}(U^c) = \{0\}$.

The previous paragraph shows that there is a countable family $\{g_n\}_n$ of continuous functions $g_n : X \to [0,1]$ with the property that, for each $x \in X$ and neighbourhood U of x, there is some $n \in \mathbb{N}$ such that $g_n(x) = 1$ and $g_n(U^c) = \{0\}$.

The product topology on $[0,1]^{\mathbb{N}}$ coincides with the subspace topology that is induced by $(\mathbb{R}^{\mathbb{N}}, \mathscr{T}^\times)$ (Proposition 1.51). The space $(\mathbb{R}^{\mathbb{N}}, \mathscr{T}^\times)$ is metrisable (Proposition 1.57), and therefore so is the subspace $[0,1]^{\mathbb{N}}$. Define $f : X \to [0,1]^{\mathbb{N}}$ by

$$f(x) = (g_n(x))_n.$$

Each coordinate function g_n is continuous, and so, because $[0,1]^{\mathbb{N}}$ has the product topology, the function f is continuous (Proposition 1.92). The function f is also injective, for if $x,y \in X$ are distinct, then they are separated by disjoint open sets U and V, and so there is a function g_n which maps x to 1 and y to 0, which implies that $f(x) \neq f(y)$.

The map f is a continuous bijection from X to $f(X) \subseteq [0,1]^{\mathbb{N}}$, and the subspace $f(X)$ is a subspace of a metric space, and is therefore Hausdorff. Thus, Proposition 2.9 asserts that f is a homeomorphism, which proves that the topology on X is metrisable. $\qquad\square$

The Cantor ternary function Φ is a continuous map of the Cantor ternary set $\mathscr{C}$ onto the closed unit interval $[0,1]$. There is a very interesting property that is shared by all compact metric spaces.

Proposition 2.49. *If X is a compact metric space, then there exists a continuous function $f : \mathscr{C} \to X$, where $\mathscr{C}$ is the Cantor ternary set, such that f is surjective.*

Proof. By Exercise 1.133, the compact Hausdorff spaces $\mathscr{C}$ and $\mathscr{C}^{\mathbb{N}}$ (in the product topology) are homeomorphic. Because the map $\tilde{\Phi} : \mathscr{C}^{\mathbb{N}} \to [0,1]^{\mathbb{N}}$ defined by $\tilde{\Phi}((x_n)_n) = (\Phi(x_n))_n$ is continuous and surjective, the compact metric space $[0,1]^{\mathbb{N}}$ is a continuous image $h(\mathscr{C})$ of $\mathscr{C}$ via some continuous function h. The proof of Theorem 2.48 shows that X is homeomorphic to a closed subset of $[0,1]^{\mathbb{N}}$. Therefore, without loss of generality, assume that X is a closed subset of $[0,1]^{\mathbb{N}}$ and let $L = h^{-1}(X)$, which is a closed subset of $\mathscr{C}$. If it can be shown that $L = g(\mathscr{C})$ for some continuous g, then $f = h \circ g$ would be a continuous surjection of $\mathscr{C}$ onto X.

In the product topology, $\{0,1\}^{\mathbb{N}}$ and $\{0,5\}^{\mathbb{N}}$ are obviously homeomorphic. By Exercise 1.133, the former is homeomorphic to the Cantor ternary set $\mathscr{C}$, while the latter is homeomorphic to the set $\mathscr{C}_{2/3}$ defined by

$$\mathscr{C}_{2/3} = \left\{ \sum_{k=1}^{\infty} \frac{\alpha_k}{6^k} \,\middle|\, \alpha_k \in \{0,5\} \right\} \subset [0,1].$$

Hence, $\mathscr{C}$ and $\mathscr{C}_{2/3}$ are homeomorphic, which implies that L is homeomorphic to some closed subset $K \subseteq \mathscr{C}_{2/3}$.

The set $\mathscr{C}_{2/3}$ may be viewed as the "Cantor two-thirds set", which has the property that it does not contain the midpoint between any two of its elements. Let $d_K : \mathscr{C}_{2/3} \to \mathbb{R}$ be the continuous function $d_K(x) = \sup_{y \in K} |x - y|$. Because K is compact, the supremum is achieved at some point $y_x \in K$. By the "no midpoint" geometry of $\mathscr{C}_{2/3}$, for each $x \in \mathscr{C}_{2/3}$ the point $y_x \in K$ is uniquely determined by x, and so the map $\tilde{g} : \mathscr{C}_{2/3} \to K$ defined by $\tilde{g}(x) = y_x$ is a continuous surjection. Hence, L and X are continuous images of $\mathscr{C}$. $\qquad\square$

Inspired by the metric completeness of the real numbers, one can consider the metric completeness of more general metric spaces.

Definition 2.50. Assume that (X,d) is a metric space.

1. A sequence $\{x_k\}_{k \in \mathbb{N}}$ of elements $x_k \in X$ is *convergent* to $x \in X$ if for every $\varepsilon > 0$ there exists $N_\varepsilon \in \mathbb{N}$ such that $d(x,x_n) < \varepsilon$ for all $n \geq N_\varepsilon$.
2. A sequence $\{x_k\}_{k \in \mathbb{N}}$ of elements $x_k \in X$ is a *Cauchy sequence* if for every $\varepsilon > 0$ there exists $N_\varepsilon \in \mathbb{N}$ such that $d(x_n,x_m) < \varepsilon$ for all $m,n \geq N_\varepsilon$.

The convergence of a sequence $\{x_k\}_{k \in \mathbb{N}}$ to x is denoted by $x = \lim_{k \to \infty} x_k$.

Definition 2.51. A metric space (X,d) is *complete* if for every Cauchy sequence $\{x_k\}_{k \in \mathbb{N}}$ of elements $x_k \in X$ there exists $x \in X$ such that $x = \lim_{k \to \infty} x_k$.

Because metric spaces are Hausdorff, if a Cauchy sequence in (X,d) is convergent, then the limit of this sequence is unique.

Example 2.52. $\mathbb{R}^n$ *is a complete metric space in the Euclidean metric d_2.*

Proof. Denote the canonical basis vectors of $\mathbb{R}^n$ by $e_1, \ldots, e_n$, and suppose that $\{w_k\}_k$ is a Cauchy sequence of elements in $\mathbb{R}^n$, where

$$w_k = \sum_{j=1}^{n} \alpha_j^{(k)} e_j, \quad \forall k \in \mathbb{N}.$$

Fix j. The inequality $|\alpha_j^{(k)} - \alpha_j^{(m)}| \leq d_2(w_k, w_m)$ implies that the sequence $\{\alpha_j^{(k)}\}_k$ is a Cauchy sequence in $\mathbb{R}$. Theorem 1.14 asserts that $\mathbb{R}$ is a complete metric space in the absolute-value metric; hence, there is an $\alpha_j \in \mathbb{R}$ such that $\alpha_j = \lim_{k \to \infty} \alpha_j^{(k)}$. This is true for each j, and so let $w = \sum_{j=1}^{n} \alpha_j e_j$. Since

$$d_2(w, w_k) = \sqrt{\sum_{j=1}^{n} (\alpha_j - \alpha_j^{(k)})^2},$$

the sequence $\{w_k\}_k$ converges to w. Hence, $(\mathbb{R}^n, d_2)$ is a complete metric space. $\square$

Although not every metric space need be complete, the following theorem shows that every metric space is a dense subset of some complete metric space.

Theorem 2.53. *For every metric space (X, d), there is a metric space $(\tilde{X}, \tilde{d})$ and a continuous injective function $f : X \to \tilde{X}$ such that:*

1. *$(\tilde{X}, \tilde{d})$ is a complete metric space;*
2. *$\tilde{d}(f(x), f(y)) = d(x, y)$, for all $x, y \in X$; and*
3. *$f(X)$ is a dense subset of $\tilde{X}$.*

Proof. Let $Z = \prod_{n=1}^{\infty} X$, the Cartesian product of countably many copies of X, which we think of as the space of all sequences $x = (x_n)_{n \in \mathbb{N}}$ in X. Let $C \subset Z$ be the set of all Cauchy sequences and define a relation $\sim$ on C by $(x_n)_n \sim (y_n)$ if $\lim_n d(x_n, y_n) = 0$. It is straightforward to verify that $\sim$ is an equivalence relation, and so consider the space $\tilde{X} = C/\sim$ of equivalence classes, whose elements we denote by $\dot{s}$ for each $s = (s_n)_n \in C$. Let $q : C \to \tilde{X}$ denote the quotient map $q(s) = \dot{s}$.

If $s, t \in C$, then the sequence $\{d(s_n, t_n)\}_n$ is a Cauchy sequence in $\mathbb{R}$ (Exercise 2.101). The limit $\lim_n d(s_n, t_n)$ exists, because $\mathbb{R}$ is a complete metric space, Further, suppose that $s, s', t, t' \in C$ satisfy $\dot{s} = \dot{s}'$ and $\dot{t} = \dot{t}'$. Then, for a fixed $n \in \mathbb{N}$,

$$d(s_n, t_n) \leq d(s_n, s'_n) + d(s'_n, t'_n) + d(t'_n, t_n),$$

and

$$d(s'_n, t'_n) \leq d(s'_n, s_n) + d(s_n, t_n) + d(t_n, t'_n).$$

Hence, $\lim_n d(s_n, t_n) = \lim_n d(s'_n, t'_n)$, and so the formula

$$\tilde{d}(\mathfrak{s}, \mathfrak{t}) = \lim_{n \to \infty} d(s_n, t_n)$$

yields a well-defined function $\tilde{d} : \tilde{X} \times \tilde{X} \to \mathbb{R}$. It is simple to verify that $\tilde{d}$ is in fact a metric on $\tilde{X}$.

Consider the function $\iota : X \to C$ that sends each $x \in X$ to the constant (and Cauchy) sequence $\iota(x) = (x, x, x, \dots)$, and define $f : X \to \tilde{X}$ by $f = q \circ \iota$. Observe that, for all $x, y \in X$, $\tilde{d}(f(x), f(y)) = d(x, y)$; thus, f is a continuous and injective map (Exercise 2.102). If $\mathfrak{s} \in C$ belongs to the equivalence class of $\iota(x)$, for an element $x \in X$, then $x = \lim_n s_n$ in X.

Choose any $\dot{\mathfrak{s}} \in \tilde{X}$ and $\varepsilon > 0$. If $\mathfrak{s} = (s_n)_s \in C$ is a representative of the class $\dot{\mathfrak{s}}$, then, by virtue of the fact that $\mathfrak{s}$ is a Cauchy sequence, there is an $N \in \mathbb{N}$ such that $d(s_n, s_m) < \varepsilon$ for all $n, m \geq N$. Let $x = s_N$ and consider $f(x)$. Because $\tilde{d}(f(x), \dot{\mathfrak{s}}) = \lim_{m \geq N} d(s_N, s_m) < \varepsilon$, we deduce that $B_\varepsilon(\dot{\mathfrak{s}}) \cap f(X) \neq \emptyset$. Hence, $\overline{f(X)} = \tilde{X}$.

To show that $(\tilde{X}, \tilde{d})$ is a complete metric space, let $\{\dot{\mathfrak{s}}_k\}_k$ be a Cauchy sequence in $\tilde{X}$, and suppose that $\varepsilon > 0$. Thus, there is an $N_0 \in \mathbb{N}$ for which $\tilde{d}(\dot{\mathfrak{s}}_k, \dot{\mathfrak{s}}_\ell) < \varepsilon/3$ for all $k, \ell \geq N_0$. The set $f(X)$ is dense in $\tilde{X}$, and so, for each $k \in \mathbb{N}$, there exists $x_k \in X$ such that $\tilde{d}(f(x_k), \dot{\mathfrak{s}}_k) < \frac{1}{k}$. Thus,

$$\tilde{d}(f(x_k), f(x_\ell)) \leq \tilde{d}(f(x_k), \dot{\mathfrak{s}}_k) + \tilde{d}(\dot{\mathfrak{s}}_k, \dot{\mathfrak{s}}_\ell) + \tilde{d}(\dot{\mathfrak{s}}_\ell, f(x_\ell))$$

$$\leq 1/k + \varepsilon/3 + 1/\ell.$$

If $N \geq N_0$ is such that $\frac{1}{k} < \frac{\varepsilon}{3}$ and $\frac{1}{\ell} < \frac{\varepsilon}{3}$ for all $k, \ell \geq N$, then $\tilde{d}(f(x_k), f(x_\ell)) < \varepsilon$, for all $k, \ell \geq N$, which proves that $\{f(x_k)\}_k$ is a Cauchy sequence in $\tilde{X}$. Furthermore, the equation $\tilde{d}(f(x_k), f(x_\ell)) = d(x_k, x_\ell)$ shows the sequence $\mathfrak{x} = \{x_k\}_k$ is a Cauchy sequence in X also. Thus, $\mathfrak{x} \in C$ and we may consider $\dot{\mathfrak{x}} \in \tilde{X}$. The inequalities

$$\tilde{d}(\dot{\mathfrak{x}}, \dot{\mathfrak{s}}_k) \leq \tilde{d}(f(x_k), \dot{\mathfrak{s}}_k) < \frac{1}{k},$$

demonstrate that the Cauchy sequence $\{\dot{\mathfrak{s}}_k\}_k$ converges in $\tilde{X}$ to $\dot{\mathfrak{x}}$. $\qquad\square$

Definition 2.54. Let X be a topological space.

1. A subset $G \subseteq X$ is a G_δ-set if there is a countable family $\{U_k\}_{k \in \mathbb{N}}$ of open sets $U_k \subseteq X$ such that

$$G = \bigcap_{k \in \mathbb{N}} U_k.$$

2. A subset $F \subseteq X$ is an F_σ-set if there is a countable family $\{F_k\}_{k \in \mathbb{N}}$ of closed sets $F_k \subseteq X$ such that

$$F = \bigcup_{k \in \mathbb{N}} F_k.$$

The Baire Category Theorem below is a remarkable result illustrating the nature of complete metric spaces, as well as being widely relevant in applications of topology to analysis.

Theorem 2.55 (Baire Category Theorem). *If $\{U_k\}_{k \in \mathbb{N}}$ is a sequence of open sets in a complete metric space (X,d), and if each U_k is dense in X, then the G_δ-set*

$$\bigcap_{k \in \mathbb{N}} U_k$$

is also dense in X.

Proof. Choose $x_0 \in X$ and let $\varepsilon > 0$. We aim to prove that $B_\varepsilon(x_0) \cap G \neq \emptyset$, where $G = \bigcap_{k \in \mathbb{N}} U_k$.

By hypothesis, U_1 is open and dense in V. Thus, there is a $\delta_1 \in (0,1)$ and an element $x_1 \in B_\varepsilon(x_0) \cap U_1$ such that $\overline{B_{\delta_1}(x_1)} \subseteq B_\varepsilon(x_0) \cap U_1$. Likewise, U_2 is open and dense in X, and so there is a $\delta_2 \in (0,\frac{1}{2})$ and an element $x_2 \in B_{\delta_1}(x_1) \cap U_2$ such that $\overline{B_{\delta_2}(x_2)} \subseteq B_{\delta_1}(x_1) \cap U_2$.

It is clear that this process may be continued by induction to obtain sequences $\{x_n\}_{n \in \mathbb{N}}$ in X and $\{\delta_n\}_{n \in \mathbb{N}}$ in $\mathbb{R}$ such that

$$\delta_n \in \left(0, \frac{1}{2^{n-1}}\right) \quad \text{and} \quad \overline{B_{\delta_n}(x_n)} \subseteq B_{\delta_{n-1}}(x_{n-1}) \cap U_n.$$

By construction,

$$\overline{B_{\delta_k}(x_k)} \subseteq B_{\delta_n}(x_n) \subseteq \overline{B_{\delta_n}(x_n)} \subseteq B_\varepsilon(x_0), \quad \forall k > n.$$

Fix $n \in \mathbb{N}$ and let $k, m > n$. By the inclusions above,

$$x_k, x_m \in B_{\delta_n}(x_n) \subseteq \overline{B_{\delta_n}(x_n)}.$$

Therefore, $d(v_k, v_m) < \frac{1}{2^{n-2}}$, which shows that $\{x_n\}_{n \in \mathbb{N}}$ is a Cauchy sequence. Since X is a complete metric space, there is a limit $x \in X$ to this sequence. Choose any $n \in \mathbb{N}$. If $k > n$, then

$$x_k \in \overline{B_{\delta_n}(x_n)} \subseteq (B_{\delta_{n-1}}(x_{n-1}) \cap U_n) \subseteq U_n.$$

Hence, $x \in \overline{B_{\delta_n}(x_n)} \subseteq U_n$ and $x \in \overline{B_{\delta_n}(x_n)} \subseteq B_\varepsilon(x_0)$, and so $x \in G \cap B_\varepsilon(x_0)$. $\square$

Definition 2.56. A subset $F \subset X$ is *nowhere dense* in a topological space X if the interior of the closure $\overline{F}$ of F is the empty set.

Corollary 2.57. *In a complete metric space,*

1. *the intersection of a countable family of dense G_δ-sets is a dense G_δ-set, and*
2. *the union of a countable family of nowhere dense F_σ sets is a nowhere dense F_σ set.*

Proof. Exercise 2.104. □

2.5 Connected Spaces

Definition 2.58. A *separation* of a topological space X is a pair (U,V) of open nonempty subsets $U,V \subseteq X$ such that $U \cap V = \emptyset$ and $U \cup V = X$. If X admits a separation, then X is said to be a *disconnected space*; and if no separation of X exists, then X is called a *connected space*.

The standard geometric model for the continuum $\mathbb{R}$ is that of an unbroken continuous line. The following example provides the topological justification for this model.

Example 2.59. Every closed interval $[a,b]$ of real numbers is connected.

Proof. Assume $[a,b]$ is not connected. Thus, there is a separation (U,V) of $[a,b]$. Without loss of generality, assume that $a \in U$ and define $L = \{x \in \mathbb{R} \mid [a,x] \subseteq U\}$. Some observations concerning L are: (i) $a \in L$, (ii) $b \notin L$ (because $U^c = V$ is a nonempty subset of $[a,b]$), (iii) $x \leq b$ for all $x \in L$, and (iv) if $x \in L$ and $y \in [a,x]$, then $y \in L$.

As L is a nonempty set bounded above by b, the supremum of L exists, which we denote by c. Thus, for every $\varepsilon > 0$ there is a $z \in L$ such that $c - \varepsilon < z$, which shows that $W \cap L \neq \emptyset$ for every neighbourhood W of c. That is, $c \in \overline{L}$. From $L \subseteq U$ we obtain $\overline{L} \subseteq \overline{U}$. Further, U and $U^c = V$ are open, and so U is also closed, whence $\overline{U} = U$. Therefore, $c \in U$.

We now show that $a < c < b$. First of all, because U is open, there is some $\varepsilon > 0$ for which $[a, a+\varepsilon) \subseteq U$, which implies that $a + \varepsilon \leq c$. Likewise, if it were true that $c = b$, then U would contain $[a, b-\varepsilon]$ for all $0 < \varepsilon < (b-a)$, and so U would contain $[a,b)$, thereby forcing the nonempty open set V to be $V = \{b\}$ (which is not open in $[a,b]$). Thus, $c < b$.

Now because U is open and $c \in U$, there exists $\varepsilon > 0$ such that $(c-\varepsilon, c+\varepsilon) \subseteq U$. As $c = \sup L$, there is an element $x \in L$ such that $x \in (c-\varepsilon, c] \subseteq U$. Select $z \in (c, c+\varepsilon) \subseteq U$. Thus, $[a,x] \subseteq U$ and $(x,z] \subseteq (c-\varepsilon, c+\varepsilon) \subseteq U$ imply that $[a,z] = [a,x] \cup (x,z] \subseteq U$. But $c < z \leq \sup L = c$ is a contradiction. Hence, it must be that $[a,b]$ is connected. □

To construct additional examples of connected spaces, the following result is useful.

Proposition 2.60. *If $\{X_\alpha\}_{\alpha \in \Lambda}$ is a family of connected subspaces of a topological space X, such that $\bigcap_\alpha X_\alpha \neq \emptyset$, then $\bigcup_\alpha X_\alpha$ is connected.*

Proof. Without loss of generality assume that $X = \bigcup_\alpha X_\alpha$, and suppose that U and V are open subsets of X such that $U \cap V = \emptyset$ and $U \cup V = X$. Let $U_\alpha = U \cap X_\alpha$ and $V_\alpha = V \cap X_\alpha$ so that U_α and V_α are open in X_α and satisfy $U_\alpha \cap V_\alpha = \emptyset$ and $U_\alpha \cup V_\alpha = X_\alpha$. Select $x \in \bigcap_\alpha X_\alpha$. Thus, x is an element of exactly one of U or V, say U. Because $x \in X_\alpha$ for every $\alpha \in \Lambda$, we deduce that $x \in U_\alpha$ for all α. Hence $U_\alpha \neq \emptyset$ implies $V_\alpha = \emptyset$ because X_α is connected. Thus, $U = X$ and $V = \emptyset$, which implies that X does not admit a separation. $\square$

Example 2.61. $\mathbb{R}$ *is connected.*

Proof. $\mathbb{R} = \bigcup_{k \in \mathbb{N}} [-k, k]$ and $\bigcap_{k \in \mathbb{N}} [-k, k] = [-1, 1] \neq \emptyset$. Therefore, Example 2.59 and Proposition 2.60 imply that $\mathbb{R}$ is connected. $\square$

Concerning mappings of connected spaces, we have:

Theorem 2.62 (Intermediate Value Theorem). *If X is a connected topological space and if $f : X \to \mathbb{R}$ is a continuous function, then for any $x_1, x_2 \in X$ and real number r between $f(x_1)$ and $f(x_2)$ there is an $x \in X$ with $f(x) = r$.*

Proof. Without loss of generality we may assume $f(x_1) \leq f(x_2)$. Choose $r \in [f(x_1), f(x_2)]$. If, contrary to what we aim to prove, $r \notin f(X)$, then $f(X) \subset (-\infty, r) \cup (r, \infty)$. The open sets $U = f^{-1}(-\infty, r)$ and $V = f^{-1}(r, \infty)$ satisfy $U \cap V = \emptyset$, $U \cup V = X$, $x_1 \in U$, and $x_2 \in V$. Thus, (U, V) is a separation of X, in contradiction to the connectivity of X. $\square$

In a similar vein:

Proposition 2.63. *If $f : X \to Y$ is a continuous function and if X is connected, then $f(X)$ is a connected subspace of Y.*

Proof. Exercise 2.107. $\square$

The connectivity of the closed interval $[0, 1]$ gives a useful way to determine whether a space is connected.

Proposition 2.64 (A Connectivity Criterion). *If X a topological space X has the property that for every pair $x_0, x_1 \in X$ there is a continuous function $f : [0, 1] \to X$ such that $f(0) = x_0$ and $f(1) = x_1$, then X is a connected space.*

Proof. Assume, contrary to what we aim to prove, that X is disconnected. Thus, there is a separation (U, V) of X. Select $x_0 \in U$ and $x_1 \in V$, and let $f : [0, 1] \to X$ be a continuous function such that $f(0) = x_0$ and $f(1) = x_1$. Hence, $(f^{-1}(U), f^{-1}(V))$ is a separation of $[0, 1]$, in contradiction to the connectivity of $[0, 1]$. Hence, it must be that X is connected. $\square$

Proposition 2.64 does not characterise connected spaces in that there exist connected topological spaces X for which the hypothesis of Proposition 2.64 is not satisfied. Therefore, spaces that satisfy this connectivity criterion are called path connected.

Definition 2.65. A topological space X is *path connected* if X satisfies the hypothesis of Proposition 2.64.

Example 2.66. *If $n \geq 2$, then $\mathbb{R}^n \setminus \{0\}$ is a path-connected subspace of $\mathbb{R}^n$.*

Proof. Let $x_0, x_1 \in \mathbb{R}^n$. If, considered as vectors, x_0 and x_1 are linearly independent, then no nontrivial linear combination of x_0 and x_1 yields the zero vector. Hence, the map $f : [0,1] \to \mathbb{R}^n$ defined by $f(t) = (1-t)x_0 + tx_1$ is continuous, satisfies $f(0) = x_0$ and $f(1) = x_1$, and has range contained in $\mathbb{R}^n \setminus \{0\}$. On the other hand, if x_0 and x_1 are linearly dependent—that is, $x_1 = \lambda x_0$ for some $\lambda \in \mathbb{R}$—then there is a nonzero $z \in \mathbb{R}^n$ such that z is linearly independent of x_0 and, thus, of x_1. So, by what has already been proved, there are continuous $g, h : [0,1] \to \mathbb{R}^n \setminus \{0\}$ such that $g(0) = x_0$, $g(1) = z$, $h(0) = z$, and $h(1) = x_1$. Now let $f : [0,1] \to \mathbb{R}^n \setminus \{0\}$ be given by $f(t) = g(2t)$, if $t \in [0, 1/2]$, and by $f(t) = h(2t-1)$, if $t \in [1/2, 1]$. Then f is continuous (Exercise 1.127) and satisfies $f(0) = x_0$ and $f(1) = x_1$. □

At the other end of the connectivity spectrum lies the notion of a totally disconnected space.

Definition 2.67. A topological space X is *totally disconnected* if for every pair of distinct $x, y \in X$ there exist subsets $U, V \subseteq X$ such that

1. $x \in U$ and $y \in V$,
2. U and V are both open and closed in X, and
3. $U \cap V = \emptyset$.

To better understand the definition above, first note that a space A is disconnected if A can be written as $A = U \cup V$, for some nonempty open disjoint subsets $U, V \subset A$; in this case, U and V are necessarily both closed and open. Therefore, if X is totally disconnected, and if x and y are distinct elements of X, then there is no connected subset A of X that contains both x and y.

Example 2.68. *The Cantor set is totally disconnected.*

Proof. The Cantor ternary function Φ is monotone increasing and is a continuous surjection of the Cantor set $\mathscr{C}$ onto $[0,1]$ (Proposition 1.86). Suppose that $x, y \in \mathscr{C}$ are such that $x < y$. Recall that the open set $[0,1] \setminus \mathscr{C}$ is a countable union of pairwise disjoint open intervals. Of these intervals, select one, say (a, b), for which $x \leq a$ and $b \leq y$. Now select $r, s \in (a, b)$ such that $r < s$, and consider the open subsets U and V of $\mathscr{C}$ given by $U = \Phi^{-1}([0, r))$ and $V = \Phi^{-1}((s, 1])$. Because Φ is monotone increasing, we deduce that $x \in U$, $y \in V$, and $U \cap V = \emptyset$; and because Φ is constant on every open interval in $[0,1] \setminus \mathscr{C}$ (Corollary 1.87), we observe that $\Phi^{-1}([0, r)) = \Phi^{-1}([0, r])$ and $\Phi^{-1}((s, 1]) = \Phi^{-1}([s, 1])$. Hence, U and V are also closed sets, by the continuity of Φ. □

2.6 Stone-Čech Compactification

Earlier, in Proposition 2.22, we noted that every non-compact, locally compact Hausdorff space X can be embedded into a compact Hausdorff space $\tilde{X}$ by adding a single point (the so-called point at infinity). The goal of the present section is to show that another embedding of X into a compact Hausdorff space is possible, and this embedding has the advantage that every bounded continuous real-valued function on X extends to a continuous function on the larger space.

The first issue to address is setting the precise meaning of "compactification" of a topological space.

Definition 2.69. A *compactification* of a topological space X is a pair $(K, \iota_{X,K})$ consisting of

1. a compact space K, and
2. a continuous function $\iota_{X,K} : X \to K$ such that

   a. $\iota_{X,K}$ is a homeomorphism of X and $\iota_{X,K}(X)$, and
   b. $\iota_{X,K}(X)$ is dense in K.

By identifying X with its image $\iota_{X,K}(X)$ in K, one can view K as being a compact space that contains X as a dense subspace.

The definition of compactification does not ask, for example, that X (or $\iota_{X,K}(X)$) be open in K. However, if X is locally compact and K is Hausdorff, then such will be the case.

Proposition 2.70. *If a locally compact space X is dense in a Hausdorff space Y, then X is an open subset of Y.*

Proof. As in Proposition 1.64, the closure in X of a subset $A \subseteq X$ will be denoted by $\overline{A}^X$, and the closure in Y of a subset $B \subseteq Y$ will be denoted by $\overline{B}^Y$.

Select $x_0 \in X$. Because X is locally compact, there are an open set U in X and a compact subset K of X such that $x_0 \in U \subseteq K$. Therefore, $\overline{U}^X$ is a compact subset of K and, hence, is compact in X. If one takes an open cover $\{V_\alpha\}_{\alpha \in \Lambda}$ in Y of the set $\overline{U}^X$, then $\{X \cap V_\alpha\}_{\alpha \in \Lambda}$ is an open cover of $\overline{U}^X$ in X. Hence, by compactness, finitely many $X \cap V_\alpha$ cover $\overline{U}^X$, and so finitely many V_α cover $\overline{U}^X$, which proves that $\overline{U}^X$ is a compact subset of Y. Because Y is Hausdorff, compact subsets of Y are closed in Y (Proposition 2.8); therefore, $\overline{U}^X$ is a closed subset of Y. The set $\overline{U}^X$ contains U and is a closed subset of Y; thus, $\overline{U}^X \supseteq \overline{U}^Y$. On the other hand, $\overline{U}^X \subseteq \overline{U}^Y$, by virtue of $X \subseteq Y$. Hence, $\overline{U}^X = \overline{U}^Y$.

The set U has the form $U = X \cap V$, for some open set V in Y. Therefore, $\overline{U}^Y = \overline{X \cap V}^Y = \overline{V}^Y$ (because X is dense in Y). Moreover, $\overline{U}^Y = \overline{U}^X \subseteq X$ implies that $\overline{V}^Y \subseteq X$. Therefore, $V \subset X$ and, hence, $U = X \cap V = V$. This proves that, for every $x_0 \in X$, there is an open set V in Y such that $x_0 \in V \subseteq X$. In other words, X is an open subset of Y. □

The following theorem asserts the existence of a compactification K of a locally compact Hausdorff space X in which every bounded continuous function $X \to \mathbb{C}$ extends to a continuous function $K \to \mathbb{C}$.

Theorem 2.71 (Stone-Čech Compactification: Existence). *If X is a locally compact Hausdorff space, then there exists a compactification $(\beta X, \iota_X)$ of X such that*

1. *βX is Hausdorff, and*
2. *for every bounded continuous function $f : X \to \mathbb{C}$ there exists a unique continuous $\tilde{f} : \beta X \to \mathbb{C}$ such that $\tilde{f} \circ \iota_X(x) = f(x)$, for all $x \in X$.*

Proof. The first part of the proof is devoted to the construction of the compact Hausdorff space $\beta(X)$. To this end, let $C_b(X)$ denote the set of all continuous functions $f : X \to \mathbb{C}$ such that $\sup_{x \in X} |f(x)| < \infty$. For each such $f \in C_b(X)$, let $\|f\| = \sup_{x \in X} |f(x)|$ and consider the compact Hausdorff subspace K_f of $\mathbb{C}$ given by $K_f = \{\zeta \in \mathbb{C} \,|\, |\zeta| \le \|f\|\}$. Endow the product space

$$Z = \prod_{f \in C_b(X)} K_f$$

with the product topology. By Tychonoff's Theorem, Z is a compact Hausdorff space.

Define $\iota_{X,Z} : X \to Z$ by

$$\iota_{X,Z}(x) = \prod_{f \in C_b(X)} f(x),$$

for $x \in X$. Each component map $f : X \to K_f$ is continuous, and so the map $\iota_{X,Z}$ is also continuous (Proposition 1.92). Furthermore, the local compactness of X implies (by Exercise 2.92) that there exists an element $f \in C_b(X)$ such that $f(x) \ne f(y)$, for any two distinct $x, y \in X$. Hence, the function $\iota_{X,Z}$ is necessarily injective.

To prove that $\iota_{X,Z}$ is a homeomorphism between X and its range $\iota_{X,Z}(X)$ in Z, it is sufficient to prove that $\iota_{X,Z}$ maps open sets in X to open sets in $\iota_{X,Z}(X)$. Therefore, suppose that $U \subseteq X$ is an open set, select $z_0 \in \iota_{X,Z}(X)$, and let $x_0 \in U$ be the unique element for which $z_0 = \iota_{X,Z}(x_0)$. Consider the function $g_0 : \{x_0\} \to \mathbb{C}$ given by $g_0(x_0) = 1$. The point set $\{x_0\}$ is compact and is contained in the open set U. Hence, by the Tietze Extension Theorem (Theorem 2.43), there is a bounded continuous function $g : X \to \mathbb{C}$ such that $g(x_0) = 1$ and $g(U^c) = \{0\}$. The canonical projection $p_g : Z \to K_g \subset \mathbb{C}$ is continuous (Proposition 1.92); therefore, the set $V = p_g^{-1}(\mathbb{C} \setminus \{0\})$ is open in Z. Let $W = V \cap \iota_{X,Z}(X)$, which is open in $\iota_{X,Z}(X)$ and contains z_0 (because $p_g(z_0) \ne 0$). If $z \in W$, then there is a unique $x \in X$ with $z = \iota_{X,Z}(x)$. Thus, as $0 \ne p_g(x) = g(x)$, we deduce that $x \notin U^c$; hence, $x \in U$. Therefore, $z = \iota_{X,Z}(x) \in \iota_{X,Z}(U)$. In other words, for each $z_0 \in \iota_{X,Z}(U)$ there is an open set W in $\iota_{X,Z}(X)$ such that $z_0 \in W \subseteq \iota_{X,Z}(U)$. Thus, $\iota_{X,Z}(U)$ is open in $\iota_{X,Z}(X)$, which completes the proof that X and $\iota_{X,Z}(X)$ are homeomorphic.

Denote the closure $\overline{\iota_{X,Z}(X)}$ of $\iota_{X,Z}(X)$ in Z by βX, and let ι_X denote $\iota_{X,Z}$. Observe that βX is a compact Hausdorff space and contains $\iota_X(X)$ as a dense subspace. Thus, $(\beta X, \iota_X)$ is a compactification of X.

Suppose now that $f : X \to \mathbb{C}$ is a bounded continuous function, and define $\tilde{f} :$ $\beta(X) \to \mathbb{C}$ by $\tilde{f} = p_{f|\beta(X)}$. Note that, if $x \in X$, then $\tilde{f} \circ \iota_X(x) = f(x)$. To show the uniqueness of the extension $\tilde{f}$, suppose that $\tilde{f}$ and $\tilde{\tilde{f}}$ are two extensions of f. If $z \in$ Z, then there is a net $\{\iota_X(x_\alpha)\}_\alpha$ convergent to z. Thus, $\tilde{f} \circ \iota_X(z) = \lim_\alpha f(x_\alpha)$ and $\tilde{\tilde{f}} \circ \iota_X(z) = \lim_\alpha f(x_\alpha)$. Because convergent nets have unique limits in the Hausdorff space $\mathbb{C}$, we deduce that $\tilde{f} = \tilde{\tilde{f}}$. $\square$

For the moment, the following definition of the Stone-Čech compactification of a locally compact space will suffice.

Definition 2.72. The *Stone-Čech compactification* of a locally compact Hausdorff space X is the compactification $(\beta X, \iota_X)$ constructed in the proof of Theorem 2.71.

It is natural to wonder about the conclusion of Theorem 2.71 in the case where the locally compact Hausdorff space X is compact.

Proposition 2.73. *If X is a compact Hausdorff space, then $\beta X \simeq X$.*

Proof. By construction, βX is the closure of $\iota_X(X)$ in the compact product space Z. Furthermore, $\iota_X(X)$ and X are homeomorphic with respect to the subspace topology of $\iota_X(X)$. Because $\iota_X(X)$ is a compact subspace of a Hausdorff space, the set $\iota_X(X)$ is closed in Z (Proposition 2.9). Hence, $\beta X = \overline{\iota_X(X)} = \iota_X(X) \simeq X$. $\square$

The following proposition captures a very elegant feature of the Stone-Čech compactification.

Proposition 2.74. *If X and Y are locally compact Hausdorff spaces, and if $h : X \to$ Y is a continuous function, then there is a continuous function $H : \beta X \to \beta Y$ such that the following commutative diagram holds:*

$$
\begin{array}{ccc}
X & \xrightarrow{\;h\;} & Y \\
{\scriptstyle\iota_X}\downarrow & & \downarrow{\scriptstyle\iota_Y} \\
\beta X & \xrightarrow[\;H\;]{} & \beta Y.
\end{array}
$$

That is, $H \circ \iota_X = \iota_Y \circ h$.

Proof. For a fixed $g \in C_b(Y)$, the function $g \circ h : X \to \mathbb{C}$ is bounded and continuous. Therefore, by Theorem 2.71, there is a continuous function $\hat{g} : \beta X \to \mathbb{C}$ such that $\hat{g}(\iota_X(x)) = g(h(x))$, for all $x \in X$. Let Z_Y be the compact product space constructed in the proof of Theorem 2.71 and which contains βY as a subspace. Define a function $H : \beta X \to Z_Y$ by

$$
H(\omega) = \left(\hat{g}(\omega)\right)_{g \in C_b(Y)}, \quad \text{for all } \omega \in \beta X.
$$

By definition of H, it is clear that $H \circ \iota_X = \iota_Y \circ h$. To show that H is continuous, observe that $p_g \circ H = \hat{g}$, where p_g denotes the canonical projection $Z_Y \to \mathbb{C}$ of Z_Y onto its g-th coordinate. Because $\hat{g}$ is continuous, Proposition 1.92 implies that H is continuous.

To this point, we know only that H maps βX into Z_Y. However,

$$H(\beta X) = H\left(\overline{\iota_X(X)}\right) \subseteq \overline{H(\iota_X(X))} = \overline{\iota_Y \circ h(X)} \subseteq \overline{\iota_Y \circ h(Y)} = \beta Y$$

shows that the range of H is indeed within βY. (Note that the first of the inclusions above is a consequence of the continuity of H.)   □

A variant of Proposition 2.74 is the following result, with essentially the same proof.

Proposition 2.75. *Assume that $(K, \iota_{X,K})$ is a Hausdorff compactification of a locally compact Hausdorff space X with the property that for every bounded continuous function $f : X \to \mathbb{C}$ there exists a unique continuous $\tilde{f} : K \to \mathbb{C}$ such that $\tilde{f} \circ \iota_{X,K} = f$. If L is an arbitrary compact Hausdorff space and if $h : X \to L$ is a continuous function, then there exists a continuous function $H : K \to L$ such that $H \circ \iota_{X,K} = h$.*

Proof. Because βL is a subspace of the product space $Z_L = \prod_{g \in C_b(L)} K_g$, each $g \in C_b(L)$ yields a function $g \circ h : X \to \mathbb{C}$ that is bounded and continuous and which, by hypothesis, extends to a function $\hat{g} : K \to \mathbb{C}$. Following the proof of Proposition 2.74 verbatim leads to a continuous map $H_0 : K \to \beta L \subseteq Z_L$. Proposition 2.73 shows that $\beta L = \iota_L(L)$, where $\iota_L : L \to Z_L$ is a homeomorphism of L and $\iota_L(L)$. Set $H = \iota_L^{-1} \circ H_0$ to get the map $H : K \to L$ with the desired properties.   □

One final note, before continuing further, is that the method of proof employed in Theorem 2.71 and the tangential results above (Propositions 2.74 and 2.75) also yield the following noteworthy fact.

Proposition 2.76. *If X is a locally compact Hausdorff space, then there exists a set Λ such that X is homeomorphic to a subspace of the compact Hausdorff hypercube $([0,1]^\Lambda, \mathscr{T}^\times)$.*

Proof. Exercise 2.112   □

It is preferable to define the Stone-Čech compactification by a property rather than by a construction. To do so, it is necessary to understand what other compactifications of X possess properties (1) and (2) of Theorem 2.71 and to understand how these relate to βX.

Theorem 2.77 (Stone-Čech Compactification: Uniqueness). *Assume that X is a locally compact Hausdorff space and that $(K, \iota_{X,K})$ is a compactification of X in which K is Hausdorff and for every bounded continuous function $f : X \to \mathbb{C}$ there exists a unique continuous $\tilde{f} : K \to \mathbb{C}$ such that $\tilde{f} \circ \iota_{X,K}(x) = f(x)$, for all $x \in X$. Then there exists a homeomorphism $\alpha : \beta X \to K$ such that $\alpha \circ \iota_X = \iota_{X,K}$. That is,*

$$X \xrightarrow{\mathrm{id}_X} X$$

$$\iota_X \downarrow \qquad\qquad \downarrow \iota_{X,K}$$

$$\beta X \xrightarrow{\quad\alpha\quad} K.$$

Proof. If Q is a set, then id_Q shall denote the (identity) function $Q \to Q$ in which $\mathrm{id}_Q(q) = q$, for every $q \in Q$.

We apply Proposition 2.75 twice, as both compactifications K and βX satisfy the hypothesis of Proposition 2.75. In the first instance, Proposition 2.75 yields a continuous map $I_X : K \to \beta X$ such that $I_X \circ \iota_{X,K} = \iota_X$. In the second instance, Proposition 2.75 yields a continuous map $I_{X,K} : \beta X \to K$ such that $I_{X,K} \circ \iota_X = \iota_{X,K}$. Hence,

$$I_{X,K} \circ I_X \circ \iota_{X,K} = I_{X,K} \circ \iota_X = \iota_{X,K} \text{ and } I_X \circ I_{X,K} \circ \iota_X = I_X \circ \iota_{X,K} = \iota_X.$$

Therefore, because $\iota_{X,K}(X)$ is dense in K and $\iota_X(X)$ is dense in βX, the functional equations above lead to

$$I_{X,K} \circ I_X = \mathrm{id}_K \text{ and } I_X \circ I_{X,K} = \mathrm{id}_{\beta X}.$$

Hence, $I_{X,K}$ is a bijection with continuous inverse I_X. Therefore, with $\alpha = I_{X,K}$ we have a homeomorphism $\alpha : \beta X \to K$ such that $\alpha \circ \iota_X = \iota_{X,K}$. $\square$

In light of Theorem 2.77 and Proposition 2.70, the Stone-Čech compactification of a locally compact Hausdorff space X is any compact Hausdorff space K that contains X as a dense open subspace and has the property that every bounded continuous function $f : X \to \mathbb{C}$ extends to a unique continuous $\tilde{f} : K \to \mathbb{C}$.

The final general property of the Stone-Čech compactification to be remarked upon here is given by the following proposition, which asserts that βX is the largest of all possible compactifications of a given locally compact Hausdorff space X.

Proposition 2.78. *If X is a locally compact Hausdorff space and if K is a compact Hausdorff space that contains X as a dense subspace, then K is a quotient of βX.*

Proof. Exercise 2.113. $\square$

The complete determination of βX using existing knowledge about X is very difficult, if not impossible. Indeed, even if X is well understood and has good topological properties, βX can be tremendously different, as the following result indicates.

Proposition 2.79. *If X is an infinite discrete space, then βX is nonmetrisable and totally disconnected.*

Proof. Because X is an infinite set with the discrete topology, X is a locally compact Hausdorff space; moreover, without loss of generality, we may assume that X is a dense open subspace of βX.

Assume, contrary to what we aim to prove, that βX is metrisable. Let $\{x_n\}_{n \in \mathbb{N}}$ be any sequence in X consisting of infinitely many distinct elements. As X is metrisable, Theorem 2.19 asserts that the sequence $\{x_n\}_{n\in\mathbb{N}}$ admits a convergent subsequence $\{x_{n_j}\}_{j\in\mathbb{N}}$ consisting of distinct points and with limit $z \in \beta X$. Because X is discrete, the sets $A = \{x_{n_{2k}} \mid k \in \mathbb{N}\}$ and $B = \{x_{n_{2k+1}} \mid k \in \mathbb{N}\}$ are closed, and so the function $f_0 : A \cup B \to [0,1]$ in which $f_0(x) = 0$, if $x \in A$, and $f_0(x) = 1$, if $x \in B$, has a continuous extension $f : X \to [0,1]$ by Theorem 2.43. Further, f has a continuous extension $f^\beta : \beta X \to \mathbb{R}$. However, $f^\beta(z) = \lim_k f(x_{n_{2k}}) = 0$ and $f^\beta(z) = \lim_k f(x_{n_{2k+1}}) = 1$, which because of $z = \lim_j x_{n_j}$ is a contradiction of the continuity of f^β. Therefore, it cannot be true that $\{x_n\}_{n\in\mathbb{N}}$ admits a convergent subsequence, which by Theorem 2.19 implies that βX is not metrisable.

Select any nonempty proper subset Y of X. Because X has the discrete topology, both Y and $X \setminus Y$ are open in X. Therefore, the function $f : X \to [0,1]$ defined by $f(x) = 1$ if $x \in Y$ and $f(x) = 0$ if $x \notin Y$ is continuous. If $\overline{Y}^\beta \cap \overline{X \setminus Y}^\beta$ is nonempty, where $\overline{A}^\beta$ denotes the closure in βX of a subset $A \subseteq X$, then there exist $z \in \beta X$ and nets $\{y_\alpha\}_{\alpha\in\Lambda}$ and $\{x_\delta\}_{\delta\in\Delta}$ in Y and $X \setminus Y$, respectively, such that $z = \lim_\alpha y_\alpha = \lim_\delta x_\delta$. Therefore, the continuous extension f^β of f to βX has the property that $f^\beta(z) = \lim_\alpha f(y_\alpha) = 1$ and $f^\beta(z) = \lim_\delta f(x_\delta) = 0$, which is a contradiction. Hence: disjoint subsets of X have disjoint closures in βX.

Select any two distinct $z_1, z_2 \in \beta X$. Because βX is Hausdorff, there are open sets $U_1, U_2 \subseteq \beta X$ with each $x_j \in U_j$ and $U_1 \cap U_2 = \emptyset$. Because X is open in βX, both $X \cap U_1$ and $X \cap U_2$ are open sets, and thus their closures $\overline{X \cap U_1}^\beta$ and $\overline{X \cap U_2}^\beta$ are open in βX, by Exercise 2.116. Therefore, all that remains is to show that each $z_j \in \overline{X \cap U_j}^\beta$. To this end, assume that $z \in \beta X$ and U is any open set in βX that contains z. Because X is dense in βX, for every open set W containing z there is an element $x_W \in X$ such that $x_W \in U \cap W$. Hence, $\{x_W\}_W$ is a net in $X \cap U$ converging to z, and so $z \in \overline{X \cap U}^\beta$. $\square$

Corollary 2.80. $\beta\mathbb{N}$ *is separable, but not second countable.*

Proof. The countable set $\mathbb{N}$ is dense in $\beta\mathbb{N}$, and so $\beta\mathbb{N}$ is separable. If $\beta\mathbb{N}$ were second countable, then the compact Hausdorff space $\beta\mathbb{N}$ would be metrisable (Theorem 2.48), in contradiction to Proposition 2.79. $\square$

Observe that $\beta\mathbb{N}$ also furnishes us with an example that shows Theorem 2.19 need not hold beyond metric spaces. Specifically, using the proof of Proposition 2.79, no sequence in $\mathbb{N}$ admits a convergent subsequence in $\beta\mathbb{N}$.

Problems

2.81. Prove that the following statements are equivalent for a topological space X:

1. X is compact;
2. $\bigcap_{\alpha\in\Lambda} F_\alpha \neq \emptyset$ for every family $\{F_\alpha\}_{\alpha\in\Lambda}$ of closed sets $F_\alpha \subseteq X$ with the finite intersection property.

2.82. Prove that the following statements are equivalent for a topological space X:

1. X is compact;
2. $\bigcap_{\alpha \in \Lambda} \overline{E}_\alpha \neq \emptyset$ for every family $\{E_\alpha\}_{\alpha \in \Lambda}$ of sets $E_\alpha \subseteq X$ with the finite intersection property.

2.83. Prove that if F is a finite set in a Hausdorff space X, then F is a closed set.

2.84. Prove that if $Y \subseteq X$ in a Hausdorff space X, and that if $x \in Y$ and $U \subseteq X$ is a neighbourhood of x, then $U \cap Y$ is an infinite set.

2.85. Suppose that (X,d) is a compact metric space and that $U_1, \ldots, U_n$ are open subsets of X such that $\bigcup_{j=1}^{n} U_j = X$.

1. Prove that the function $f : X \to \mathbb{R}$ defined by $f(x) = \dfrac{1}{n} \sum_{j=1}^{n} \mathrm{dist}, (x, U_j^c)$ is continuous and satisfies $f(x) > 0$ for every $x \in X$.
2. Prove that if $\delta = \inf\{f(x) \,|\, x \in X\}$, then $\delta > 0$.
3. Prove that if $S \subset X$ is a nonempty set for which $\sup\{d(s_1, s_2) \,|\, s_1, s_2 \in S\} < \delta$, then $S \subseteq U_j$ for some $j \in \{1, \ldots, n\}$.

2.86. In the Euclidean metric space $(\mathbb{R}^n, d_2)$, prove that a subset $K \subset \mathbb{R}^n$ is compact if and only if K is closed and $\sup_{x \in K} d_2(x, 0) < \infty$.

2.87. Prove that the n-spheres S^n are compact.

2.88. Prove that if x and x' are limits of a convergent sequence $\{x_n\}_{n \in \mathbb{N}}$ in a Hausdorff space X, then $x' = x$.

2.89. Recall from Proposition 1.57 that the topological space $(\mathbb{R}^{\mathbb{N}}, \mathscr{T}^{\times})$ is metrisable with respect to the metric

$$\rho((x_n)_n, (y_n)_n) = \sup_n \left(\min \left(\frac{|x_n - y_n|}{n}, \frac{1}{n} \right) \right).$$

Show that $S = \{x \in \mathbb{R}^{\mathbb{N}} \,|\, \rho(x, 0) = 1\}$ is closed by not compact.

2.90. Determine whether $(\mathbb{R}^{\mathbb{N}}, \mathscr{T}^{\times})$ is a locally compact space.

2.91. Prove that in a locally compact Hausdorff space X that if $K \subseteq U$, where K is compact and U is open, then there exists an open subset $W \subseteq X$ such that $K \subseteq W \subseteq U$ and $\overline{W}$ is compact.

2.92. Assume that x and y are distinct points in a locally compact Hausdorff space X. Prove that there exists a bounded continuous function $f : X \to \mathbb{C}$ such that $f(x) \neq f(y)$.

2.93. Prove that the topology $\mathscr{T}$ on $\tilde{X}$ in Proposition 2.22 is Hausdorff.

2.94. Prove that $\tilde{\mathbb{R}}$ and the circle S^1 are homeomorphic, where $\tilde{\mathbb{R}}$ is the one-point compactification of $\mathbb{R}$.

2.95. Assume that Y is a nonempty subset of a topological space X. Suppose that $\{y_\alpha\}_{\alpha\in\Lambda}$ is a net in Y with limit $x \in X$. Prove that $x \in \overline{Y}$.

2.96. If X and Y are topological spaces, then prove that the following statements are equivalent for a function $f : X \to Y$:

1. f is continuous;
2. for every convergent net $\{x_\alpha\}_{\alpha\in\Lambda}$ in X with limit $x \in X$, $\{f(x_\alpha)\}_{\alpha\in\Lambda}$ is a convergent net in Y with limit $f(x)$.

2.97. Prove that the following statements are equivalent for a topological space X:

1. X is Hausdorff;
2. every convergent net $\{x_\alpha\}_{\alpha\in\Lambda}$ in X has a unique limit point.

2.98. Prove that a topological space X is normal if and only if (i) $\{x\}$ is a closed set, for all $x \in X$, and (ii) for every closed set F and open set U for which $F \subseteq U$ there is an open set V with $F \subseteq V \subseteq \overline{V} \subseteq U$.

2.99. Prove that every separable metric space is second countable.

2.100. Prove that the metric spaces $(\mathbb{R}^n, d_1)$ and $\mathbb{R}^n, d_\infty)$ are complete.

2.101. Prove that if (X, d) is a metric space and if $\{x_k\}_k$ and $\{y_k\}_k$ are two Cauchy sequences in X, then $\{d(x_k, y_k)\}_k$ is a Cauchy sequence in $\mathbb{R}$.

2.102. Suppose that (X_1, d_1) and (X_2, d_2) are metric spaces and that $f : X_1 \to X_2$ is an *isometry*, which is to say that $d_2(f(x), f(y)) = d_1(x, y)$ for all $x, y \in X_1$. Prove that f is continuous and injective.

2.103. Prove that in a separable metric space, every open covering of an open set admits a countable subcovering.

2.104. Prove that in a complete metric space the intersection of a countable family of dense G_δ-sets is a dense G_δ-set and that the union of a countable family of nowhere dense F_σ sets is a nowhere dense F_σ set.

2.105. Prove or find a counterexample to the following assertion: if X is a compact metric space, then X is a complete metric space.

2.106. Prove that if Y is a connected subspace of a topological space X, then $\overline{Y}$ is connected.

2.107. Prove that if $f : X \to Y$ is a continuous function and if X is connected, then $f(X)$ is a connected subspace of Y.

2.108. Prove that the n-spheres S^n are path connected.

2.109. Prove that every connected open set U in $\mathbb{R}^n$ is path connected.

2.110. Prove that every countable metric space is totally disconnected.

2.111. Show that the bounded continuous function $f(x) = \sin(1/x)$ on the locally compact space $X = (0,1)$ does not extend to a continuous function $\tilde{f}$ on the compact set $\overline{X} = [0,1]$.

2.112. Prove that if X is a locally compact Hausdorff space, then there exists a set Λ such that X is homeomorphic to a subspace of the compact Hausdorff hypercube $([0,1]^\Lambda, \mathscr{T}^\times)$.

2.113. Prove that if X is a locally compact Hausdorff space and if K is a compact Hausdorff space that contains X as a dense subspace, then K is a quotient of βX.

2.114. Show that the one-point compactification of $\mathbb{N}$ is not homeomorphic to the Stone-Čech compactification $\beta\mathbb{N}$ of $\mathbb{N}$.

2.115. Assume that X is a non-compact, locally compact space X. Prove that the Stone-Čech compactification βX of X is not metrisable.

2.116. A topological space is *extremely disconnected* if the closure of every open set is open. Show that if X is an infinite discrete space, then βX is extremely disconnected.

Part II
Measure Theory and Integration

Chapter 3
Measure Theory

If topology derives its inspiration from the qualitative features of geometry, then the subject of the present chapter, measure theory, may be thought to have its origins in the quantitative concepts of length, area, and volume. However, a careful theory of area, for example, turns out to be much more delicate than one might expect initially, as any given set may possess an irregular feature, such as having a jagged boundary or being dispersed across many subsets. Even in the setting of the real line, if one has a set E of real numbers, then in what sense can the length of the set E be defined and computed? Furthermore, to what extent can we expect the length (or area, volume) of a union $A \cup B$ of disjoint sets A and B to be the sum of the individual lengths (or areas, volumes) of A and B?

This present chapter is devoted to measure theory, which, among other things, entails a rigorous treatment of length, area, and volume. However, as with the subject of topology, the context and results of measure theory reach well beyond these basic geometric quantities.

3.1 Measurable Spaces and Functions

Definition 3.1. If X is a set, then a σ-*algebra* on X is a collection Σ of subsets of X with the following properties:

1. $X \in \Sigma$;
2. $E^c \in \Sigma$ for every $E \in \Sigma$; and
3. for every countable family $\{E_k\}_{k\in\mathbb{N}}$ of sets $E_k \in \Sigma$,

$$\bigcup_{k\in\mathbb{N}} E_k \in \Sigma.$$

© Springer International Publishing Switzerland 2016
D. Farenick, *Fundamentals of Functional Analysis*, Universitext,
DOI 10.1007/978-3-319-45633-1_3

The pair (X, Σ) is called a *measurable space*, and the elements E of Σ are called *measurable sets*.

The smallest and largest σ-algebras on a set X are, respectively, $\Sigma = \{\emptyset, X\}$ and $\Sigma = \mathscr{P}(X)$, the power set $\mathscr{P}(X)$ of X. The following definition, while abstract in essence, allows for the determination of more interesting, intermediate examples of σ-algebras.

Definition 3.2. If $\mathscr{S}$ is any collection of subsets of X, then the intersection of all σ-algebras on X that contain $\mathscr{S}$ is called the σ-*algebra generated by* $\mathscr{S}$.

It is elementary to verify that the σ-algebra generated by a collection of $\mathscr{S}$ of subsets of X is a σ-algebra in the sense of Definition 3.1.

Definition 3.3. If $(X, \mathscr{T})$ is a topological space, then the σ-algebra generated by $\mathscr{T}$ is called the σ-algebra of *Borel sets* of X.

Let us now consider functions of interest for measure theory.

Definition 3.4. If (X, Σ) is a measurable space, then a function $f : X \to \mathbb{R}$ is *measurable* if $f^{-1}(U) \in \Sigma$, for every open set $U \subseteq \mathbb{R}$.

Proposition 3.5. *If (X, Σ) is a measurable space, then the following statements are equivalent for a function $f : X \to \mathbb{R}$:*

1. $f^{-1}((\alpha, \infty)) \in \Sigma$ for all $\alpha \in \mathbb{R}$;
2. $f^{-1}([\alpha, \infty)) \in \Sigma$ for all $\alpha \in \mathbb{R}$;
3. $f^{-1}((-\infty, \alpha)) \in \Sigma$ for all $\alpha \in \mathbb{R}$;
4. $f^{-1}((-\infty, \alpha]) \in \Sigma$ for all $\alpha \in \mathbb{R}$.

Proof. To begin, observe that (2) follows from (1), because

$$f^{-1}([\alpha, \infty)) = f^{-1}\left(\bigcap_{k \in \mathbb{N}}(\alpha - \frac{1}{k}, \infty)\right) = \bigcap_{k \in \mathbb{N}}f^{-1}\left((\alpha - \frac{1}{k}, \infty)\right) \in \Sigma.$$

Statement (3) follows easily from (2), since

$$f^{-1}((-\infty, \alpha)) = f^{-1}([\alpha, \infty))^c \in \Sigma.$$

Next, we see that (3) implies (4), because

$$f^{-1}((-\infty, \alpha]) = f^{-1}\left(\bigcap_{k \in \mathbb{N}}(-\infty, \alpha + \frac{1}{k})\right) = \bigcap_{k \in \mathbb{N}}f^{-1}\left((-\infty, \alpha + \frac{1}{k})\right) \in \Sigma.$$

Statement (4) implies (1), because

$$f^{-1}((\alpha, \infty)) = f^{-1}((-\infty, \alpha])^c \in \Sigma,$$

which completes the proof.   □

An additional equivalent condition for the measurability of a function is set aside, for future reference, as the following result.

Proposition 3.6 (Criterion for Measurability). *If (X, Σ) is a measurable space, then a function $f : X \to \mathbb{R}$ is measurable if and only if $f^{-1}((\alpha, \infty)) \in \Sigma$, for all $\alpha \in \mathbb{R}$.*

Proof. By definition of measurable function, $f^{-1}((\alpha, \infty)) \in \Sigma$ for all $\alpha \in \mathbb{R}$ because each (α, ∞) is open in $\mathbb{R}$.

Conversely, assume that $f^{-1}((\alpha, \infty)) \in \Sigma$, for all $\alpha \in \mathbb{R}$. Let $U \subseteq \mathbb{R}$ be an open set. By Cantor's Lemma (Proposition 1.30), there is a family of pairwise disjoint open intervals $\{J_k\}_{k \in \mathbb{N}}$ such that $U = \bigcup_k J_k$. Because $f^{-1}(U) = \bigcup_k f^{-1}(J_k)$ and Σ is closed under countable unions, it is enough to prove that $f^{-1}(J) \in \Sigma$ for every open interval J. For open intervals of the form (α, ∞) and $(-\infty, \alpha)$, this is handled by Proposition 3.5. If one has an open interval of the form $J = (\alpha, \beta)$, then $f^{-1}(J)$ is given by $f^{-1}(J) = f^{-1}((-\infty, \alpha])^c \cap f^{-1}([\beta, \infty))^c$, which by Proposition 3.5 is the intersection of two sets in Σ.   □

Proposition 3.7. *If (X, Σ) is a measurable space, if $f, g : X \to \mathbb{R}$ are measurable functions, and if $\lambda \in \mathbb{R}$, then $f + g$, λf, $|f|$, and fg are measurable functions. If, in addition, $g(x) \neq 0$ for every $x \in X$, then f/g is measurable function.*

Proof. The equivalent criteria for measurability of Proposition 3.5 will be used in each case. We begin with a proof that $f + g$ is measurable.

Fix $\alpha \in \mathbb{R}$ and consider the set $S_\alpha = \{x \in X \mid f(x) + g(x) > \alpha\}$. Because f and g are measurable, for each $q \in \mathbb{Q}$ we have

$$\{x \in X \mid f(x) > q\} \in \Sigma \quad \text{and} \quad \{x \in X \mid q > \alpha - g(x)\} \in \Sigma.$$

Hence, as Σ is closed under intersections and countable unions,

$$\bigcup_{q \in \mathbb{Q}} (\{x \in X \mid f(x) > q\} \cap \{x \in X \mid q > \alpha - g(x)\}) \in \Sigma. \tag{3.1}$$

Let G denote the set in (3.1); we shall prove that $S_\alpha = G$. If $y \in S_\alpha$, then $f(y) > \alpha - g(y)$. In fact, by the density of $\mathbb{Q}$ in $\mathbb{R}$, there is a rational number $q_y \in \mathbb{Q}$ such that $f(y) > q_y > \alpha - g(y)$. Thus,

$$y \in \{x \in X \mid f(x) > q_y\} \cap \{x \in X \mid q_y > \alpha - g(x)\},$$

which shows that $S_\alpha \subseteq G$. Conversely, if $y \in G$, then there is a rational $q_y \in \mathbb{Q}$ such that $y \in \{x \in X \mid f(x) > q_y\} \cap \{x \in X \mid q_y > \alpha - g(x)\}$. Thus, $f(y) > q_y > \alpha - g(y)$ implies that $f(y) + g(y) > \alpha$, whence $y \in S_\alpha$ and, consequently, $G \subseteq S_\alpha$. This proves that $f + g$ is measurable.

The proof that λf is measurable is clear, and we move to the proof that $|f|$ is measurable. Note that if $\alpha \in \mathbb{R}$, then $|f|^{-1}((\alpha, \infty)) = X$ if $\alpha < 0$, and

$$|f|^{-1}((\alpha, \infty)) = f^{-1}((\alpha, \infty)) \cup f^{-1}((-\infty, -\alpha)), \text{ if } \alpha \geq 0.$$

In either case, $|f|^{-1}((\alpha, \infty)) \in \Sigma$, which proves that $|f|$ is measurable.

To prove that the product fg is measurable, first assume that $h : X \to \mathbb{R}$ is a measurable function and consider h^2. If $\alpha \in \mathbb{R}$, then $\{x \in X \mid h(x)^2 > \alpha\} = X$ if $\alpha < 0$, otherwise $\{x \in X \mid h(x)^2 > \alpha\} = |h|^{-1}\left((\sqrt{\alpha}, \infty)\right)$. In either case, the sets belong to Σ. This proves that the square of a measurable function is measurable. To conclude that fg is measurable, express fg as

$$fg = \frac{1}{4}\left((f+g)^2 - (f-g)^2\right). \tag{3.2}$$

As the sums, squares, and scalar multiples of measurable functions are measurable, equation (3.2) demonstrates that fg is measurable.

If $g(x) \neq 0$ for every $x \in X$, then $1/g$ is measurable (Exercise 3.79), which implies that the function $f/g = f \cdot (1/g)$ is measurable.   □

Using the algebraic features exhibited in Proposition 3.7, one deduces that the following functions are measurable as well.

Corollary 3.8. *Suppose that $f, g : X \to \mathbb{R}$ are measurable functions.*

1. *If $\max(f, g)$ is the function whose value at each $x \in X$ is the maximum of $f(x)$ and $g(x)$, and if $\min(f, g)$ is the function whose value at each $x \in X$ is the minimum of $f(x)$ and $g(x)$, then $\max(f, g)$ and $\min(f, g)$ are measurable.*
2. *f^+ is the function $\max(f, 0)$ and f^- is the function $-\min(f, 0)$, then f^+ and f^- are measurable.*

Proof. By Proposition 3.7, the sum, difference, and absolute value of measurable functions are measurable. Therefore, the formulae

$$\max(f, g) = 1/2(f + g + |f - g|),$$
$$\min(f, g) = 1/2(f + g - |f - g|),$$
$$f^+ = 1/2(|f| + f), \text{ and}$$
$$f^- = 1/2(|f| - f)$$

imply the asserted conclusions.   □

The purpose of the following result is to use sequences of measurable functions to determine new measurable functions.

Proposition 3.9. *Suppose that $f_k : X \to \mathbb{R}$ is a measurable function for each $k \in \mathbb{N}$. If*

$$S = \{x \in X \mid \sup_k f_k(x) \text{ exists}\},$$
$$LS = \{x \in X \mid \limsup_k f_k(x) \text{ exists}\},$$
$$I = \{x \in X \mid \inf_k f_k(x) \text{ exists}\},$$
$$LI = \{x \in X \mid \liminf_k f_k(x) \text{ exists}\}, \text{ and}$$
$$L = \{x \in X \mid \lim_k f_k(x) \text{ exists}\},$$

then each of the sets S, LS, I, LI, and L is measurable. Moreover,

1. $\sup_k f_k$ *is a measurable function on S,*
2. $\limsup_k f_k$ *is a measurable function on LS,*
3. $\inf_k f_k$ *is a measurable function on I,*
4. $\liminf_k f_k$ *is a measurable function on LI, and*
5. $\lim f_k$ *is a measurable function on L.*

Proof. The set $f_k^{-1}((-\infty, q))$ is measurable for every $k \in \mathbb{N}$ and $q \in \mathbb{Q}$; therefore, so is

$$\bigcup_{q \in \mathbb{Q}} \bigcap_{k \in \mathbb{N}} f_k^{-1}((-\infty, q)) = S.$$

Consider now the function $\sup_k f_k$ defined on the (measurable) set S with values in $\mathbb{R}$. For every $\alpha \in \mathbb{R}$,

$$\{x \in S \mid \sup_k f_k(x) > \alpha\} = \bigcup_{k \in \mathbb{N}} \{x \in S \mid f_k(x) > \alpha\} \in \Sigma(S).$$

Hence, $\sup_k f_k$ is measurable as a function $S \to \mathbb{R}$.

The proofs that I is a measurable set and that $\inf_k f_k$ is a measurable function $I \to \mathbb{R}$ are handled in a similar fashion. For example, in this case, I is given by

$$I = \bigcup_{q \in \mathbb{Q}} \bigcap_{k \in \mathbb{N}} f_k^{-1}((q, \infty)).$$

For each $k \in \mathbb{N}$ consider the measurable function $g_k : S \to \mathbb{N}$ defined by

$$g_k(x) = \sup_{n \geq k} f_n(x), \quad x \in S.$$

For every $x \in LS$, $\limsup_k f_k$ is precisely $\inf_k g_k$. Moreover, by the discussion of the previous paragraph, $\inf g_k$ is a measurable function on the (measurable) set

$$\bigcup_{q \in \mathbb{Q}} \bigcap_{k \in \mathbb{N}} g_k^{-1}((q, \infty)) = LS.$$

Hence, as a function $LS \to \mathbb{R}$, $\limsup_k f_k$ is measurable.

The proofs that LI is a measurable set and that $\liminf_k f_k$ is a measurable function $LI \to \mathbb{R}$ are similarly handled.

Consider the measurable set $E = LS \cap LI$ and let $h : E \to \mathbb{R}$ be the function

$$h(x) = \limsup_k f_k(x) - \liminf_k f_k(x).$$

Note that h is measurable and that

$$L = \{x \in X \mid \limsup_k f_k(x) = \liminf_k f_k(x)\} = h^{-1}(\{0\}),$$

which is a measurable set because

$$h^{-1}(\{0\}) = E \setminus \left(h^{-1}(-\infty, 0) \cup h^{-1}(0, \infty) \right).$$

Finally, for every $\alpha \in \mathbb{R}$,

$$\{x \in L \mid \lim_k f_k(x) > \alpha\} = L \cap \{x \in L \mid \limsup_k f_k(x) > \alpha\} \in \Sigma(L).$$

Therefore, $\lim_k f_k$ is measurable as a function from L to $\mathbb{R}$.   □

Definition 3.10. If X is a set, then the *characteristic function* of a subset $E \subseteq X$ is the function $\chi_E : X \to \mathbb{R}$ defined by

$$\chi_E(x) = 1, \text{ if } x \in E, \text{ and } \chi_E(x) = 0, \text{ if } x \notin E.$$

From the definition above, the following proposition is immediate:

Proposition 3.11. *If (X, Σ) is a measurable space and if $E \subseteq X$, then the characteristic function $\chi_E : X \to \mathbb{R}$ is a measurable function if and only if $E \in \Sigma$.*

Characteristic functions can be used to restrict or extend the domain of functions (Exercise 3.82).

Definition 3.12. If (X, Σ) is a measurable space, then a *simple function* is a measurable function $\varphi : X \to \mathbb{R}$ such that φ assumes at most a finite number of values in $\mathbb{R}$.

Suppose that φ is a simple function on a measurable space (X, Σ). If $\varphi(X) = \{\alpha_1, \ldots, \alpha_n\} \subset \mathbb{R}$, then let $E_k = \varphi^{-1}(\{\alpha_k\})$ (which is a measurable set, as φ is a measurable function) so that

$$\varphi = \sum_{k=1}^{n} \alpha_k \chi_{E_k}$$

represents φ as a linear combination of the characteristic functions χ_{E_k}.

Definition 3.13. A sequence $\{f_k\}_{k \in \mathbb{N}}$ of real-valued functions f_k on a set X is a *monotone increasing sequence* if $f_k(x) \leq f_{k+1}(x)$ for every $k \in \mathbb{N}$ and every $x \in X$.

The analysis of measurable functions depends, to a very large extent, on the following approximation theorem.

Theorem 3.14 (Approximation of Nonnegative Measurable Functions). *For every nonnegative measurable function f on a measurable space (X, Σ), there is a monotone increasing sequence $\{\varphi_k\}_{k \in \mathbb{N}}$ of nonnegative simple functions φ_k on X such that*

$$\lim_{k \to \infty} \varphi_k(x) = f(x),$$

for every $x \in X$.

Proof. Let $\kappa_n : [0,n] \to \mathbb{Z}$ be the function whose value at t is the unique $j \in \mathbb{Z}$ for which $t \in [\frac{j}{2^n}, \frac{j+1}{2^n})$, and define $\omega_n : \mathbb{R} \to \mathbb{Q}$ by $\omega_n(t) = \kappa_n(t)/2^n$, if $t \in [0,n]$, and by $\omega_n(t) = 0$ if $t \in (n, \infty)$. The functions ω_n satisfy $\omega_n(t) \le t$ for every $t \in [0, \infty)$ and $\omega_n(t) \le \omega_{n+1}(t)$ for all $n \in \mathbb{N}$ and $t \in [0, \infty)$. Now let $\varphi_n = \omega_n \circ f$. Thus, $\{\varphi_n\}_n$ is a monotone increasing sequence of nonnegative functions, each with finite range. For each $x \in X$ there is some $n \in \mathbb{N}$ for which $f(x) \in [0, n)$. Thus, for every $k > n$,

$$f(x) - \varphi_k(x) < \frac{1}{2^k},$$

which proves that $\lim_{n \to \infty} \varphi_n(x) = f(x)$. All that remains is to verify that φ_n is measurable. To this end, select $n \in \mathbb{N}$ and let

$$E_n = f^{-1}([n, \infty)) \quad \text{and} \quad E_{nj} = f^{-1}\left(\left[\frac{j-1}{2^n}, \frac{j}{2^n}\right)\right), \quad \text{for } 1 \le j \le 2^n n.$$

These are measurable sets and

$$\varphi_n = \sum_{j=1}^{2^n n} \frac{j-1}{2^n} \chi_{E_{nj}} + n \chi_{E_n}.$$

Hence, φ_n is a simple function. $\qquad\square$

By decomposing a real-valued function f into a difference its positive and negative parts, namely $f = f^+ - f^-$, where

$$f^+ = \frac{|f| + f}{2} \quad \text{and} \quad f^- = \frac{|f| - f}{2}, \tag{3.3}$$

we obtain the following approximation result for arbitrary measurable functions.

Corollary 3.15. *If (X, Σ) is a measurable space and if $f : X \to \mathbb{R}$ is a measurable function, then there is a sequence $\{\psi_k\}_{k \in \mathbb{N}}$ of simple functions $\psi_k : X \to \mathbb{R}$ such that*

$$\lim_{k \to \infty} \psi_k(x) = f(x), \quad \forall x \in X.$$

3.2 Measure Spaces

Before continuing further, the values $-\infty$ and $+\infty$ will be added to the arithmetic system of $\mathbb{R}$. Formally, the *extended real number system* are the elements of the set denoted by $[-\infty, +\infty]$ and defined by $\{-\infty\} \cup \mathbb{R} \cup \{+\infty\}$. (Here, $-\infty$ and $+\infty$ are meant to denote the "ends" of the real axis.) The arithmetic of $[-\infty, +\infty]$ is prescribed by the following laws:

1. $r \cdot s$ and $r + s$ are the usual product and sum in $\mathbb{R}$, for all $r, s \in \mathbb{R}$;
2. $0 \cdot (-\infty) = 0 \cdot (+\infty) = 0$;
3. $r \cdot (-\infty) = -\infty$ and $r \cdot (+\infty) = +\infty$, for all $r \in \mathbb{R}$ with $r > 0$ and for $r = +\infty$;

4. $r \cdot (-\infty) = +\infty$ and $r \cdot (+\infty) = -\infty$, for all $r \in \mathbb{R}$ with $r < 0$ and for $r = -\infty$;
5. $r + (-\infty) = -\infty$ and $r + (+\infty) = +\infty$, for all $r \in \mathbb{R}$.

The sum of $-\infty$ and $+\infty$ is not defined in the extended real number system, which is a small fact that will be of note in our study of signed measures in Section 3.7.

Henceforth, $[0, \infty]$ denotes the subset of the extended real numbers given by

$$[0, \infty] = [0, \infty) \cup \{+\infty\}.$$

The terminology below concerning families of sets will be used extensively, beginning with the definition of measure in Definition 3.17.

Definition 3.16. A family $\{X_\alpha\}_{\alpha \in \Lambda}$ of subsets of a given set X is a *family of pairwise disjoint sets* if $X_\alpha \cap X_\beta = \emptyset$ for all $\alpha, \beta \in \Lambda$ such that $\alpha \neq \beta$.

Definition 3.17. A *measure* on a measurable space (X, Σ) is a function $\mu : \Sigma \to [0, +\infty]$ such that $\mu(\emptyset) = 0$ and

$$\mu\left(\bigcup_{k \in \mathbb{N}} E_k\right) = \sum_{k \in \mathbb{N}} \mu(E_k), \tag{3.4}$$

for every sequence $\{E_k\}_{k \in \mathbb{N}}$ of pairwise disjoint sets $E_k \in \Sigma$. Furthermore,

1. if $\mu(X) < \infty$, then μ is said to be a *finite measure*, and
2. if $\mu(X) = 1$, then μ is said to be a *probability measure*.

The (X, Σ, μ) is called a *measure space*.

Measures are not easy to construct or determine in general, but there are some very simple examples nevertheless.

Example 3.18. *Consider the measurable space (X, Σ) in which X is an uncountable infinite set and Σ is the σ-algebra of all subsets $E \subseteq X$ that have the property that E or E^c is countable (see Exercise 3.71). If $\mu : \Sigma \to [0, +\infty]$ is defined by*

$$\mu(E) = 0 \ \text{if } E \text{ is countable, and } \mu(E) = 1 \ \text{if } E^c \text{ is countable,}$$

then μ is a measure on (X, Σ).

Example 3.19 (Dirac Measures). *If Σ is a σ-algebra of subsets of X in which $\{x\} \in \Sigma$ for every $x \in X$, then for each $x \in X$ the function $\delta_x : \Sigma \to [0, 1]$ given by*

$$\delta_x(E) = 1 \ \text{if } x \in E, \ \text{and } \delta_x(E) = 0 \ \text{if } x \notin E,$$

is a probability measure on (X, Σ). The measures δ_x are called Dirac measures *or* point mass measures.

Example 3.20 (Counting Measure). *Consider the measurable space* $(\mathbb{N}, \mathscr{P}(\mathbb{N}))$, *where* $\mathscr{P}(\mathbb{N})$ *is the power set of* $\mathbb{N}$. *If* $\mu : \Sigma \to [0, +\infty]$ *is the function defined by*

$$\mu(E) = \text{the cardinality of } E,$$

then μ *is a measure on* $(\mathbb{N}, \mathscr{P}(\mathbb{N}))$ *and is called* counting measure.

We turn now to some general properties of measures and measure spaces.

Proposition 3.21 (Monotonicity of Measure). *Let* (X, Σ, μ) *denote a measure space. Suppose that* $E, F \in \Sigma$ *are such that* $E \subseteq F$. *Then* $\mu(E) \leq \mu(F)$. *Furthermore, if* $\mu(F) < \infty$, *then* $\mu(F \backslash E) = \mu(F) - \mu(E)$.

Proof. Because $E \subseteq F$, we may express F as $F = E \cup (E^c \cap F)$, which is a union of disjoint sets E and $E^c \cap F$, each of which belongs to Σ. Hence, $\mu(F) = \mu(E) + \mu(E^c \cap F) \geq \mu(E)$.   □

Proposition 3.22 (Continuity of Measure). *Let* (X, Σ, μ) *denote a measure space. Suppose that* $\{A_k\}_{k \in \mathbb{N}}$ *and* $\{E_k\}_{k \in \mathbb{N}}$ *are sequences of sets* $E_k \in \Sigma$.

1. If $A_k \subseteq A_{k+1}$, *for all* $k \in \mathbb{N}$, *then*

$$\mu\left(\bigcup_{k \in \mathbb{N}} A_k\right) = \lim_{k \to \infty} \mu(A_k). \tag{3.5}$$

2. If $E_k \supseteq E_{k+1}$, *for all* $k \in \mathbb{N}$, *and if* $\mu(E_1) < \infty$, *then*

$$\mu\left(\bigcap_{k \in \mathbb{N}} E_k\right) = \lim_{k \to \infty} \mu(E_k). \tag{3.6}$$

Proof. (1) Equation (3.5) plainly holds if $\mu(A_k) = \infty$ for at least one k; hence, assume that $\mu(A_k) < \infty$ for all $k \in \mathbb{N}$. The sequence $\{A_k\}_{k \in \mathbb{N}}$ is nested and ascending, and so it is simple to produce from it a sequence of pairwise disjoint sets $G_k \in \Sigma$ by taking set differences: that is, define G_1 to be A_1 and let

$$G_k = A_k \backslash A_{k-1}, \quad \forall k \geq 2.$$

Observe that $\mu(A_k) < \infty$ implies that $\mu(G_k) = \mu(A_k) - \mu(A_{k-1})$, by Proposition 3.21. Furthermore, the sets G_k are pairwise disjoint. Because $A_k = \bigcup_{n=1}^{k} G_n$, we have

$$\bigcup_{k \in \mathbb{N}} A_k = \bigcup_{k \in \mathbb{N}} G_k.$$

Thus, by the countable additivity of μ on disjoint unions,

$$\mu\left(\bigcup_{k\in\mathbb{N}} A_k\right) = \mu\left(\bigcup_{k\in\mathbb{N}} G_k\right)$$
$$= \sum_{k\in\mathbb{N}} \mu(G_k)$$
$$= \mu(A_1) + \lim_{n\to\infty} \sum_{k=2}^{n} [\mu(A_k) - \mu(A_{k-1})]$$
$$= \mu(A_1) + \left[\lim_{n\to\infty} \mu(A_n)\right] - \mu(A_1),$$

which establishes formula (3.5).

(2) The sequence $\{E_k\}_{k\in\mathbb{N}}$ is nested and descending, and so it is simple to produce from it a sequence of pairwise disjoint sets $F_k \in \Sigma$ by taking set differences: that is, let

$$F_k = E_k \backslash E_{k+1}, \quad \forall k \in \mathbb{N}.$$

Observe that $E_k = E_{k+1} \cup F_k$ and that $E_{k+1} \cap F_k = \emptyset$. Thus, by the countable additivity of μ on disjoint unions,

$$\mu(E_k) = \mu(E_{k+1}) + \mu(F_k), \quad \forall k \in \mathbb{N}.$$

Because

$$\left(\bigcap_{k\in\mathbb{N}} E_k\right) \cap \left(\bigcup_{k\in\mathbb{N}} F_k\right) = \emptyset,$$

and

$$E_1 = \left(\bigcap_{k\in\mathbb{N}} E_k\right) \cup \left(\bigcup_{k\in\mathbb{N}} F_k\right),$$

the countable additivity of μ on disjoint unions yields

$$\mu(E_1) = \mu\left(\bigcap_{k\in\mathbb{N}} E_k\right) + \mu\left(\bigcup_{k\in\mathbb{N}} F_k\right)$$
$$= \mu\left(\bigcap_{k\in\mathbb{N}} E_k\right) + \sum_{k\in\mathbb{N}} \mu(F_k)$$
$$= \mu\left(\bigcap_{k\in\mathbb{N}} E_k\right) + \lim_{n\to\infty} \sum_{k=1}^{n} [\mu(E_k) - \mu(E_{k+1})]$$
$$= \mu\left(\bigcap_{k\in\mathbb{N}} E_k\right) + \mu(E_1) - \lim_{n\to\infty} \mu(E_{n+1}),$$

which establishes formula (3.6).   □

As an application of the continuity of measure, the following result shows that if a measurable function f on a finite measure space is unbounded, then the set on which the values of f are very large has arbitrarily small measure.

Proposition 3.23. *If (X, Σ, μ) is a finite measure space and if $f : X \to \mathbb{R}$ is measurable, then for each $\varepsilon > 0$ there is an $n \in \mathbb{N}$ such that*

$$\mu(\{x \in X \mid |f(x)| > n\}) < \varepsilon.$$

Proof. Let $E_n = \{x \in X \mid |f(x)| > n\}$, for each $n \in \mathbb{N}$. Note that $\mu(E_1) \leq \mu(X) < \infty$ and $E_{n+1} \supseteq E_n$ for every n. Hence, by Proposition 3.22, if $E = \bigcap_{n \in \mathbb{N}} E_n$, then $\mu(E) = \lim_{n \to \infty} \mu(E_n)$. Now because, in this particular case, $E = \emptyset$ and thus $\mu(E) = 0$, we deduce that for each $\varepsilon > 0$ there is an $n \in \mathbb{N}$ such that $\mu(E_n) < \varepsilon$. $\square$

One might not have a sequence of pairwise disjoint sets at hand. Nevertheless, it is possible to obtain an estimate on the measure of their union.

Proposition 3.24 (Countable Subadditivity of Measure). *Let (X, Σ, μ) denote a measure space. Suppose that $\{E_k\}_{k \in \mathbb{N}}$ is any sequence of sets $E_k \in \Sigma$. Then,*

$$\mu\left(\bigcup_{k \in \mathbb{N}} E_k\right) \leq \sum_{k \in \mathbb{N}} \mu(E_k). \tag{3.7}$$

Proof. For each $k \in \mathbb{N}$, let

$$F_k = E_k \setminus \left(\bigcup_{j=1}^{k-1} E_j\right).$$

Note that the sequence $\{F_k\}_{k \in \mathbb{N}}$ consists of pairwise disjoint elements of Σ and that each $F_k \subseteq E_k$. Thus, $\mu(F_k) \leq \mu(E_k)$, by Proposition 3.21. Also,

$$\bigcup_{k \in \mathbb{N}} E_k = \bigcup_{k \in \mathbb{N}} F_k.$$

Thus,

$$\mu\left(\bigcup_{k \in \mathbb{N}} E_k\right) = \mu\left(\bigcup_{k \in \mathbb{N}} F_k\right) = \sum_{k \in \mathbb{N}} \mu(F_k) \leq \sum_{k \in \mathbb{N}} \mu(E_k),$$

which proves inequality (3.7). $\square$

There is a rather significant difference between those measure spaces (X, Σ, μ) in which $\mu(X)$ is finite and those for which $\mu(X) = \infty$. A hybrid between these two alternatives occurs with the notion of a σ-finite space.

Definition 3.25. A measure space (X, Σ, μ) is σ-*finite* if there is a sequence $\{X_n\}_{n \in \mathbb{N}}$ of measurable sets $X_n \in \Sigma$ such that $\mu(X_n) < \infty$ for every n and $X = \bigcup_{n \in \mathbb{N}} X_n$.

3.3   Outer Measures

Having examined to this point some properties of measures, we turn now to the issue of constructing measures. This will be done by first defining an outer measure.

Definition 3.26. If X is a set, then a function $\mu^* : \mathscr{P}(X) \to [0, \infty]$ on the power set $\mathscr{P}(X)$ of X is an *outer measure* on X if

1. $\mu^*(\emptyset) = 0$,
2. $\mu^*(S_1) \leq \mu^*(S_2)$, if $S_1 \subseteq S_2$, and
3. $\mu^* \left(\bigcup_{k \in \mathbb{N}} S_k \right) \leq \sum_{k \in \mathbb{N}} \mu^*(S_k)$ for every sequence $\{S_k\}_{k \in \mathbb{N}}$ of subsets $S_k \subseteq X$.

An outer measure is generally not a measure. And note that the domain of an outer measure is the power set $\mathscr{P}(X)$, rather than some particular σ-algebra of subsets of X.

Definition 3.27. A *sequential cover* of X is a collection $\mathscr{O}$ of subsets of X with the properties that $\emptyset \in \mathscr{O}$ and for every $S \subseteq X$ there is a countable subcollection $\{I_k\}_{k \in \mathbb{N}} \subseteq \mathscr{O}$ such that

$$S \subseteq \bigcup_{k \in \mathbb{N}} I_k.$$

Sequential covers lead to outer measures as follows.

Proposition 3.28. *Assume that $\mathscr{O}$ is a sequential cover of a set X. If $\lambda : \mathscr{O} \to [0, \infty)$ is any function for which $\lambda(\emptyset) = 0$, then the function $\mu^* : \mathscr{P}(X) \to [0, \infty]$ defined by*

$$\mu^*(S) = \inf \left\{ \sum_{k=1}^{\infty} \lambda(I_k) \, | \, \{I_k\}_{k \in \mathbb{N}} \subseteq \mathscr{O} \text{ and } S \subseteq \bigcup_{k \in \mathbb{N}} I_k \right\} \tag{3.8}$$

is an outer measure on X.

Proof. Clearly $\mu^*(\emptyset) = 0$. If $S \subseteq T$, then any $\{I_k\}_{k \in \mathbb{N}} \subseteq \mathscr{O}$ that covers the set T also covers the set S, and so $\mu^*(S) \leq \mu^*(T)$. Thus, all that remains is to verify that μ^* is countable subadditive.

To this end, suppose that $\{S_k\}_{k \in \mathbb{N}}$ is a sequence of subsets $S_k \subseteq X$. Since we aim to show that

$$\mu^* \left(\bigcup_{k \in \mathbb{N}} S_k \right) \leq \sum_{k \in \mathbb{N}} \mu^*(S_k),$$

only the case where the sum $\sum_k \mu^*(S_k)$ converges need be considered. For this case, suppose that $\varepsilon > 0$. For each $k \in \mathbb{N}$ there is a countable family $\{I_{kj}\}_{j \in \mathbb{N}} \subseteq \mathcal{O}$ such that $S_k \subseteq \bigcup_j I_{kj}$ and

$$\sum_j \lambda(I_{kj}) \leq \mu^*(S_k) + \frac{\varepsilon}{2^k}.$$

Thus, $\{I_{kj}\}_{k,j \in \mathbb{N}}$ forms a countable subcollection of sets from $\mathcal{O}$ that cover $\bigcup_k S_k$ and satisfies

$$\mu^*\left(\bigcup_{k \in \mathbb{N}} S_k\right) \leq \sum_k \sum_j \lambda(I_{kj}) \leq \sum_k \left(\mu^*(S_k) + \frac{\varepsilon}{2^k}\right) \leq \sum_{k \in \mathbb{N}} \mu^*(S_k) + \varepsilon.$$

As $\varepsilon > 0$ is chosen arbitrarily, μ^* is indeed countably subadditive.   □

The value of an outer measure is two-fold: (i) it is frequently easier to define an outer measure on the power set of X than it is to define a measure on some σ-algebra of subsets of X (indeed, determining nontrivial σ-algebras on X is in itself a nontrivial task), and (ii) if one has an outer measure at hand, then there is a σ-algebra Σ of subsets of X for which the restriction of μ^* to Σ is a measure on (X, Σ). This latter fact is the content of the following theorem.

Theorem 3.29 (Carathéodory). *If μ^* is an outer measure on a set X, then*

1. the collection $\mathfrak{M}_{\mu^*}(X)$ of all subsets $E \subseteq X$ for which

$$\mu^*(S) = \mu^*(E \cap S) + \mu^*(E^c \cap S), \quad \forall S \subseteq X,$$

is a σ-algebra, and

2. the function $\mu : \mathfrak{M}_{\mu^*}(X) \to [0, \infty]$ defined by $\mu(E) = \mu^*(E)$, $E \in \mathfrak{M}_{\mu^*}(X)$, is a measure on the measurable space $(X, \mathfrak{M}_{\mu^*}(X))$.

The criterion (1) in Theorem 3.29 for membership in $\mathfrak{M}_{\mu^*}(X)$ is called the *Carathéodory criterion*. The proof of Theorem 3.29 requires the following lemma.

Lemma 3.30. *If $E_1, \ldots, E_n \in \mathfrak{M}_{\mu^*}(X)$, then*

$$\bigcup_{k=1}^n E_k \in \mathfrak{M}_{\mu^*}(X).$$

Moreover, if $E_1, \ldots, E_n \in \mathfrak{M}_{\mu^*}(X)$ are pairwise disjoint, then

$$\mu^*\left(S \cap \left[\bigcup_{k=1}^n E_k\right]\right) = \sum_{k=1}^n \mu^*(S \cap E_k) \quad \forall S \subseteq X. \tag{3.9}$$

Proof. It is sufficient to consider the case $n = 2$, as the remaining cases follow by induction on n.

We shall prove that $E_1 \cup E_2 \in \mathfrak{M}_{\mu^*}(X)$, for all $E_1, E_2 \in \mathfrak{M}_{\mu^*}(X)$ Let $S \subseteq X$ be arbitrary and note that $S \cap (E_1 \cup E_2)$ can be written as

$$S \cap (E_1 \cup E_2) = (S \cap E_1) \cup (S \cap E_2) = (S \cap E_1) \cup ([S \cap E_1^c] \cap E_2).$$

Likewise,

$$S \cap (E_1 \cup E_2)^c = S \cap (E_1^c \cap E_2^c) = (S \cap E_1^c) \cap E_2^c.$$

Thus,

$$\mu^*(S) \leq \mu^*(S \cap (E_1 \cup E_2)) + \mu^*(S \cap (E_1 \cup E_2)^c)$$

$$= \mu^*\big((S \cap E_1) \cup ([S \cap E_1^c] \cap E_2)\big) + \mu^*\big([S \cap E_1^c] \cap E_2^c\big)$$

$$\leq \mu^*(S \cap E_1) + \mu^*\big([S \cap E_1^c] \cap E_2)\big) + \mu^*\big([S \cap E_1^c] \cap E_2^c\big)$$

$$= \mu^*(S \cap E_1) + \mu^*(S \cap E_1^c)$$

$$= \mu^*(S),$$

where the final two equalities are because of $E_2 \in \mathfrak{M}_{\mu^*}(X)$ and $E_1 \in \mathfrak{M}_{\mu^*}(X)$, respectively. Hence,

$$\mu^*(S) = \mu^*(S \cap (E_1 \cup E_2)) + \mu^*(S \cap (E_1 \cup E_2)^c), \quad \forall S \subseteq X.$$

This proves that $E_1 \cup E_2 \in \mathfrak{M}_{\mu^*}(X)$.

Next, let $E_1, E_2 \subseteq X$ be disjoint elements of $\mathfrak{M}_{\mu^*}(X)$. If $S \subseteq X$, then

$$\begin{aligned} [S \cap (E_1 \cup E_2)] \cap E_2 &= S \cap E_2 \quad \text{and} \\ [S \cap (E_1 \cup E_2)] \cap E_2^c &= S \cap E_1. \end{aligned} \tag{3.10}$$

Thus, by using (3.10) together with the fact that $E_2 \in \mathfrak{M}_{\mu^*}(X)$, we obtain

$$\mu^*(S \cap E_1) + \mu^*(S \cap E_2) = \mu^*(S \cap (E_1 \cup E_2)),$$

which completes the proof.   □

We are now equipped to prove Theorem 3.29.

Proof. To prove (1), namely that $\mathfrak{M}_{\mu^*}(X)$ is a σ-algebra, recall that a subset $E \subseteq X$ is an element of $\mathfrak{M}_{\mu^*}(X)$ if and only if $E^c \in \mathfrak{M}_{\mu^*}(X)$. Hence, $\mathfrak{M}_{\mu^*}(X)$ is closed under complements. Further, the empty set $\emptyset$ clearly belongs to $\mathfrak{M}_{\mu^*}(X)$. Thus, all that remains is to prove that $\mathfrak{M}_{\mu^*}(X)$ is closed under countable unions.

Lemma 3.30 states that $\mathfrak{M}_{\mu^*}(X)$ is closed under finite unions. To get the same result for finite intersections, note that

$$E_1, E_2 \in \mathfrak{M}_{\mu^*}(X) \implies E_1^c, E_2^c \in \mathfrak{M}_{\mu^*}(X)$$

$$\implies E_1^c \cup E_2^c \in \mathfrak{M}_{\mu^*}(X)$$

$$\implies (E_1^c \cup E_2^c)^c \in \mathfrak{M}_{\mu^*}(X)$$

$$\implies (E_1^c)^c \cap (E_2^c)^c = E_1 \cap E_2 \in \mathfrak{M}_{\mu^*}(X).$$

That is, $E_1 \cap E_2 \in \mathfrak{M}_{\mu^*}(X)$. By induction, $E_1 \cap \cdots \cap E_n \in \mathfrak{M}_{\mu^*}(X)$, for all $E_1, \ldots, E_n \in \mathfrak{M}_{\mu^*}(X)$.

Now let $\{A_k\}_{k\in\mathbb{N}}$ be a sequence for which $A_k \in \mathfrak{M}_{\mu^*}(X)$ for all $k \in \mathbb{N}$. Let $E_0 = \emptyset$ and

$$E_k = A_k \setminus \bigcup_{j=1}^{k-1} A_j, \quad \forall k \in \mathbb{N}.$$

As $\mathfrak{M}_{\mu^*}(X)$ is closed under finite unions and intersections, $E_k \in \mathfrak{M}_{\mu^*}(X)$ for all $k \in \mathbb{N}$. Furthermore, by Exercise 3.76, $\{E_k\}_{k\in\mathbb{N}}$ is a sequence of pairwise disjoint sets for which

$$\bigcup_{k\in\mathbb{N}} E_k = \bigcup_{k\in\mathbb{N}} A_k.$$

Let

$$E = \bigcup_{k\in\mathbb{N}} E_k \quad \text{and} \quad F_n = \bigcup_{k=1}^{n} E_k, \quad \forall n \in \mathbb{N}.$$

Because $F_n \subseteq E$, we have that $E^c \subseteq F_n^c$. The sets F_n are elements of $\mathfrak{M}_{\mu^*}(X)$; thus, for any subset $S \subseteq X$,

$$\mu^*(S) = \mu^*(S \cap F_n) + \mu^*(S \cap F_n^c)$$
$$\geq \mu^*(S \cap F_n) + \mu^*(S \cap E^c).$$

Equation (3.9) of Lemma 3.30 yields

$$\mu^*(S \cap F_n) = \sum_{k=1}^{n} \mu^*(S \cap E_k).$$

Thus, this equation and the inequality $\mu^*(S) \geq \mu^*(S \cap F_n) + \mu^*(S \cap E^c)$ imply that

$$\mu^*(S) \geq \sum_{k=1}^{n} \mu^*(S \cap E_k) + \mu^*(S \cap E^c), \quad \forall n \in \mathbb{N}.$$

Therefore, by making use of the fact that μ^* is countably subadditive,

$$\mu^*(S) \geq \sum_{k=1}^{\infty} \mu^*(S \cap E_k) + \mu^*(S \cap E^c)$$

$$\geq \mu^*(S \cap E) + \mu^*(S \cap E^c)$$

$$\geq \mu^*(S).$$

Hence, $\mu^*(S) = \mu^*(S \cap E) + \mu^*(S \cap E^c)$, which proves that $E \in \mathfrak{M}_{\mu^*}(X)$.

To prove (2), namely that μ^* restricted to $\mathfrak{M}_{\mu^*}(X)$ is a measure, note first that $\mu(\emptyset) = 0$ and that the range of μ is obviously all of $[0, \infty]$.

Suppose now that $\{E_k\}_{k \in \mathbb{N}}$ is a sequence in $\mathfrak{M}_{\mu^*}(X)$ of pairwise disjoint sets and let $E = \bigcup_k E_k$. We aim to prove that $\mu(E) = \sum_k \mu(E_k)$. Outer measure is countably subadditive; thus,

$$\mu(E) = \mu^*(E) \leq \sum_{k=1}^{\infty} \mu^*(E_k) = \sum_{k=1}^{\infty} \mu(E_k).$$

Let $S \subseteq X$ be arbitrary. By Lemma 3.30,

$$\mu^*\left(S \cap \left[\bigcup_{k=1}^{n} E_k\right]\right) = \sum_{k=1}^{n} \mu^*(S \cap E_k) \quad \text{for every } n \in \mathbb{N}.$$

In particular, for $S = X$, this yields, for every $n \in \mathbb{N}$,

$$\mu\left(\bigcup_{k=1}^{n} E_k\right) = \mu^*\left(\bigcup_{k=1}^{n} E_k\right) = \sum_{k=1}^{n} \mu^*(E_k) = \sum_{k=1}^{n} \mu(E_k).$$

Thus,

$$\sum_{k=1}^{\infty} \mu(E_k) \geq \mu\left(\bigcup_{k=1}^{\infty} E_k\right) \geq \mu\left(\bigcup_{k=1}^{n} E_k\right) = \sum_{k=1}^{n} \mu(E_k),$$

for every $n \in \mathbb{N}$, and so $\mu\left(\bigcup_{k=1}^{\infty} E_k\right) = \sum_{k=1}^{\infty} \mu(E_k)$. □

One useful consequence is the following simple result.

Proposition 3.31. *Suppose that $E, F \in \mathfrak{M}(X)$. If $E \subseteq F$ and if $\mu^*(F) < \infty$, then $\mu^*(F \backslash E) = \mu^*(F) - \mu^*(E)$.*

Proof. Write F as $F = (F \backslash E) \cup E$, which is a disjoint union of elements of $\mathfrak{M}(X)$. Both $\mu^*(F)$ and $\mu^*(F \backslash E)$ are finite. Thus, $\mu^*(F) = \mu^*(F \backslash E) + \mu^*(E)$, by (3.9) [with $S = X$]. □

The next definition and proposition indicate that sets that have zero outer measure are measurable.

Definition 3.32. If μ^* is an outer measure on X, then a subset $S \subset \mathbb{R}$ is μ^*-*null* if $\mu^*(S) = 0$.

Proposition 3.33. *If μ^* is an outer measure on X and if $E \subseteq X$ is μ^*-null, then $E \in \mathfrak{M}_{\mu^*}(X)$.*

Proof. Let $E \subseteq X$ be a μ^*-null set. If $S \subseteq X$, then $E \cap S \subseteq E$ and so $0 \leq \mu^*(E \cap S) \leq \mu^*(E) = 0$. Hence, by the subadditivity of outer measure,

$$\mu^*(S) \leq \mu^*(E \cap S) + \mu^*(E^c \cap S) = 0 + \mu^*(E^c \cap S) \leq \mu^*(S).$$

That is, $\mu^*(S) = \mu^*(E \cap S) + \mu^*(E^c \cap S)$ for every $S \subseteq X$. □

What other subsets $E \subseteq X$ will belong to the σ-algebra $\mathfrak{M}_{\mu^*}(X)$? The answer to this question depends, of course, on the character of the outer measure μ^*. A useful answer in the setting of metric spaces is given by Proposition 3.35 below, for which following definition will be required.

Definition 3.34. If (X,d) is a metric space and if A and B are nonempty subsets of X, then the *distance between A and B* is the quantity denoted by $\mathrm{dist}\,(A,B)$ and defined by

$$\mathrm{dist}\,(A,B) = \inf\{d(a,b) \,|\, a \in A,\, b \in B\}.$$

If, in a metric space (X,d), the distance between subsets A and B is positive, then A and B are disjoint and $\mu^*(A \cup B) \leq \mu^*(A) + \mu^*(B)$. If equality is achieved in all such cases, then the induced σ-algebra $\mathfrak{M}_{\mu^*}(X)$ will contain the Borel sets of X.

Proposition 3.35. *If an outer measure μ^* on a metric space (X,d) has the properties that $\mu^*(X) < \infty$ and that*

$$\mu^*(A \cup B) = \mu^*(A) + \mu^*(B),$$

for all subsets $A, B \subseteq X$ for which $\mathrm{dist}\,(A,B) > 0$, then every Borel set of X belongs to the σ-algebra $\mathfrak{M}_{\mu^*}(X)$ induced by μ^*.

Proof. By the Carathéodory criterion of Theorem 3.29, our objective is to show that, for every open subset $U \subseteq X$, the equation

$$\mu^*(S) = \mu^*(S \cap U) + \mu^*(S \cap U^c)$$

holds for all $S \subseteq X$.

To this end, select a nonempty subset S of X. If $S \cap U = \emptyset$, then the equation $\mu^*(S) = \mu^*(S \cap U) + \mu^*(S \cap U^c)$ holds trivially. Thus, assume that $S \cap U \neq \emptyset$, and for each $n \in \mathbb{N}$ let

$$S_n = \left\{ x \in U \cap S \,\middle|\, \mathrm{dist}\,(\{x\}, U^c) \geq \frac{1}{n} \right\}.$$

Observe that $S_n \subseteq S_{n+1}$ for all $n \in \mathbb{N}$ and that $U \cap S = \bigcup_{n \in \mathbb{N}} S_n$. By the hypothesis on μ^*, the distance inequalities $\operatorname{dist}(S_n, S \cap U^c) \geq \operatorname{dist}(S_n, U^c) \geq \frac{1}{n} > 0$ imply that

$$\mu^*(S) \geq \mu^*((S \cap U^c) \cup S_n) = \mu^*(S \cap U^c) + \mu^*(S_n).$$

Because $\mu^*(X) < \infty$ and because the sets S_n form an ascending sequence, the limit $\lim_{n \to \infty} \mu^*(S_n)$ exists and is bounded above by $\mu^*(S \cap U)$. If it were known that $\lim_{n \to \infty} \mu^*(S_n) = \mu^*(S \cap U)$, then the inequality above would lead to

$$\mu^*(S) \geq \mu^*(S \cap U^c) + \mu^*(S \cap U),$$

which, when coupled with the inequality $\mu^*(S) \leq \mu^*(S \cap U^c) + \mu^*(S \cap U)$ arising from the subadditivity of μ^*, would imply $\mu^*(S) = \mu^*(S \cap U) + \mu^*(S \cap U^c)$. Therefore, all that remains is to prove that $\lim_{n \to \infty} \mu^*(S_n) = \mu^*(S \cap U)$.

For every $n \in \mathbb{N}$, let $A_n = S_{n+1} \setminus S_n$. If $m, n \in \mathbb{N}$ satisfy $|m - n| \geq 2$, then the distance between A_m and A_n is positive, and so $\mu^*(A_m \cup A_n) = \mu^*(A_m) + \mu^*(A_n)$. Therefore, by induction,

$$\sum_{k=1}^{n} \mu^*(A_{2k}) = \mu^* \left(\bigcup_{k=1}^{n} A_{2k} \right) \leq \mu^*(S_{2n+1}) \leq \mu^*(S \cap U) < \infty.$$

Hence, the series $\sum_{k=1}^{\infty} \mu^*(A_{2k})$ converges. Likewise, $\sum_{k=1}^{\infty} \mu^*(A_{2k+1})$ converges, and so the series $\sum_{k=1}^{\infty} \mu^*(A_k)$ converges. Therefore, by the countable subadditivity of μ^*,

$$\mu^*(S_n) \leq \mu^*(S \cap U) \leq \mu^*(S_n) + \sum_{k=n+1}^{\infty} \mu^*(A_k),$$

and so

$$|\mu^*(S \cap U) - \mu^*(S_n)| \leq \sum_{k=n+1}^{\infty} \mu^*(A_k).$$

The convergence of $\sum_{k=1}^{\infty} \mu^*(A_k)$ yields $\lim_{n \to \infty} |\mu^*(S \cap U) - \mu^*(S_n)| = 0$. $\square$

3.4 Lebesgue Measure

The original motivation for the development of measure theory was to put the notion of length, area, volume, and so forth on rigorous mathematical footing, with the understanding that the sets to be measured may not be intervals, rectangles, or boxes. The measures that captures length, area, and volume are called Lebesgue measures.

Proposition 3.36. *The collection*

$$\mathscr{O}_n = \left\{ \prod_{i=1}^n (a_i, b_i) \,|\, a_i, b_i \in \mathbb{R}, \, a_i < b_i \right\}$$

is a sequential cover of $\mathbb{R}^n$.

Proof. Let $S \subseteq \mathbb{R}^n$. For each $x \in S$ there is a neighbourhood U_x of x of the form $U_x = \prod_{i=1}^n (a_i, b_i)$. Let $V = \bigcup_{x \in S} U_x$, which is an open set. By Proposition 1.26, the set $\mathscr{B}$ of all finite open intervals with rational end points is a basis for the topology of $\mathbb{R}$. Thus,

$$\mathscr{B}_n = \left\{ \prod_{i=1}^n (p_i, q_i) \,|\, p_i, q_i \in \mathbb{Q}, \, p_i < q_i \right\}$$

is a basis for the topology of $\mathbb{R}^n$. By Proposition 1.24, every open set is a union of basic open sets. Thus, since $\mathscr{B}_n$ is countable, there is a countable family $\{I_k\}_{k \in \mathbb{N}} \subseteq \mathscr{B}_n \subseteq \mathscr{O}_n$ such that $V = \bigcup_{k \in \mathbb{N}} I_k$, whence $S \subseteq \bigcup_{k \in \mathbb{N}} I_k$. $\qquad\square$

Definition 3.37. *Lebesgue outer measure on* $\mathbb{R}^n$ *is the function* m^* *on* $\mathscr{P}(\mathbb{R}^n)$ defined by

$$m^*(S) = \inf \left\{ \sum_{k=1}^{\infty} \lambda(I_k) \,|\, \{I_k\}_{k \in \mathbb{N}} \subseteq \mathscr{O}_n \text{ and } S \subseteq \bigcup_{k \in \mathbb{N}} I_k \right\},$$

where $\mathscr{O}_n$ is the sequential cover of $\mathbb{R}^n$ given by Proposition 3.36 and the function λ is defined by

$$\lambda \left(\prod_{i=1}^n (a_i, b_i) \right) = \prod_{i=1}^n (b_i - a_i).$$

Observe that if $E \subseteq \mathbb{R}^n$ is an open box in $\mathbb{R}^n$ (that is, $E \in \mathscr{O}_n$), then $\lambda(E)$ is the volume of E and $m^*(E) = \lambda(E)$.

The first proposition shows that, in the case $n = 1$, m^* is a length function for all finite intervals, open or otherwise.

Proposition 3.38. *If $a, b \in \mathbb{R}$ are such that $a < b$, then*

$$m^* ([a,b]) = m^* ((a,b]) = m^* ([a,b)) = m^* ((a,b)) = b - a.$$

Proof. Because m^* is an outer measure, $m^*(S_1) \leq m^*(S_2)$ if $S_1 \subseteq S_2$. Therefore, $m^* ((a,b)) \leq m^* ((a,b]) \leq m^* ([a,b])$ and $m^* ((a,b)) \leq m^* ([a,b)) \leq m^* ([a,b])$. Since by definition, $m^* ((a,b)) = b - a$, it is enough to prove that $m^*([a,b]) = b - a$. To this end, observe that, for every $\varepsilon > 0$, $[a,b] \subset (a - \varepsilon, b + \varepsilon)$. Because this open interval covers $[a,b]$, we have that $m^*([a,b]) \leq \lambda ((a - \varepsilon, b + \varepsilon)) = b - a + 2\varepsilon$. As this is true for every $\varepsilon > 0$, one concludes that $m^* ([a,b]) \leq b - a = m^* ((a,b))$. $\quad\square$

Similarly, one has:

Proposition 3.39. *If $E \in \mathcal{O}_n$, then $m^*(\overline{E}) = m^*(E)$.*

Proof. Exercise 3.88.

The notion of μ^*-null set, for an outer measure μ^* on a set X, was introduced earlier. To simplify the terminology here, we shall say a subset $S \subseteq \mathbb{R}$ is a *null set* if its Lebesgue outer measure $m^*(S)$ is 0. Thus, from Proposition 3.33, every null set $S \subseteq \mathbb{R}^n$ is necessarily Lebesgue measurable.

Example 3.40 (Some Null Sets). *The following subsets of $\mathbb{R}^n$ are null sets:*

1. *every finite or countably infinite set;*
2. *every countable union of null sets;*
3. *every subset of a null set;*
4. *the Cantor ternary set in $\mathbb{R}$.*

Proof. The details of these examples are left as an exercise (Exercise 3.89), but the case of the Cantor set is described here.

The Cantor ternary set $\mathscr{C}$ is given by $\mathscr{C} = \bigcap_{n \in \mathbb{N}} \mathscr{C}_n$, where each $\mathscr{C}_n$ is a union of 2^n pairwise disjoint closed intervals $F_{n,j}$ of length $(1/3)^n$. Thus,

$$m^*(\mathscr{C}) \leq m^*(\mathscr{C}_n) = m^* \left(\bigcup_{j=1}^{2^n} F_{n,j} \right) \leq \sum_{j=1}^{2^n} m^*(F_{n,j}) = \left(\frac{2}{3} \right)^n.$$

As the inequality above holds for all $n \in \mathbb{N}$, $m^*(\mathscr{C}) = 0$. $\quad\square$

Proposition 3.41. *If c is the cardinality of the continuum, then the cardinality of $\mathfrak{M}(\mathbb{R})$ is 2^c (the cardinality of the power set of $\mathbb{R}$).*

Proof. The Cantor ternary set $\mathscr{C}$ has the cardinality if the continuum (Proposition 1.83) and every subset of $\mathscr{C}$ is Lebesgue measurable. Hence, the cardinality of $\mathfrak{M}(\mathbb{R})$ is the cardinality of the power set of $\mathbb{R}$. $\quad\square$

In addition to null sets, every open set is a Lebesgue measurable set.

Proposition 3.42. *Every open set in $\mathbb{R}^n$ is Lebesgue measurable.*

Proof. If $W = (a,b)$ and $U = (p,q)$ are open intervals, and if $a < p < b < q$, then $W \cap U$ is the interval $I = (p,b)$ with length $\lambda(I) = (b-p)$ and $W \cap U^c$ is an interval $J = (a,p)$ with length $\lambda(J) = (p-a)$. Thus,

$$b-a = m^*(W) = m^*(I) + m^*(J) = m^*(W \cap U) + m^*(W \cap U^c).$$

The equation above holds in cases where the inequalities $a < p < b < q$ are not satisfied, because either one of $W \cap U$ or $W \cap U^c$ is empty, or $W \cap U$ and $W \cap U^c$ are nonempty disjoint open intervals whose lengths sum to $b - a$.

A similar feature holds in $\mathbb{R}^n$. If $W = \prod_{j=1}^{n}(a_j, b_j)$ and $U = \prod_{j=1}^{n}(p_j, q_j)$ are elements of $\mathcal{O}_n$, then either one of $W \cap U$ or $W \cap U^c$ is empty, or $W \cap U$ and $W \cap U^c$ are nonempty disjoint elements of $\mathcal{O}_n$ whose volumes sum to $\prod_{j=1}^{n}(b_j - a_j)$. Hence,

$$m^*(W) = m^*(W \cap U) + m^*(W \cap U^c)$$

for all $W, U \in \mathcal{O}_n$.

To prove that every open set in $\mathbb{R}^n$ is Lebesgue measurable, assume that $V \subseteq \mathbb{R}^n$ is an open set. Because $\mathbb{R}^n$ has a countable basis for its topology, every open set is a countable union of open sets. Therefore, we may assume without loss of generality that V is a basic open set: $V = \prod_{j=1}^{n}(a_j, b_j)$, for some $a_j, b_j \in \mathbb{Q}$. Let $S \subseteq \mathbb{R}^n$ be arbitrary and assume that $\varepsilon > 0$. Select a covering $\{U_k\}_k \subset \mathcal{O}_n$ of S such that $\sum_k \lambda(U_k) \le m^*(S) + \varepsilon$. Because

$$S \cap V \subseteq \bigcup_k (U_k \cap V) \text{ and } S \cap V^c \subseteq \bigcup_k (U_k \cap V^c),$$

we have that

$$m^*(S \cap V) + m^*(S \cap V^c) \le \sum_k \left(m^*(U_k \cap V) + m^*(U_k \cap V^c) \right)$$

$$= \sum_k m^*(U_k)$$

$$\le m^*(S) + \varepsilon.$$

As $\varepsilon > 0$ is arbitrary, we deduce that $m^*(S) = m^*(S \cap V) + m^*(S \cap V^c)$, which proves that the open set V is Lebesgue measurable. $\square$

Corollary 3.43. *Every Borel subset of $\mathbb{R}^n$ is Lebesgue measurable.*

If E and F are Lebesgue-measurable sets, then it is necessarily true that

$$m(E \cup F) + m(E \cap F) = m(E) + m(F).$$

Proposition 3.44 below extends this property to outer measure of arbitrary sets, but at the expense of weakening the equality above to an inequality.

Proposition 3.44. *For any subsets* $A, B \subseteq \mathbb{R}^n$,

$$m^*(A \cup B) + m^*(A \cap B) \leq m^*(A) + m^*(B).$$

Proof. Let $\varepsilon > 0$ be given, and let $\{I_k\}_k$ and $\{J_i\}_i$ be coverings of A and B, respectively, by open boxes I_k and J_i such that

$$\sum_k \ell(I_k) \leq m^*(A) + \varepsilon \quad \text{and} \quad \sum_i \ell(J_i) \leq m^*(B) + \varepsilon.$$

Let $U = \bigcup_k I_k$ and $V = \bigcup_i J_i$. Thus, $A \subseteq U$ and $B \subseteq V$, and $A \cup B \subseteq U \cup V$ and $A \cap B \subseteq U \cap V$. Because U and V are open sets, they are Lebesgue measurable and, hence,

$$m^*(A \cup B) + m^*(A \cap B) \leq m^*(U \cup V) + m^*(U \cap V) = m(U \cup V) + m(U \cap V)$$

$$= m(U) + m(V) \leq \sum_k m(I_k) + \sum_i m(J_i) \leq m^*(A) + m^*(B) + 2\varepsilon.$$

Because $\varepsilon > 0$ is arbitrary, we have $m^*(A \cup B) + m^*(A \cap B) \leq m^*(A) + m^*(B)$. □

The notion of σ-finite measure space was introduced in Definition 3.25 as a hybrid of finite measure space and infinite measure space. Lebesgue measure on $\mathbb{R}^n$ is a concrete example of a σ-finite space.

Proposition 3.45. *The measure space* $(\mathbb{R}^n, \mathfrak{M}(\mathbb{R}^n), m)$ *is* σ-*finite.*

Proof. If $K_j = \prod_1^n [-j, j]$ for each $j \in \mathbb{N}$, then K_j is measurable of finite measure $m(K_j) = (2j)^n$, and $\mathbb{R}^n = \bigcup_{j \in \mathbb{N}} K_j$. □

Every Borel subset of $\mathbb{R}^n$ is Lebesgue measurable, and Borel sets are determined by open subsets. Therefore, it seems natural to expect that the measures of arbitrary Lebesgue-measurable sets can be approximated by the measures of open and/or closed sets—this is the notion of regularity. The idea of translation invariance of measure is related to the fact, for example, that if one moved an n-cube C in $\mathbb{R}^n$ to some other position in space, the volume of C would not change.

A tool in analysing the regularity and translation invariance of Lebesgue measure is the following proposition.

Proposition 3.46. *The following statements are equivalent for a subset* $E \subseteq \mathbb{R}^n$.

1. E is a Lebesgue-measurable set.
2. For every $\varepsilon > 0$ there is an open set $U \subseteq \mathbb{R}^n$ such that $E \subseteq U$ and $m^*(U \backslash E) < \varepsilon$.
3. For every $\varepsilon > 0$ there is a closed set $F \subseteq \mathbb{R}^n$ such that $F \subseteq E$ and $m^*(E \backslash F) < \varepsilon$.

Proof. The logic of proof is slightly unusual in that following implications will be established: (1)$\Rightarrow$(2) and (1)$\Rightarrow$(3), then followed by (3)$\Rightarrow$(1) and (2)$\Rightarrow$(1).

To prove that (1) implies (2), suppose that $E \subseteq \mathbb{R}^n$ is Lebesgue measurable and let $\varepsilon > 0$ be given. The cases where $m^*(E)$ is finite or infinite will be treated separately.

In the first case, assume that $m^*(E) < \infty$. By definition, there is countable covering $\{I_k\}_{k \in \mathbb{N}} \subset \mathcal{O}_n$ of E such that

$$\sum_{k=1}^{\infty} \lambda(I_k) < m^*(E) + \varepsilon.$$

Let $U = \bigcup_k I_k$, which is an open (and, hence, Lebesgue measurable) set containing E. Note that

$$m^*(E) \leq m^*(U) \leq \sum_{k=1}^{\infty} \lambda(I_k) < m^*(E) + \varepsilon.$$

Because $m^*(U) < \infty$ and $E \subseteq U$ is a containment of Lebesgue-measurable sets, Proposition 3.31 states that

$$m^*(U \backslash E) = m^*(U) - m^*(E) \leq \sum_{k=1}^{\infty} \lambda(I_k) - m^*(E) < \varepsilon,$$

which proves (2) in the case where $m^*(E) < \infty$.

Assume now that $m^*(E) = \infty$. Define $E_k = E \cap ([-k, k]^n)$, for each $k \in \mathbb{N}$. Hence,

$$m^*(E_k) \leq (2k)^n \text{ and } E = \bigcup_{k \in \mathbb{N}} E_k.$$

Because $m^*(E_k) < \infty$, the first case implies there are open sets $U_k \subseteq \mathbb{R}^n$ such that $E_k \subseteq U_k$ and $m^*(U_k \backslash E_k) < \frac{\varepsilon}{2^{k+1}}$. Let $U = \bigcup_k U_k$, which is open and contains E. Thus,

$$U \backslash E \subseteq \bigcup_{k \in \mathbb{N}} U_k \backslash E_k$$

and

$$m^*(U \backslash E) \leq m^* \left(\bigcup_{k \in \mathbb{N}} U_k \backslash E_k \right) \leq \sum_{k=1}^{\infty} m^*(U_k \backslash E_k) \leq \frac{1}{2} \sum_{k=1}^{\infty} \frac{\varepsilon}{2^k} < \varepsilon,$$

which proves (2) in the case where $m^*(E) = \infty$.

For the proof of (1) implies (3), suppose that $E \subseteq \mathbb{R}^n$ is Lebesgue measurable and let $\varepsilon > 0$ be given. As E is Lebesgue measurable, so is E^c. Apply (1)$\Rightarrow$(2) to

E^c to conclude that there is an open set U such that $E^c \subseteq U$ and $m^*(U \backslash E^c) < \varepsilon$. Let $F = U^c$, which is a closed set contained in E. Thus,

$$m^*(E \backslash F) = m^*(E \cap F^c) = m^*(E \cap U) = m^*(U \backslash E^c) < \varepsilon,$$

thereby proving that (1) implies (3).

To prove that (3) implies (1), assume hypothesis (3) and let $\varepsilon > 0$ be given. By hypothesis, there is a closed set F such that $F \subseteq E$ and $m^*(E \backslash F) < \epsilon$. Let $S \subseteq \mathbb{R}^n$ be any set. Note that $(S \cap E) \cap F = S \cap F$ and $(S \cap E) \cap F^c \subseteq E \cap F^c$; hence,

$$
\begin{aligned}
m^*(S \cap E) &= m^*((S \cap E) \cap F) + m^*((S \cap E) \cap F^c) \\[6pt]
&\leq m^*(S \cap F) + m^*(E \cap F^c) \\[6pt]
&= m^*(S \cap F) + m^*(E \backslash F) \\[6pt]
&\leq m^*(S \cap F) + \varepsilon .
\end{aligned}
\tag{3.11}
$$

The inclusion $F \subseteq E$ implies that

$$m^*(S \cap E^c) \leq m^*(S \cap F^c). \tag{3.12}$$

Therefore, (3.11) and (3.12) combine to produce

$$
\begin{aligned}
m^*(S) &\leq m^*(S \cap E) + m^*(S \cap E^c) \\[6pt]
&\leq \varepsilon + m^*(S \cap F) + m^*(S \cap F^c) \\[6pt]
&= \varepsilon + m^*(S) .
\end{aligned}
\tag{3.13}
$$

(The final equality arises from the fact that F—being closed—is Lebesgue measurable.) As ε is arbitrary, the inequalities (3.13) imply that $m^*(S) = m^*(S \cap E) + m^*(S \cap E^c)$. That is, E is Lebesgue measurable.

Lastly, the proof of (2) implies (1) is similar to the proof of the (3)$\Rightarrow$(1) and is, therefore, omitted. $\square$

Proposition 3.47 (Regularity of Lebesgue Measure). *Lebesgue measure m on $\mathbb{R}^n$ has the following properties:*

1. *$m(K) < \infty$ for every compact subset $K \subset \mathbb{R}^n$;*
2. *$\mu(E) = \inf\{\mu(U) \mid U \subseteq \mathbb{R}^n \text{ is open and } E \subseteq U\}$ for every measurable set E;*
3. *$\mu(E) = \sup\{\mu(K) \mid K \text{ is compact and } K \subseteq E\}$, for every measurable set E.*

Proof. Assume that $K \subset \mathbb{R}^n$ is compact. For each $x \in K$ there is an open box $W_x \in \mathcal{O}_n$ of volume 1 such that $x \in W_x$. From the open cover $\{W_x\}_{x \in K}$ of the compact set K

extract a finite subcover $\{W_{x_j}\}_{j=1}^{\ell}$ and deduce that

$$m(K) \le \sum_{j=1}^{\ell} m(W_{x_j}) = \ell < \infty.$$

Next, assume that $E \subseteq \mathbb{R}^n$ is a Lebesgue-measurable set and that $\varepsilon > 0$. By Proposition 3.46, there is an open set $U \subseteq \mathbb{R}$ such that $E \subseteq U$ and $m(U \setminus E) < \varepsilon$. Thus,

$$m(U) = m(E) + m(U \setminus E) \le m(E) + \varepsilon.$$

Hence, $\mu(E) = \inf\{\mu(U) \mid U \subseteq \mathbb{R}^n \text{ is open and } E \subseteq U\}$.

Now assume that E is a Lebesgue-measurable set such that the closure $\overline{E}$ of E is compact. Let $\varepsilon > 0$ be given. By the previous paragraph there is an open set U containing $\overline{E} \setminus E$ such that $m(U) < m(\overline{E} \setminus E) + \varepsilon$. Let $K = \overline{E} \cap U^c$, which is a closed subset of the compact set $\overline{E}$; hence, K is compact. Furthermore, if $x \in K$, then $x \in \overline{E}$ and $x \notin \overline{E} \cap E^c$, which is to say that $x \in E$. Thus, $K \subseteq E$. Because

$$m(\overline{E}) - m(K) = m(\overline{E}) - \big(m(\overline{E}) - m(U)\big) = m(U) < m(\overline{E}) - m(E) + \varepsilon,$$

we deduce that $m(E) < m(K) + \varepsilon$ and $\mu(E) = \sup\{\mu(K) \mid K \text{ is compact and } K \subseteq E\}$.

For each $k \in \mathbb{N}$, the set $B_k = \prod_{1}^{n} [-k, k]$ is compact. If $E_k = E \cap B_k$, then $\{E_k\}_{k \in \mathbb{N}}$ is an ascending sequence of sets such that $E = \bigcup_{k \in \mathbb{N}} E_k$. Thus, by continuity of measure, $m(E) = \lim_{k \to \infty} m(E_k)$. Choose any positive $r \in \mathbb{R}$ such that $r < m(E)$. Thus, there is a $k \in \mathbb{N}$ such that $r < m(E_k) < m(E)$. Because $\overline{E}_k$ is compact, the previous paragraph shows that there is a compact subset K of E_k such that $r < m(K)$. Now since $E_k \subseteq E$, K is also a subset of E. As the choice of $r < m(E)$ is arbitrary, this shows that $\mu(E) = \sup\{\mu(K) \mid K \text{ is compact and } K \subseteq E\}$. $\square$

If $x \in \mathbb{R}^n$ and $S \subseteq \mathbb{R}^n$, then $x + S$ denotes the subset of $\mathbb{R}^n$ defined by

$$x + S = \{x + y \mid y \in S\}.$$

Proposition 3.48 (Translation Invariance of Lebesgue Measure). *If $E \subset \mathbb{R}^n$ is Lebesgue measurable and if $x \in \mathbb{R}$, then $x + E$ is Lebesgue measurable and*

$$m(x + E) = m(E). \tag{3.14}$$

Proof. If $I \in \mathcal{O}_n$ is the open box $I = \prod_{j=1}^{n} (a_j, b_j)$, then $x + I$ and I have the same volume. Thus, for any subset $S \in \mathbb{R}^n$ and $x \in \mathbb{R}^n$, the outer measures of S and $x + S$

coincide. Therefore, we aim to prove that $x + E$ is a Lebesgue-measurable set if E is a Lebesgue-measurable set. To this end, we shall employ Proposition 3.46.

Let $\varepsilon > 0$. Because E is measurable, Proposition 3.46 states that there is an open set $U \subseteq \mathbb{R}^n$ such that $E \subseteq U$ and $m^*(U \backslash E) < \varepsilon$. Thus, there is a countable covering of $U \backslash E$ by open boxes I_k such that

$$\sum_{k \in \mathbb{N}} \lambda(I_k) < \varepsilon.$$

For each k, $x + I_k$ is an open box of volume $\lambda(x + I_k) = \lambda(I_k)$. Furthermore, because U is a countable union of basic opens (all of which are open boxes), the set $x + U$ is open, the inclusion $x + E \subseteq r + U$ is clear, and

$$(x + U) \backslash (x + E) = \{x + y \mid y \in U \backslash E\} = x + (U \backslash E) \subseteq \bigcup_{k \in \mathbb{N}} (x + I_k).$$

Thus,

$$m^*((x + U) \backslash (x + E)) \leq \sum_{k \in \mathbb{N}} \lambda(x + I_k) = \sum_{k \in \mathbb{N}} \lambda(I_k) < \varepsilon.$$

Hence, $x + E$ satisfies the hypothesis of Proposition 3.46, thereby completing the proof that $x + E$ is Lebesgue measurable. $\qquad \square$

It is natural to wonder whether every subset of $\mathbb{R}$ is Lebesgue measurable. That is not the case, as the following theorem shows. Because the proof of the theorem below requires the Axiom of Choice, the result is existential rather than constructive.

Theorem 3.49 (Vitali). *There is a subset $\mathcal{V}$ of $\mathbb{R}$ such that $\mathcal{V}$ is not Lebesgue measurable.*

Proof. Consider the relation $\sim$ on $\mathbb{R}$ defined by $x \sim y$ if and only if $y - x \in \mathbb{Q}$. It is not difficult to verify that $\sim$ is an equivalence relation, and so the equivalence classes $\dot{x}$ of $x \in \mathbb{R}$ form a partition of $\mathbb{R}$. Note that $\dot{x} = x + \mathbb{Q}$, for each $x \in \mathbb{R}$.

For each $x \in (-1, 1)$, let $A_x = \dot{x} \cap (-1, 1)$. Of course, if $x_1, x_2 \in (-1, 1)$, then either $A_{x_1} = A_{x_2}$ or $A_{x_1} \cap A_{x_2} = \emptyset$. By the Axiom of Choice, there is a set $\mathcal{V}$ such that, for every $x \in (-1, 1)$, $\mathcal{V} \cap A_x$ is a singleton set.

The set $\mathbb{Q} \cap (-2, 2)$ is countable; hence, we may write

$$\mathbb{Q} \cap (-2, 2) = \{q_k \mid k \in \mathbb{N}\}.$$

For each $k \in \mathbb{N}$, consider $q_k + \mathcal{V}$. Suppose that $x \in (q_k + \mathcal{V}) \cap (q_m + \mathcal{V})$, for some $k, m \in \mathbb{N}$. Then there are $c_k, c_m \in \mathcal{V}$ such that $q_k + c_k = q_m + c_m$; that is, $c_k - c_m = q_m - q_k \in \mathbb{Q}$, which implies that $c_k \in A_{c_m}$. As $\mathcal{V} \cap A_{c_m}$ is a singleton set, it must be that $c_k = c_m$ and $q_k = q_m$. Hence, $\{q_k + \mathcal{V}\}_{k \in \mathbb{N}}$ is a countable family of pairwise disjoint sets, each of which is obviously contained in the open interval $(-3, 3)$.

Let $x \in (-1, 1)$ and consider A_x. By construction of $\mathcal{V}$, there is precisely one element $y \in (-1, 1)$ that is common to both A_x and $\mathcal{V}$. Thus, x and y are equivalent, which is to say that $x - y \in \mathbb{Q}$. Because $x, y \in (-1, 1)$, $x - y \in (-2, 2)$; hence, $x - y = q_k$, for some $k \in \mathbb{N}$. Therefore, $x \in q_k + \mathcal{V}$.

The arguments above establish that

$$(-1,1) \subset \bigcup_{k \in \mathbb{N}} (q_k + \mathcal{V}) \subset (-3,3). \tag{3.15}$$

If $\mathcal{V}$ were Lebesgue measurable, then each $q_k + \mathcal{V}$ would be Lebesgue measurable, by Proposition 3.48, and $m(q_k + \mathcal{V})$ would equal $m(\mathcal{V})$. Therefore, if $\mathcal{V}$ were Lebesgue measurable, then

$$m\left(\bigcup_{k \in \mathbb{N}} (q_k + \mathcal{V})\right) = \sum_{k \in \mathbb{N}} m(q_k + \mathcal{V}) = \sum_{k \in \mathbb{N}} m(\mathcal{V})$$

would hold. Furthermore, computation of Lebesgue measure in (3.15) would yield

$$2 < \sum_{k=1}^{\infty} m(\mathcal{V}) < 6. \tag{3.16}$$

But there is no real number $m(\mathcal{V})$ for which (3.16) can hold. Therefore, it cannot be that $\mathcal{V}$ is a Lebesgue-measurable set. $\qquad \square$

Corollary 3.50. *Outer measure m^* on $\mathbb{R}$ is not countably additive. That is, there is a sequence $\{S_k\}_{k \in \mathbb{N}}$ of pairwise disjoint subsets $S_k \subseteq \mathbb{R}$ such that*

$$m^*\left(\bigcup_{k \in \mathbb{N}} S_k\right) < \sum_{k \in \mathbb{N}} m^*(S_k).$$

Proof. Let $S_k = q_k + \mathcal{V}$, as in the proof of Theorem 3.49. Because m^* is countably subadditive and because $(-1,1) \subseteq \bigcup_{k \in \mathbb{N}} (q_k + \mathcal{V})$,

$$2 \leq m^*\left(\bigcup_{k \in \mathbb{N}} S_k\right).$$

Therefore, because $m^*(q_k + \mathcal{V}) = m^*(\mathcal{V})$, we have that. $m^*(S_k) = m^*(S_1)$, for all $k \in \mathbb{N}$, and the inequality above shows that $m^*(S_1) \neq 0$. Thus,

$$\sum_{k \in \mathbb{N}} m^*(S_k) = \infty.$$

On the other hand,

$$\bigcup_{k \in \mathbb{N}} S_k \subset (-3,3) \implies m^*\left(\bigcup_{k \in \mathbb{N}} S_k\right) \leq 6.$$

Hence,

$$m^* \left(\bigcup_{k \in \mathbb{N}} S_k \right) < \sum_{k \in \mathbb{N}} m^*(S_k),$$

as claimed.   □

Vitali's Theorem produces a nonmeasurable subset of $(-1, 1)$; the argument can be modified to produce a nonmeasurable subset of any measurable set of positive measure.

Theorem 3.51. *If $E \in \mathfrak{M}(\mathbb{R})$ and if $m(E) > 0$, then there is a subset $\mathcal{V} \subset E$ such that $\mathcal{V} \notin \mathfrak{M}(\mathbb{R})$.*

Proof. Let $A_k = E \cap [-k, k]$ for each $k \in \mathbb{N}$. The sequence $\{A_k\}_{k \in \mathbb{N}}$ is an ascending sequence in $\mathfrak{M}(\mathbb{R})$ with union E. Hence, by continuity of measure,

$$0 < m(E) = \lim_{k \to \infty} m(A_k),$$

and so $m(A_{k_0}) > 0$ for some $k_0 \in \mathbb{N}$. Now apply the argument of Theorem 3.49 using $E \cap [-k_0, k_0]$ in place of $(-1, 1)$ to determine a nonmeasurable subset $\mathcal{V}$ of $E \cap [-k_0, k_0]$.   □

The Borel sets and null sets determine the structure of Lebesgue-measurable sets.

Proposition 3.52. *The following statements are equivalent for a subset $E \subseteq \mathbb{R}$:*

1. *E is a Lebesgue-measurable set;*
2. *there exist $B, E_0 \subseteq \mathbb{R}$ such that:*

   a. *B is a Borel set,*
   b. *E_0 is a null set,*
   c. *$E_0 \cap B = \emptyset$, and*
   d. *$E = B \cup E_0$.*

Proof. Exercise 3.91.   □

Proposition 3.52 shows how Borel sets can be used to characterise Lebesgue-measurable sets. Much less obvious is the following theorem, which indicates that these two σ-algebras are in fact distinct.

Theorem 3.53 (Suslin). *There exist Lebesgue-measurable subsets of $\mathbb{R}$ that are not Borel sets. In fact, there are Lebesgue-measurable subsets of the Cantor ternary set that are not Borel sets.*

Proof. Let $\tilde{\Phi}$ denote an extension of the Cantor ternary function (see Proposition 1.86) $\Phi : [0, 1] \to [0, 1]$ to a function $\mathbb{R} \to [0, 1]$ by setting $\tilde{\Phi} = 0$ on $(-\infty, 0)$, $\tilde{\Phi} = \Phi$ on $[0, 1]$, and $\tilde{\Phi} = 1$ on $(1, \infty)$. Let $f : \mathbb{R} \to \mathbb{R}$ be given by

$$f(x) = \tilde{\Phi}(x) + x, \quad \forall x \in \mathbb{R}.$$

Observe that f is continuous and monotone increasing.

Define a collection Σ of subsets of $\mathbb{R}$ as follows:

$$\Sigma = \{S \subseteq \mathbb{R} \mid f(S) \in \mathfrak{B}(\mathbb{R})\}.$$

We now show that Σ is a σ-algebra of subsets of $\mathbb{R}$. Because $\mathbb{R} = f(\mathbb{R})$, $\mathbb{R} \in \Sigma$. Moreover, if $A \in \Sigma$, then $f(A^c) \cap f(A) = \emptyset$ and $\mathbb{R} = f(\mathbb{R}) = f(A) \cup f(A^c)$ imply that $f(A^c) = f(A)^c$, whence $A^c \in \Sigma$. That is, Σ is closed under complementation. Now suppose that $\{A_k\}_{k \in \mathbb{N}} \subset \Sigma$; then

$$f\left(\bigcup_{k \in \mathbb{N}} A_k\right) = \bigcup_{k \in \mathbb{N}} f(A_k) \in \mathfrak{B}(\mathbb{R}).$$

Hence, Σ is a σ-algebra.

If $p, q \in \mathbb{Q}$ and $p < q$, then the continuity of f and the fact that f is monotone increasing leads to $f((p,q)) = (f(p), f(q))$. Therefore, $(p,q) \in \Sigma$ for all $p, q \in \mathbb{Q}$. Because Σ is a σ-algebra and Σ contains the base for the topology on $\mathbb{R}$, Σ necessarily contains the Borel sets of $\mathbb{R}$. Hence,

$$f(B) \in \mathfrak{B}(\mathbb{R}), \quad \forall B \in \mathfrak{B}(\mathbb{R}). \tag{3.17}$$

In particular, if $\mathscr{C}$ is the Cantor ternary set, then $f(\mathscr{C})$ is a Borel set. We now show that $f(\mathscr{C})$ has positive measure.

To this end note that $[0,1] \setminus \mathscr{C}$ is a union of countably many pairwise disjoint intervals (a_k, b_k), where $a_k, b_k \in \mathscr{C}$ for all $k \in \mathbb{N}$. Proposition 1.86 shows that Φ is constant on each such open interval. Therefore,

$$2 = m([0,2]) = m(f([0,1)))$$

$$= m(f(\mathscr{C} \cup ([0,1] \setminus \mathscr{C})))$$

$$= m(f(\mathscr{C})) + m(f([0,1] \setminus \mathscr{C}))$$

$$= m(f(\mathscr{C})) + \sum_{k \in \mathbb{N}} m((a_k + \Phi(a_k), b_k + \Phi(b_k)))$$

$$= m(f(\mathscr{C})) + \sum_{k \in \mathbb{N}} (b_k - a_k)$$

$$= m(f(\mathscr{C})) + m([0,1] \setminus \mathscr{C})$$

$$= m(f(\mathscr{C})) + 1.$$

Thus, $m(f(\mathscr{C})) = 1 > 0$ and so, by Theorem 3.51, $f(\mathscr{C})$ contains a subset $\mathscr{V}$ that is not Lebesgue measurable. Let $Q = f^{-1}(\mathscr{V})$. Because f is an injective function, the preimage Q of $\mathscr{V}$ under f must be contained in $\mathscr{C}$. Thus, Q is a null set and, hence, is a Lebesgue-measurable set. However, Q is not a Borel set. (If Q were a Borel set, then inclusion 3.17 would imply that $f(Q) = \mathcal{V}$ would be a Borel set—but it is not.) Hence, $Q \in \mathfrak{M}(\mathbb{R})$ and $Q \notin \mathfrak{B}(\mathbb{R})$.   □

Similar results hold in higher dimensions.

Theorem 3.54. *Not every subset of $\mathbb{R}^n$ is Lebesgue measurable, and there exist Lebesgue-measurable subsets of $\mathbb{R}^n$ that are not Borel measurable.*

Proof. Exercise 3.92.   □

3.5   Atomic and Non-Atomic Measures

There are a variety of ways to distinguish between qualitative properties of measures, and in this section we consider atomic measures and their polar opposites, non-atomic measures.

Definition 3.55. Assume that (X, Σ, μ) is a measure space.

1. A measurable subset $E \subseteq X$ is an *atom* for μ if $\mu(E) > 0$ and one of $\mu(E \cap F)$ or $\mu(E \cap F^c)$ is 0, for every $F \in \Sigma$.
2. The measure μ on (X, Σ) is *atomic* if every measurable set of positive measure contains an atom for μ.
3. The measure μ is *non-atomic* if μ has no atoms.

Thus, counting measure on $\mathbb{N}$ is atomic, whereas Lebesgue measure on $\mathbb{R}$ is non-atomic (Exercises 3.93 and 3.94). Every measure can be decomposed uniquely as a sum of two such measures, as shown by Proposition 3.57 below. The proof will make use of the following concept of singularity.

Definition 3.56. If μ and $\tilde{\mu}$ are measures on (X, Σ), then μ is *singular with respect to* $\tilde{\mu}$ for each $E \in \Sigma$ there exists a set $F \in \Sigma$ with the properties that

1. $F \subseteq E$,
2. $\mu(E) = \mu(F)$, and
3. $\tilde{\mu}(F) = 0$.

The notation $\mu \mathscr{S} \tilde{\mu}$ indicates that μ is singular with respect to $\tilde{\mu}$, and if both $\mu \mathscr{S} \tilde{\mu}$ and $\tilde{\mu} \mathscr{S} \mu$ occur, then μ and $\tilde{\mu}$ are said to be *mutually singular*.

Proposition 3.57. *Every measure μ on a measurable space (X, Σ) has the form $\mu = \mu_a + \mu_{na}$, for some mutually singular atomic measure μ_a and non-atomic measure μ_{na} on (X, Σ). Moreover, if $\tilde{\mu}_a$ and $\tilde{\mu}_{na}$ are atomic and non-atomic measures on (X, Σ) such that $\mu = \tilde{\mu}_a + \tilde{\mu}_{na}$, then $\tilde{\mu}_a = \mu_a$ and $\tilde{\mu}_{na} = \mu_{na}$.*

Proof. Let $\mathscr{D}$ be the family of all countable unions of sets that are atoms for μ. For each $E \in \Sigma$, define

$$\mu_a(E) = \sup\{\mu(E \cap D) \mid D \in \mathscr{D}\}$$

$$\mu_{\text{na}}(E) = \sup\{\mu(E \cap N) \mid \mu_a(N) = 0\}.$$

Observe that μ_a and μ_{na} are measures on (X, Σ) and satisfy $\mu = \mu_a + \mu_{\text{na}}$. If $\mu_a(N) = 0$, then $\mu(N \cap D) = 0$ for all $D \in \mathscr{D}$; hence, $\mu_{\text{na}}(D) = 0$ for all $D \in \mathscr{D}$, and so $\mu_a \mathscr{S} \mu_{\text{na}}$. By definition of μ_{na}, $\mu_{\text{na}}(D) = \mu_{\text{na}}(D \cap N)$ and $\mu_a(N) = 0$ imply that $\mu_a(D \cap N) = 0$; hence, $\mu_{\text{na}} \mathscr{S} \mu_a$.

To show that μ_a is atomic, suppose $E \in \Sigma$ such that $\mu_a(E) \neq 0$. Because $\mu_a(E) = 0$ when $\mu(E) = 0$, we deduce that $\mu(E) \neq 0$. Furthermore, by the definition of μ_a, $\mu_a(E) \neq 0$ implies there is some $D \in \mathscr{D}$ such that $\mu(E \cap D) \neq 0$. By the definition of $\mathscr{D}$, we can write $D = \bigcup_{n \in \mathbb{N}} D_n$, where each D_n is an atom for μ, and so $\mu(E \cap D_n) \neq 0$ for some $n \in \mathbb{N}$. Since $E \cap D_n$ is an atom for μ such that $\mu_a(E \cap D_n) \neq 0$, and because $\mu_a(E) = 0$ when $\mu(E) = 0$, we see that $E \cap D_n$ is an atom for μ_a. Thus, μ_a is an atomic measure.

To show that μ_{na} is non-atomic, suppose that $\mu_{\text{na}}(E) \neq 0$. Therefore, $\mu(E \cap N) \neq 0$ for some $N \in \Sigma$ with $\mu_a(N) = 0$. The set $E \cap N$ is not an atom for μ, because $\mu_a(E \cap N) \neq 0$ if $E \cap N$ were an atom. Since $\mu(E \cap N) \neq 0$ and because $E \cap N$ is not an atom for μ, there exists $F \in \Sigma$ such that $\mu(E \cap N \cap F) \neq 0$ and $\mu((E \cap N) \setminus F) \neq 0$. Hence, $\mu_{\text{na}}(E \cap F) \neq 0$ and $\mu_{\text{na}}(E \setminus F) \neq 0$, implying that μ_{na} is non-atomic.

The proof of the uniqueness of the decomposition is left as Exercise 3.95. $\square$

3.6 Measures on Locally Compact Hausdorff Spaces

If one considers the Borel sets of a topological space X, then it is natural to expect that certain topological features of X play a role in the measure theory of X. But for this to occur, the particular measure under consideration needs to be aware of the topology. One class of measures that is sensitive to topology is the class of regular measures.

Definition 3.58. Let $(X, \mathscr{T})$ be a topological space and consider a measurable space (X, Σ) in which Σ contains the σ-algebra $\mathfrak{B}(X)$ of Borel sets of X. A measure μ on (X, Σ) is said to be a *regular measure* if

1. $\mu(K) < \infty$ for every compact subset $K \subseteq X$,
2. $\mu(E) = \inf\{\mu(U) \mid U$ is open and $E \subseteq U\}$, for every $E \in \Sigma$, and
3. $\mu(U) = \sup\{\mu(K) \mid K$ is compact and $K \subseteq U\}$, for every open set U.

Observe that Proposition 3.47 asserts that Lebesgue measure is regular.

The third property above for the measure of an open set extends to arbitrary measurable sets of finite measure.

Proposition 3.59. *Assume that μ is a regular measure on (X, Σ), where X is a topological space and where Σ contains the Borel sets of X. If $E \in \Sigma$ satisfies $\mu(E) < \infty$, then*

$$\mu(E) = \sup\{\mu(K) \,|\, K \text{ is compact and } K \subseteq E\}.$$

Proof. Assume that $E \in \Sigma$ has finite measure and let $\varepsilon > 0$. Because $\mu(E) = \inf\{\mu(U) \,|\, U \text{ is open and } E \subseteq U\}$, there is an open set $U \subseteq X$ such that $E \subseteq U$ and $\mu(U) < \mu(E) + \varepsilon/2$. Hence, $\mu(U) < \infty$ and $\mu(U \setminus E) < \varepsilon/2$. Because U is open and has finite measure, the same type of argument shows that there is a compact set A with $A \subseteq U$ and $\mu(U) < \mu(A) + \varepsilon/2$. Lastly, since $\mu(U \setminus E) < \varepsilon/2$, there is an open set W with $U \setminus E \subseteq W$ and $\mu(W) < \varepsilon/2$. The set W^c is closed and is contained in $U^c \cup E$. Thus, $K = A \cap W^c$ is a closed subset of a compact set and is, hence, compact. Further,

$$K = A \cap W^c \subseteq A \cap (U^c \cup E) = (A \cap U^c) \cup (A \cap E) = A \cap E \subseteq E$$

and

$$\mu(E) \leq \mu(U)$$

$$< \mu(A) + \varepsilon/2$$

$$= \mu(A \cap W) + \mu(A \cap W^c) + \varepsilon/2$$

$$< \varepsilon/2 + \mu(K) + \varepsilon/2.$$

Hence, $K \subseteq E$ and $\mu(E) < \mu(K) + \varepsilon$ implies that $\mu(E)$ is the least upperbound of all real numbers $\mu(K)$ in which K is a compact subset of E.   □

Proposition 3.59 admits a formulation for σ-finite spaces, which will be of use in our analysis of L^p-spaces.

Proposition 3.60. *If (X, Σ, μ) is a σ-finite measure space in which X is a topological space, Σ contains the Borel sets of X, and μ is regular, then*

$$\mu(E) = \sup\{\mu(K) \,|\, K \text{ is compact and } K \subseteq E\}$$

for every $E \in \Sigma$.

Proof. Exercise 3.96.   □

Continuous functions are, from the point of view of analysis, fairly well understood. In comparison, measurable functions appear to be harder to grasp because of the existential nature of measurability. Therefore, in this light, the following two theorems are striking, for they show that, under the appropriate conditions, measurable functions within ε of being continuous.

Theorem 3.61. *Assume that μ is a regular finite measure on (X, Σ), where X is a compact Hausdorff space and where Σ contains the Borel sets of X. If $f : X \to \mathbb{R}$ is a* *bounded measurable function, then for every $\varepsilon > 0$ there exist a continuous function* $g : X \to \mathbb{R}$ *and a compact set K such that $g_{|K} = f_{|K}$ and $\mu(K^c) < \varepsilon$.*

Proof. To begin with, assume f is a simple function with range $\{\alpha_1, \ldots, \alpha_n\}$, where $\alpha_1, \ldots, \alpha_n \in [0,1]$ are distinct real numbers. Let $E_j = f^{-1}(\{\alpha_j\})$, which is a measurable set; note that $E_j \cap E_i = \emptyset$ if $i \neq j$. Let $\varepsilon > 0$ be given. By Proposition 3.59, for each j there is a compact subset $K_j \subseteq E_j$ with $\mu(K_j) + \frac{\varepsilon}{n} > \mu(E_j)$. Thus, $\mu(E_j \setminus K_j) < \frac{\varepsilon}{n}$ for every j. Because $E_j \cap E_i = \emptyset$ for $i \neq j$, if $K = \bigcup\limits_{j=1}^{n} K_j$, then

$$\mu(K^c) = \mu\left(\bigcup_{j=1}^{n} E_j \setminus K_j\right) = \sum_{j=1}^{n} \mu(E_j \setminus K_j) < \varepsilon.$$

The restriction $f_{|K}$ of f to the compact set K is plainly continuous. Since X is normal, the Tietze Extension Theorem asserts that $f_{|K}$ has a continuous extension $g : X \to [0,1]$.

Assume now that f is an arbitrary measurable function with $0 \leq f(x) \leq 1$ for every $x \in X$. By Proposition 3.14 there is a monotone increasing sequence $\{\varphi_n\}_{n \in \mathbb{N}}$ of nonnegative simple functions φ_n on X such that $\lim\limits_{n \to \infty} \varphi_n(x) = f(x)$ for every $x \in X$. In fact, because $f(X) \subseteq [0,1]$, the convergence of $\{\varphi_n\}_n$ to f is uniform on X (Exercise 3.83). By the previous paragraph, for each $n \in \mathbb{N}$ there is a compact set K_n such that $\mu(K_n^c) < \varepsilon/2^n$ and $\varphi_{n|K_n}$ is continuous. Observe that $\varphi_1(x) + \sum\limits_{n=2}^{N}(\varphi_n(x) - \varphi_{n-1}(x)) = \varphi_N(x)$, and so $\varphi_1 + \sum\limits_{n=2}^{\infty}(\varphi_n - \varphi_{n-1}) = f$. Because $\varphi_1 + \sum\limits_{n=2}^{N}(\varphi_n - \varphi_{n-1})$ is continuous on $K = \bigcap\limits_{n=1}^{\infty} K_n$, and because $\varphi_1 + \sum\limits_{n=2}^{\infty}(\varphi_n - \varphi_{n-1})$ converges uniformly to f, the measurable function f is continuous on K. By the Tietze Extension Theorem, $f_{|K}$ has a continuous extension $g : X \to [0,1]$. Because

$$\mu(K^c) \leq \sum_{n=1}^{\infty} \mu(K_n^c) < \varepsilon,$$

this completes the proof of the theorem in the case where $f(X) \subseteq [0,1]$.

For the case of general f, select $\alpha \in \mathbb{R}$ such that $\alpha f(X) \subseteq [-1, 1]$, and decompose αf as $(\alpha f)^+ - (\alpha f)^-$, where $(\alpha f)^+$ and $(\alpha f)^-$ are measurable functions with ranges contained in $[0,1]$. Thus, the case of general f follows readily from the case of f with $f(X) \subseteq [0,1]$.   □

Theorem 3.62 (Lusin).   *Assume that μ is a regular measure on (X, Σ), where X is a locally compact Hausdorff space and where Σ contains the Borel sets of X. If $f : X \to \mathbb{R}$ is a measurable function with the property that $f_{|E^c} = 0$ for some $E \in \Sigma$ with finite measure, then for every $\varepsilon > 0$ there exists a continuous and bounded* *function $g : X \to \mathbb{R}$ such that*

$$\mu\left(\{x \in X \mid f(x) \neq g(x)\}\right) < \varepsilon.$$

Proof. By hypothesis, $f_{|E^c} = 0$; thus, $E_n = \{x \in X \mid |f(x)| > n\}$ is a subset of E for every $n \in \mathbb{N}$. Because $\mu(E) < \infty$, Proposition 3.23 implies that $\mu(E_n) < \varepsilon/6$ for some $n \in \mathbb{N}$. Hence, if $F = E \cap E_n^c$, which is a set of finite measure, then $f_{|F}$ is bounded.

By Proposition 3.59, there is a compact subset $Y \subseteq F$ such that $\mu(F \setminus Y) < \varepsilon/6$. Consider the bounded measurable function $f_{|Y}$. By Theorem 3.61, there is a compact subset $K \subseteq Y$ and a continuous function $g_0 : Y \to \mathbb{R}$ such that $g_{0|K} = f_{|K}$ and $\mu(Y \setminus K) < \varepsilon/6$. Thus,

$$E \setminus K = E_n \cup (F \setminus Y) \cup (Y \setminus K)$$

yields $\mu(E \setminus K) < \varepsilon/2$.

The regularity of μ again implies the existence of an open set $U \subseteq X$ for which $E \subseteq U$ and $\mu(U) < \mu(E) + \varepsilon/2$. Hence, $\mu(U)$ is finite and $\mu(U \setminus E) < \varepsilon/2$. Because K is compact and $K \subseteq U$, Theorem 2.43 asserts that g_0 admits a continuous and bounded extension $g : X \to \mathbb{R}$ such that $g(x) = 0$ for all $x \notin U$. Therefore, $0 = g_{|U^c} = f_{|U^c}$ and, hence,

$$\mu\left(\{x \in X \mid f(x) \neq g(x)\}\right) = \mu\left(\{x \in E \mid f(x) \neq g(x)\}\right) + \mu\left(\{x \in E^c \mid f(x) \neq g(x)\}\right)$$

$$< \varepsilon/2 + \varepsilon/2 = \varepsilon,$$

which completes the proof.   □

3.7 Signed and Complex Measures

Extending the notions of length, area, volume, and other arbitrary measures to real- and complex-valued quantities results in the concepts of signed measure and complex measure.

Definition 3.63. A function $\omega : \Sigma \to [-\infty, +\infty]$ on a measurable space (X, Σ) is called a *signed measure* if $\omega(\emptyset) = 0$ and

$$\omega\left(\bigcup_{k=1}^{\infty} E_k\right) = \sum_{k=1}^{\infty} \omega(E_k),$$

for every sequence $\{E_k\}_{k \in \mathbb{N}}$ of pairwise disjoint sets $E_k \in \Sigma$.

The definition of signed measure entails some subtleties. First of all, arithmetic in the extended real number system $[-\infty, +\infty]$ does not admit sums of the form $(-\infty) + (+\infty)$ or $(+\infty) + (-\infty)$, which implies that for each $E \in \Sigma$ at most one of $\omega(E)$ or $\omega(E^c)$ can have an infinite value (because $\omega(X) = \omega(E) + \omega(E^c)$). In particular, this means that if there exists a measurable set E with $\omega(E) = +\infty$, then necessarily $\omega(X) = +\infty$; and if there exists a measurable set E with $\omega(E) = -\infty$, then $\omega(X) = -\infty$ necessarily. Therefore, $\omega(X)$ can achieve at most one of the values $-\infty$ or $+\infty$. If μ does not achieve either of these infinite values, then ω is said to be a *finite signed measure*. The triple (X, Σ, ω) is called a *signed measure space*.

Definition 3.64. If (X, Σ, ω) is a signed measure space, and if $P, N \in \Sigma$, then

1. P is said to be *positive* with respect to ω if $\omega(E \cap P) \geq 0$ for every $E \in \Sigma$, and
2. N is said to be *negative* with respect to ω if $\omega(E \cap N) \leq 0$ for every $E \in \Sigma$.

Interestingly, a signed measure partitions a signed measure space into a positive part and a negative part, as shown by the Hahn Decomposition Theorem below.

Theorem 3.65 (Hahn Decomposition of Signed Measures).   *If (X, Σ, ω) is a signed measure space, then there exist $P, N \in \Sigma$ such that*

1. *P is positive with respect to ω and N is negative with respect to ω,*
2. *$P \cap N = \emptyset$, and*
3. *$X = P \cup N$.*

Proof. We may assume without loss of generality that $-\infty$ is not one of the values assumed by ω. Let $\alpha = \inf\{\omega(E) \mid E \in \Sigma \text{ is a negative set}\}$. (Because $\emptyset$ is a negative set, the infimum is defined.) Let $\{E_k\}_{k \in \mathbb{N}}$ be a sequence of measurable sets for which $\alpha = \lim_k \omega(E_k)$. For each k let $N_k = E_k \setminus \left(\bigcup_{j=1}^{k-1} E_k \right)$ so that $\{N_k\}_{k \in \mathbb{N}}$ is a sequence of pairwise disjoint negative sets such that $\alpha = \inf_k \omega(N_k)$. Thus, with $N = \bigcup_{k=1}^{\infty} N_k$, we have for every $j \in \mathbb{N}$ that $\omega(N) = \sum_{k=1}^{\infty} \omega(N_k) \leq \omega(N_j)$. Hence, $\omega(N) = \alpha$ and N is a negative set. Because $-\infty$ is not in the range of ω, it must be that $\omega(N) \in \mathbb{R}$. Hence, α is the minimum measure of all negative subsets of X.

Let $P = N^c$. Assume, contrary to what we aim to prove, that P is not a positive set. Thus, there exists a measurable subset $E \subset P$ such that $\omega(E) < 0$. The set E is not negative because, if it were, then $N \cup E$ would also be a negative set of measure $\omega(N \cup E) = \alpha + \omega(E) < \alpha$, which contradicts the fact that α is the minimum measure of all negative subsets of X. Hence, E must possess a measurable subset F of positive measure. Let $n_1 \in \mathbb{N}$ denote the smallest positive integer for which there exists a measurable subset $F_1 \subset E$ of measure $\omega(F_1) \geq 1/n_1$. Since $E \setminus F_1$ and F_1 are disjoint and have union E, $\omega(E) = \omega(E \setminus F_1) + \omega(F_1)$. That is, $\omega(E \setminus F_1) = \omega(E) - \omega(F_1) \leq \omega(E) - n_1^{-1} < \omega(E)$. For the very same reasons given earlier, the set $E \setminus F_1$ cannot be negative; thus, $E \setminus F_1$ contains a measurable subset of positive measure. Let $n_2 \in \mathbb{N}$

denote the smallest positive integer for which there exists a measurable subset $F_2 \subset (E \setminus F_1)$ of measure $\omega(F_2) \geq 1/n_2$. Repeating this argument inductively produces a subset $\{n_k\}_{k \in \mathbb{N}} \subseteq \mathbb{N}$ and a sequence $\{F_k\}_{k \in \mathbb{N}}$ of pairwise disjoint measurable subsets $F_k \subset E$ such that the set $F = \bigcup_{k=1}^{\infty} F_k$ satisfies

$$\omega(F) = \sum_{k=1}^{\infty} \omega(F_k) \geq \sum_{k=1}^{\infty} \frac{1}{n_k} > 0.$$

Therefore, the subset $G = E \setminus F$ of E satisfies $\omega(G) \leq \omega(G) + \omega(F) = \omega(E) < 0$. Since $-\infty$ is not in the range of ω, $\omega(G)$ is a negative real number, and so

$$0 < \sum_{k=1}^{\infty} \frac{1}{n_k} \leq \sum_{k=1}^{\infty} \omega(F_k) = \omega(F) = \omega(E) - \omega(G) < |\omega(G)| < \infty$$

implies that $\lim_k n_k^{-1} = \lim_k \omega(F_k) = 0$. Therefore, if Q is a measurable subset of G, then

$$Q \subseteq G = E \cap F^c = E \cap \left(\bigcap_{k=1}^{\infty} F_k^c \right) = \bigcap_{k=1}^{\infty} (E \setminus F_k)$$

implies that $Q \subseteq E \setminus F_k$ for every $k \in \mathbb{N}$. If it were true that $\omega(Q) > 0$, then for some $j \in \mathbb{N}$ we would have $\omega(Q) > \frac{1}{n_j - 1}$, which is to say that Q is a subset of $E \setminus F_j$ of measure $\omega(Q) > \frac{1}{n_j - 1} > \frac{1}{n_j}$, in contradiction to the property of n_j being the smallest positive integer for which $E \setminus F_j$ has a subset A of measure $\omega(A) > \frac{1}{n_j}$. Hence, $\omega(Q) \leq 0$ and the fact that Q is an arbitrary measurable subset of G implies that G is a negative set. But $G \subseteq P$ implies that $G \cap N = \emptyset$ and so the negative subset $G \cup N$ satisfies $\omega(G \cup N) < \alpha$, which is in contradiction to the fact that α is the minimum measure of all negative subsets of X. Therefore, it must be that P is a positive set. $\square$

The sets P and N that arise in Theorem 3.65 are said to be a *Hahn decomposition* of the signed measure space (X, Σ, ω). While this decomposition need not be unique, Exercise 3.99 shows that if (P_1, N_1) and (P_2, N_2) are two Hahn decompositions of a signed measure space (X, Σ, ω), then

$$\omega(E \cap P_1) = \omega(E \cap P_2) \quad \text{and} \quad \omega(E \cap N_1) = \omega(E \cap N_2)$$

for all $E \in \Sigma$. Therefore, the functions $\omega_+, \omega_- : \Sigma \to [0, +\infty]$ defined by

$$\omega_+(E) = \omega(E \cap P) \quad \text{and} \quad \omega_-(E) = -\omega(E \cap N) \tag{3.18}$$

are measures on (X, Σ) and are independent of the choice of Hahn decomposition (P, N) of (X, Σ, ω). Note, also, that at least one of ω_+ and ω_- is a finite measure. These observations give rise to the next theorem.

Theorem 3.66 (Jordan Decomposition Theorem). *For every signed measure ω on a measurable space (X, Σ), there exist measures ω_+ and ω_- on (X, Σ) such that*

1. *at least one of ω_+ and ω_- is a finite measure, and*
2. *$\omega(E) = \omega_+(E) - \omega_-(E)$, for every $E \in \Sigma$.*

Furthermore, if γ, δ are measures on (X, Σ), where at least one of which is finite, and if $\omega(E) = \gamma(E) - \delta(E)$ for every $E \in \Sigma$, then $\omega_+(E) \leq \gamma(E)$ and $\omega_-(E) \leq \delta(E)$, for all $E \in \Sigma$.

Proof. Exercise 3.100.   □

Turning now to complex measures, the definition below departs from the definitions of measure and signed measure in that it is assumed from the outset that the measure is finite.

Definition 3.67. A function $\nu : \Sigma \to \mathbb{C}$ on a measurable space (X, Σ) is called a *complex measure* if $\nu(\emptyset) = 0$ and

$$\nu\left(\bigcup_{k=1}^{\infty} E_k\right) = \sum_{k=1}^{\infty} \nu(E_k),$$

for every sequence $\{E_k\}_{k \in \mathbb{N}}$ of pairwise disjoint sets $E_k \in \Sigma$.

By decomposing a complex measure ν into its real and imaginary parts $\Re\nu$ and $\Im\nu$, two finite signed measures are obtained, each of which is a difference of finite measures. Hence, there are finite measures μ_j on (X, Σ), for $j = 1, \ldots, 4$, such that

$$\nu = (\mu_1 - \mu_2) + i(\mu_3 - \mu_4).$$

By considering the function $E \mapsto |\nu(E)|$, something close to a measure is obtained—but the triangle inequality makes this function countably subadditive rather than additive on sequences of pairwise disjoint sets. Therefore, to obtain a measure from a complex measure requires slightly more effort.

Definition 3.68. In a measurable space (X, Σ), a *measurable partition* of a measurable set $E \subseteq X$ is a family $\mathscr{P}_E$ of countably many subsets $A \in \Sigma$ such that $A \subseteq E$ for all $A \in \mathscr{P}_E$, $\bigcup_{A \in \mathscr{P}_E} A = E$, and $A \cap B = \emptyset$ whenever $A, B \in \mathscr{P}_E$ are distinct.

Proposition 3.69. *If ν is a complex measure on a measurable space (X, Σ) and if $|\nu| : \Sigma \to [0, \infty]$ is defined by*

$$|\nu|(A) = \sup\left\{\sum_{E \in \mathscr{P}_A} |\nu(E)| \,\Big|\, \mathscr{P}_A \text{ is a measurable partition of } A\right\},$$

then $|\nu|$ is a finite measure on (X, Σ).

Proof. Because $\mathscr{P}_\emptyset = \{A, B\}$, where $A = B = \emptyset$, is measurable partition of the empty set $\emptyset$, we have that $\nu(\emptyset) = \nu(\emptyset) + \nu(\emptyset)$ in $\mathbb{C}$ and so $\mu(\emptyset) = 0$.

To prove that $|\nu|$ is countably additive, let $\{E_k\}_{k \in \mathbb{N}}$ be a sequence of pairwise disjoint measurable sets and let $E = \bigcup_{k \in \mathbb{N}} E_k$. For each k, consider an arbitrary measurable partition $\{F_{kj}\}_{j \in \mathbb{N}}$ of E_k; thus, $\sum_j |\nu(F_{kj})| \leq |\nu|(E_k)$. Because $\{F_{kj}\}_{k,j \in \mathbb{N}}$ is an arbitrary measurable partition of E,

$$\sum_{k=1}^{\infty} \sum_{j=1}^{\infty} |\nu(F_{kj})| \leq |\nu|(E).$$

For each k, $|\nu|(E_k)$ is the supremum of $\sum_j |\nu(F_{kj})|$ over all measurable partitions $\{F_{kj}\}_{j \in \mathbb{N}}$ of E_k, and therefore the inequality above yields $\sum_{k=1}^{\infty} |\nu|(E_k) \leq |\nu|(E)$.

Conversely, select an arbitrary measurable partition $\{A_\ell\}_{\ell \in \mathbb{N}}$ of E. Because the sets $\{E_k\}_{k \in \mathbb{N}}$ are pairwise disjoint, $\{A_\ell \cap E_k\}_{k \in \mathbb{N}}$ is a partition of A_ℓ for every $\ell \in \mathbb{N}$, and $\{A_\ell \cap E_k\}_{\ell \in \mathbb{N}}$ is a partition of E_k for every $k \in \mathbb{N}$. Thus,

$$\sum_{\ell=1}^{\infty} |\nu(A_\ell)| \leq \sum_{\ell=1}^{\infty} \sum_{k=1}^{\infty} |\nu(A_\ell \cap E_k)| = \sum_{k=1}^{\infty} \sum_{\ell=1}^{\infty} |\nu(A_\ell \cap E_k)| \leq \sum_{k=1}^{\infty} |\nu(E_k)|,$$

and so $|\nu|(E) \leq \sum_{k=1}^{\infty} |\nu(E_k)|$. Hence, $|\nu|$ is countably additive.

As indicated previously, there are finite measures μ_j on (X, Σ), for $j = 1, \ldots, 4$, such that $\nu = (\mu_1 - \mu_2) + i(\mu_3 - \mu_4)$. Thus, for any measurable set $E \in \Sigma$, $|\nu(E)| \leq \sum_{j=1}^{4} \mu_j(E)$. Therefore, if $\mathscr{P}_X$ is a partition of X, then

$$\sum_{E \in \mathscr{P}_X} |\nu(E)| \leq \sum_{E \in \mathscr{P}_X} \sum_{j=1}^{4} \mu_j(E) = \sum_{j=1}^{4} \sum_{E \in \mathscr{P}_X} \mu_j(E) = \sum_{j=1}^{4} \mu_j(X) < \infty.$$

Hence, $|\nu|(X) \leq \sum_{j=1}^{4} \mu_j(X)$, which proves that $|\nu|$ is a finite measure. $\square$

Definition 3.70. In Proposition 3.69 above, the measure $|\nu|$ on (X, Σ) induced by the complex measure ν is called the *total variation of* ν.

Problems

3.71. Show that the collection Σ of all subsets E of an infinite set X for which E or the complement E^c of E is countable is a σ-algebra.

3.72. Prove that if Λ is a family of σ-algebras on a subset X, then $\bigcap_{\Sigma \in \Lambda} \Sigma$ is a σ-algebra.

3.73. Assume that Σ is a σ-algebra of subsets of X. Show that, for each $E \in \Sigma$, the collection $\Sigma(E)$ of subsets of X defined by

$$\Sigma(E) = \{E \cap A \mid A \in \Sigma\}$$

is a σ-algebra on E.

3.74. Let Σ be a σ-algebra of subsets of a nonempty set X, and let $E_k \in \Sigma$ for $k \in \mathbb{N}$. Define

$$\limsup E_k = \bigcap_{k \geq 1} \left(\bigcup_{n \geq k} E_n \right),$$
$$\liminf E_k = \bigcup_{k \geq 1} \left(\bigcap_{n \geq k} E_n \right).$$

Prove the following statements.

1. $\limsup E_k$ and $\liminf E_k$ belong to Σ.
2. If $E_1 \subseteq E_2 \subseteq E_3 \subseteq \ldots$, then $\limsup E_k = \bigcup_k E_k = \liminf E_k$

3.75. Let E_k denote the closed interval $E_k = [0, 1 + \frac{(-1)^k}{k}]$. Determine the sets $\limsup E_k$ and $\liminf E_k$. (Suggestion: consider the cases k even and k odd separately.)

3.76. Let X be a nonempty set X and let $\{A_k\}_{k \in \mathbb{N}}$ be a sequence of subsets of X. Define $E_0 = \emptyset$ and, for $n, m \in \mathbb{N}$,

$$E_m = \bigcup_{k=1}^{m} A_k, \quad F_n = A_n \setminus E_{n-1}.$$

Prove the following statements.

1. $\{E_n\}_n$ is a monotone increasing sequence of sets (that is, $E_n \subseteq E_{n+1}$ for all n).
2. $\{F_n\}_n$ is a sequence of pairwise disjoint sets.
3. $\bigcup_n E_n = \bigcup_n F_n = \bigcup_n A_n$.

3.77. Prove that if a σ-algebra Σ on an infinite set X has infinitely many elements, then Σ is uncountable.

3.78. Prove that if $(X, \mathscr{T})$ is a topological space, and if $\Sigma_{\mathscr{T}}$ is the σ-algebra generated by $\mathscr{T}$, then, with respect to the measurable space $(X, \Sigma_{\mathscr{T}})$, every continuous function $f : X \rightarrow \mathbb{R}$ is a measurable function.

3.79. Suppose that (X, Σ) is a measurable space and that $h : X \rightarrow \mathbb{R}$ is a measurable function for which $h(x) \neq 0$, for all $x \in X$. Prove that the function $1/h$ is measurable.

3.80. Prove that if (X, Σ) is a measurable space and if $E \subseteq X$, then the characteristic function $\chi_E : X \rightarrow \mathbb{R}$ is a measurable function if and only if $E \in \Sigma$.

3.81. Let U be a nonempty subset of $\beta \mathbb{N}$ (see Section 2.6), and consider the characteristic function χ_U. Prove that χ_U is continuous if and only if both U and $\overline{U}$ are open in $\beta \mathbb{N}$.

3.82. Assume that (X, Σ) is a measurable space and that $E \in \Sigma$.

1. If $f : E \rightarrow \mathbb{R}$ is a measurable function relative to the measurable space $(E, \Sigma(E))$, then prove that the extension $\tilde{f} : X \rightarrow \mathbb{R}$ of f defined by $\tilde{f} = f \chi_E$ is a measurable function with respect to the measurable space (X, Σ).
2. Conversely, if $\tilde{f} : X \rightarrow \mathbb{R}$ is a measurable function with respect to the measurable space (X, Σ), and if $f = \tilde{f}|_E$ (the restriction of $\tilde{f}$ to E), then prove that $f : E \rightarrow R$ is a measurable function with respect to the measurable space $(E, \Sigma(E))$.

3.83. If $f : X \rightarrow [0, 1]$ is a measurable function, then prove that there is a monotone-increasing sequence of nonnegative simple functions $\varphi_n : X \rightarrow [0, 1]$ such that $\lim_{n \to \infty} \varphi_n(x) = f(x)$ uniformly—that is, for every $\varepsilon > 0$ there is an $N_\varepsilon \in \mathbb{N}$ such that $|f(x) - \varphi_n(x)| < \varepsilon$ for all $n \geq N_\varepsilon$ and all $x \in X$.

3.84. Let X be an infinite set and let Σ be the σ-algebra in Exercise 3.71. Define a function $\mu : \Sigma \rightarrow [0, \infty]$ by $\mu(E) = 0$ if $E \in \Sigma$ is countable and $\mu(E) = 1$ if $E \in \Sigma$ is uncountable. Show that μ is a measure on (X, Σ).

3.85. Consider the measurable space $(\mathbb{N}, \mathscr{P}(\mathbb{N}))$, where $\mathscr{P}(\mathbb{N})$ is the power set of $\mathbb{N}$. Prove that the function $\mu : \Sigma \rightarrow [0, \infty]$ defined by

$$\mu(E) = \text{the cardinality of } E$$

defines a measure on $(\mathbb{N}, \mathscr{P}(\mathbb{N}))$.

3.86. A function $\mu : \Sigma \rightarrow [0, \infty)$ on a measurable space (X, Σ) is *finitely additive* if, for all finite sub-collections $\{E_k\}_{k=1}^n$ of pairwise disjoint measurable sets E_k,

$$\mu \left(\bigcup_{k=1}^n E_k \right) = \sum_{k=1}^n \mu(E_k).$$

Prove that if a finitely additive function μ also satisfies $\lim_k \mu(A_k) = 0$, for every descending sequence $A_1 \supseteq A_2 \supseteq A_3 \supseteq \cdots$ of sets $A_j \in \Sigma$ in which $\bigcap_{k=1}^{\infty} A_k = \emptyset$, then μ is in fact a measure on (X, Σ).

3.87. Assume that μ is a measure on a measurable space (X, Σ). Prove that

$$\mu(E \cup F) + \mu(E \cap F) = \mu(E) + \mu(F),$$

for all $E, F \in \Sigma$.

3.88. Prove that if $E \in \mathcal{O}_n$, then $m^*(\overline{E}) = m^*(E)$, where

$$\mathcal{O}_n = \{ \prod_{i=1}^{n} (a_i, b_i) \mid a_i, b_i \in \mathbb{R}, \, a_i \le b_i \}.$$

3.89. Prove that each of the following subsets of $\mathbb{R}^n$ is a null set.

1. Every finite or countably infinite set.
2. Every countable union of null sets.
3. Every subset of a null set.

3.90. Prove that if $E \subseteq \mathbb{R}^n$ is Lebesgue measurable such that $m(E) > 0$, then E contains a nonmeasurable subset.

3.91. Prove that the following statements are equivalent for a subset $E \subseteq \mathbb{R}$:

1. E is a Lebesgue-measurable set;
2. there exist $B, E_0 \subseteq \mathbb{R}$ such that:

   a. B is a Borel set,
   b. E_0 is a null set,
   c. $E_0 \cap B = \emptyset$, and
   d. $E = B \cup E_0$.

3.92. Prove that there exist subsets S of $\mathbb{R}^n$ that are not Lebesgue measurable, and that there exist Lebesgue-measurable subsets E of $\mathbb{R}^n$ that are not Borel measurable.

3.93. Determine the atoms for counting measure on $\mathbb{N}$.

3.94. Prove that Lebesgue measure on $\mathbb{R}^n$ is non-atomic.

3.95. Suppose that $\mu = \mu_a + \mu_{na} = \tilde{\mu}_a + \tilde{\mu}_{na}$ are two decompositions of a measure μ on (X, Σ) as the sum of an atomic measure and a non-atomic measure, where $\mu_a \mathscr{S} \mu_{na}$, $\mu_{na} \mathscr{S} \mu_a$, $\tilde{\mu}_a \mathscr{S} \tilde{\mu}_{na}$, and $\tilde{\mu}_{na} \mathscr{S} \tilde{\mu}_a$.

1. Show that $\mu_{na} \mathscr{S} \tilde{\mu}_a$ $\tilde{\mu}_a \mathscr{S} \mu_{na}$
2. Show that $\tilde{\mu}_{na}(E) - \mu_{na}(E) \ge 0$ and $\mu_a(E) - \tilde{\mu}_a(E) \ge 0$ dor every $E \in \Sigma$
3. Show that $\tilde{\mu}_a = \mu_a$ and $\tilde{\mu}_{na} = \mu_{na}$.

3.96. Prove that if (X, Σ, μ) is a σ-finite measure space in which X is a topological space, Σ contains the Borel sets of X, and μ is regular, then

$$\mu(E) = \sup\{\mu(K) \,|\, K \text{ is compact and } K \subseteq E\}$$

for every $E \in \Sigma$.

3.97. Let Σ denote the Borel sets of $X = [0, 1]$ and define a function $\mu : \Sigma \to [0, \infty]$ by $\mu(E) = m(E)$, if $0 \notin E$, and $\mu(E) = \infty$, if $0 \in E$.

1. Prove that μ is a measure on (X, Σ).
2. Prove that (X, Σ, μ) is not a σ-finite measure space.

3.98. Show that, in a signed measure space (X, Σ, ω), the union and intersection of finitely many positive sets are positive sets, and that the union and intersection of finitely many negative sets are negative sets.

3.99. Suppose that (P_1, N_1) and (P_2, N_2) are Hahn decompositions of a signed measure space (X, Σ, ω). Prove that, for every $E \in \Sigma$,

$$\omega(E \cap P_1) = \omega(E \cap P_1 \cap P_2) = \omega(E \cap P_2).$$

3.100. Assume that (X, Σ, ω) is a signed measure space with Hahn decomposition (P, N). Show that the functions ω_+ and ω_- defined by

$$\omega_+(E) = \omega(E \cap P) \quad \text{and} \quad \omega_-(E) = -\omega(E \cap N),$$

for $E \in \Sigma$ are measures on (X, Σ) with the following properties:

1. at least one of ω_+ and ω_- is a finite measure;
2. $\omega(E) = \omega_+(E) - \omega_-(E)$, for every $E \in \Sigma$;
3. if γ, δ are measures on (X, Σ), where at least one of which is finite, and if $\omega(E) = \gamma(E) - \delta(E)$ for every $E \in \Sigma$, then $\omega_+(E) \leq \gamma(E)$ and $\omega_-(E) \leq \delta(E)$, for all $E \in \Sigma$.

Chapter 4
Integration

This chapter is devoted to the main results concerning the Lebesgue integral. There are many reasons for considering a more robust theory of integration than that afforded by the classical theories of Cauchy and Riemann, and one of the most compelling of these reasons arises from the inclination to view the integral as a continuous linear transformation from a vector space of functions into the real or complex numbers. Continuity of integration, in this regard, may be considered to be the property that the integral of a convergent sequence of functions f_n is the limit of the integrals of f_n. Limiting properties such as these are possible with the Lebesgue integral, but not with the Riemann integral in general, and this is one reason why the Lebesgue integral plays a central role in functional analysis.

The classical approach to the integration of continuous real-valued functions on a closed interval $[a,b] \subset \mathbb{R}$ is based on a partitioning of the domain, $[a,b]$, into n-subintervals, and to then use Riemann sums to approximate the integral. Lebesgue's approach to the integration of continuous real-valued functions is to partition the range of a function rather than its domain. This simple idea extends beyond continuous functions to measurable functions, and leads to a remarkably powerful theory of integration.

4.1 Integration of Nonnegative Functions

Definition 4.1. Suppose that (X, Σ, μ) is a measure space, and that $\varphi : X \to \mathbb{R}$ is a simple function with range $\{\alpha_1, \ldots, \alpha_n\} \subseteq [0, \infty)$. Let $E_k = \varphi^{-1}(\{\alpha_k\})$, for each k.

1. The *canonical form* of φ is the representation of φ given by

$$\varphi = \sum_{k=1}^{n} \alpha_k \, \chi_{E_k}. \tag{4.1}$$

© Springer International Publishing Switzerland 2016
D. Farenick, *Fundamentals of Functional Analysis*, Universitext,
DOI 10.1007/978-3-319-45633-1_4

2. The *Lebesgue integral*, or simply the *integral*, of φ is the quantity in the extended nonnegative real number system $[0, \infty]$ denoted by $\int_X \varphi \, d\mu$ and defined by

$$\int_X \varphi \, d\mu = \sum_{k=1}^{n} \alpha_k \, \mu(E_k). \tag{4.2}$$

Although ∞ is one value that the Lebesgue integral of φ could take , if $\mu(X)$ is finite, then so is $\int_X \varphi \, d\mu$, for every nonnegative simple function φ.

The first proposition about integration states that the definition of the integral of φ does not depend on whether or not φ is expressed in canonical form.

Proposition 4.2. *If (X, Σ, μ) is a measure space and if $\varphi : X \to \mathbb{R}$ is a nonnegative simple function of the form*

$$\varphi = \sum_{k=1}^{m} \beta_k \, \chi_{F_k},$$

where $F_1, \ldots, F_m$ are pairwise disjoint measurable sets for which $F_1 \cup \cdots \cup F_m = X$, then

$$\int_X \varphi \, d\mu = \sum_{k=1}^{m} \beta_k \, \mu(F_k).$$

Proof. If (4.1) denotes the canonical form of φ,

$$\sum_{k=1}^{n} \alpha_k \, \chi_{E_k} = \sum_{k=1}^{m} \beta_k \, \chi_{F_k}.$$

For all pairs $(k, j) \in \{1, \ldots, n\} \times \{1, \ldots, m\}$, consider $A_{kj} = E_k \cap F_j$. The sequence $\{A_{kj}\}_{k,j}$ consists of pairwise disjoint measurable sets whose union is X. If $x \in A_{kj}$, then $\varphi(x) = \alpha_k = \beta_j$. Thus, the sets $\{\alpha_1, \ldots, \alpha_n\}$ and $\{\beta_1, \ldots, \beta_m\}$ coincide. Moreover,

$$\begin{aligned}
\int_X \varphi \, d\mu = \sum_{k=1}^{n} \alpha_k \, \mu(E_k) &= \sum_{k=1}^{n} \alpha_k \sum_{j=1}^{m} \mu(E_k \cap F_j) \\
&= \sum_{k=1}^{n} \sum_{j=1}^{m} \alpha_k \, \mu(A_{kj}) \\
&= \sum_{j=1}^{m} \sum_{k=1}^{n} \beta_j \, \mu(A_{kj}) \\
&= \sum_{j=1}^{m} \beta_j \, \mu(F_j),
\end{aligned}$$

which completes the proof.   □

Corollary 4.3. *Suppose that (X, Σ, μ) is a measure space, and that $\varphi, \psi : X \to \mathbb{R}$ are nonnegative simple functions.*

1. If $\psi(x) \leq \varphi(x)$, for all $x \in X$, then

$$\int_X \psi \, d\mu \leq \int_X \varphi \, d\mu.$$

2. If $\alpha, \beta \in \mathbb{R}$ are nonnegative, then $\alpha\varphi + \beta\psi$ is a nonnegative simple function and

$$\int_X (\alpha\varphi + \beta\psi) \, d\mu = \alpha \int_X \varphi \, d\mu + \beta \int_X \psi \, d\mu.$$

Proof. Exercise 4.62. $\qquad\qquad\qquad\qquad\qquad\qquad\qquad\qquad\qquad\qquad\qquad$ $\square$

Theorem 3.14 states that nonnegative measurable functions can be realised as the limit (pointwise) of a monotone increasing sequence of simple functions. Thus, simple functions lead the way to the definition of integral for nonnegative functions.

Definition 4.4. If (X, Σ, μ) is a measure space and $f : X \to \mathbb{R}$ is a nonnegative measurable function, then the *Lebesgue integral*, or *integral*, of f is the quantity in $[0, \infty]$ denoted by $\int_X f \, d\mu$ and defined by

$$\int_X f \, d\mu = \sup \left\{ \int_X \varphi \, d\mu \mid \varphi \text{ is simple, and } 0 \leq \varphi(x) \leq f(x), \text{ for all } x \in X \right\}.$$

At times one prefers to integrate a function over a measurable subset of X rather than over the whole space.

Definition 4.5. Let (X, Σ, μ) be a measure space and $f : X \to \mathbb{R}$ be a nonnegative measurable function. If $E \in \Sigma$, then the *Lebesgue integral of f over E* is the quantity in $[0, \infty]$ denoted by $\int_E f \, d\mu$ and defined by $\int_E f \, d\mu = \int_X \chi_E f \, d\mu$, where χ_E is the characteristic function of E.

By defining the integral of a nonnegative measurable function as in Definition 4.4, one has the following continuity property.

Theorem 4.6 (Monotone Convergence Theorem). *If $\{f_k\}_{k \in \mathbb{N}}$ is a monotone increasing sequence of nonnegative measurable functions on a measure space (X, Σ, μ) such that $\lim_k f_k(x)$ exists for all $x \in X$, then*

$$\lim_{k \to \infty} \int_X f_k \, d\mu = \int_X (\lim_{k \to \infty} f_k) \, d\mu.$$

Proof. Let $f : X \to \mathbb{R}$ denote the limiting function: $f(x) = \lim_k f_k(x)$, $x \in X$. By Theorem 3.9, f is measurable. Because $f_k(x) \leq f_{k+1}(x) \leq f(x)$, for all $x \in X$,

Exercise 4.63 shows that

$$\int_X f_k \, d\mu \; \leq \; \int_X f_{k+1} \, d\mu \; \leq \; \int_X f \, d\mu, \; \forall k \in \mathbb{N}.$$

Thus,

$$\lim_{k\to\infty} \int_X f_k \, d\mu \; \leq \; \int_X (\lim_{k\to\infty} f_k) \, d\mu.$$

Conversely, let $\varphi : X \to \mathbb{R}$ be a simple function such that $0 \leq \varphi(x) \leq f(x)$, for every $x \in X$. Assume that the canonical form of φ is

$$\varphi = \sum_{k=1}^{n} \alpha_k \, \chi_{E_k}.$$

Fix any $\gamma \in (0, 1)$ and consider the set

$$F_k = \{x \in X \, | \, f_k(x) \geq \gamma \varphi(x)\}.$$

Because $F_k = (f_k - \gamma\varphi)^{-1}([0, \infty))$, F_k is a measurable set. By definition of F_k,

$$\gamma\varphi(x) \; \leq \; f_k(x) \; \leq \; f(x) \quad \forall x \in F_k.$$

Hence, again by Exercise 4.63,

$$\int_X \chi_{F_k} \gamma \varphi_k \, d\mu \; \leq \; \int_X \chi_{F_k} f_k \, d\mu \; \leq \; \int_X f_k \, d\mu. \tag{4.3}$$

Moreover, because the sequence of functions f_k is monotone increasing, the sets F_k are monotone increasing. In fact,

$$\bigcup_{k\in\mathbb{N}} F_k = X. \tag{4.4}$$

To prove (4.4), select $x \in X$. Because $\gamma < 1$, we have that $\gamma\varphi(x) < \varphi(x) \leq f(x)$. Furthermore, $\lim_k f_k(x) = f(x)$ and $\{f_k\}_{k\in\mathbb{N}}$ is a monotone-increasing sequence of functions; thus, there is a $k_0 \in \mathbb{N}$ such that $\gamma\varphi(x) < f_{k_0}(x) \leq f(x)$, which implies that $x \in F_{k_0}$. Hence, $\bigcup_{k\in\mathbb{N}} F_k = X$.

The continuity of μ (Proposition 3.22) and equation (4.4) yield $\mu(X) = \lim_{k\to\infty} \mu(F_k)$, and so $\mu(E_j) = \lim_{k\to\infty} \mu(E_j \cap F_k)$, for each $j = 1, \dots, n$. Furthermore,

$$\gamma \int_X \varphi \, d\mu = \gamma \sum_{j=1}^{n} \alpha_j \mu(E_j) = \gamma \sum_{j=1}^{n} \alpha_j \lim_{k\to\infty} \mu(E_j \cap F_k) = \gamma \lim_{k\to\infty} \sum_{j=1}^{n} \alpha_j \mu(E_j \cap F_k)$$

$$= \gamma \lim_{k\to\infty} \int_X \chi_{F_k} \varphi \, d\mu \le \lim_{k\to\infty} \int_X \chi_{F_k} f_k \, d\mu \le \lim_{k\to\infty} \int_X f_k \, d\mu .$$

(The final two inequalities above are on account of (4.3).) Therefore,

$$\int_X \varphi \, d\mu = \lim_{\gamma \to 1^-} \gamma \int_X \varphi \, d\mu \le \lim_{k\to\infty} \int_X f_k \, d\mu .$$

Hence,

$$\sup \left\{ \int_X \varphi \, d\mu \mid \varphi \text{ is simple, and } 0 \le \varphi(x) \le f(x), \text{ for all } x \in X \right\} \le \lim_{k\to\infty} \int_X f_k \, d\mu .$$

That is,

$$\int_X \lim_{k\to\infty} f_k \, d\mu \le \lim_{k\to\infty} \int_X f_k \, d\mu ,$$

thereby completing the proof of the Monotone Convergence Theorem. $\qquad\square$

The Monotone Convergence Theorem does not apply to monotone decreasing sequences, as demonstrated by the example below.

Example 4.7. *There is a monotone-decreasing sequence $\{g_k\}_{k\in\mathbb{N}}$ of Lebesgue-measurable nonnegative functions $g_k : \mathbb{R} \to \mathbb{R}$ such that $\lim_k g_k(x)$ exists for all $x \in \mathbb{R}$, but*

$$\lim_{k\to\infty} \int_{\mathbb{R}} g_k \, dm \ne \int_{\mathbb{R}} (\lim_{k\to\infty} g_k) \, dm.$$

Proof. Let $g_k = \chi_{[k,\infty)}$. For each $x \in \mathbb{R}$, $g_1(x) \ge g_2(x) \cdots \ge 0$ and $\lim_k g_k(x) = 0$. Thus,

$$\int_{\mathbb{R}} (\lim_{k\to\infty} g_k) \, dm = \int_{\mathbb{R}} 0 \, dm = 0.$$

On the other hand, $\int_{\mathbb{R}} g_k \, dm = \infty$ for every $k \in \mathbb{N}$. $\qquad\square$

Example 4.7 indicates that the Monotone Convergence Theorem does not extend to convergent sequences that are not monotone increasing. Nevertheless, there is a useful partial result, known as Fatou's Lemma.

Theorem 4.8 (Fatou's Lemma). *Suppose that (X, Σ, μ) is a measure space and that $f_n : X \to \mathbb{R}$ is a measurable nonnegative function for each $n \in \mathbb{N}$. If $\liminf_n f_n(x)$*

exists, for all $x \in X$, then

$$\int_X (\lim_{n\to\infty} \inf f_n)\, d\mu \leq \lim_{n\to\infty} \inf \int_X f_n\, d\mu. \tag{4.5}$$

Proof. For each $k \in \mathbb{N}$, let $g_k(x) = \inf\{f_m(x) \mid m \geq k\}$, for every $x \in X$. By Theorem 3.9, g_k and $\liminf_{n\to\infty} f_n$ are measurable. Moreover, $\{g_k\}_{k\in\mathbb{N}}$ is a monotone-increasing sequence of nonnegative functions such that

$$\lim_{k\to\infty} g_k(x) = \lim_{n\to\infty} \inf f_n(x), \quad \forall x \in X.$$

Therefore, by the Monotone Convergence Theorem,

$$\lim_{k\to\infty} \int_X g_k\, d\mu = \int_X (\lim_{n\to\infty} \inf f_n)\, d\mu.$$

On the other hand, $g_k(x) \leq f_k(x)$, for all $x \in X$, and $\{g_k\}_{k\in\mathbb{N}}$ monotone increasing implies that

$$\int_X g_k\, d\mu \leq \int_X f_m\, d\mu \quad \forall m \geq k,$$

and so

$$\lim_{k\to\infty} \int_X g_k\, d\mu \leq \lim_{n\to\infty} \inf \int_X f_n\, d\mu.$$

This completes the proof of inequality (4.5). $\qquad\qquad\qquad\qquad\qquad\qquad\square$

Corollary 4.9. *If $\{f_k\}_{k\in\mathbb{N}}$ is a sequence of nonnegative measurable functions on a measure space (X, Σ, μ) such that $\lim_k f_k(x)$ exists, for all $x \in X$, then*

$$\int_X (\lim_{k\to\infty} f_k)\, d\mu \leq \lim_{k\to\infty} \inf \int_X f_k\, d\mu.$$

Basic algebraic properties of the integral can now be investigated,

Theorem 4.10. *Suppose that (X, Σ, μ) is a measure space and that f and g are nonnegative measurable functions $X \to \mathbb{R}$. If $\alpha, \beta \in [0, \infty)$, then*

$$\int_X (\alpha f + \beta g)\, d\mu = \alpha \int_X f\, d\mu + \beta \int_X g\, d\mu.$$

Proof. By Proposition 3.14, there are monotone increasing sequences $\{\varphi_k\}_{k\in\mathbb{N}}$ and $\{\psi_k\}_{k\in\mathbb{N}}$ of nonnegative simple functions such that, for every $x \in X$,

$$\lim_{k\to\infty} \varphi_k(x) = f(x) \quad \text{and} \quad \lim_{k\to\infty} \psi_k(x) = g(x).$$

By the Monotone Convergence Theorem,

$$\int_X f \, d\mu = \lim_{k \to \infty} \int_X \varphi_k \, d\mu \quad \text{and} \quad \int_X g \, d\mu = \lim_{k \to \infty} \int_X \psi_k \, d\mu. \qquad (4.6)$$

Let $\vartheta_k = \alpha \varphi_k + \beta \psi_k$, for every $k \in \mathbb{N}$. Then $\{\vartheta_k\}_{k \in \mathbb{N}}$ is a monotone-increasing sequence of simple functions such that

$$\int_X \vartheta_k \, d\mu = \int_X \varphi_k \, d\mu + \int_X \psi_k \, d\mu$$

and

$$\lim_{k \to \infty} \vartheta_k(x) = (\alpha f + \beta g)(x), \quad \forall x \in X.$$

The Monotone Convergence Theorem yields

$$\int_X (\alpha f + \beta g) \, d\mu = \lim_{k \to \infty} \int_X \vartheta_k \, d\mu,$$

which is simplified to

$$\int_X (\alpha f + \beta g) \, d\mu = \alpha \int_X f \, d\mu + \beta \int_X g \, d\mu,$$

by the equations in (4.6). $\qquad \square$

By induction, Theorem 4.10 extends to finite linear combinations of nonnegative functions via nonnegative coefficients.

Corollary 4.11. *Suppose that (X, Σ, μ) is a measure space and that $f_1, \ldots, f_n$ are nonnegative measurable functions $X \to \mathbb{R}$. If $\alpha_1, \ldots, \alpha_n \in [0, \infty)$, then*

$$\int_X \left(\sum_{j=1}^n \alpha_j f_j \right) d\mu = \sum_{j=1}^n \int_X \alpha_j f_j \, d\mu.$$

Another very useful fact is the next result.

Proposition 4.12. *Suppose that (X, Σ, μ) is a measure space and that f is a nonnegative measurable function $X \to \mathbb{R}$. If $E_1, \ldots, E_n \in \Sigma$ are pairwise disjoint, then*

$$\int_{E_1 \cup \cdots \cup E_n} f \, d\mu = \sum_{j=1}^n \int_{E_j} f \, d\mu.$$

Proof. Let $E = E_1 \cup \cdots \cup E_n$ and note that $\chi_E f = \sum_{j=1}^{n} \chi_{E_j} f$. Hence, by Corollary 4.11,

$$\int_X \chi_E f \, d\mu = \sum_{j=1}^{n} \int_X \chi_{E_j} f \, d\mu. \text{ That is, } \int_{E_1 \cup \cdots \cup E_n} f \, d\mu = \sum_{j=1}^{n} \int_{E_j} f \, d\mu. \qquad \square$$

Corollary 4.13. *Suppose that* (X, Σ, μ) *is a measure space and that* f *is a nonnegative measurable function* $X \to \mathbb{R}$. *If* $E \in \Sigma$ *is such that* $\mu(E) = 0$, *then*

$$\int_E f \, d\mu = 0 \quad \text{and} \quad \int_X f \, d\mu = \int_{X \setminus E} f \, d\mu.$$

Proof. Because $\mu(E) = 0$, $\mu(E \cap F) = 0$ for every $F \in \Sigma$. Hence, if φ is a simple function in canonical form $\varphi = \sum_{k=1}^{n} \alpha_k \chi_{E_k}$, then

$$\int_E \varphi \, d\mu = \sum_{k=1}^{n} \alpha_k \mu(E \cap E_k) = 0.$$

Thus, by the definition of integral of f as the supremum of a set of integrals of simple functions, one concludes immediately that

$$\int_E f \, d\mu = 0.$$

This fact, together with Theorem 4.12, shows that

$$\int_X f \, d\mu = \int_{X \setminus E} f \, d\mu + \int_E f \, d\mu = \int_{X \setminus E} f \, d\mu,$$

which completes the proof. $\square$

4.2   Integrable Functions

Definition 4.14. If X is a set, and if $f : X \to \mathbb{R}$ is a function, then the *Jordan decomposition of* f is the equation

$$f = f^+ - f^-, \tag{4.7}$$

where

$$f^+ = \frac{|f| + f}{2} \quad \text{and} \quad f^- = \frac{|f| - f}{2}. \tag{4.8}$$

Observe that f^+ and f^- are nonnegative functions such that

$$f^+ f^- = 0 \text{ and } f^+ + f^- = |f|.$$

Moreover, if (X, Σ) is a measurable space, and if $f : X \to \mathbb{R}$ is a measurable function, then f^+ and f^- are measurable functions.

Using the Jordan decomposition, the integral of any measurable function can now be defined.

Definition 4.15. If (X, Σ, μ) is a measure space, and if $f : X \to \mathbb{R}$ is a measurable function, then the *Lebesgue integral*, or simply the *integral*, of f is the quantity in the extended real number system $[-\infty, \infty]$ denoted by $\int_X f \, d\mu$ and defined by

$$\int_X f \, d\mu = \int_X f^+ \, d\mu - \int_X f^- \, d\mu,$$

where $f = f^+ - f^-$ is the Jordan decomposition of f.

Our interest is with the class of functions for which $\int_X f^+ \, d\mu - \int_X f^- \, d\mu$ is a (finite) real number.

Definition 4.16. If (X, Σ, μ) is a measure space, and if $f : X \to \mathbb{R}$ is a measurable function, then f is an *integrable function* if

$$\int_X f^+ \, d\mu < \infty \text{ and } \int_X f^- \, d\mu < \infty.$$

Proposition 4.17. *Assume that (X, Σ, μ) is a measure space and $f, g : X \to \mathbb{R}$ are integrable functions. For all $\alpha, \beta \in \mathbb{R}$, the function $\alpha f + \beta g$ is integrable, and*

$$\int_X (\alpha f + \beta g) \, d\mu = \alpha \int_X f \, d\mu + \beta \int_X g \, d\mu.$$

Proof. Let $h_1, h_2 : X \to \mathbb{R}$ be nonnegative, integrable functions, and let $h = h_1 - h_2$. It may be that $h^+ \neq h_1$ and $h^- \neq h_2$; nevertheless, it is true that

$$\int_X h \, d\mu = \int_X h_1 \, d\mu - \int_X h_2 \, d\mu. \tag{4.9}$$

To prove this, express h in its Jordan decomposition: $h = h^+ - h^-$. Because $h^+ - h^- = h_1 - h_2$, we have that $h^+ + h_2 = h^- + h_1$. By additivity of the integral for nonnegative functions (Theorem 4.10),

$$\int_X h^+ \, d\mu + \int_X h_2 \, d\mu = \int_X h^- \, d\mu + \int_X h_1 \, d\mu.$$

As all four integrals above are finite, by the hypothesis that h, h_1, and h_2 are integrable,

$$\int_X h^+ d\mu - \int_X h^- d\mu = \int_X h_1 d\mu - \int_X h_2 d\mu,$$

which proves equation (4.9).

The argument above applies to $f + g$ via

$$f + g = (f^+ + g^+) - (f^- + g^-).$$

(This need not be the Jordan decomposition $(f + g)^+ - (f + g)^-$.) Lastly, now that we know that

$$\int_X (f + g) d\mu = \int_X f d\mu + \int_X g d\mu,$$

the case of $\alpha f + \beta g$ is handled by expressing each scalar in its Jordan decomposition: $\alpha = \alpha^+ - \alpha^-$ and $\beta = \beta^+ - \beta^-$. $\square$

By Proposition 4.17, the set of all integrable functions has the structure of a vector space.

Corollary 4.18. *If (X, Σ, μ) is a measure space, then the set $\mathscr{L}^1_{\mathbb{R}}(X, \Sigma, \mu)$ of all integrable functions $f : X \to \mathbb{R}$ is a vector space over the field $\mathbb{R}$, and the map $f \mapsto \int_X f d\mu$ is a linear transformation of $\mathscr{L}^1_{\mathbb{R}}(X, \Sigma, \mu)$ onto $\mathbb{R}$.*

The next result gives a characterisation of integrability in terms of the absolute-value function.

Proposition 4.19. *Assume that (X, Σ, μ) is a measure space and $f : X \to \mathbb{R}$ is a measurable function. Then f is integrable if and only if $|f|$ is integrable. Moreover, if f is integrable, then the following triangle inequality holds:*

$$\left| \int_X f d\mu \right| \leq \int_X |f| d\mu.$$

Proof. Assume that f is integrable. Thus, $\int_X f^+ d\mu$ and $\int_X f^- d\mu$ are finite. Because $|f| = f^+ + f^-$, Theorem 4.17 shows that

$$\int_X |f| d\mu = \int_X (f^+ + f^-) d\mu = \int_X f^+ d\mu + \int_X f^- d\mu < \infty.$$

That is, $|f|$ is integrable.

Conversely, assume that $|f|$ is integrable. The equation $|f| = f^+ + f^-$ yields $f^+(x) \leq |f|(x)$ and $f^-(x) \leq |f|(x)$, for all $x \in X$. Thus,

$$\int_X f^+ \, d\mu \leq \int_X |f| \, d\mu < \infty \quad \text{and} \quad \int_X f^- \, d\mu \leq \int_X |f| \, d\mu < \infty.$$

Hence, f is integrable.

To prove the triangle inequality, assume that f is integrable. The triangle inequality in real numbers α and β is $|\alpha + \beta| \leq |\alpha| + |\beta|$. Applying this to integrals leads to

$$\left| \int_X f \, d\mu \right| = \left| \int_X f^+ \, d\mu - \int_X f^- \, d\mu \right| \leq \left| \int_X f^+ \, d\mu \right| + \left| \int_X f^- \, d\mu \right| \leq \int_X |f| \, d\mu,$$

which completes the proof. $\square$

Proposition 4.19 above demonstrates the integrability condition is similar to that of absolute convergence in the theory of series. Indeed, if one considers the measure space $(\mathbb{N}, \mathscr{P}(\mathbb{N}), \mu)$, where μ is the counting measure, then a function $f : \mathbb{N} \to \mathbb{R}$ has Lebesgue integral

$$\int_{\mathbb{N}} f \, d\mu = \sum_{k=1}^{\infty} f(k),$$

and so f is integrable if and only if

$$\sum_{k=1}^{\infty} |f(k)| < \infty.$$

Theorem 4.20 (Dominated Convergence Theorem). *Suppose that (X, Σ, μ) is a measure space, and that $\{f_k\}_{k \in \mathbb{N}}$ is a sequence of measurable functions $f_k : X \to \mathbb{R}$ such that $\lim_{k \to \infty} f_k(x)$ exists, for all $x \in X$. If there is a nonnegative integrable function $g : X \to \mathbb{R}$ such that $|f_k(x)| \leq g(x)$, for every $k \in \mathbb{N}$ and $x \in X$, and if $f : X \to \mathbb{R}$ is the measurable function defined by $f(x) = \lim_{k \to \infty} f_k(x)$, for every $x \in X$, then f is integrable, and*

$$\lim_{k \to \infty} \int_X f_k \, d\mu = \int_X f \, d\mu \quad \text{and} \quad \lim_{k \to \infty} \int_X |f_k - f| \, d\mu = 0.$$

Proof. Because $f(x) = \lim_k f_k(x)$, for all $x \in X$, f is measurable and $|f(x)| \leq g(x)$ for every $x \in X$. Thus,

$$\int_X |f| \, d\mu \leq \int_X g \, d\mu < \infty,$$

which implies that f is integrable. For each $k \in \mathbb{N}$, let $h_k = 2g - |f_k - f|$. Note that the triangle inequality in real numbers gives $|f_k - f|(x) \leq 2g(x)$, and so the functions h_k are nonnegative. Moreover, $\lim_k |f_k - f|(x) = 0$; thus, $\lim_k h_k = 2g$. Fatou's Lemma yields

$$
\begin{aligned}
\int_X 2g \, d\mu &= \int_X (\lim_{k \to \infty} \inf h_k) \, d\mu \\
&\leq \lim_{k \to \infty} \inf \int_X h_k \, d\mu \\
&= \int_X 2g \, d\mu + \lim_{k \to \infty} \inf \left(- \int_X |f_k - f| \, d\mu \right) \\
&= \int_X 2g \, d\mu - \lim_{k \to \infty} \sup \int_X |f_k - f| \, d\mu \, .
\end{aligned}
$$

Therefore,

$$
0 \leq \lim_{k \to \infty} \sup \int_X |f_k - f| \, d\mu \leq 0. \tag{4.10}
$$

If a sequence $\{\alpha_k\}_k$ of positive numbers does not converge to 0, then it must necessarily be that $\lim \sup_k \alpha_k > 0$. Thus, inequality (4.10) implies that

$$
0 \leq \lim_{k \to \infty} \int_X |f_k - f| \, d\mu \leq 0.
$$

For each $k \in \mathbb{N}$,

$$
\left| \int_X f_k \, d\mu - \int_X f \, d\mu \right| \leq \int_X |f_k - f| \, d\mu \, .
$$

Hence, as $k \to \infty$ we obtain

$$
\lim_{k \to \infty} \int_X f_k \, d\mu = \int_X f \, d\mu \, ,
$$

thereby completing the proof.   □

Some of the convergence results in integration can be reformulated for convergence on complements of sets of measure zero.

Definition 4.21. If $f, g : X \to \mathbb{R}$ are measurable functions on a measure space (X, Σ, μ), then $f = g$ *almost everywhere* if

$$
\mu \left(\{ x \in X \, | \, f(x) \neq g(x) \} \right) = 0.
$$

That is, f and g are equal almost everywhere if the set on which they are not equal is a set of measure zero. This is of significance in integration because of Corollary 4.13, which states that if (X, Σ, μ) is a measure space and if f is a nonnegative measurable function $X \to \mathbb{R}$, then, for any $E \in \Sigma$ such that $\mu(E) = 0$,

$$\int_E f \, d\mu = 0 \quad \text{and} \quad \int_X f \, d\mu = \int_{X \setminus E} f \, d\mu .$$

By passing to differences of positive functions, the line above extends to any real-valued integrable function.

The point is this: sets of measure zero have no role in the value of the integral. That is, $\int_X f \, d\mu = \int_X g \, d\mu$, if $f = g$ almost everywhere.

Definition 4.22. If (X, Σ, μ) is a measure space and if $f_k : X \to \mathbb{R}$ is measurable function, for all $k \in \mathbb{N}$, then the sequence $\{f_k\}_{k \in \mathbb{N}}$ *converges almost everywhere* if

$$\mu \left(\{ x \in X \mid \lim_{k \to \infty} f_k(x) \text{ does not exist} \} \right) = 0 .$$

The following result, which is a partial converse to the Monotone Convergence Theorem, demonstrates one way in which almost everywhere convergence can arise.

Proposition 4.23. *Suppose that (X, Σ, μ) is a measure space, and that $\{f_k\}_{k \in \mathbb{N}}$ is a monotone increasing sequence of nonnegative integrable functions $f_k : X \to \mathbb{R}$. If $\lim_{k \to \infty} \int_X f_k \, d\mu$ exists, then there is a measurable set $D \subseteq X$ such that $\mu(D) = 0$, $\lim_{k \to \infty} f_k(x)$ exists for all $x \in X \setminus D$, and*

$$\lim_{k \to \infty} \int_X f_k \, d\mu = \int_{X \setminus D} (\lim_{k \to \infty} f_k) \, d\mu .$$

Proof. Because $\{f_k\}_{k \in \mathbb{N}}$ is a monotone-increasing sequence,

$$\int_X f_k \, d\mu \leq \int_X f_{k+1} \, d\mu ,$$

for every $k \in \mathbb{N}$. The hypothesis implies the existence of a real number $W > 0$ such that $\int_X f_k \, d\mu \leq W$ for every k.

Let $D = \{ x \in X \mid \lim_k f_k(x) \text{ does not exist} \}$. Then $D = X \setminus L$, where L is the set of points $x \in X$ for which $\lim_k f_k(x)$ exists. By Theorem 3.9, $L \in \Sigma$ and, hence, $D \in \Sigma$. Let $\varepsilon > 0$ be arbitrary and define, for each $k \in \mathbb{N}$, $D_k = \{ x \in X \mid f_k(x) > W/\varepsilon \}$. Because the sequence $\{f_k\}_k$ is monotone increasing, so is the sequence $\{D_k\}_k$. If $x \in D$, then there is a $k_0 \in \mathbb{N}$ such that $f_{k_0}(x) > W/\varepsilon$, and so $x \in D_{k_0}$. Hence,

$$D \subseteq \bigcup_{k \in \mathbb{N}} D_k . \tag{4.11}$$

Therefore, the continuity of the measure μ (Proposition 3.22) applied to the inclusion (4.11) yields

$$\mu(D) \leq \mu\left(\bigcup_{k\in\mathbb{N}} D_k\right) = \lim_{k\to\infty}\mu(D_k).$$

But, for each $k \in \mathbb{N}$,

$$\mu(D_k) = \int_X \chi_{D_k}\, d\mu \leq \frac{\varepsilon}{W}\int_X \chi_{D_k} f_k\, d\mu \leq \frac{\varepsilon}{W}\int_X f_k\, d\mu \leq \varepsilon.$$

Thus, $\mu(D) = 0$.

Let $f : X \to \mathbb{R}$ be given by $f(x) = 0$ for $x \in D$ and $f(x) = \lim_{k\to\infty} f_k(x)$ for $x \notin D$. The sequence $\{\chi_{X\setminus D} f_k\}_{k\in\mathbb{N}}$ is monotone increasing and $f = \lim_k \chi_{X\setminus D} f_k$. Thus, f is measurable and, by the Monotone Convergence Theorem,

$$\lim_{k\to\infty}\int_X \chi_{X\setminus D} f_k\, d\mu = \int_X f\, d\mu.$$

For every nonnegative measurable function g, $\int_D g\, d\mu = 0$, because $\mu(D) = 0$. Thus,

$$\lim_{k\to\infty}\int_X f_k\, d\mu = \lim_{k\to\infty}\left(\int_{X\setminus D} f_k\, d\mu + \int_D f_k\, d\mu\right) = \lim_{k\to\infty}\int_{X\setminus D} f_k\, d\mu$$

$$= \lim_{k\to\infty}\int_X \chi_{X\setminus D} f_k\, d\mu = \int_X f\, d\mu = \int_{X\setminus D}(\lim_{k\to\infty} f_k)\, d\mu,$$

which completes the proof.   □

It is somewhat clumsy to make explicit reference to the set D of measure zero in the integral

$$\int_{X\setminus D}(\lim_{k\to\infty} f_k)\, d\mu,$$

especially as null sets do not contribute to the value of the integral. Therefore, a notational convention is adopted: if $\{f_k\}_{k\in\mathbb{N}}$ converges almost everywhere, then

$$\int_X (\lim_{k\to\infty} f_k)\, d\mu$$

is understood to represent the quantity

$$\int_{X\setminus D}(\lim_{k\to\infty} f_k)\, d\mu,$$

where $D \subset X$ is the set (of measure zero) of points for which $\{f_k(x)\}_{k\in\mathbb{N}}$ has no limit.

With this notational convention, the Monotone Convergence Theorem and the Dominated Convergence Theorem have versions for almost-everywhere convergence. For example, the Dominated Convergence Theorem is formulated as follows.

Theorem 4.24. *Suppose that (X, Σ, μ) is a measure space, and that $\{f_k\}_{k \in \mathbb{N}}$ is a sequence of integrable functions such that $\lim\limits_{k \to \infty} f_k(x)$ exists almost everywhere. If there is a nonnegative integrable function $g : X \to \mathbb{R}$ such that, for every $k \in \mathbb{N}$ and $x \in X$,*

$$|f_k(x)| \le g(x),$$

then

$$\lim_{k \to \infty} \int_X f_k \, d\mu = \int_X \left(\lim_{k \to \infty} f_k \right) d\mu.$$

Proof. Exercise 4.70.   □

4.3   Complex-Valued Functions and Measures

The goal of this section is to outline how integration of real-valued functions extends to integration of complex-valued functions by using the real and imaginary parts of a complex-valued function. The main item of note is that the Dominated Convergence Theorem extends to complex-valued functions.

The role of signed and complex measures in integration theory is less prominent than the role played ordinary measures. However, our study of Banach space duality will require integration to be extended so as to encompass both complex-valued functions and complex measures. Therefore, this section concludes with a brief description of how such an extension is formulated and achieved.

Definition 4.25. If (X, Σ) is a measurable space, then a function $f : X \to \mathbb{C}$ is said to be *complex measurable* if $f^{-1}(U) \in \Sigma$, for all open subsets $U \subseteq \mathbb{C}$.

If $f : X \to \mathbb{C}$, then let $\bar{f}$ and $|f|$ denote the functions defined by $\bar{f}(x) = \overline{f(x)}$ and $|f|(x) = |f(x)|$, for all $x \in X$. As with complex numbers, a function f can be expressed in its real and imaginary parts:

$$f = \Re f + i \Im f, \quad \text{where} \quad \Re f = \frac{1}{2}(f + \bar{f}) \text{ and } \Im f = \frac{1}{2i}(f - \bar{f}).$$

Note that $\Re f$ and $\Im f$ are real-valued functions on X.

Proposition 4.26. *If (X, Σ) is a measurable space, then a function $f : X \to \mathbb{C}$ is complex measurable if and only if $\Re f$ and $\Im f$ are measurable functions $X \to \mathbb{R}$.*

Proof. Exercise 4.71.   □

Definition 4.27. If (X, Σ, μ) is a measure space, then $f : X \to \mathbb{C}$ is *integrable* if $\Re f$ and $\Im f$ are integrable functions $X \to \mathbb{R}$, and the *integral* of f is defined by

$$\int_X f\, d\mu = \int_X \Re f\, d\mu + i \int_X \Im f\, d\mu.$$

Note that $\Re f$ and $\Im f$ are integrable only if $\Re f$ and $\Im f$ are measurable. In such cases, f is automatically complex measurable (by Proposition 4.26), and this is why complex measurability is not mentioned in the definition of integrability.

Proposition 4.28. *If (X, Σ, μ) is a measure space, then $f : X \to \mathbb{C}$ is integrable if and only if $|f|$ is integrable, in which case*

$$\left| \int_X f\, d\mu \right| \leq \int_X |f|\, d\mu.$$

Proof. Write f in terms of its real and imaginary parts: $f = \Re f + i\Im f$. If f is integrable, then each of $\Re f$ and $\Im f$ is integrable (by definition). Hence, $|\Re f|$ and $|\Im f|$ are integrable. Because $|f| \leq |\Re f| + |\Im f|$, one concludes that $|f|$ is integrable. Conversely, assume that $|f|$ is integrable. Since $|\Re f| \leq |f|$ and $|\Im f| \leq |f|$, both $|\Re f|$ and $|\Im f|$ are integrable. Thus, f is integrable.

To show the triangle inequality, assume that f is integrable. For any complex number ζ, a complex number ω satisfies $|\omega| \leq |\zeta|$ if and only if $e^{i\theta} \Re \omega \leq |\zeta|$ for all $\theta \in [0, 2\pi]$. Note that $e^{i\theta} \Re f = \Re(e^{i\theta} f) \leq |e^{i\theta} f|$. Thus,

$$e^{i\theta} \Re \left(\int_X f\, d\mu \right) = \int_X \Re(e^{i\theta} f)\, d\mu \leq \int_X |e^{i\theta} f|\, d\mu = \int_X |f|\, d\mu,$$

for all $\theta \in [0, 2\pi]$. Hence, $\left| \int_X f\, d\mu \right| \leq \int_X |f|\, d\mu.$ □

Finally, the Dominated Convergence Theorem extends to complex-valued functions.

Theorem 4.29 (Dominated Convergence Theorem: Complex Case). *Suppose that (X, Σ, μ) is a measure space, and that $\{f_k\}_{k \in \mathbb{N}}$ is a sequence of measurable functions $f_k : X \to \mathbb{C}$ such that $\lim_{k \to \infty} f_k(x)$ exists, for all $x \in X$. If there is a nonnegative integrable function $g : X \to \mathbb{R}$ such that $|f_k(x)| \leq g(x)$, for every $k \in \mathbb{N}$ and $x \in X$, and if $f : X \to \mathbb{C}$ is the measurable function defined by $f(x) = \lim_{k \to \infty} f_k(x)$, for every $x \in X$, then f is integrable, and*

$$\lim_{k \to \infty} \int_X f_k\, d\mu = \int_X f\, d\mu \text{ and } \lim_{k \to \infty} \int_X |f_k - f|\, d\mu = 0.$$

If ω is a signed measure on a measurable space (X, Σ), then expressing ω in its Jordan Decomposition (see Theorem 3.66) as a difference of two measures, ω_+ and ω_-, allows one to define the integral of a measurable function $f : X \to \mathbb{C}$ with respect to the signed measure ω by

$$\int_X f\, d\omega = \int_X f\, d\omega_+ - \int_X f\, d\omega_-,$$

as each of $\int_X f\, d\omega_+$ and $\int_X f\, d\omega_-$ is well defined.

Definition 4.30. If ω is a signed measure on (X, Σ) with Jordan decomposition $\omega = \omega_+ - \omega_-$, then a measurable function $f : X \to \mathbb{C}$ is *integrable* if

$$\int_X |f|\, d\omega_+ \text{ and } \int_X |f|\, d\omega_-$$

are finite.

In defining $|\omega|$ by $|\omega| = \omega_+ + \omega_-$, we obtain a measure $|\omega|$ on (X, Σ) and for each integrable function f, we have a triangle inequality:

$$\left| \int_X f\, d\omega \right| = \left| \int_X f\, d\omega_+ - \int_X f\, d\omega_- \right| \le \int_X |f|\, d\omega_+ + \int_X |f|\, d\omega_- = \int_X |f|\, d|\omega|.$$

One can approach integration with respect to a complex measure ν in a similar way, by first considering the signed measures $\Re\nu$ and $\Im\nu$ induced by the real and imaginary parts of ν. In so doing, two finite signed measures are obtained, each of which is a difference of finite measures. Hence, there are finite measures μ_j on (X, Σ), for $j = 1, \ldots, 4$, such that

$$\nu = (\mu_1 - \mu_2) + i(\mu_3 - \mu_4).$$

The integral of a measurable function $f : X \to \mathbb{C}$ with respect to the complex measure ν is defined by

$$\int_X f\, d\nu = \int_X f\, d(\mu_1 - \mu_2) + i \int_X f\, d(\mu_3 - \mu_4).$$

Definition 4.31. If ν is a complex measure on (X, Σ), expressed as $\nu = (\mu_1 - \mu_2) + i(\mu_3 - \mu_4)$, where $\mu_1 - \mu_2$ and $\mu_3 - \mu_4$ are Jordan decompositions of the real and imaginary parts of ν, respectively, then a measurable function $f : X \to \mathbb{C}$ is *integrable* if

$$\int_X |f|\, d\mu_j < \infty$$

for each $j = 1, \ldots, 4$.

A version of the triangle inequality is noted in Exercise 4.72.

We will not have any significant need of such integrals in this book, except for in the discussion of duality, where the use of the integral $\int_X f\,dv$ with respect to a complex measure v is convenient.

4.4 Continuity

A starting point for our discussion of continuity is the following important observation:

Proposition 4.32. *If (X, Σ, μ) is a measurable space and $g : X \to \mathbb{R}$ is a nonnegative measurable function, then the function $v_g : \Sigma \to \mathbb{R}$ defined by*

$$v_g(E) = \int_E g\,d\mu, \ E \in \Sigma,$$

is a measure on (X, Σ).

Proof. Exercise 4.73.   □

Proposition 4.32 shows that there are an abundance of measures one can introduce on a measurable space, and that finite measures v_g are achieved from using nonnegative integrable functions g, even if μ itself is not finite. In this case, as shown by the next result, the pair of measures μ and v_g exhibit a continuity feature (see Proposition 4.37 as well).

Proposition 4.33. *If $f : X \to \mathbb{C}$ is an integrable function on a measure space (X, Σ, μ), then for every $\varepsilon > 0$ there exists a $\delta > 0$ such that $\int_E |f|\,d\mu < \varepsilon$ for all measurable sets $E \subseteq X$ with $\mu(E) < \delta$.*

Proof. For each $n \in \mathbb{N}$, let $E_n = \{x \in X \mid |f(x)| < n\}$ and define $g_n : X \to \mathbb{R}$ by

$$g_n = |f|\chi_{E_n} + n\chi_{E_n^c}.$$

Thus, $\{g_n\}_{n\in\mathbb{N}}$ is a monotone increasing sequence of nonnegative measurable functions with $\lim_n g_n(x) = |f(x)|$ for every $x \in X$. Therefore, by the Monotone Convergence Theorem, $\{\int_X g_n\,d\mu\}_{n\in\mathbb{N}}$ is a monotone increasing sequence in $\mathbb{R}$ with limit $\int_X |f|\,d\mu$. Hence, if $\varepsilon > 0$, then there is a $K \in \mathbb{N}$ such that

$$\int_X |f|\,d\mu < \int_X g_K\,d\mu + \frac{\varepsilon}{3}.$$

Thus,

$$\int_X |f|\,d\mu < \left(\int_{E_K} g_K\,d\mu + \int_{E_K^c} g_K\,d\mu \right) + \frac{\varepsilon}{3},$$

which implies that

$$\int_E |f|\,d\mu < \left(\int_{E\cap E_K} g_K\,d\mu + \int_{E\cap E_K^C} g_K\,d\mu\right) + \frac{\varepsilon}{3} \le K\mu(E) + K\mu(E) + \frac{\varepsilon}{3}.$$

Therefore, if $\delta = \varepsilon/(3K)$ and $\mu(E) < \delta$, then $\int_E |f|\,d\mu < \varepsilon$. $\qquad\square$

For integrable function f on a measure space (X, Σ, μ), the induced measure $\nu_{|f|}$ on (X, Σ) is finite and has the continuity property exhibited in Proposition 4.33 above. This property of a pair of such measures can be characterised in the abstract in the following way.

Proposition 4.34. *The following statements are equivalent for measures μ and ν, where ν is finite, on a measurable space (X, Σ):*

1. *for every $\varepsilon > 0$ there exists a $\delta > 0$ such that $\nu(E) < \varepsilon$ for each $E \in \Sigma$ that satisfies $\mu(E) < \delta$;*
2. *$\nu(E) = 0$ for each $E \in \Sigma$ that satisfies $\mu(E) = 0$.*

Proof. The proof of (1) $\Rightarrow$ (2) is left as an exercise. To prove the converse, we shall prove the contrapositive: if there exists a $\varepsilon > 0$ such that for each $\delta > 0$ there exists a set $E_\delta \in \Sigma$ with $\mu(E_\delta) < \delta$ and $\nu(E_\delta) \ge \varepsilon$, then there is also a set $E \in \Sigma$ for which $\mu(E) = 0$ and $\nu(E) > 0$. To this end, assume such a $\varepsilon > 0$ exists and for each $k \in \mathbb{N}$ let $E_k \in \Sigma$ satisfy $\mu(E_k) < 2^{-k}$ and $\nu(E_k) \ge \varepsilon$. For every $n \in \mathbb{N}$, let $A_n = \bigcup_{k \ge n} E_k$ so that $\{A_n\}_{n\in\mathbb{N}}$ is a descending sequence of measurable sets. Observe that

$$\mu(A_n) \le \sum_{k=n}^{\infty} \mu(E_k) \le \sum_{k=n}^{\infty} 2^{-k} = 2^{-n-1} < 2^{-n} \quad \text{and} \quad \nu(A_n) \ge \nu(E_n) \ge \varepsilon. \quad \text{In particular,}$$

$\mu(A_1) < \infty$ and, by hypothesis, $\nu(A_1) < \infty$. Therefore, if $E = \bigcap_{n\in\mathbb{N}} A_n$, then

$$\mu(E) = \lim_{n\to\infty} \mu(A_n) \le \lim_{n\to\infty} 2^{-n} = 0 \quad \text{and} \quad \nu(E) = \lim_{n\to\infty} \nu(A_n) \ge \varepsilon,$$

by continuity of measure (Proposition 3.22). $\qquad\square$

The second of the equivalent conditions in Proposition 4.34 is called absolute continuity, the relevance of which will be explained by the Radon-Nikodým Theorem.

Definition 4.35. If (X, Σ) is a measurable space and if μ and ν are measures on (X, Σ), then ν is *absolutely continuous* with respect to μ if $\nu(E) = 0$ for every $E \in \Sigma$ for which $\mu(E) = 0$.

Theorem 4.36 (Radon-Nikodým). *Suppose that μ and ν are finite measures on a measurable space (X, Σ). If ν is absolutely continuous with respect to μ, then there is a nonnegative measurable function $g : X \to \mathbb{R}$ such that*

$$\nu(E) = \int_E g\,d\mu, \quad \forall E \in \Sigma.$$

Proof. Let $\mathscr{G}$ be the set of all measurable functions $f : X \to [0,\infty)$ for which $\int_E f\,d\mu \le \nu(E)$ for every $E \in \Sigma$. The set $\mathscr{G}$ is nonempty (because $0 \in \mathscr{G}$) and $\max(f_1,f_2) \in \mathscr{G}$ for all $f_1,f_2 \in \mathscr{G}$. Because $\nu(X) < \infty$, the supremum α of the set $\{\int_X f\,d\mu \,|\, f \in \mathscr{G}\}$ exists. In particular, for each $n \in \mathbb{N}$ there exists $f_n \in \mathscr{G}$ such that $\int_X f_n\,d\mu > \alpha - \frac{1}{n}$. Define a sequence $\{g_n\}_{n\in\mathbb{N}}$ by $g_n = \max(f_1,\dots,f_n)$ and note that $\{g_n\}_{n\in\mathbb{N}}$ is monotone increasing and $\lim_{n\to\infty} \int_X g_n\,d\mu = \alpha$. Therefore, by Proposition 4.23, there exists a measurable function $g : X \to [0,\infty]$ such that $g(x) = \lim_n g_n(x)$ for almost every $x \in X$ and $\int_X g\,d\mu = \alpha$.

Because $\int_E g\,d\mu = \sup_n \int_E g_n\,d\mu \le \nu(E)$ for every $E \in \Sigma$, the function $\tilde{\nu} : \Sigma \to \mathbb{R}$ defined by

$$\tilde{\nu}(E) = \nu(E) - \int_E g\,d\mu$$

is a finite measure. We aim to show that $\tilde{\nu}$ is identically zero. Therefore, assume that $\tilde{\nu}$ is not the zero measure. Thus, $\tilde{\nu}(X) > 0$. Fix $\varepsilon > 0$ such that $\tilde{\nu}(X) > \varepsilon\mu(X)$ and let $\omega = \tilde{\nu} - \varepsilon\mu$. Because ω is a difference of finite measures, ω is a signed measure. Furthermore, $\omega(X) > 0$.

The Hahn Decomposition (Theorem 3.65) asserts that there exist $P,N \in \Sigma$ such that P is positive with respect to ω, N is negative with respect to ω, $P \cap N = \emptyset$, and $X = P \cup N$. Thus, for $E \in \Sigma$, $0 \le \omega(E \cap P)$ implies that $\varepsilon\mu(E \cap P) \le \tilde{\nu}(E \cap P)$. Hence,

$$\nu(E) = \tilde{\nu}(E) + \int_E g\,d\mu \ge \varepsilon\tilde{\nu}(E \cap P) + \int_E g\,d\mu$$

$$\ge \varepsilon\tilde{\nu}(E \cap P) + \int_E g\,d\mu = \int_E (g + \varepsilon\chi_P)\,d\mu.$$

The inequality above shows that $(g + \varepsilon\chi_P) \in \mathscr{G}$. However,

$$\int_X (g + \varepsilon\chi_P)\,d\mu > \int_X g\,d\mu$$

contradicts the fact that $\int_X g\,d\mu = \alpha$. Therefore, it must be that $\tilde{\nu}(E) = 0$ for every $E \in \Sigma$. $\qquad\square$

4.5   Fundamental Theorem of Calculus

As in calculus, Lebesgue integrals display certain continuity and differential properties that facilitate calculations. The first of these properties is the following familiar continuity feature.

Proposition 4.37. *If $f : [a,b] \to \mathbb{C}$ is integrable with respect to Lebesgue measure, then the function $F : [a,b] \to \mathbb{C}$ defined by*

$$F(x) = \int_{[a,x]} f \, dm,$$

for $x \in [a,b]$, is continuous on $[a,b]$.

Proof. Choose $x_0 \in [a,b]$ and let $\varepsilon > 0$. By Proposition 4.33, there exists a $\delta > 0$ such that $\int_E |f| \, dm < \varepsilon$ if $E \subseteq [a,b]$ is a measurable set with $m(E) < \delta$. If $x \in [a,b]$ satisfies $|x - x_0| < \delta$ and if $x_0 \leq x$, then

$$|F(x) - F(x_0)| = \left| \int_{[a,x]} f \, dm - \int_{[a,x_0]} f \, dm \right| = \left| \int_{[x_0,x]} f \, dm \right| < \int_{[x_0,x]} |f| \, dm < \varepsilon.$$

The same inequality holds if $x \leq x_0$ by replacing $[x_0,x]$ with $[x,x_0]$. Hence, for each $\varepsilon > 0$, there exists $\delta > 0$ such that $|F(x) - F(x_0)| < \varepsilon$ whenever $x \in [a,b]$ satisfies $|x - x_0| < \delta$.   □

Continuity of the integrand at a point leads to differentiability of the integral at that same point.

Proposition 4.38. *If $f : [a,b] \to \mathbb{R}$ is integrable with respect to Lebesgue measure, and if f is continuous at $x_0 \in (a,b)$, then the function $F : [a,b] \to \mathbb{R}$ defined by*

$$F(x) = \int_{[a,x]} f \, dm,$$

for $x \in [a,b]$, is differentiable at x_0 and

$$\frac{dF}{dx}(x_0) = f(x_0).$$

Proof. Let $\varepsilon > 0$. The continuity of f at x_0 implies the existence of a $\delta > 0$ with the properties that $(x_0 - \delta, x_0 + \delta) \subset (a,b)$, and that $|f(x) - f(x_0)| < \varepsilon$, for all $|x - x_0| < \delta$. If $\delta > h > 0$ (the case where h is negative is similar), then

$$|F(x_0+h) - F(x_0) - hf(x_0)| = \left| \int_{[x_0,x_0+h]} f\,dm - hf(x_0) \right|$$

$$= \left| \int_{[x_0,x_0+h]} f\,dm - \int_{[x_0,x_0+h]} f(x_0)\,dm \right|$$

$$\leq \int_{[x_0,x_0+h]} |f - f(x_0)|\,dm$$

$$< \varepsilon h.$$

(For the final inequality: $|x - x_0| \leq h < \delta$ implies that $|f(x) - f(x_0)| < \varepsilon$.) Dividing the inequalities above by h yields shows that each $\varepsilon > 0$ there is a $\delta > 0$ such that

$$|h| < \delta \quad \Longrightarrow \quad \left| \frac{F(x_0+h) - F(x_0)}{h} - f(x_0) \right| < \varepsilon.$$

That is, F is differentiable at x_0 and $\dfrac{dF}{dx}(x_0) = f(x_0)$. $\square$

The results above allow for the calculation of integrals in the style of Cauchy, via antiderivatives.

Theorem 4.39 (Fundamental Theorem of Calculus). *If $F : [a,b] \to \mathbb{R}$ is differentiable on (a,b), and if dF/dx is integrable and continuous on (a,b), then*

$$\int_{[a,\xi]} \frac{dF}{dx}\,dm = F(\xi) - F(a), \tag{4.12}$$

for every $\xi \in [a,b]$.

Proof. Let $G : [a,b] \to \mathbb{R}$ be the function

$$G(\xi) = \int_{[a,\xi]} \frac{dF}{dx}\,dm,$$

for $\xi \in [a,b]$. Note that $G(a) = 0$ and that $\frac{dG}{dx} = \frac{dF}{dx}$ at every point of (a,b) (by the hypothesis that $\frac{dF}{dx}$ is continuous on (a,b) and by Proposition 4.38). Let $H = G - F$ and fix $\xi \in (a,b)$. By the Mean Value Theorem in differential calculus, there is a point $x_0 \in (a,b)$ such that

$$\frac{H(\xi) - H(a)}{\xi - a} = \frac{dH}{dx}(x_0).$$

However, $\frac{dH}{dx} = \frac{dG}{dx} - \frac{dF}{dx}$ implies that $\frac{dH}{dx}$ is identically zero on (a,b). Thus,

$$0 = H(\xi) - H(a) = G(\xi) - F(\xi) - (G(a) - F(a))$$

$$= G(\xi) - F(\xi) + F(a)$$

$$= -F(\xi) + F(a) + \int_{[a,\xi]} \frac{dF}{dx} \, dm.$$

That is, equation (4.12) holds.   □

Corollary 4.40. *If* $F : [a,b] \to \mathbb{R}$ *is differentiable on* (a,b) *and if* dF/dx *is continuous on* (a,b), *then*

$$\int_{[a,b]} \frac{dF}{dx} \, dm = F(b) - F(a). \tag{4.13}$$

Moreover, if $f : [a,b] \to \mathbb{R}$ *is continuous, then*

$$\int_{[a,b]} f \, dm = \int_a^b f(x) \, dx, \tag{4.14}$$

where the integral on the right is the Cauchy–Riemann integral.

Proof. Equation (4.13) follows from Theorem 4.39.

Assume now that f is continuous. Define F by

$$F(x) = \int_{[a,x]} f \, dm, \quad x \in [a,b].$$

Then F is differentiable on (a,b) and $dF/dx = f$ (by Proposition 4.38). Hence,

$$F(x) = \int_{[a,x]} \frac{dF}{dx} \, dm = \int_{[a,b]} f \, dx = F(b) - F(a).$$

Equation (4.13) for the Lebesgue integral is precisely what the Fundamental Theorem of Calculus yields for the Cauchy–Riemann integral, namely that

$$\int_a^b \frac{dF}{dx} \, dx = F(b) - F(a).$$

This completes the proof.   □

We note below that the Fundamental Theorem of Calculus does not admit an "almost everywhere" version.

Example 4.41. *The Cantor ternary function* Φ *is differentiable at almost every point of* $[0,1]$, *and* $\frac{d\Phi}{dx} = 0$ *almost everywhere, yet*

$$\int_{[0,1]} \frac{d\Phi}{dx} \, dm \neq \Phi(1) - \Phi(0).$$

Proof. The function Φ is constant on any open interval $(\alpha, \beta) \subset [0,1] \backslash \mathscr{C}$. Hence, the derivative of Φ is zero on (α, β). As the Cantor set $\mathscr{C}$ is a set of measure zero, Φ is differentiable at almost every point of $[0,1]$. Therefore, the integral of $\frac{d\Phi}{dx}$ is zero, whereas $\Phi(1) - \Phi(0) = 1 - 0 = 1$, and so the Fundamental Theorem of Calculus does not hold. $\qquad\qquad\square$

4.6 Series

Proposition 4.42. *On a measure space* (X, Σ, μ), *assume that* $\{u_n\}_{n \in \mathbb{N}}$ *is a sequence of nonnegative integrable functions such that* $\sum_{n=1}^{\infty} u_n(x)$ *converges for all* $x \in X$. *If* $\sum_{n=1}^{\infty} u_n$ *is integrable, then*

$$\int_X \left(\sum_{n=1}^{\infty} u_n \right) d\mu = \sum_{n=1}^{\infty} \int_X u_n \, d\mu. \tag{4.15}$$

Proof. Let $g = \sum_{n=1}^{\infty} u_n$. If $f_k = \sum_{n=1}^{k}$, for each $k \in \mathbb{N}$, then $f_k(x) \leq g(x)$ for all $x \in X$. Because g is integrable, the Dominated Convergence Theorem states that

$$\int_X (\lim_{k \to \infty} f_k) \, d\mu = \lim_{k \to \infty} \int_X f_k \, d\mu,$$

which is precisely equation (4.15). $\qquad\qquad\square$

A partial converse to Proposition 4.42 is:

Proposition 4.43. *If* (X, Σ, μ) *is a measure space and if* $\{u_n\}_{n \in \mathbb{N}}$ *is a sequence of nonnegative integrable functions such that* $\sum_{n=1}^{\infty} u_n$ *is integrable, then* $\sum_{n=1}^{\infty} u_n(x)$ *converges for almost all* $x \in X$ *and*

$$\int_X \left(\sum_{n=1}^{\infty} u_n \right) d\mu = \sum_{n=1}^{\infty} \int_X u_n \, d\mu.$$

Proof. Apply the Monotone Convergence Theorem and Proposition 4.23 to the partial sums of the series. □

The following example demonstrates how Propositions 4.42 and 4.43 can be invoked for calculations.

Example 4.44. *If $\log x$ denotes the natural logarithm function, and if $f : [0,1] \to \mathbb{R}$ is defined by*

$$f(0) = 0; \quad f(1) = 1; \quad f(x) = \frac{x \log x}{x - 1} \quad \forall x \in (0,1),$$

then f is continuous, and

$$\int_0^1 f(x)\,dx = 1 - \sum_{n=2}^{\infty} \frac{1}{n^2(n-1)}.$$

Proof. Note that f is continuous on $(0,1)$ and that

$$\lim_{x \to 0^+} f(x) = 0 \quad \text{and} \quad \lim_{x \to 1^-} f(x) = 1.$$

Thus, f is continuous on $[0,1]$ and therefore

$$\int_0^1 f(x)\,dx = \int_{[0,1]} f\,dm.$$

Let $f_k = \chi_{[\frac{1}{k}, 1-\frac{1}{k}]} f$, for all $k \in \mathbb{N}$. The sequence $\{f_k\}_{k \in \mathbb{N}}$ of nonnegative measurable functions is monotone increasing and $\lim_k f_k = f$. By the Monotone Convergence Theorem,

$$\int_{[0,1]} f\,dm = \lim_{k \to \infty} \int_{[0,1]} f_k\,dm = \lim_{k \to \infty} \int_{\frac{1}{k}}^{1-\frac{1}{k}} \frac{x \log x}{x - 1}\,dx.$$

Change variables: let $\omega = 1 - x$ to obtain

$$\int_{\frac{1}{k}}^{1-\frac{1}{k}} \frac{x \log x}{x - 1}\,dx = \int_{\frac{1}{k}}^{1-\frac{1}{k}} \left(\frac{1-\omega}{\omega} \right) \log(1-\omega)\,d\omega.$$

The function $\log(1 - \omega)$ has a power series expansion that converges (uniformly, although we need only pointwise) on $[\frac{1}{k}, 1 - \frac{1}{k}]$. Thus,

$$\left(\frac{1-\omega}{\omega}\right)\log(1-\omega) = \frac{1-\omega}{\omega}\sum_{n=1}^{\infty}\frac{\omega^n}{n}$$

$$= 1 - \sum_{n=2}^{\infty}\omega^{n-1}\left(\frac{1}{n-1}-\frac{1}{n}\right)$$

$$= 1 - \sum_{n=2}^{\infty}\frac{\omega^{n-1}}{(n-1)n}.$$

By Proposition 4.42, this final series can be integrated term by term, which yields

$$\int_{\frac{1}{k}}^{1-\frac{1}{k}}\frac{1-\omega}{\omega}\log(1-\omega)\,d\omega = \left(1-\frac{2}{k}\right) - \sum_{n=2}^{\infty}\frac{1}{(n-1)n}\int_{\frac{1}{k}}^{1-\frac{1}{k}}\omega^n\,d\omega$$

$$= \left(1-\frac{2}{k}\right) - \sum_{n=2}^{\infty}\frac{(1-\frac{1}{k})^n-(\frac{1}{k})^n}{n^2(n-1)}.$$

Thus,

$$\int_{[0,1]} f\,dm = \lim_{k\to\infty}\left(1-\frac{2}{k}-\sum_{n=2}^{\infty}\frac{(1-\frac{1}{k})^n-(\frac{1}{k})^n}{n^2(n-1)}\right)$$

$$= 1 - \sum_{n=2}^{\infty}\frac{1}{n^2(n-1)},$$

as claimed. $\qquad\qquad\square$

4.7 Integral Inequalities

If x and y are elements in a real or complex vector space V, then the line segment $L_{x,y}$ in V that joins x with y has parametric form

$$L_{x,y} = \{\lambda x + (1-\lambda)y \mid \lambda \in [0,1]\}.$$

This is, of course, only one of infinitely many ways to parameterise the line segment $L_{x,y}$, but it is perhaps the simplest one. In particular, if J is an interval of real numbers, then $L_{x,y} \subseteq J$ for all $x,y \in J$.

Definition 4.45. If $J \subseteq \mathbb{R}$ is an interval, then a function $\vartheta : J \to \mathbb{R}$ is:

1. a *convex* function if, for all $x, y \in J$ and $\lambda \in (0, 1)$,

$$\vartheta\left(\lambda x + (1 - \lambda)y\right) \leq \lambda\vartheta(x) + (1 - \lambda)\vartheta(y);$$

2. a *concave* function if, for all $x, y \in J$ and $\lambda \in (0, 1)$,

$$\lambda\vartheta(x) + (1 - \lambda)\vartheta(y) \leq \vartheta\left(\lambda x + (1 - \lambda)y\right).$$

Note that a function ϑ is convex if and only if $-\vartheta$ is concave. The general shape of the graph of a convex function on an interval $J \subseteq \mathbb{R}$ is "concave up", whereas the shape of concave functions is "concave down". This is made precise by the following proposition.

Proposition 4.46. *If $J \subset \mathbb{R}$ is an open interval, and $\vartheta : J \to \mathbb{R}$ has a continuous nonnegative second derivative at every point of J, then ϑ is a convex function.*

Proof. Assume that $d^2\vartheta/dt^2$ is nonnegative on J. Then, $d\vartheta/dt$ is monotone increasing on J. To prove that ϑ is convex, choose any $x, y \in J$ and $\lambda \in (0, 1)$. Let $\zeta = \lambda x + (1 - \lambda)y$. By the Fundamental Theorem of Calculus, and by the fact that $d\vartheta/dt$ is monotone increasing,

$$\vartheta(\zeta) - \vartheta(x) = \int_x^\zeta \left(\frac{d\vartheta}{dt}\right) dt \leq \left(\frac{d\vartheta}{dt}[\zeta]\right)(\zeta - x).$$

Likewise,

$$\vartheta(y) - \vartheta(\zeta) = \int_\zeta^y \left(\frac{d\vartheta}{dt}\right) dt \geq \left(\frac{d\vartheta}{dt}[\zeta]\right)(y - \zeta).$$

Because $\zeta - x = (1 - \lambda)(y - x)$ and $y - \zeta = \lambda(y - x)$, the second terms in each of the two inequalities above can be expressed in terms of $(y - x)$, leading to

$$\vartheta(\zeta) \leq \vartheta(x) + \left(\frac{d\vartheta}{dt}[\zeta]\right)(1 - \lambda)(y - x)$$

$$\vartheta(\zeta) \leq \vartheta(y) - \left(\frac{d\vartheta}{dt}[\zeta]\right)\lambda(y - x).$$

Hence,

$$\lambda\vartheta(\zeta) \leq \lambda\vartheta(x) + \left(\frac{d\vartheta}{dt}[\zeta]\right)\lambda(1 - \lambda)(y - x)$$

$$(1 - \lambda)\vartheta(\zeta) \leq (1 - \lambda)\vartheta(y) - \left(\frac{d\vartheta}{dt}[\zeta]\right)\lambda(1 - \lambda)(y - x).$$

Adding these two inequalities leads to

$$\vartheta(\lambda x + (1-\lambda)y) \leq \lambda\vartheta(x) + (1-\lambda)\vartheta(y).$$

This proves that ϑ is a convex function. $\square$

Proposition 4.46 allows one to readily produce examples of convex functions. Among the most important convex functions are the ones below.

Corollary 4.47. *The following functions ϑ are convex:*

1. $\vartheta(t) = e^{\alpha t}$ *on* $\mathbb{R}$, *for any* $\alpha \in \mathbb{R}$;
2. $\vartheta(t) = t^p$ *on* $(0,\infty)$, *where* $p \in \mathbb{R}$ *is such that* $p \geq 1$;
3. $\vartheta(t) = -\log t$ *on* $(0,\infty)$.

Proposition 4.48. *If* $\vartheta : J \to \mathbb{R}$ *is a convex function, and if* $x,y,z \in J$ *satisfy* $x < y < z$, *then*

$$\frac{\vartheta(y) - \vartheta(x)}{y - x} \leq \frac{\vartheta(z) - \vartheta(y)}{z - y}. \tag{4.16}$$

Proof. There is a unique $\lambda \in (0,1)$ such that $y = \lambda x + (1-\lambda)z$. Thus,

$$z = \frac{y - \lambda x}{1 - \lambda} \quad \text{and} \quad z - y = \frac{\lambda}{1 - \lambda}(y - x).$$

Because ϑ is a convex function, $\vartheta(y) \leq \lambda\vartheta(x) + (1-\lambda)\vartheta(z)$, and so

$$\lambda(\vartheta(z) - \vartheta(x)) \leq \vartheta(z) - \vartheta(y).$$

That is,

$$\vartheta(z) - \vartheta(x) \leq \frac{1}{\lambda}(\vartheta(z) - \vartheta(y)).$$

Hence,

$$\frac{\vartheta(y) - \vartheta(x)}{y - x} \leq \frac{\lambda\vartheta(x) + (1-\lambda)\vartheta(z) - \vartheta(x)}{y - x}$$

$$= \frac{(1-\lambda)(\vartheta(z) - \vartheta(x))}{y - x}$$

$$\leq \frac{\frac{(1-\lambda)}{\lambda}(\vartheta(z) - \vartheta(x))}{y - x}$$

$$= \frac{\vartheta(z) - \vartheta(y)}{\frac{\lambda}{1-\lambda}(y - x)}$$

$$= \frac{\vartheta(z) - \vartheta(y)}{z - y}.$$

This completes the proof. $\square$

If one views each side of inequality (4.16) as a difference quotient, then inequality (4.16) says that the derivative of ϑ (if it exists) is an increasing function.

The defining condition for a convex function is an inequality. Therefore, it is not surprising that convex functions lead to a variety of inequalities, and one of the most fundamental and generic of such inequalities is the following inequality of Jensen.

Theorem 4.49 (Jensen's Inequality). *Suppose that* (X, Σ, μ) *is a measure space such that* $\mu(X) = 1$. *If* $f : X \to [a', b'] \subset (a, b)$ *is a measurable function, then, for every convex function* $\vartheta : (a, b) \to \mathbb{R}$,

$$\vartheta\left(\int_X f \, d\mu\right) \le \int_X \vartheta \circ f \, d\mu.$$

Proof. Note that $\int_X f \, d\mu \in (a, b)$ because

$$a < a' \le f(x) \le b' < b, \quad \forall x \in X,$$

implies that

$$a = \int_X a \, d\mu < \int_X f \, d\mu < \int_X b \, d\mu = b.$$

Let $\zeta = \int_X f \, d\mu$ and let

$$\beta = \sup_{z \in (a, \zeta)} \frac{\vartheta(\zeta) - \vartheta(z)}{\zeta - z}.$$

Hence,

$$\vartheta(z) \le \vartheta(\zeta) + \beta(z - \zeta), \quad \forall z \in (a, \zeta).$$

Because ϑ is a convex function, Proposition 4.48 implies that

$$\beta \le \frac{\vartheta(y) - \vartheta(\zeta)}{y - \zeta}, \quad \forall y \in (\zeta, b).$$

Thus,

$$\vartheta(y) \ge \beta(y - \zeta) + \vartheta(\zeta), \quad \forall y \in (\zeta, b).$$

Conclusion:

$$\vartheta(t) \ge \vartheta(\zeta) + \beta(t - \zeta), \quad \forall t \in (a, b).$$

In particular,

$$\vartheta(f(x)) - \vartheta(\zeta) - \beta f(x) + \beta \zeta \ge 0, \quad \forall x \in X. \tag{4.17}$$

On passing to integrals, and noting that $\mu(X) = 1$ and $\zeta = \int_X f \, d\mu$, inequality (4.17) yields

$$\int_X \vartheta \circ f \, d\mu - \vartheta \left(\int_X f \, d\mu \right) - \beta \zeta + \beta \zeta \geq 0.$$

That is,

$$\vartheta \left(\int_X f \, d\mu \right) \leq \int_X \vartheta \circ f \, d\mu,$$

which completes the proof. $\square$

Some very basic yet far reaching inequalities may be derived from Jensen's inequality. Among the most elementary of these are the arithmetic-geometric mean inequality and Young's inequality.

Proposition 4.50. *Suppose that* $\alpha_1, \ldots, \alpha_n$ *are positive real numbers.*

1. Arithmetic-Geometric Mean Inequality:

$$(\alpha_1 \cdots \alpha_n)^{1/n} \leq \frac{1}{n} (\alpha_1 + \cdots + \alpha_n).$$

2. Young's Inequality: If $p_1, \ldots, p_n$ *are positive and* $\frac{1}{p_1} + \cdots + \frac{1}{p_n} = 1$, *then*

$$\alpha_1 \cdots \alpha_n \leq \frac{1}{p_1} \alpha_1^{p_1} + \cdots + \frac{1}{p_n} \alpha_n^{p_n}.$$

Proof. For the proof of the Arithmetic-Geometric Mean Inequality, let $X = \{1, 2, \ldots, n\}$, $\Sigma = \mathscr{P}(X)$, and μ be

$$\mu(E) = \frac{1}{n} |E|, \quad \forall E \subseteq X,$$

where $|E|$ denotes the cardinality of E. Thus, $\mu(X) = 1$.

Let $\beta_1, \ldots, \beta_n \in \mathbb{R}$ be such that $e^{\beta_k} = \alpha_k$, for each k. If $f : X \to \mathbb{R}$ is defined by $f(k) = \beta_k$, for $k \in X$, then

$$\int_X f \, d\mu = \sum_{k=1}^{n} f(k) \mu(\{k\}) = \frac{1}{n} \sum_{k=1}^{n} \beta_k.$$

The function $\vartheta(t) = e^t$ is convex on $\mathbb{R}$ and

$$\int_X \vartheta \circ f \, d\mu = \frac{1}{n} \sum_{k=1}^{n} e^{f(k)} = \frac{1}{n} \sum_{k=1}^{n} e^{\beta_k}.$$

Jensen's inequality states that

$$\vartheta\left(\int_X f\,d\mu\right) \le \int_X \vartheta \circ f\,d\mu.$$

Hence,

$$(\alpha_1\alpha_2\cdots\alpha_n)^{1/n} = e^{\beta_1/n + \cdots + \beta_n/n} \le \frac{1}{n}\sum_{k=1}^n e^{\beta_k} = \frac{1}{n}\sum_{k=1}^n \alpha_k.$$

This completes the proof.

The proof of Young's inequality is left to the reader as Exercise 4.87.   □

Proposition 4.50 is a rather straightforward application of Jensen's inequality. A somewhat more sophisticated application of Jensen's inequality leads to the fundamental inequalities of Hölder and Minkowksi (Theorems 4.53 and 4.54).

Definition 4.51. If (X, Σ, μ) is a measure space and if $p \ge 1$, then a measurable function $f : X \to \mathbb{C}$ is said to be *p-integrable* if f^p is integrable.

Definition 4.52. Two positive real numbers $p, q \in \mathbb{R}$ are said to be *conjugate* if

$$\frac{1}{p} + \frac{1}{q} = 1.$$

If p and q are conjugate real numbers, then q to p is uniquely determined by p, and $p > 1$ and $q > 1$. The notion of "conjugate positive numbers" is nothing more than the association of a pair of convex coefficients, namely $\frac{1}{p}$ and $\frac{1}{q}$, with the positive real numbers p and q.

Theorem 4.53 (Hölder's Inequality). *Suppose that p and q are conjugate real numbers. If (X, Σ, μ) is a measure space and if $f, g : X \to \mathbb{C}$ are such that f is p-integrable and g is q-integrable, then $fg : X \to \mathbb{C}$ is integrable and*

$$\int_X |fg|\,d\mu \le \left(\int_X |f|^p\,d\mu\right)^{1/p}\left(\int_X |g|^q\,d\mu\right)^{1/q} \tag{4.18}$$

Proof. Because f^p and g^q are integrable, so are $|f|^p$ and $|g|^q$. Let $F, G : X \to \mathbb{R}$ be the functions defined by

$$F(x) = \frac{|f(x)|}{\left(\int_X |f|^p\,d\mu\right)^{1/p}} \quad \text{and} \quad G(x) = \frac{|g(x)|}{\left(\int_X |g|^q\,d\mu\right)^{1/q}}.$$

Thus,

$$\int_X F^p\,d\mu = \int_X G^q\,d\mu = 1.$$

By Young's Inequality (Proposition 4.50), for each $x \in X$,

$$F(x)G(x) \leq \frac{1}{p}F(x)^p + \frac{1}{q}G(x)^q.$$

Hence,

$$\int_X FG\,d\mu \leq \frac{1}{p}\int_X F^p\,d\mu + \frac{1}{q}\int_X G^q\,d\mu = \frac{1}{p} + \frac{1}{q} = 1.$$

That is,

$$\frac{\displaystyle\int_X |fg|\,d\mu}{\left(\int_X |f|^p\,d\mu\right)^{1/p}\left(\int_X |g|^q\,d\mu\right)^{1/q}} \leq 1,$$

which completes the proof. $\square$

Theorem 4.54 (Minkowski's Inequality).   *Suppose that $p \geq 1$. If (X, Σ, μ) is a measure space and if $f, g : X \to \mathbb{C}$ are p-integrable, then $f + g$ is p-integrable and*

$$\left(\int_X |f+g|^p\,d\mu\right)^{1/p} \leq \left(\int_X |f|^p\,d\mu\right)^{1/p} + \left(\int_X |g|^p\,d\mu\right)^{1/p} \qquad (4.19)$$

Proof. The theorem is true for $p = 1$ because the sum of integrable functions is integrable and because inequality (4.19) is simply a consequence of the triangle inequality in real and complex numbers.

Assume, therefore, that $p > 1$ and consider the function $\vartheta : \mathbb{R}^+ \to \mathbb{R}^+$ defined by $\vartheta(t) = t^p$. Because ϑ is convex,

$$\vartheta\left(\frac{1}{2}|f(x)| + \frac{1}{2}|g(x)|\right) \leq \frac{1}{2}\vartheta(|f(x)|) + \frac{1}{2}\vartheta(|g(x)|), \quad \forall x \in X.$$

Hence,

$$\left(\frac{1}{2}\right)^p (|f| + |g|)^p \leq \frac{1}{2}|f|^p + \frac{1}{2}|g|^p.$$

Because the sum of integrable functions is integrable, the inequality above shows that $(|f| + |g|)^p$ is integrable. But, by the triangle inequality, $|f + g|^p \leq (|f| + |g|)^p$; hence, $f + g$ is p-integrable.

Let $q \in \mathbb{R}^+$ be conjugate to p. Thus, $p = (p-1)q$ and

$$\left((|f| + |g|)^{p-1}\right)^q = (|f| + |g|)^p.$$

Hence, $(|f| + |g|)^{p-1}$ is q-integrable. Therefore, one can apply Hölder's Inequality to obtain

$$\int_X |f|(|f| + |g|)^{p-1}\, d\mu \leq \left(\int_X |f|^p\, d\mu \right)^{1/p} \left(\int_X (|f| + |g|)^{(p-1)q}\, d\mu \right)^{1/q}$$

and

$$\int_X |g|(|f| + |g|)^{p-1}\, d\mu \leq \left(\int_X |g|^p\, d\mu \right)^{1/p} \left(\int_X (|f| + |g|)^{(p-1)q}\, d\mu \right)^{1/q}.$$

Because $|f|^p + |g|^p = |f|(|f| + |g|)^{p-1} + |g|(|f| + |g|)^{p-1}$, summing the two inequalities above yields a new inequality whose left-hand side is

$$\int_X (|f| + |g|)^p\, d\mu$$

and whose right-hand side is

$$\left(\int_X (|f| + |g|)^{(p-1)q}\, d\mu \right)^{1/q} \left[\left(\int_X |f|^p\, d\mu \right)^{1/p} + \left(\int_X |g|^p\, d\mu \right)^{1/p} \right].$$

Divide the new inequality through by $\left(\int_X (|f| + |g|)^{(p-1)q}\, d\mu \right)^{1/q}$ and use that $p = (p-1)q$ to obtain

$$\left(\int_X (|f| + |g|)^p\, d\mu \right)^{1-1/q} \leq \left(\int_X |f|^p\, d\mu \right)^{1/p} + \left(\int_X |g|^p\, d\mu \right)^{1/p}.$$

Because $\frac{1}{p} = 1 - \frac{1}{q}$ and $|f + g|^p \leq (|f| + |g|)^p$, the inequality above implies inequality (4.19). $\qquad\square$

Definition 4.55. Assume that $p \in \mathbb{R}$ satisfies $p \geq 1$. A sequence $\{\alpha_k\}_{k \in \mathbb{N}}$ in $\mathbb{C}$ is *p-summable* if

$$\sum_{k \in \mathbb{N}} |\alpha_k|^p < \infty.$$

The Hölder and Minkowski inequalities also have formulations for sequences of complex numbers.

Theorem 4.56. *Suppose that $\{\alpha_k\}_{k \in \mathbb{N}}$ and $\{\beta_k\}_{k \in \mathbb{N}}$ are sequences in $\mathbb{C}$. Assume that $p \geq 1$.*

1. (Hölder) If p and q are conjugate real numbers, and if $\{\alpha_k\}_{k \in \mathbb{N}}$ is p-summable and $\{\beta_k\}_{k \in \mathbb{N}}$ is q-summable, then $\{\alpha_k \beta_k\}_{k \in \mathbb{N}}$ is summable and

$$\sum_{k\in\mathbb{N}} |\alpha_k \beta_k| \le \left(\sum_{k\in\mathbb{N}} |\alpha_k|^p\right)^{1/p} \left(\sum_{k\in\mathbb{N}} |\beta_k|^q\right)^{1/q}.$$

2. (Minkowski) *If $\{\alpha_k\}_{k\in\mathbb{N}}$ and $\{\beta_k\}_{k\in\mathbb{N}}$ are p-summable, then $\{\alpha_k + \beta_k\}_{k\in\mathbb{N}}$ is p-summable and*

$$\left(\sum_{k\in\mathbb{N}} |\alpha_k + \beta_k|^p\right)^{1/p} \le \left(\sum_{k\in\mathbb{N}} |\alpha_k|^p\right)^{1/p} + \left(\sum_{k\in\mathbb{N}} |\beta_k|^p\right)^{1/p}.$$

Proof. Apply Theorems 4.53 and 4.54 to the case where the measure space (X, Σ, μ) is given by $X = \mathbb{N}$, $\Sigma = \mathscr{P}(\mathbb{N})$, and μ is counting measure. □

Note that the Hölder and Minkowski inequalities are nontrivial even in the cases where the sequences $\{\alpha_k\}_{k\in\mathbb{N}}$ and $\{\beta_k\}_{k\in\mathbb{N}}$ have only finitely many nonzero elements.

Appendix: The Issue of How to Integrate

Some concepts in mathematics develop rapidly, while other ideas take many decades to mature and settle. The theory of integration is of the latter type.

In elementary calculus, integration is usually carried out in the style of Cauchy, mostly for the purpose of making explicit calculations. However, the more theoretical aspects of analysis require a theory of integration in which certain properties of integrals—especially properties of limits—are known to hold. For such purposes, the integration theories of Cauchy and, as well, Riemann, do not do quite all that we require of them. These difficulties, while recognised by many outstanding mathematicians of the late nineteenth-century, were not so easily overcome. Nevertheless, after many new and innovative results had been obtained by several mathematicians, a satisfactory theory of integration eventually emerged in the early twentieth-century in a series of works by Henri Lebesgue.

Riemann's Approach

Assume that $a, b \in \mathbb{R}$ satisfy $a < b$ and that $f : [a, b] \to \mathbb{R}$ is a bounded function. That is, there is a positive real number M for which $|f(x)| \le M$, for all $x \in [a, b]$. If $n \in \mathbb{N}$, then an *n-partition* of $[a, b]$ is a collection of n finite intervals of the form $(\zeta_{k-1}, \zeta_k]$, where $k = 1, \ldots, n$ and

$$a = \zeta_0 < \zeta_1 < \ldots < \zeta_n = b. \tag{4.20}$$

Let $\zeta = (\zeta_0, \zeta_1, \ldots, \zeta_n)$ and denote the partition (4.20) of $[a, b]$ by $\mathscr{P}_\zeta$. As f is bounded, the real numbers m_k and M_k, defined below, exist for all $1 \leq k \leq n$:

$$m_k = \inf\{f(x) \,|\, x \in (\zeta_{k-1}, \zeta_k]\} \quad \text{and} \quad M_k = \sup\{f(x) \,|\, x \in (\zeta_{k-1}, \zeta_k]\}.$$

Thus, every n-partition $\mathscr{P}_\zeta$ of $[a, b]$ induces a *lower Riemann sum* s_ζ, namely

$$s_\zeta = \sum_{k=1}^{n} m_k(\zeta_k - \zeta_{k-1}),$$

and an *upper Riemann sum* S_ζ, which in this case is

$$S_\zeta = \sum_{k=1}^{n} M_k(\zeta_k - \zeta_{k-1}).$$

The *lower and upper Riemann integrals* of f are defined, respectively, as follows:

$$\underline{\int_a^b} f(x)\,dx = \sup\left\{s_\zeta \,|\, \mathscr{P}_\zeta \text{ is an } n\text{-partition of } [a, b],\, n \in \mathbb{N}\right\}$$

$$\overline{\int_a^b} f(x)\,dx = \inf\left\{S_\zeta \,|\, \mathscr{P}_\zeta \text{ is an } n\text{-partition of} [a, b],\, n \in \mathbb{N}\right\}$$

Definition 4.57. A bounded function $f : [a, b] \to \mathbb{R}$ is *Riemann integrable* if

$$\underline{\int_a^b} f(x)\,dx = \overline{\int_a^b} f(x)\,dx. \tag{4.21}$$

In this case, the *Riemann integral* of f is the value of (4.21) and is denoted by

$$\int_a^b f(x)\,dx.$$

It is not difficult to see that any function $f : [a, b] \to \mathbb{R}$ that assumes only one value—that is, any constant function f—is Riemann integrable. However, as soon as we consider functions that assume two values, things go wrong.

Proposition 4.58. *There is a bounded function f on $[0, 1]$ that assumes only two values yet fails to be Riemann integrable.*

Proof. Consider the function $f : [0, 1] \to \mathbb{R}$ given by $f = \chi_{\mathbb{Q} \cap [0,1]}$. That is, $f(q) = 1$ for every rational $q \in [0, 1]$ and $f(r) = 0$ for every irrational $r \in [0, 1]$. As f assumes only two values, f is bounded. Take any n-partition $\mathscr{P}_\zeta$ of $[0, 1]$. In each interval

$(\zeta_{k-1},\zeta_k]$ there are both rationals and irrationals, and so $m_k = 0$ and $M_k = 1$ on each $(\zeta_{k-1},\zeta_k]$. Hence,

$$\underline{\int_a^b} f(x)\,dx = 0 \quad \text{and} \quad \overline{\int_a^b} f(x)\,dx = 1.$$

Therefore, f is not Riemann integrable.   □

What is the problem with the function f in Proposition 4.58 ? Well, for one thing, no matter has small the interval $(\zeta_{k-1},\zeta_k]$ may be, the oscillation of f on $(\zeta_{k-1},\zeta_k]$ (namely, between 0 and 1) is large relative to the length of $(\zeta_{k-1},\zeta_k]$. Bounded functions with uncontrollable oscillations generally fail to be Riemann integrable. On the other hand, the oscillations of continuous functions diminish as the size of an interval $(\zeta_{k-1},\zeta_k]$ decreases, and so continuous functions are Riemann integrable. (This last explanation is not, of course, a proof.)

The next result points toward the more serious issue of limits.

Proposition 4.59. *There exists a sequence of Riemann-integrable functions* $f_k :$ $[0,1] \to \mathbb{R}$ *such that*

1. $0 \le f_k(x) \le f_{k+1}(x) \le 1$, *for all* $x \in [0,1]$ *and all* $k \in \mathbb{N}$,
2. *the Riemann integrals of* f_k *converge to 0, that is,*

$$\lim_{k\to\infty} \int_0^1 f_k(x)\,dx = 0, \text{ and}$$

3. *the function* $f(x) = \lim_k f_k(x)$, $x \in [0,1]$ *is not Riemann integrable.*

Proof. Enumerate $\mathbb{Q} \cap [0,1]$ as $\{q_n\}_{n\in\mathbb{N}}$. For each $k \in \mathbb{Q}$, let $f_k = \chi_{q_1,\dots,q_k}$. Because f_k is continuous and equal to zero except at the k points $q_1,\dots,q_k$, it is simple to verify that f_k is Riemann integrable and

$$\int_0^1 f_k(x)\,dx = 0, \quad \forall k \in \mathbb{N}.$$

It is also clear that $f_k(x) \le f_{k+1}(x)$, for all x and all k, and that the limiting function f is the characteristic function $\chi_{\mathbb{Q}\cap[0,1]}$, which is not Riemann integrable (by the proof of Proposition 4.58).   □

The point of the Proposition 4.59 is that the "hoped for" formula

$$\lim_{k\to\infty} \int_a^b f_k(x)\,dx = \int_a^b \lim_{k\to\infty} f_k(x)\,dx \tag{4.22}$$

does not hold.

Lebesgue's Approach

To overcome the troubles suggested by Propositions 4.58 and 4.59, Lebesgue thought to partition the "y-axis" rather than the "x-axis". That is, suppose that $f : [a,b] \to \mathbb{R}$ is a bounded function with range in the closed interval $[c,d]$. Let $\mathscr{P}_\eta$ be an n-partition of $[c,d]$ and consider the sets

$$E_k = \{x \in [a,b] \,|\, f(x) \in (\eta_{k-1}, \eta_k]\}, \quad 1 \le k \le n.$$

Here, $E_1, \ldots, E_n$ are pairwise disjoint and their union is $[a,b]$. Assume, further, that there is a "length function" m defined on subsets of $[a,b]$ such that when m is evaluated at any interval (x,y), m gives the actual length of an interval (x,y): $m((x,y)) = y - x$. With these assumptions, let

$$\ell_k = \inf\{f(x) \,|\, x \in E_k\} \quad \text{and} \quad L_k = \sup\{f(x) \,|\, x \in E_k\},$$

and

$$t_\eta = \sum_{k=1}^{n} \ell_k \, m(E_k) \quad \text{and} \quad T_\eta = \sum_{k=1}^{n} L_k \, m(E_k).$$

Define lower and upper "Lebesgue" integrals, respectively, by

$$\underline{\int_a^b} f \, dm = \sup\left\{t_\eta \,|\, \mathscr{P}_\eta \text{ is an } n\text{-partition of } [c,d], \, n \in \mathbb{N}\right\}$$

$$\overline{\int_a^b} f \, dm = \inf\left\{T_\eta \,|\, \mathscr{P}_\eta \text{ is an } n\text{-partition of } [c,d], \, n \in \mathbb{N}\right\}$$

and say that f is Lebesgue integrable if

$$\underline{\int_a^b} f \, dm = \overline{\int_a^b} f \, dm.$$

The value of partitioning $[c,d]$ rather than $[a,b]$ is that now oscillations of f over an interval of $[a,b]$ have no impact. However, one now requires the length function m, and this function m has to be defined in such a way that $m(E)$ can be evaluated even if E is a quite complicated subset. Thus, measure theory arose.

Lebesgue's Criterion for Riemann Integration

Although the following theorem of Lebesgue will not be proved here, it is certainly striking in that it shows that a bounded function f is Riemann integrable if and only if f is continuous almost-everywhere.

Theorem 4.60. *A bounded function $f : [a,b] \to \mathbb{R}$ is Riemann integrable if and only if the set $D \subset [a,b]$ of points of discontinuity of f is a null set.*

The Improper Riemann Integral

Another difference between Riemann and Lebesgue integration may be found in the notion of an improper Riemann integral.

Definition 4.61. If $f : (0,\infty) \to \mathbb{R}$ is a Riemann-integrable function on $[a,b]$, for every $0 < a < b$, then the *improper Riemann integral* of f over $(0,\infty)$ is the quantity denoted by $\int_0^\infty f(x)\,dx$ and defined by

$$\int_0^\infty f(x)\,dx = \lim_{n\to\infty} \int_{1/n}^n f(x)\,dx.$$

The theory of improper Riemann integrals is a continuous analogue of the theory of conditionally convergent series. For example, in calculus it is shown that

$$\int_0^\infty \frac{\sin x}{x}\,dx = \frac{\pi}{2} \quad \text{(as an improper Riemann integral)}.$$

However, the measurable function $f(x) = \frac{\sin x}{x}$ on $(0,\infty)$ is not (Lebesgue) integrable on $(0,\infty)$. To prove this, one may argue by contradiction. Assume that f *is* integrable on $(0,\infty)$. Thus, by Theorem 4.19, so is $|f|$:

$$|f|(x) = \frac{|\sin x|}{x}.$$

Let $h_k = \chi_{(\pi,k\pi]}|f|$, for each $k \in \mathbb{N}$. Then, $\{h_k\}_{k\in\mathbb{N}}$ is a monotone-increasing sequence of measurable nonnegative functions. In fact, each h_k is integrable and $|f|(x) = \lim_k h_k(x)$ for every $x \in (\pi,\infty)$. Thus, by the Monotone Convergence Theorem,

$$\lim_{k\to\infty} \int_{(\pi,\infty)} h_k\,dm = \int_{(\pi,\infty)} |f|\,dm < \infty.$$

We shall derive a contradiction by demonstrating that

$$\lim_{k\to\infty} \int_{(\pi,\infty)} h_k\,dm = +\infty.$$

To do this, decompose the interval $(\pi, k\pi]$ as union of pairwise disjoint subintervals $(j\pi, (j+1)\pi]$, for $j = 1, \ldots, k-1$. Note that $h_k(x) > |\sin x|/((j+1)\pi)$ for all $x \in (j\pi, (j+1)\pi]$. Thus,

$$\int_{(\pi,\infty)} h_k \, dm = \sum_{j=1}^{k-1} \int_{(j\pi,(j+1)\pi]} f_k \, dm$$

$$\geq \sum_{j=1}^{k-1} \int_{(j\pi,(j+1)\pi]} \frac{|\sin x|}{(j+1)\pi} \, dx$$

$$= \frac{2}{\pi} \sum_{j=1}^{k-1} \frac{1}{(j+1)},$$

which diverges as $k \to \infty$.

As this discussion above points out, Lebesgue's approach to integration is analogous to the theory of absolutely convergent series, whereas Riemann's approach is more like the theory of conditionally convergent series.

Problems

4.62. Assume that (X, Σ, μ) is a measure space and that $\varphi, \psi : X \to \mathbb{R}$ are nonnegative simple functions.

1. If $\psi(x) \leq \varphi(x)$, for all $x \in X$, then prove that

$$\int_X \psi \, d\mu \leq \int_X \varphi \, d\mu.$$

2. If $\alpha, \beta \in \mathbb{R}$ are nonnegative, then prove that $\alpha\varphi + \beta\psi$ is a nonnegative simple function and that

$$\int_X (\alpha\varphi + \beta\psi) \, d\mu = \alpha \int_X \varphi \, d\mu + \beta \int_X \psi \, d\mu.$$

4.63. If $f, g : X \to \mathbb{R}$ are nonnegative and measurable, and if $f(x) \leq g(x)$, for all $x \in X$, then prove that

$$\int_X f \, d\mu \leq \int_X g \, d\mu.$$

4.64. Suppose that $f : X \to \mathbb{R}$ is measurable, nonnegative, and

$$\int_X f \, d\mu = 0.$$

Prove that $f = 0$ almost everywhere.

4.65. Assume that (X, Σ, μ) is a measure space and that $f_n : X \to \mathbb{R}$ is a nonnegative measurable function for each $n \in \mathbb{N}$. Assume that

$$f(x) = \sum_{n=1}^{\infty} f_n(x), \quad \forall x \in X.$$

If the series converges for every $x \in X$, then prove that

$$\int_X f \, d\mu = \sum_{n=1}^{\infty} \int_X f_n \, d\mu.$$

4.66. Assume that (X, Σ, μ) is a measure space and that $f : X \to \mathbb{R}$ is a nonnegative measurable function. Define a function $\nu : \Sigma \to [0, \infty]$ by

$$\nu(E) = \int_E f \, d\mu, \quad \forall E \in \Sigma.$$

Prove that ν is a measure on (X, Σ) and that $\nu(E) = 0$ for every $E \in \Sigma$ for which $\mu(E) = 0$.

4.67. Assume that (X, Σ, μ) is a measure space and that $f_k : X \to \mathbb{R}$ is a nonnegative measurable function for each $k \in \mathbb{N}$. Suppose that, for every $x \in X$, $\lim_k f_k(x)$ exists and let $f = \lim_k f_k$. If $f(x) \geq f_k(x)$, for every $x \in X$ and every $k \in \mathbb{N}$, then use Fatou's Lemma to prove that

$$\int_X f \, d\mu = \lim_{k \to \infty} \int_X f_k \, d\mu.$$

4.68. Assume that (X, Σ, μ) is a measure space and that $f : X \to \mathbb{R}$ is an integrable function. If $\mu(X) < \infty$, then prove that for every $\varepsilon > 0$ there is a simple function $\varphi : X \to \mathbb{R}$ such that

$$\int_X |f - \varphi| \, d\mu < \varepsilon.$$

Is the assumption that $\mu(X)$ be finite a necessary assumption ? Explain.

4.69. Assume that (X, Σ, μ) is a measure space and that $f, f_k : X \to \mathbb{R}$ are measurable functions such that

$$\lim_{k \to \infty} \left(\sup_{x \in X} |f_k(x) - f(x)| \right) = 0.$$

(That is, $f_k \to f$ uniformly on X.) If $\mu(X) < \infty$ and each f_k is integrable, then prove that f is integrable and that

$$\int_X f\,d\mu = \lim_{k\to\infty} \int_X f_k\,d\mu.$$

4.70. Assume that (X, Σ, μ) is a measure space and that $f_k : X \to \mathbb{R}$ is an integrable function, for each $k \in \mathbb{N}$. Furthermore, assume that

$$\lim_{k\to\infty} f_k(x) \quad \text{exists almost everywhere}.$$

If there is a nonnegative integrable function $g : X \to \mathbb{R}$ such that, for every $k \in \mathbb{N}$,

$$|f_k(x)| \leq g(x) \quad \forall x \in X,$$

then prove that

$$\lim_{k\to\infty} \int_X f_k\,d\mu = \int_X (\lim_{k\to\infty} f_k)\,d\mu.$$

4.71. If (X, Σ) is a measurable space, then prove that a function $f : X \to \mathbb{C}$ is complex measurable if and only if $\Re f$ and $\Im f$ are measurable functions $X \to \mathbb{R}$.

4.72. Prove that if v is a complex measure on (X, Σ), and if $f : X \to \mathbb{C}$ is an integrable function, then

$$\left| \int f\,dv \right| \leq \int_X |f|\,d|v|,$$

where $|v|$ is the total variation of v (see Definition 3.70).

4.73. Prove that if (X, Σ, μ) is a measurable space and $g : X \to \mathbb{R}$ is a nonnegative measurable function, then the function $v_g : \Sigma \to \mathbb{R}$ defined by

$$v_g(E) = \int_E g\,d\mu, \ E \in \Sigma,$$

is a measure on (X, Σ).

4.74. Compute the integral of Cantor's ternary function Φ over $[0, 1]$.

4.75. Compute $\displaystyle\int_E f\,dm$ for each of the following sets E and functions f:

1. $E = [0, 1)$ and $f(x) = (1 - x^2)^{-1/2}$;
2. $E = (0, 1]$ and $f(x) = x\log x$;
3. $E = (0, \infty)$ and $f(x) = e^{-x}$;

4. $E = (0, \infty)$ and $f(x) = ([x]!)^{-1}$, where $[x]$ denotes the largest integer n for which
   $n \leq x$;
5. $E = \mathbb{R}$ and $f(x) = (1 + x^2)^{-1}$.

4.76. Prove that

$$\lim_{n \to \infty} \int_0^n \left(1 + \frac{x}{n}\right)^n e^{-2x}\, dx \quad \text{exists}$$

and evaluate the limit.

4.77. Prove that

$$\lim_{n \to \infty} \int_0^n \left(1 - \frac{x}{n}\right)^n e^{x/2}\, dx \quad \text{exists}$$

and evaluate the limit.

4.78. Let u_n be the characteristic function of $(0, \frac{1}{n}]$, for each $n \in \mathbb{N}$. Show that $\sum_{n=1}^{\infty} u_n(x)$ converges for all $x \in \mathbb{R}$, but $\sum_{n=1}^{\infty} u_n$ fails to be integrable.

4.79. Assume that (X, Σ, μ) is a measure space. Prove or disprove the following assertion: If $u_n : X \to \mathbb{R}$ is integrable, for each $n \in \mathbb{N}$, and if $\sum_{n=1}^{\infty} u_n$ converges absolutely on X to a function $u : X \to \mathbb{R}$, then u is integrable and

$$\int_X \left(\sum_{n=1}^{\infty} u_n\right) d\mu = \sum_{n=1}^{\infty} \int_X u_n\, d\mu.$$

4.80. Consider the function $f : [0, 1] \to \mathbb{R}$ given by the series

$$f(x) = \sum_{n=0}^{\infty} \frac{x}{(1 + x)^n}.$$

1. Show that $f(x) = x + 1$, if $x \in (0, 1]$, and that $f(0) = 0$.
2. Does the series converge uniformly on $[0, 1]$?
3. Is it true that

$$\int_0^1 \left(\sum_{n=0}^{\infty} \frac{x}{(1 + x)^n}\right) dx = \sum_{n=0}^{\infty} \int_0^1 \frac{x}{(1 + x)^n}\, dx?$$

4.81. Prove that the Radon-Nikodým Theorem (Theorem 4.36) holds for σ-finite measures on a measurable space (X, Σ).

4.82. Prove that if $\{\vartheta_k\}_{k\in\mathbb{N}}$ is a sequence of convex functions on an open interval $J \subseteq \mathbb{R}$ such that $\sup_k \vartheta_k(x)$ exists for all $x \in J$, then $\sup_k \vartheta_k$ is a convex function.

4.83. Prove that every convex function is continuous.

4.84. Assume that $J \subset \mathbb{R}$ is an open interval and that $\vartheta : J \to \mathbb{R}$ has a continuous second derivative $d^2\vartheta/dt^2$ at every point of J. Prove that is ϑ is a convex function, then $d^2\vartheta/dt^2$ is nonnegative on J.

4.85. Assume that J is an open interval and that $\vartheta : J \to \mathbb{R}$ is a convex function. Prove that if $t_1,\ldots,t_n \in [0,1]$ satisfy $t_1 + \cdots + t_n = 1$, and if $x_1,\ldots,x_n \in J$, then

$$\vartheta\left(\sum_{k=1}^n t_k x_k\right) \leq \sum_{k=1}^n t_k \vartheta(x_k).$$

4.86. A function $\vartheta : J \to \mathbb{R}$ is *strictly convex* if

$$\vartheta(\lambda x + (1-\lambda)y) < \lambda\vartheta(x) + (1-\lambda)\vartheta(y), \quad \forall \lambda \in (0,1) \text{ and } \forall x \neq y.$$

1. Prove that $\vartheta(t) = e^{\alpha t}$ is strictly convex on $\mathbb{R}$ for every nonzero $\alpha \in \mathbb{R}$.
2. Prove that $\vartheta(t) = t^p$ on $(0,\infty)$ for every $p > 1$.
3. Prove that, for positive real numbers $\alpha_1,\ldots,\alpha_n$,

$$(\alpha_1 \cdots \alpha_n)^{1/n} = \frac{1}{n}(\alpha_1 + \cdots + \alpha_n)$$

   if and only if $\alpha_1 = \cdots = \alpha_n$.

4.87. Let $\alpha_1,\ldots,\alpha_n$ be positive real numbers and let $p_1,\ldots p_n$ be positive real numbers that satisfy $\frac{1}{p_1} + \cdots + \frac{1}{p_n} = 1$.

1. Prove Young's inequality:

$$\alpha_1 \ldots \alpha_n \leq \sum_{k=1}^n \frac{\alpha_k^{p_k}}{p_k}.$$

2. Characterise the cases of equality in Young's inequality.

4.88. Suppose that (X, Σ, μ) is a finite measure space and that $f : X \to \mathbb{C}$ is integrable. Furthermore, suppose that $U \subseteq \mathbb{C}$ is an open set for which $E = f^{-1}(E)$ has positive measure. Prove that $\frac{1}{\mu(X)} \int_E f \, d\mu \in U$.

Part III
Banach Spaces

Chapter 5
Banach Spaces

Various collections of functions carry the structure of a vector space, and functional analysis is devoted to the study of such vector spaces, mainly from an analytic rather than linear algebraic perspective. When working with vector spaces, one is required to specify the underlying base field. In analysis, the natural choices are the fields $\mathbb{R}$ and $\mathbb{C}$, which are preferable to the field $\mathbb{Q}$ or some finite field $\mathbb{F}$, because of the completeness properties enjoyed by the real and complex fields. However, because the field $\mathbb{C}$ is algebraically closed, whereas $\mathbb{R}$ is not, there is a richer and more widely used theory in the case of complex vector spaces. Thus, the base field for all vector spaces under consideration is assumed, with very few exceptions, to be the field $\mathbb{C}$ of complex numbers.

By equipping a vector space with additional structure, such as a topology, linear transformations of the space and the vector space itself are poised to be studied from the point of view of analysis.

5.1 Normed Vector Spaces

The most basic vehicle for introducing a topological structure to vector spaces is through the use of a norm.

Definition 5.1. A *norm* on a (complex) vector space V is a function $\|\cdot\| : V \to \mathbb{R}$ such that, for all $v, w \in V$ and $\alpha \in \mathbb{C}$,

1. $\|v\| \geq 0$, and $\|v\| = 0$ only if $v = 0$,
2. $\|\alpha v\| = |\alpha| \|v\|$,
3. $\|v + w\| \leq \|v\| + \|w\|$.

© Springer International Publishing Switzerland 2016
D. Farenick, *Fundamentals of Functional Analysis*, Universitext,
DOI 10.1007/978-3-319-45633-1_5

The inequality $\|v + w\| \leq \|v\| + \|w\|$ is called the *triangle inequality*. Observe that a norm $\|\cdot\|$ on a normed vector space V induces a metric topology, which is also called a *norm topology*, on V via the metric $d : V \times V \to \mathbb{R}$ defined by

$$d(v_1, v_2) = \|v_1 - v_2\|, \quad v_1, v_2 \in V.$$

Proposition 5.2. *If $\|\cdot\|$ is a norm on V, then a basis $\mathscr{B}$ for the norm topology $\mathscr{T}$ on V is given by*

$$\mathscr{B} = \{B_r(v) \mid v \in V, r \in \mathbb{R}, r > 0\},$$

where $B_r(v) = \{w \in V \mid \|w - v\| < r\}$.

Proposition 5.2 above is a reformulation, for normed vector spaces rather than arbitrary metric spaces, of Proposition 1.39.

In equipping a vector space with a norm—and, hence, a metric topology—one can ask whether the vector space operations of scalar multiplication and vector addition are continuous.

Proposition 5.3. *If V is a normed vector space, then the maps $a : V \times V \to V$ and $m : \mathbb{C} \times V \to V$ defined by*

$$a(v_1, v_2) = v_1 + v_2 \quad and \quad m(\alpha, v) = \alpha v, \forall \alpha \in \mathbb{C}, v, v_1, v_2 \in V,$$

are continuous.

Proof. Exercise 5.99. $\square$

Example 5.4. *The equation $\|\xi\| = \sqrt{\displaystyle\sum_{j=1}^{n} |\xi_j|^2}$, for*

$$\xi = \begin{bmatrix} \xi_1, \\ \vdots \\ \xi_n \end{bmatrix} \in \mathbb{C}^n,$$

defines a norm—called the Euclidean norm—*on $\mathbb{C}^n$.*

Proof. The only nontrivial verification is that of the triangle inequality. If $\xi, \eta \in \mathbb{C}^n$, then

$$\|\xi + \eta\|^2 = \|\xi\|^2 + 2\Re\left(\sum_{j=1}^{n} \xi_j \overline{\eta}_j\right) + \|\eta\|^2.$$

The complex version of the classical Cauchy-Schwarz inequality (1.2) is

$$\left| \sum_{j=1}^{n} \xi_j \overline{\eta}_j \right| \leq \sqrt{\left(\sum_{j=1}^{n} |\xi_j|^2 \right) \left(\sum_{j=1}^{n} |\eta_j|^2 \right)}.$$

Because $\Re \zeta \leq |\zeta|$ for every complex number ζ, the Cauchy-Schwarz inequality above yields $2 \Re \left(\sum_{j=1}^{n} \xi_j \overline{\eta}_j \right) \leq 2 \|\xi\| \, \|\eta\|$. Hence, $\|\xi + \eta\|^2 \leq (\|\xi\| + \|\eta\|)^2$, which proves the triangle inequality.   □

Example 5.5. *Assume that μ is a finite Borel measure on a compact Hausdorff space X and let $C(X)$ be the vector space of all continuous maps $f : X \to \mathbb{C}$. Then each of the equations*

$$\|f\|_\infty = \int |f| \, d\mu$$

$$\|f\|_1 = \max_{x \in X} |f(x)|$$

defines norms $\| \cdot \|_1$ and $\| \cdot \|_\infty$ on $C(X)$.

Proof. Because X is compact and $|f| \in C(X)$ for each $f \in C(X)$, the maximum modulus of each $f \in C(X)$ is achieved at some point $x \in X$ (Proposition 2.13). Furthermore, bounded continuous functions are integrable on a finite measure space. Hence, the definitions of $\| \cdot \|_1$ and $\| \cdot \|_\infty$ make sense for the vector space $C(X)$.

As before, it is only the triangle inequality that is in need of verification, as the other requirements for a norm are clearly met by each of $\| \cdot \|_1$ and $\| \cdot \|_\infty$. The triangle inequality $\|f + g\|_1 \leq \|f\|_1 + \|g\|_1$ is a consequence of the triangle inequality for integrals–namely $\left| \int h \, d\mu \right| \leq \int |h| \, d\mu$–and the triangle inequality for complex numbers, while the triangle inequality $\|f + g\|_\infty \leq \|f\|_\infty + \|g\|_\infty$ is a consequence of the triangle inequality for complex numbers and for the "max" function.   □

There are many norms of interest on $\mathbb{C}^n$. For example, by taking $X = \{1, \dots, n\}$ and μ to be counting measure on X, Example 5.5 implies that

$$\|\xi\|_1 = \max_{1 \leq j \leq n} |\xi_j|$$

$$\|\xi\|_\infty = \sum_{j=1}^{n} |\xi_j|$$

are norms on $\mathbb{C}^n$. (This can be easily verified directly as well.) To distinguish the Euclidean norm from the two norms above, we write $\|\xi\|_2$ for the Euclidean norm $\| \cdot \|$ of Example 5.4

Example 5.5 demonstrates how natural it is to consider different norms on a given vector space. In moving from one norm to another, the shape of the closed unit ball will change, and it may happen that V is complete in one norm but not in another norm. With equivalent norms, this pathology does not arise.

Definition 5.6. If $\| \cdot \|$ and $\| \cdot \|'$ are norms on a vector space V, then $\| \cdot \|$ and $\| \cdot \|'$ are *equivalent norms* if there are positive constants c and C such that

$$c\|v\| \le \|v\|' \le C\|v\|,$$

for every vector $v \in V$.

If one uses the notation $\| \cdot \| \sim \| \cdot \|'$ to mean that there are positive constants c and C such that $c\|v\| \le \|v\|' \le C\|v\|$ for every $v \in V$, then it is easily proved that $\sim$ is an equivalence relation on the set of all norms on a given vector space V.

Proposition 5.7. *If $\mathscr{T}$ and $\mathscr{T}'$ are the norm topologies on a vector space V induced, respectively, by equivalent norms $\| \cdot \|$ and $\| \cdot \|'$ on V, then $\mathscr{T} = \mathscr{T}'$.*

Proof. Exercise 5.102.   □

In finite-dimensional vector spaces, all norms are equivalent.

Proposition 5.8. *All norms on a finite-dimensional vector space are equivalent.*

Proof. Fix a linear basis $\{v_1, \ldots, v_n\}$ of V and let $\| \cdot \|_2$ be defined by

$$\left\| \sum_{j=1}^{n} \alpha_j v_j \right\|_2 = \sqrt{\sum_{j=1}^{n} |\alpha_j|^2}.$$

By Example 5.4, $\| \cdot \|_2$ is a norm on V. Because equivalence of norms is an equivalence relation, it is sufficient to show that if $\| \cdot \|$ is a norm on V, then $\| \cdot \|_2 \sim \| \cdot \|$.

To this end, let $C = \sqrt{\sum_{j=1}^{n} \|v_j\|^2}$. Consider $S = \{\xi \in \mathbb{C}^n \,|\, \|\xi\|_2 = 1\}$ and the function $f : S \to \mathbb{R}$ defined by $f(\xi) = \left\| \sum_{j=1}^{n} \xi_j v_j \right\|$. We claim that f is continuous. Fix $\xi \in S$ and consider a neighbourhood $W \subseteq \mathbb{R}$ of $f(\xi)$. Thus, there is a $\varepsilon > 0$ such that $(f(\xi) - \varepsilon, f(\xi) + \varepsilon) \subseteq W$. Let $V \subset \mathbb{C}^n$ be the set of all $\eta \in \mathbb{C}^n$ for which $\|\xi - \eta\|_2 < \varepsilon/C$. Thus, $U = S \cap V$ is a neighbourhood of ξ in S and if $\eta \in U$, then

$$|f(\xi) - f(\eta)| = \left\| \left\| \sum_{j=1}^{n} \xi_j v_j \right\| - \left\| \sum_{j=1}^{n} \eta_j v_j \right\| \right\|$$

$$\leq \left\| \sum_{j=1}^{n} (\xi_j - \eta_j) v_j \right\| \qquad \text{[by Exercise 5.104]}$$

$$\leq \sum_{j=1}^{n} |\xi_j - \eta_j| \, \|v_j\|$$

$$\leq \sqrt{\sum_{j=1}^{n} |\xi_j - \eta_j|^2} \sqrt{\sum_{j=1}^{n} \|v_j\|^2} \qquad \text{[by Cauchy-Schwarz]}$$

$$= \|\xi - \eta\|_2 C$$

$$< \varepsilon,$$

which implies that $f(\eta) \in W$. Hence, f is continuous at every $\xi \in S$, which implies f is a continuous function.

Because $|\zeta|^2 = |\Re \eta|^2 + |\Im \zeta|^2$ for every $\zeta \in \mathbb{C}$, S may be identified with the Euclidean sphere S^{2n-1} in $\mathbb{R}^{2n}$. Thus, S is a compact set and, hence, the continuous map f achieves its minimum value c at some element of S.

Now if $v = \sum_j \alpha_j v_j \in V$ for some $\alpha_1, \ldots, \alpha_n \in \mathbb{C}$ not all zero and if the coordinates of $\xi \in S$ are $\xi_j = \alpha_j / (\sum_k |\alpha_k|^2)^{1/2}$, then $c \leq f(\xi)$. Thus,

$$c \left(\sum_{j=1}^{n} |\alpha_j|^2 \right)^{1/2} \leq \left\| \sum_{j=1}^{n} \alpha_j v_j \right\| \leq \sum_{j=1}^{n} |\alpha_j| \, \|v_j\| \leq C \left(\sum_{j=1}^{n} |\alpha_j|^2 \right)^{1/2}.$$

That is, $c\|v\|_2 \leq \|v\| \leq C\|v\|_2$ for every $v \in V$, which proves that $\|\cdot\|_2 \sim \|\cdot\|$.   □

A slightly more general concept than that of a norm is the notion of a seminorm.

Definition 5.9. A *seminorm* on a vector space V is a function $\rho : V \to \mathbb{R}$ such that, for all $v, w \in V$ and $\alpha \in \mathbb{E}$,

1. $\rho(v) \geq 0$,
2. $\rho(\alpha v) = |\alpha| \rho(v)$,
3. $\rho(v + w) \leq \rho(v) + \rho(w)$.

Seminorms differ from norms in the following way: with a seminorm ρ, the equation $\rho(v) = 0$ can hold for nonzero v; however, with a norm $\|\cdot\|$, the equation $\|v\| = 0$ holds only for $v = 0$. Nevertheless, one can obtain a norm from a seminorm as follows.

Proposition 5.10. *If ρ is a seminorm on a vector space, and if $\sim$ is the relation on V defined by*

$$v \sim w \quad \text{if} \quad \rho(v-w) = 0 \,,$$

then $\sim$ is an equivalence relation. Moreover, if the equivalence classes of elements of V are denoted by

$$\dot{v} = \{w \in V \,|\, w \sim v\} \,,$$

then:

1. the set $V/\sim$ of equivalence classes is a vector space under the operations

$$\dot{v} + \dot{w} = (v \dot{+} w) \,, v, w \in V \,,$$
$$\alpha \dot{v} = (\dot{\alpha v}) \,, \quad \alpha \in \mathbb{C}, v \in V \,;$$

2. the function $\|\cdot\| : V/\sim \to \mathbb{R}$ defined by

$$\|\dot{v}\| = \rho(v) \,, \quad v \in V \,,$$

is a norm on $V/\sim$.

Proof. Exercise 5.106. $\qquad\qquad\qquad\qquad\qquad\qquad\qquad\qquad\qquad\qquad\qquad\qquad$ □

Through the use of a norm, one can introduce the notions of convergent sequences and series, as well as the idea of completeness.

Definition 5.11. A normed vector space $(V, \|\cdot\|)$ is a *Banach space* if (V, d) is a complete metric space, where d is the metric $d(v, w) = \|v - w\|$.

Normed vector spaces of finite dimension provide the simplest examples of Banach spaces.

Proposition 5.12. *Every finite-dimensional normed vector space is a Banach space.*

Proof. By Proposition 5.8 and its proof, if $\{v_1, \ldots, v_n\}$ is a basis for a normed vector space V, then there are positive constants c and C such that

$$c \left(\sum_{j=1}^{n} |\alpha_j|^2 \right)^{1/2} \leq \left\| \sum_{j=1}^{n} \alpha_j v_j \right\| \leq C \left(\sum_{j=1}^{n} |\alpha_j|^2 \right)^{1/2}, \qquad (5.1)$$

for all $\alpha_1, \ldots, \alpha_n \in \mathbb{C}$. Assume now that $\{w_k\}_k$ is a Cauchy sequence of elements in V. Write

$$w_k = \sum_{j=1}^{n} \alpha_j^{(k)} v_j \,, \quad \forall k \in \mathbb{N} \,.$$

Then inequality (5.1) implies that for each $j = 1, \ldots, n$, the sequence $\{\alpha_j^{(k)}\}_k$ converges in $\mathbb{C}$ to some α_j (because $\mathbb{C}$ is complete). Let $w = \sum_{j=1}^{n} \alpha_j v_j$. Inequality (5.1) implies, again, that the sequence $\{w_k\}_k$ converges in V to w. Hence, V is a Banach space. □

Definition 5.13. If X and Y are nonempty subsets of a vector spaces V, then $X + Y$ denotes the set of all vectors of the form $x + y$, where $x \in X$ and $y \in Y$.

The completeness axiom for Banach spaces has numerous consequences, such as the following topological property exhibited by compact sets.

Proposition 5.14. *Suppose that K and C are subsets of a Banach space V such that K is compact and C is closed. If $K \cap C = \emptyset$, then there exists $\varepsilon > 0$ such that*

$$(K + B_\varepsilon(0)) \cap C = \emptyset.$$

Proof. If the conclusion is not true, then there are $v_n \in K$ and $w_n \in V$ of norm $\|w_n\| < \frac{1}{n}$ such that $v_n + w_n \in C$, for every $n \in \mathbb{N}$. Since K is compact, $\{v_n\}_{n \in \mathbb{N}}$ admits a convergent subsequence $\{v_{n_k}\}_{k \in \mathbb{N}}$ with limit $v \in K$. Note that $\|v - (v_{n_k} + w_{n_k})\| \leq \|v - v_{n_k}\| + \frac{1}{n_k}$; thus, $v \in C$, as C is closed. But this contradicts $K \cap C = \emptyset$; therefore, the conclusion must hold. □

The notion of basis for a vector space, which is a common feature of linear algebra, has a limited role in Banach space theory because the defining conditions are purely algebraic and do not take into account the topological or analytic properties of the space. Nevertheless, it is interesting to examine the facts concerning linear bases, as in Proposition 5.16 below, for instance.

Recall that if S is a nonempty subset of a vector space V, then:

1. the elements of S are *linearly independent* if, for every finite subset $\{v_1, \ldots, v_n\} \subseteq S$, the equation $\sum_{j=1}^{n} \alpha_j v_j = 0$ holds, for $\alpha_1, \ldots, \alpha_n \in \mathbb{C}$, only if each $\alpha_j = 0$; and

2. the *span* of S is the set of all linear combinations $\sum_{j=1}^{n} \alpha_j v_j$, for all finitely many $v_1, \ldots, v_n \in S$.

The important point to keep in mind is that linear combinations are finite sums, even though in analysis one considers infinite sums.

Definition 5.15. A *linear basis* of a vector space V is a subset $\mathfrak{B} \subset V$ such that the elements of $\mathfrak{B}$ are linearly independent and span V.

Recall from linear algebra that if a vector space V admits a finite linear basis $\mathfrak{B}$, then every linear basis of V has cardinality equal to that of $\mathfrak{B}$ and this integer is called the *dimension* of V. In particular, V is finite dimensional if V has a finite linear basis, and V is infinite dimensional if V has no finite linear basis. In the case of infinite-dimensional Banach spaces, the question of existence of linear bases is addressed by the following proposition. The proposition also shows that the completeness of Banach spaces (as metric spaces) forces linear bases of Banach spaces to be of sufficiently large cardinality.

Proposition 5.16. *If V is an infinite-dimensional Banach space, then V has a linear basis and every linear basis of V is an uncountable set.*

Proof. By hypothesis, there is an infinite set $\mathfrak{B}_0$ of linearly independent vectors in V. Let $\mathfrak{S}$ be the set of all subsets $Y \subset V$ of linearly independent vectors for which $Y \supseteq \mathfrak{B}_0$, and define a partial order $\preceq$ on $\mathfrak{S}$ by set inclusion: that is, $X \preceq Y$ if and only if $X, Y \in \mathfrak{S}$ satisfy $X \subseteq Y$.

Let $\mathfrak{C} \subseteq \mathfrak{S}$ be a linearly ordered subset and consider the set $Y = \bigcup_{E \in \mathfrak{C}} E$. To show that the elements of Y are linearly independent, select $y_1, \dots, y_n \in Y$. Thus, there exist sets $E_1, \dots, E_n \in \mathfrak{C}$ such that $y_j \in E_j$ for all j. Because $\mathfrak{C}$ is linearly ordered, either $E_1 \subseteq E_2$ or $E_2 \subseteq E_1$. Thus, there is an $i_2 \in \{1, 2\}$ such that $y_1, y_2 \in E_{i_2}$. Likewise, $E_{i_2} \subseteq E_3$ or $E_3 \subseteq E_{i_2}$; hence, there is an $i_3 \in \{1, 2, 3\}$ such that $y_1, y_2, y_3 \in E_{i_3}$. Continuing by induction we obtain an integer $i_n \in \{1, \dots, n\}$ for which $y_1, \dots, y_n \in E_{i_n}$. As E_{i_n} consists of linearly independent vectors, we deduce that $y_1, y_2, \dots, y_n \in Y$ are linearly independent. Furthermore, $Y \supseteq E \supseteq \mathfrak{B}_0$ for every $E \in \mathfrak{C}$, and so Y is an element of $\mathfrak{S}$ and is an upperbound in $\mathfrak{S}$ of the linearly ordered set $\mathfrak{C}$. Hence, by Zorn's Lemma (Theorem 1.8), $\mathfrak{S}$ has a maximal element $\mathfrak{B}$.

To show that $\mathfrak{B}$ is a linear basis of V, consider the linear submanifold $M = \mathrm{Span}\,\mathfrak{B}$. If $M \neq V$, then there is a unit vector $v \in V$ with $v \notin M$, which implies that v is linearly independent of every vector in M and, in particular, of every vector in $\mathfrak{B}$. Thus, $\tilde{\mathfrak{B}} = \mathfrak{B} \cup \{v\}$ is a linearly independent set that properly contains $\mathfrak{B}$. But $\mathfrak{B} \preceq \tilde{\mathfrak{B}}$ and $\mathfrak{B} \neq \tilde{\mathfrak{B}}$ contradict the maximality of $\mathfrak{B}$ in $\mathfrak{S}$. Hence, it must be that $M = V$, which implies that $\mathfrak{B}$ is a linear basis of V.

Now suppose that $\mathfrak{B}$ is an arbitrary linear basis of V and assume, contrary to what we aim to prove, that $\mathfrak{B}$ is a countable set $\mathfrak{B} = \{v_n \,|\, n \in \mathbb{N}\}$. Thus, if $V_n = \mathrm{Span}\{v_1, \dots, v_n\}$, then

$$V = \bigcup_{n \in \mathbb{N}} V_n.$$

Each V_n has finite dimension and is, therefore, closed in the metric topology of V. We claim that V_n is nowhere dense—that is, that the interior U_n of V_n is empty. To this end, select $v \in U_n$ and $\varepsilon > 0$ such that $B_\varepsilon(v) \subset U_n$. On the line segment $\{(1-t)v + tv_{n+1} \,|\, t \in [0,1]\}$ connecting v and v_{n+1}, select $t \in (0,1)$ sufficiently small such that $\|(1-t)v + tv_{n+1} - v\| = t\|v_{n+1} - v\| < \varepsilon$. Hence, there is a $w_n \in B_\varepsilon(v)$ such that $v_{n+1} \in \mathrm{Span}\{v, w_n\} \subseteq V_n$, which is a contradiction. Therefore, it must be that $U_n = \emptyset$ for every $n \in \mathbb{N}$, which implies that each V_n is nowhere dense. By Corollary 2.57 of the Baire Category Theorem (Theorem 2.55), a countable union of nowhere dense sets in a complete metric space is nowhere dense. Hence, V is nowhere dense, which is impossible because V is both open and closed in V. Thus, no linear basis of V can be countable. $\qquad\square$

5.2 Subspaces, Quotients, and Bases

Because Banach spaces are vector spaces, their subspaces are of interest. However, the presence of an underlying topology means that one needs to distinguish subspaces that are closed in the topology from those that are not. Thus, the following terminology is adopted.

Definition 5.17. Suppose that V is a Banach space and that $L \subseteq V$.

1. L is a *linear submanifold*, or *linear manifold*, of V if
$$\alpha_1 w_1 + \alpha_2 w_2 \in L, \quad \forall \alpha_1, \alpha_2 \in \mathbb{C}, w_1, w_2 \in L.$$

2. L is a *subspace* of V if L is a linear submanifold of V and L is a closed set in the topology of V.

If L is a subspace of a Banach space V, then the elements of the quotient vector space V/L are denoted by $\dot{v}$, for $v \in V$, and are given by
$$\dot{v} = \{w \in V \mid v - w \in L\}.$$

Proposition 5.18. *If L is a subspace of a Banach space V, then the function $\|\cdot\|$ on the quotient space V/L, defined by*
$$\|\dot{v}\| = \inf\{\|v - y\| \mid y \in L\}, \tag{5.2}$$
is a norm on V/L such that, with respect to this norm, V/L is a Banach space.

Proof. Observe that, by definition, $\|\dot{v}\| \leq \|v\|$ for every $v \in V$. To show $\|\dot{v}\| = 0$ only if $\dot{v} = \dot{0}$, suppose that $\|\dot{v}\| = 0$. Thus, there is a sequence of vectors $y_n \in L$ such that $\|v - y_n\| \to 0$. Because L is closed and v is the limit of the sequence $\{y_n\}_{n \in \mathbb{N}}$, v must belong to L, which yields $\dot{v} = \dot{0}$. It is readily apparent that $\|\alpha \dot{v}\| = |\alpha| \|\dot{v}\|$ and $\|\dot{v} + \dot{w}\| \leq \|\dot{v}\| + \|\dot{w}\|$, and so the function $\|\cdot\|$ defined by equation (5.2) is indeed a norm on the vector space V/L.

Now suppose that $\{\dot{v}_k\}_{k \in \mathbb{N}}$ be a Cauchy sequence in V/L, and select a subsequence $\{\dot{v}_{k_j}\}_{j \in \mathbb{N}}$ with the property that $\|\dot{v}_{k_j} - \dot{v}_{k_{j-1}}\| < 2^{-j}$ for every $j \in \mathbb{N}$. Therefore, for each j there exists $y_j \in (\dot{v}_{k_j} - \dot{v}_{k_{j-1}})$ with $\|y_j\| \leq \|\dot{v}_{k_j} - \dot{v}_{k_{j-1}}\| < 2^{-j}$. Hence, $\sum_{j=2}^{\infty} \|y_j\|$ is convergent in $\mathbb{R}$, which implies that $\sum_{j=2}^{\infty} y_j$ is convergent in V to some vector y. Consider $\dot{y} + \dot{v}_{k_1} \in V/L$. For any $j \in \mathbb{N}$,

$$\left\| \dot{v}_{k_j} - (\dot{y} + \dot{v}_{k_1}) \right\| = \left\| (\dot{v}_{k_j} - \dot{v}_{k_1}) - \dot{y} \right\| = \left\| \sum_{i=2}^{j} (\dot{v}_{k_i} - \dot{v}_{k_{i-1}}) - \dot{y} \right\|$$

$$= \left\| \sum_{i=2}^{j} \dot{y}_i - \dot{y} \right\| \leq \left\| \sum_{i=2}^{j} y_i - y \right\|.$$

Because $\sum_{i=2}^{\infty} y_i$ converges to $y \in V$, the sequence $\{\dot{v}_{k_j}\}_j$ converges to $\dot{y} + \dot{v}_{k_1}$ in V/L. However, as $\{\dot{v}_{k_j}\}_j$ is a convergent subsequence of the Cauchy sequence $\{\dot{v}_k\}_{k \in \mathbb{N}}$, the sequence $\{\dot{v}_k\}_k$ must also converge in V/L to $\dot{y} + \dot{v}_{k_1}$. □

The next two lemmas give important analytical information about infinite-dimensional Banach spaces, and we shall make use of them frequently in this study.

Lemma 5.19 (Increasing Subspace Chain Lemma). *If V is an infinite-dimensional Banach space and if $\{M_n\}_{n \in \mathbb{N}}$ is a sequence of subspaces such that $M_n \subset M_{n+1}$ (proper containment), then for each $\delta \in (0, 1)$ there is a sequence of vectors $v_n \in V$ such that*

1. $v_n \in M_{n+1}$ and $v_n \notin M_n$,
2. $\|v_n\| = 1$,
3. $\|v_n - u\| \geq \delta$ for all vectors $u \in M_n$, and
4. $\|v_k - v_j\| \geq \delta$ for all $j, k \in \mathbb{N}$ such that $k \neq j$.

Proof. Fix $n \in \mathbb{N}$. Because the containment $M_n \subset M_{n+1}$ is proper, the quotient space M_{n+1}/M_n is nonzero; hence, there is a vector of norm δ. Since $\delta < 1$, there are vectors $f_n \in M_n$ and $g_n \in M_{n+1}$ such that $\|\dot{g}_n\| = \delta$ and $\|g_n - f_n\| = 1$. Let $v_n = g_n - f_n$. Because $\dot{v}_n = \dot{g}_n$ and $\delta = \|\dot{v}_n\| = \inf\{\|v_n - f\| \mid f \in M_n\}$, we deduce that $\|v_n - u\| \geq \delta$ for all vectors $u \in M_n$.

Having produced such a unit vector $v_n \in M_{n+1}$ for each $n \in \mathbb{N}$, now select $j, k \in \mathbb{N}$ with $k \neq j$. Without loss of generality, assume that $j < k$. Thus, $v_j \in M_{j+1} \subseteq M_k$ and, therefore, $\|v_k - v_j\| \geq \delta$. □

The second lemma is proved by precisely the same type of argument.

Lemma 5.20 (Decreasing Subspace Chain Lemma). *If V is an infinite-dimensional Banach space and if $\{M_n\}_{n \in \mathbb{N}}$ is a sequence of subspaces such that $M_n \supset M_{n+1}$ (proper containment), then for each $\delta \in (0, 1)$ there is a sequence of vectors $v_n \in V$ such that*

1. $v_n \in M_n$ and $v_n \notin M_{n+1}$,
2. $\|v_n\| = 1$,
3. $\|v_n - u\| \geq \delta$ for all vectors $u \in M_{n+1}$, and
4. $\|v_k - v_j\| \geq \delta$ for all $j, k \in \mathbb{N}$ such that $k \neq j$.

A notable consequence of Lemma 5.19 is:

Proposition 5.21. *If V is an infinite-dimensional Banach space, then there exists a sequence $\{v_n\}_{n \in \mathbb{N}}$ of (distinct) unit vectors $v_n \in V$ such that $\{v_n\}_{n \in \mathbb{N}}$ has no convergent subsequences.*

Proof. Select $\delta \in (0, 1)$. By Lemma 5.19 there is a sequence $\{v_n\}_{n \in \mathbb{N}}$ of (distinct) unit vectors $v_n \in V$ such that $\|v_k - v_j\| \geq \delta$ for every $j \neq k$. Thus, $\{v_n\}_{n \in \mathbb{N}}$ has no Cauchy subsequences and, therefore, $\{v_n\}_{n \in \mathbb{N}}$ has no convergent subsequences. □

Proposition 5.21 is noted here for future reference. The first use of the result is in the following characterisation of the compactness of the closed unit ball in terms of dimension.

Proposition 5.22. *The closed unit ball in a Banach space V is compact if and only if V has finite dimension.*

Proof. Suppose that V has dimension $n \in \mathbb{N}$. By Proposition 5.8, the closed unit balls of V and $\ell^\infty(n)$ are homeomorphic. Because the unit ball of $\ell^\infty(n)$ is a product of n copies of the closed unit disc $\overline{\mathbb{D}} = \{z \in \mathbb{C} \mid |z| \leq 1\}$, and because $\overline{\mathbb{D}}$ is compact, the closed unit balls of $\ell^\infty(n)$ and V are compact.

If V has infinite dimension, then Proposition 5.21 shows that there exists a sequence $\{v_n\}_{n\in\mathbb{N}}$ of (distinct) unit vectors $v_n \in V$ such that $\{v_n\}_{n\in\mathbb{N}}$ has no convergent subsequences. Because sequences in compact metric spaces admit convergence subsequences (Proposition 2.19), the unit sphere is not a compact set, nor is any closed set that contains the closed unit sphere. Hence, the closed unit ball of V is compact only if V has finite dimension. □

5.3 Banach Spaces of Continuous Functions

Recall from Definition 2.20 that a topological space X is *locally compact* if, for every $x \in X$, there is an open set $U \subset X$ containing x such that $\overline{U}$ is compact. Assuming X is locally compact, let $C_b(X)$ denote the set of all functions $f : X \to \mathbb{C}$ that are continuous and bounded. Thus, for every $f \in C_b(X)$ means that there is an $R > 0$ such that $|f(x)| < R$ for all $x \in X$.

Theorem 5.23. *If X is a locally compact space, then $C_b(X)$ is a Banach space, where the vector space operations are given by the usual pointwise operations, and where the norm of $f \in C_b(X)$ is defined by*

$$\|f\| = \sup_{x\in X} |f(x)|. \tag{5.3}$$

Proof. It is elementary that $C_b(X)$ is a vector space and that (5.3) defines a norm on $C_b(X)$. Thus, it remains only to show that every Cauchy sequence in $C_b(X)$ is convergent in $C_b(X)$.

Let $\{f_k\}_{k\in\mathbb{N}} \subset C_b(X)$ denote a Cauchy sequence. For each $x \in X$,

$$|f_n(x) - f_m(x)| \leq \sup_{y\in X} |f_n(y) - f_m(y)| = \|f_n - f_m\|.$$

Since $\{f_k\}_{k\in\mathbb{N}}$ is a Cauchy sequence in $C_b(X)$, $\{f_k(x)\}_{k\in\mathbb{N}}$ is a Cauchy sequence in $\mathbb{C}$ for each $x \in X$. Because $\mathbb{C}$ is complete, $\lim_k f_k(x)$ exists for every $x \in X$. Therefore, define $f : X \to \mathbb{C}$ by $f(x) = \lim_k f_k(x)$, for each $x \in X$. We aim to show (i) that f is continuous and bounded, and (ii) that $\{f_k\}_{k\in\mathbb{N}}$ converges to f in $C_b(X)$.

Let $\varepsilon > 0$. Because $\{f_k\}_{k\in\mathbb{N}}$ is a Cauchy sequence, there exists $N_\varepsilon \in \mathbb{N}$ such that $\|f_n - f_m\| < \varepsilon$ for all $n, m \geq N_\varepsilon$. Assume that $n \geq N_\varepsilon$. Choose any $x \in X$; thus,

$$|f(x) - f_n(x)| \leq |f(x) - f_m(x)| + |f_m(x) - f_n(x)|$$

$$\leq |f(x) - f_m(x)| + \|f_m - f_n\|.$$

As the inequalities above are true for all $m \in \mathbb{N}$,

$$|f(x) - f_n(x)| \leq \inf_{m\in\mathbb{N}} (|f(x) - f_m(x)| + \|f_m - f_n\|)$$

$$\leq 0 + \varepsilon.$$

This right-hand side of the inequality above is independent of the choice of $x \in X$. Hence, if $n \geq N_\varepsilon$ is fixed, then $f - f_n$ is a bounded function $X \to \mathbb{C}$ and

$$\sup_{x\in X} |f(x) - f_n(x)| \leq \varepsilon.$$

Since f is uniformly within ε of a continuous function, f is continuous at each $x \in X$. Furthermore, since the sum of bounded functions is bounded, $f_n + (f - f_n) = f$ is bounded. This proves that $f \in C_b(X)$. Finally, since $f \in C_b(X)$ satisfies $\|f - f_n\| \leq \varepsilon$ for all $n \geq N_\varepsilon$, the Cauchy sequence $\{f_k\}_{k\in\mathbb{N}}$ converges in $C_b(X)$ to $f \in C_b(X)$. □

Notational Convention For a compact space X, we denote $C_b(X)$ by $C(X)$.

If X is compact and $f \in C(X)$, then $f(X)$ is a compact subset of $\mathbb{C}$; hence, f is bounded and attains its supremum at some point of X. Therefore, the norm $\|f\|$ is given by

$$\|f\| = \max_{x\in X} |f(x)|.$$

Because $\mathbb{C}$ has multiplication as well as addition, we may multiply $f, g \in C_b(X)$ to produce a function $fg : X \to \mathbb{C}$ whose value $(fg)[x]$ at each $x \in X$ is defined by

$$(fg)[x] = f(x)g(x).$$

It is not difficult to see that $fg \in C_b(X)$ and that $\|fg\| \leq \|f\|\,\|g\|$.

Definition 5.24. An *associative algebra*—or, more simply, an *algebra*—is a complex vector space A endowed by with a product (or multiplication) operation such that, for all $a, b, c \in A$ and all $\alpha \in \mathbb{C}$,

$$(a+b)c = (ac + bc), \quad a(b+c) = ab + ac, \quad a(bc) = (ab)c,$$

and

$$(\alpha a)(b) = a(\alpha b) = \alpha(ab).$$

Furthermore, if $ab = ba$, for all $a, b \in A$, then A is called an *abelian* algebra, and if there is an element $1 \in A$ such that $a1 = 1a = a$, for every $a \in A$, then A is said to be a *unital* algebra and 1 is the *multiplicative identity* of A.

It is not difficult to show that, if 1 and $1'$ are multiplicative identities for an algebra A, then $1' = 1$.

Definition 5.25. A *Banach algebra* is a complex associative algebra A together with a norm $\| \cdot \|$ on A such that

1. $\|xy\| \leq \|x\| \|y\|$, for all $x, y \in A$, and
2. A is a Banach space under the norm $\| \cdot \|$.

Furthermore, if A is a unital algebra, then A is a *unital Banach algebra* if $\|1\| = 1$.

Thus, if X is a compact space, then $C(X)$ is a Banach algebra. The (constant) function that sends each $x \in X$ to $1 \in \mathbb{C}$ is denoted by "1" and it serves as the multiplicative identity for $C(X)$ in the sense that $f1 = f$ for every $f \in C(X)$.

Definition 5.26. A *uniform algebra* on a compact space X is a subset $A \subseteq C(X)$ such that:

1. A is a Banach subalgebra of $C(X)$;
2. $1 \in A$;
3. A separates the points of X—that is, if $x_1, x_2 \in X$ are distinct, then there exists a function $f \in A$ such that $f(x_1) \neq f(x_2)$.

Discussion of uniform algebras makes sense only if X is Hausdorff:

Proposition 5.27. $C(X)$ *is a uniform algebra on a compact space X if and only if X is Hausdorff.*

Proof. Suppose that A is any uniform algebra on X and that $x_1, x_2 \in X$ are distinct. By hypothesis, there is a function $f \in A$ such that $f(x_1) \neq f(x_2)$. In $\mathbb{C}$ there are disjoint open sets V_1 and V_2 that contain $f(x_1)$ and $f(x_2)$, respectively. Thus, by continuity of f, $U_1 = f^{-1}(V_1)$ and $U_2 = f^{-1}(V_2)$ disjoint open sets in X that contain x_1 and x_2, respectively, which proves that X is a Hausdorff space.

Conversely, suppose that X is Hausdorff. Because X is a normal space (Proposition 2.34), if $x_0, x_1 \in X$ are distinct, then Urysohn's Lemma applied to the point sets $\{x_0\}$ and $\{x_1\}$ (which are closed because X is Hausdorff) yields a function $f \in C(X)$ such that $f(x_0) = 0 \neq 1 = f(x_1)$. Hence, $C(X)$ is a uniform algebra. □

The elements of $C(X)$ are complex-valued functions. Therefore, for each $f \in C(X)$ one can consider the continuous function $\overline{f} : X \to \mathbb{C}$ defined by

$$\overline{f}(x) = \overline{f(x)}, \quad \forall x \in X.$$

Definition 5.28. A nonempty subset $S \subseteq C(X)$ is *selfadjoint* if $\overline{f} \in S$ for every $f \in S$.

The Stone-Weierstrass Theorem, Theorem 5.30 below, asserts that if A is a selfadjoint uniform algebra of continous functions $f : X \to \mathbb{C}$, where X is a compact Hausdorff space, then $A = C(X)$.

Lemma 5.29. *Suppose X is compact and that $A \subseteq C(X)$ is a unital, closed subalgebra of $C(X)$. If A is selfadjoint, and if $f, f_1, \ldots, f_n \in A$ are real-valued functions, then A also contains the following real-valued functions:*

(i) $|f|$;
(ii) $\min(f_1, \ldots, f_n)$;
(iii) $\max(f_1, \ldots, f_n)$; *and*
(iii) f^+ *and* f^-, *where* $f^+ = \frac{1}{2}(|f| + f)$ *and* $f^- = \frac{1}{2}(|f| - f)$.

Proof. If $f = 0$, then $|f| = 0$ and so $|f| \in A$ trivially. Assume, therefore, that $f \neq 0$; by normalising, we may assume without loss of generality that $\|f\| = 1$, which implies that $f(x) \in [-1, 1]$ for all $x \in X$. By Newton's Binomial Theorem,

$$\sqrt{1-t} = 1 - \frac{t}{2} + \sum_{n=2}^{\infty} (-1)^n \frac{1 \cdot 3 \cdots (2n-3)}{2^n n!} t^n,$$

which converges absolutely on $[-1, 1]$ and uniformly on compact subintervals of $(-1, 1)$. For notational convenience, let φ denote the function on $[-1, 1]$ given by $\varphi(t) = \sqrt{1-t}$ and write the power series expansion above of φ as

$$\varphi(t) = \sum_{n=0}^{\infty} \alpha_n t^n.$$

For each $\delta \in (0, 1)$ let $g_\delta \in C(X)$ be given by $g_\delta(x) = \delta + (1 - \delta)f(x)^2$; that is, $g_\delta = \delta + (1 - \delta)f^2$, where $\delta \in A$ is the constant function $x \mapsto \delta$. Because A is an algebra, $f^2 \in A$ and $g_\delta \in A$. Furthermore, $f(x)^2 \in [0, 1]$ for all $x \in X$, and so $0 \leq g_\delta \leq 1$ and $0 \leq 1 - g_\delta = (1 - \delta)(1 - f^2) \leq 1 - \delta$. That is,

$$1 - g_\delta(x) \in [0, 1 - \delta], \quad \forall x \in X.$$

Fix $k \in \mathbb{N}$ and define $f_{\delta,k}$ by

$$f_{\delta,k} = \sum_{n=0}^{k} \alpha_n (1 - g_\delta)^n,$$

where $\alpha_0, \ldots, \alpha_k \in \mathbb{R}$ are the coefficients in the power series expansion of φ. Thus, $f_{\delta,k} \in A$ and

$$\|f_{\delta,k} - (g_\delta)^{1/2}\| = \max_{x \in X} \left| \sum_{n=0}^{k} \alpha_n (1 - g_\delta(x))^n - \varphi(1 - g_\delta(x)) \right|$$

$$\leq \max_{t \in [0, 1-\delta]} \left| \sum_{n=0}^{k} \alpha_n t^n - \varphi(t) \right|.$$

By Newton's Binomial Theorem, this final limit tends to zero as $k \to \infty$. Hence, $(g_\delta)^{1/2} \in A$ (as A is norm closed). Note that $\|f^2 - g_\delta\| = \delta\|1 + f^2\| \to 0$ as $\delta \to 0$; that is, $g_\delta \to f^2$ uniformly on X as $\delta \to 0$. The function $\psi(t) = \sqrt{t}$ is uniformly continuous on the compact set $[0, 1]$, and so $\psi \circ g_\delta \to \psi \circ f^2$ uniformly on X as $\delta \to 0$. Because $\psi \circ g_\delta = (g_\delta)^{1/2} \in A$ and $\psi \circ f^2 = |f|$, the limit $\|(g_\delta)^{1/2} - |f|\| \to 0$ implies that $|f| \in A$. This completes the proof that $|f| \in A$ for every real-valued $f \in A$.

As a consequence of the arguments above, if $f_1, f_2 \in A$ are real-valued, then the continuous functions $\frac{1}{2}(f_1 + f_2 + |f_1 - f_2|)$ and $\frac{1}{2}(f_1 + f_2 - |f_1 - f_2|)$ are elements of A. That is, $\max(f_1, f_2) \in A$ and $\min(f_1, f_2) \in A$. By induction,

$$\max(f_1, \ldots, f_m) = \max(\max(f_1, \ldots, f_{n-1}), f_n)$$

and

$$\min(f_1, \ldots, f_m) = \min(\min(f_1, \ldots, f_{n-1}), f_n)$$

are elements of A. Likewise, $f^+, f^- \in A$. □

Theorem 5.30 (Stone-Weierstrass Theorem). *If X is a compact Hausdorff space and if A is a selfadjoint uniform algebra on X, then $A = C(X)$.*

Proof. First of all, because A is selfadjoint, A is spanned by real-valued functions. Indeed, if $f \in A$, then $\Re f = \frac{1}{2}(f + \bar{f})$ and $\Im f = \frac{1}{2i}(f - \bar{f})$ are real-valued elements of A and $f = \Re f + i\Im f$. Therefore it is sufficient to prove that $f \in A$ for every real-valued function $f \in C(X)$.

To this end, assume that $f \in C(X)$ and let $\varepsilon > 0$ be arbitrary. Fix $x_0 \in X$ and select any $x_1 \in X$ for which $x_1 \neq x_0$. Because A separates points, there is a function $h \in A$ such that $h(x_1) \neq h(x_0)$. At least one of $\Re h(x_1) \neq \Re h(x_0)$ or $\Im h(x_1) \neq \Im h(x_0)$ holds, and so we may assume, because A is self-adjoint, that h is a real-valued function. Consider now the real-valued function $g_{x_1} \in C(X)$ defined by

$$g_{x_1}(y) = f(x_0) + (f(x_1) - f(x_0)) \left[\frac{h(y) - h(x_0)}{h(x_1) - h(x_0)} \right], \quad \forall y \in X.$$

In particular, for $y = x_0$ and $y = x_1$ we obtain $g_{x_1}(x_0) = f(x_0)$ and $g_{x_1}(x_1) = f(x_1)$. Now this construction holds as long as $x_1 \neq x_0$; we define g_{x_0} to be f. What has been proved, then, is the following assertion: given a fixed $x_0 \in X$, there exists, for every $x \in X$, a real-valued $g_x \in A$ such that $g_x(x_0) = f(x_0)$ and $g_x(x) = f(x)$.

Continuing with the assumption that $x_0 \in X$ is fixed, note that $g_x - f \in C(X)$ for every $x \in X$. Hence, if $W_x \subseteq \mathbb{C}$ is the open set $W_x = \{z \in \mathbb{C} \,|\, \Re z < \varepsilon\}$, then

$$U_x = (g_x - f)^{-1}(W_x) = \{y \in X \,|\, g_x(y) - f(y) < \varepsilon\}$$

is open in X. Furthermore, $g_x(x) = f(x)$ implies that $x \in U_x$. Hence, $\{U_x\}_{x \in X}$ is an open cover X. The compactness of X implies that this covering admits a finite subcover, say, $U_{x_1}, \ldots, U_{x_n}$. The functions g_{x_j} that define these n open sets determine another element of A: namely, $\min\{g_{x_1},\ldots,g_{x_n}\}$, by Lemma 5.29. Because all of this has depended on the fixed element $x_0 \in X$, this minimum function shall be denoted by h_{x_0}. That is, $h_{x_0} = \min\{g_{x_1},\ldots,g_{x_n}\}$ and h_{x_0} has the property

$$h_{x_0}(y) < f(y) + \varepsilon, \quad \forall y \in X. \tag{5.4}$$

Now allow x_0 to vary throughout X, producing for each point x_0 the continuous function $h_{x_0} \in A$ described above. For each $x_0 \in X$, consider the open subset V_{x_0} of X defined by

$$V_{x_0} = \{y \in X \mid h_{x_0}(y) - f(y) > -\varepsilon\}.$$

Since $h_{x_0}(x_0) - f(x_0) = 0$, $x_0 \in V_{x_0}$. Moreover,

$$h_{x_0}(y) > f(y) - \varepsilon, \quad \forall y \in V_{x_0}.$$

Therefore, $\{V_{x_0}\}_{x_0 \in X}$ is an open cover of X and, by the compactness of X, there is a finite subcover: $V_{x_1}, \ldots, V_{x_m}$. Let $\kappa = \max\{h_{x_1},\ldots,h_{x_m}\}$, which is an element of A by Lemma 5.29. Thus, for each $j = 1,\ldots,m$,

$$h_j(y) \geq \kappa(y) > f(y) - \varepsilon, \quad \forall y \in X. \tag{5.5}$$

Combining inequalities (5.4) and (5.5) leads to

$$f(y) - \varepsilon < \kappa(y) < f(y) + \varepsilon, \quad \forall y \in X.$$

That is, $\|f - \kappa\| < \varepsilon$. $\qquad\qquad\qquad\qquad\qquad\qquad\qquad\qquad\qquad\qquad\qquad\qquad\square$

Corollary 5.31 (Classical Weierstrass Approximation Theorem). *If $f : [a,b] \to \mathbb{C}$ is a continuous function, then for every $\varepsilon > 0$ there is a polynomial p with complex coefficients such that $|f(t) - p(t)| < \varepsilon$, for all $t \in [a,b]$.*

Proof. Let A be the closure in $C([a,b])$ of the ring $\mathbb{C}[t]$ of polynomials in one indeterminate t. Thus, A is a norm-closed subalgebra of $C([a,b])$ and $1 \in A$. Moreover A separates the points of $[a,b]$, for if $x_1, x_2 \in [a,b]$ are distinct, then $q(x_1) = 0$ and $q(x_2) \neq 0$ for the element $q \in A$ given by $q(t) = t - x_1$. Therefore, the Stone-Weierstrass Theorem yields $A = C([a,b])$. In particular, by the construction of A, if $f \in C([a,b]$ and if $\varepsilon > 0$, then there is a polynomial p such that $\|f - p\| < \varepsilon$. $\qquad\square$

In the case of non-compact, locally compact spaces, there are functions that, in certain respects, mimic continuous functions on compact spaces.

Recall from Definition 2.40 that the support of a continuous function $f : X \to \mathbb{C}$ on a topological space X is the set $\operatorname{supp} f \subseteq X$ defined by

$$\operatorname{supp} f = \overline{\{x \in X \mid f(x) \neq 0\}}.$$

Definition 5.32. If X is a locally compact Hausdorff space, then let

$$C_c(X) = \{f : X \to \mathbb{C} \,|\, f \text{ is continuous and supp } f \text{ is compact}\},$$

the set of all continuous complex-valued functions on X having compact support.

Proposition 5.33. *If X is a locally compact space, then $C_c(X)$ is a subspace of $C_b(X)$.*

Proof. Exercise 5.110.   □

   Another useful Banach space of continuous functions is the space consisting of continuing functions that vanish at infinity.

Definition 5.34. Suppose that X is locally compact space. A continuous function $f : X \to \mathbb{C}$ *vanishes at infinity* if the set

$$\{x \in X \,|\, |f(x)| \geq \varepsilon\}$$

is compact in X for every $\varepsilon > 0$.

   The final result of this section makes note of the fact that $C_0(X)$ has some additional algebraic structure, and that it is a closed set in the topology of $C_b(X)$.

Proposition 5.35. *If X is a locally compact space, then*

1. $C_0(X)$ is a subspace of $C_b(X)$, and
2. $fg \in C_0(X)$, for all $f \in C_0(X)$ and $g \in C_b(X)$.

Proof. Exercise 5.111.   □

5.4   Banach Spaces of *p*-Integrable Functions

Proposition 5.36. *Suppose that (X, Σ, μ) is a measure space, and that $p \geq 1$. If*

$$\mathscr{L}^p(X, \Sigma, \mu) = \{f : X \to \mathbb{C} \,|\, f \text{ is p-integrable}\},$$

then $\mathscr{L}^p(X, \Sigma, \mu)$ is a complex vector space. Furthermore, if $\rho : \mathscr{L}^p(X, \Sigma, \mu) \to \mathbb{R}$ is given by

$$\rho(f) = \left(\int_X |f|^p \, d\mu \right)^{1/p}, \tag{5.6}$$

for all $f \in \mathscr{L}^p(X, \Sigma, \mu)$, then ρ is a seminorm on $\mathscr{L}^p(X, \Sigma, \mu)$.

Proof. It is clear that $\alpha f \in \mathscr{L}^p(X, \Sigma, \mu)$, for every $\alpha \in \mathbb{C}$ and $f \in \mathscr{L}^p(X, \Sigma, \mu)$. If $f, g \in \mathscr{L}^p(X, \Sigma, \mu)$, then $f + g \in \mathscr{L}^p(X, \Sigma, \mu)$, by Minkowski's inequality. Hence, $\mathscr{L}^p(X, \Sigma, \mu)$ is a vector space.

To verify that ρ is a seminorm, the only nontrivial fact to confirm is the triangle inequality holds. To this end, Minkowski's inequality yields:

$$\rho(f+g) = \left(\int_X |f+g|^p \, d\mu\right)^{1/p}$$

$$\leq \left(\int_X |f|^p \, d\mu\right)^{1/p} + \left(\int_X |g|^p \, d\mu\right)^{1/p}$$

$$= \rho(f) + \rho(g).$$

Hence, ρ is a seminorm.   □

The seminorm ρ of Proposition 5.36 need not be a norm. For example, if f is the characteristic function on the set $\mathbb{Q}$ of rational numbers, and if m denotes Lebesgue measure on $\mathbb{R}$, then $f \neq 0$, yet

$$\rho(f) = \left(\int_\mathbb{R} |f|^p \, dm\right)^{1/p} = m(\mathbb{Q})^{1/p} = 0.$$

On the other hand, Proposition 5.10 demonstrates that a *bona fide* normed vector space can be obtained by passing to equivalence classes.

Definition 5.37. If $p \geq 1$, and if (X, Σ, μ) is a measure space, then $L^p(X, \Sigma, \mu)$ denotes the normed vector space

$$L^p(X, \Sigma, \mu) = \mathscr{L}^p(X, \Sigma, \mu)/\sim,$$

where $\sim$ is the equivalence relation $f \sim g$ if $\rho(f-g) = 0$ and where ρ is the seminorm in (5.6). The vector space $L^p(X, \Sigma, \mu)$ is called an L^p-*space*.

Conceptual and Notational Convention The vector space $L^p(X, \Sigma, \mu)$ is a vector space of equivalence classes of p-integrable functions $f : X \to \mathbb{C}$. Thus, one properly denotes the elements of $L^p(X, \Sigma, \mu)$ by $\dot{f}$, where $f \in \mathscr{L}^p(X, \Sigma, \mu)$. However, it is a standard practice to denote the elements of $L^p(X, \Sigma, \mu)$ as simply f rather than $\dot{f}$. Nevertheless, we have adopted the notation $\mathscr{L}^p(X, \Sigma, \mu)$ for the purpose of designating functions, and so, in the interests of clarity, we retain the use of the notation $\dot{f}$ for the equivalence class of $f \in \mathscr{L}^p(X, \Sigma, \mu)$ in the normed vector space $L^p(X, \Sigma, \mu)$.

The following result gives rise to another class of Banach spaces.

Theorem 5.38 (Riesz). $L^p(X, \Sigma, \mu)$ *is a Banach space, for every* $p \geq 1$.

Proof. Suppose that $\{\dot{f}_k\}_{k\in\mathbb{N}}$ is a Cauchy sequence in $L^p(X, \Sigma, \mu)$, for some sequence of p-integrable functions $f_k \in \mathscr{L}^p(X, \Sigma, \mu)$. Because this sequence is Cauchy, one can extract from it a subsequence $\{\dot{f}_{k_j}\}_{j\in\mathbb{N}}$ such that

$$\|\dot{f}_{k_{j+1}} - \dot{f}_{k_j}\| < \left(\frac{1}{2}\right)^j, \quad \forall j \in \mathbb{N}.$$

For every $i \in \mathbb{N}$, let $g_i \in \mathscr{L}^p(X, \Sigma, \mu)$ be given by

$$g_i = \sum_{j=1}^{i} |f_{k_{j+1}} - f_{k_j}|.$$

Observe that $\{g_i\}_{i \in \mathbb{N}}$ is a monotone-increasing sequence and that

$$\|g_i\| \leq \sum_{j=1}^{i} \| |f_{k_{j+1}} - f_{k_j}| \| \leq \sum_{j=1}^{i} 2^{-j} \leq \sum_{j=1}^{\infty} 2^{-j} = 1.$$

Thus,

$$\int_X g_i^p \, d\mu \leq 1, \quad \forall i \in \mathbb{N}.$$

The converse to the Monotone Convergence Theorem (Theorem 4.23) implies that $\lim_i g_i(x)^p$ exists for almost all $x \in X$; thus, $\lim_i g_i(x)$ exists for almost all $x \in X$ and the limit function—call it g—is p-integrable. Let $L \subseteq X$ denote the set of points x in which $\lim_i g_i(x)$ exists; thus,

$$\lim_{i \to \infty} g_i(x) = \lim_{i \to \infty} \sum_{j=1}^{i} |f_{k_{j+1}}(x) - f_{k_j}(x)|. \tag{5.7}$$

If $f : L \to \mathbb{C}$ denotes the function defined by

$$f(x) = f_{k_1}(x) + \sum_{j=1}^{\infty} \left(f_{k_{j+1}}(x) - f_{k_j}(x) \right),$$

then series above converges absolutely, by (5.7), for every $x \in L$. Extend f to all of X by setting $f(x) = 0$ if $x \in X \backslash L$. The $(i-1)$-th partial sum of f is precisely f_{k_i}, and

$$|f_{k_i}| = \left| f_{k_1} + \sum_{j=1}^{i-1} (f_{k_{j+1}} - f_{k_j}) \right| \leq |f_{k_1}| + g_{i-1} \leq 2g.$$

Therefore, $|f_{k_i}|^p \leq 2^p g^p$ for all $i \in \mathbb{N}$. As g^p is integrable and $\lim_i f_{k_i}(x)^p = f(x)^p$ for all $x \in L$, the Dominated Convergence Theorem asserts that f^p is integrable. This proves that $f \in \mathscr{L}^p(X, \Sigma, \mu)$.

What remains is to show $\lim_k \|f - f_k\| = 0$. To this end, note that

$$|f - f_{k_i}|^p \leq (|f| + |f_{k_i}|)^p \leq 2^p g^p \quad \forall i \in \mathbb{N}.$$

Therefore, by the Dominated Convergence Theorem, $|f - f_{k_i}|^p$ is integrable for every i and the limit function—in this case 0—satisfies

$$\lim_{i \to \infty} \int_X |0 - |f - f_{k_i}|^p| \, d\mu = 0;$$

that is,

$$\lim_{i \to \infty} \int_X |f - f_{k_i}|^p \, d\mu = 0.$$

Hence, $\|\dot{f} - \dot{f}_{k_i}\| \to 0$ as $i \to \infty$. To show that the entire Cauchy sequence $\{\dot{f}_k\}_{k \in \mathbb{N}}$ converges in $L^p(X, \Sigma, \mu)$ to $\dot{f}$, let $\varepsilon > 0$. Since $\{\dot{f}_k\}_{k \in \mathbb{N}}$ is a Cauchy sequence, there exists $N_\varepsilon \in \mathbb{N}$ such that $\liminf_{i \in \mathbb{N}} \|\dot{f}_{k_i} - \dot{f}_{N_\varepsilon}\| < \varepsilon$. Because $f = \lim_i f_{k_i}$ almost everywhere, Fatou's Lemma yields the sought-for conclusion: namely, for any $m \geq N_\varepsilon$,

$$\|\dot{f} - \dot{f}_m\|^p = \int_X |f - f_m|^p \, d\mu$$

$$\leq \liminf_{i \in \mathbb{N}} \int_X |f_{k_i} - f_m|^p \, d\mu \qquad \text{(Fatou's Lemma)}$$

$$= \liminf_{i \in \mathbb{N}} \|\dot{f}_{k_i} - \dot{f}_m\|^p$$

$$\leq \liminf_{i \in \mathbb{N}} \|\dot{f}_{k_i} - \dot{f}_{N_\varepsilon}\|^p$$

$$< \varepsilon^p.$$

This completes the proof that $L^p(X, \Sigma, \mu)$ is a Banach space.   □

By specialising to counting measure on $\mathbb{N}$, one obtains the Banach spaces of $\ell^p(\mathbb{N})$ of p-summable sequences of complex numbers:

$$\ell^p(\mathbb{N}) = \left\{ \mathfrak{a} = \{\alpha_k\}_{k \in \mathbb{N}} \, | \, \alpha_k \in \mathbb{C}, \text{ for all } k \in \mathbb{N}, \text{ and } \sum_{k=1}^{\infty} |\alpha_k|^p < \infty \right\}.$$

In this case, the seminorm ρ of Proposition 5.36 is in fact a norm on this space.

Corollary 5.39. *If $p \geq 1$, then $\ell^p(\mathbb{N})$ is a Banach space with respect to the norm*

$$\|\mathfrak{a}\| = \left(\sum_{k \in \mathbb{N}} |\alpha_k|^p \right)^{1/p}.$$

Notational Convention If $n \in \mathbb{N}$, then $\ell^p(n)$ denotes the normed vector space of sequences

$$\mathfrak{a} = \{\alpha_k\}_{k=1}^n$$

of complex numbers, considered as a finite-dimensional subspace of $\ell^p(\mathbb{N})$.

Which measurable functions belong to $\mathscr{L}^p(X, \Sigma, \mu)$? In the case of simple functions, the criterion is quite basic.

Lemma 5.40. *If φ is a simple function on a measure space (X, Σ, μ), then φ is p-integrable if and only if $\mu\left(\varphi^{-1}(\mathbb{C} \setminus \{0\})\right) < \infty$.*

Proof. Exercise 5.112. □

The following proposition is a type of approximation result.

Proposition 5.41. *The linear submanifold $\{\dot{\varphi} \mid \varphi \text{ is simple and } p\text{-integrable}\}$ is dense in $L^p(X, \Sigma, \mu)$, for every $p \geq 1$.*

Proof. If $f \in \mathscr{L}^p(X, \Sigma, \mu)$, then $\overline{f}$ is also p-integrable, which implies that both the real and imaginary parts of f are p-integrable. Moreover, every real-valued p-integrable function is a difference of nonnegative p-integrable functions, by equation (3.3). Therefore, it is sufficient to prove that if f is a nonnegative p-integrable function and if $\varepsilon > 0$, then there is a p-integrable simple function φ such that $\int_X |f - \varphi|^p \, d\mu < \varepsilon^p$.

Assuming f is nonnegative, there exists, by Theorem 3.14, a sequence of simple functions $\varphi_n : X \to \mathbb{R}$ such that $0 \leq \varphi_n(x) \leq f(x)$ and $\lim_n \varphi_n(x) = f(x)$ for every $x \in X$. Therefore, each φ_n is p-integrable. Moreover, because $0 \leq f(x) - \varphi_n(x) \leq f(x)$ for all $x \in X$ and $n \in \mathbb{N}$, each $f - \varphi_n$ is p-integrable and $\lim_n (f(x) - \varphi_n(x))^p = 0$. Hence, by the Dominated Convergence Theorem, $\lim_n \int_X (f - \varphi_n)^p d\mu = 0$. In particular, given $\varepsilon > 0$, there exists an $n \in \mathbb{N}$ such that $\int_X |f - \varphi_n|^p \, d\mu < \varepsilon^p$, which shows that $\|\dot{f} - \dot{\varphi}_n\| < \varepsilon$. □

The latter part of the proof of Proposition 5.41 does not use the property that the functions φ_n are simple; indeed, the argument shows that the following useful and rather strong approximation property holds.

Proposition 5.42. *If $p \geq 1$, if $f \in \mathscr{L}^p(X, \Sigma, \mu)$ is nonnegative, and if $\{f_n\}_{n \in \mathbb{N}}$ is a sequence of measurable functions for which $0 \leq f_n(x) \leq f(x)$ and $\lim\limits_{n \to \infty} f_n(x) = f(x)$ for all $x \in X$, then each $f_n \in \mathscr{L}^p(X, \Sigma, \mu)$ and $\lim\limits_{n \to \infty} \|\dot{f} - \dot{f}_n\| = 0$.*

We conclude this section with another approximation result, which addresses the cases of primary interest in analysis.

If $f \in C_c(X)$, then the support of f has finite measure, because regular measures take on finite values on compact sets. Therefore, if $M = \max_{x \in X} |f(x)|$, then

$$\int_X |f|^p \, d\mu = \int_{\text{supp}f} |f|^p \, d\mu \leq M^p \mu\,(\text{supp}f) < \infty.$$

That is, $f \in \mathscr{L}^p(X, \Sigma, \mu)$ and, thus, determines an element $\dot{f} \in L^p(X, \Sigma, \mu)$. It is in this sense, then, that in Theorem 5.43 below the vector space $C_c(X)$ is viewed as linear submanifold of $L^p(X, \Sigma, \mu)$.

Theorem 5.43. *If μ is a regular measure on a locally compact Hausdorff space X, and if Σ is a σ-algebra that contains the Borel sets of X, then $C_c(X)$ is dense in $L^p(X, \Sigma, \mu)$.*

Proof. Suppose that $\varepsilon > 0$ and choose a measurable set $E \in \Sigma$ of finite measure. By the regularity of μ and Proposition 3.59, there are $K, U \subseteq X$ such that K is closed, U is open, $K \subseteq E \subseteq U$, $\mu(E \backslash K) < \frac{\varepsilon^p}{2^p+1}$, and $\mu(U \backslash E) < \frac{\varepsilon^p}{2^p+1}$. Thus,

$$\mu(U \backslash K) = \mu(U \backslash E) + \mu(E \backslash K) < \varepsilon^p / 2^p.$$

The function $\chi_{E|K} : K \to [0, 1]$ is continuous and has, by the Tietze Extension Theorem (Theoerm 2.43), an extension to a continuous function $h : X \to \mathbb{C}$ of compact support such that $\text{supp}\, h \subset \overline{U}$ and $\max_{x \in X} |h(x)| = 1$. Thus,

$$\|\dot{\chi}_E - \dot{h}\|^p = \int_X |\chi_E - h|^p \, d\mu$$

$$= \int_K |\chi_E - h|^p \, d\mu + \int_{U \backslash K} |\chi_E - h|^p \, d\mu + \int_{U^c} |\chi_E - h|^p \, d\mu$$

$$= \int_{U \backslash K} |\chi_E - h|^p \, d\mu \leq \int_{U \backslash K} (1 + 1)^p \, d\mu = 2^p \mu(U \backslash K) < \varepsilon^p.$$

That is, $\|\dot{\chi}_E - \dot{h}\| < \varepsilon$, which proves that the characteristic elements $\dot{\chi}_E$ are in the closure of $C_c(X)$ in $L^p(X, \Sigma, \mu)$.

Suppose next that φ is an arbitrary p-integrable simple function: $\varphi = \sum_{k=1}^{n} \alpha_k \chi_{E_k}$, where each $\alpha_k \neq 0$. By Lemma 5.40, each E_k has finite measure. Let $\varepsilon > 0$ and $M = \max_k |\alpha_k|$. For each k there exists $g_k \in C_c(X)$ such that $\|\dot{\chi}_{E_k} - \dot{g}_k\| < \varepsilon/(nM)$. Therefore, with $g = \alpha_1 g_1 + \cdots + \alpha_n g_n \in C_c(X)$ we have that

$$\|\dot{\varphi} - \dot{g}\| \leq \sum_{k=1}^{n} |\alpha_k| \, \|\dot{\chi}_{E_k} - \dot{g}_k\| < \varepsilon.$$

This proves that every simple function is in the closure of $C_c(X)$ in $L^p(X, \Sigma, \mu)$. By Proposition 5.41, we deduce that the closure of $C_c(X)$ is $L^p(X, \Sigma, \mu)$. $\square$

5.5   Essentially-Bounded Measurable Functions

Having spent some effort in studying the Banach spaces $L^p(X, \Sigma, \mu)$, for $p \in \mathbb{R}$ with $p \geq 1$, this section is devoted to what might be viewed as the case $p = \infty$.

Definition 5.44. If (X, Σ, μ) is a measure space, and if $f : X \to \mathbb{C}$ is a measurable function, then the *essential range of* f is the set ess-ran$f$ of all $\lambda \in \mathbb{C}$ for which

$$\mu\left(f^{-1}(U)\right) > 0,$$

for every neighbourhood $U \subseteq \mathbb{C}$ of λ.

An alternate description of the essential range is as follows.

Proposition 5.45. *If* $f : X \to \mathbb{C}$ *is a measurable function, then*

$$\text{ess-ran} f = \bigcap_{E \in \Sigma, \mu(E^c)=0} \overline{f(E)}.$$

Proof. If $\lambda \in$ ess-ran$f$ and if $E \in \Sigma$ is such that $\mu(E^c) = 0$, then necessarily $\lambda \in \overline{f(E)}$. If not, then there is an open set U containing λ that is disjoint from the closed set $\overline{f(E)}$, and so $f^{-1}(U) \subset E^c$; however, E^c has measure zero while $f^{-1}(U)$ has positive measure (since $\lambda \in$ ess-ran$f$), which is a contradiction. Hence, ess-ran$f \subseteq \overline{f(E)}$ for every measurable set E with $\mu(E^c) = 0$.

Conversely, if $\lambda \notin$ ess-ran$f$, then $\mu(f^{-1}(U)) = 0$ for some neighbourhood U of λ. Let $E = f^{-1}(U)^c$. We claim that $\lambda \notin \overline{f(E)}$. If, on the contrary, λ were contained in $\overline{f(E)}$, then $f(E) \cap U$ would be non-empty, in contradiction to $E \cap E^c = \emptyset$. Hence, $\overline{f(E)} \subseteq$ ess-ran$f$ for every $E \in \Sigma$ such that $\mu(E^c) = 0$.   □

Definition 5.46. If (X, Σ, μ) is a measure space, then the *essential supremum* of a measurable function $f : X \to \mathbb{C}$ is the quantity

$$\text{ess-sup } f = \sup\{|\lambda| \,|\, \lambda \in \text{ess-ran} f\}.$$

If the essential supremum of f is finite, then f is said to be *essentially bounded*.

If ess-sup$f = M < \infty$, then the set $\{x \in X \,|\, |f(x)| > M\}$ has measure zero; thus, it is natural to use the term "essentially bounded" in describing the function f, even though f may very well be an unbounded function.

The following proposition is essentially (ha!) self-evident.

Proposition 5.47. *If* (X, Σ, μ) *is a measure space, then the set* $\mathcal{L}^\infty(X, \Sigma, \mu)$ *of all essentially bounded measurable functions* $X \to \mathbb{C}$ *is a complex vector space. Furthermore, the function* $\rho : \mathcal{L}^\infty(X, \Sigma, \mu) \to \mathbb{R}$ *defined by*

$$\rho(f) = \text{ess-sup } |f|,$$

is a seminorm on $\mathcal{L}^\infty(X, \Sigma, \mu)$.

Definition 5.48. If $\sim$ denotes the equivalence relation on $\mathscr{L}^\infty(X,\Sigma,\mu)$ defined by $f \sim g$ if and only if ess-sup $|f-g| = 0$, and if

$$L^\infty(X,\Sigma,\mu) = \mathscr{L}^\infty(X,\Sigma,\mu)/\sim,$$

then $L^\infty(X,\Sigma,\mu)$ is called an L^∞-space.

By definition, if $f \in \mathscr{L}^\infty(X,\Sigma,\mu)$, then the norm of $\dot{f} \in L^\infty(X,\Sigma,\mu)$ is given by

$$\|\dot{f}\| = \text{ess-sup } |f|.$$

Theorem 5.49. $L^\infty(X,\Sigma,\mu)$ *is a Banach space.*

Proof. Assume that $\{f_n\}_{n\in\mathbb{N}}$ is a Cauchy sequence in $L^\infty(X,\Sigma,\mu)$, where each $f_n \in \mathscr{L}^\infty(X,\Sigma,\mu)$. For all $k,n,m \in \mathbb{N}$, let

$$E_k = \{x \in X \,|\, |f_k(x)| > \text{ess-ran} f\} \text{ and}$$
$$F_{n,m} = \{x \in X \,|\, |f_n(x)-f_m(x)| > \text{ess-ran}\,(f_n - f_m)\}.$$

The measurable sets E_k and $F_{n,m}$ have measure zero, and therefore so does G, where

$$G = \left(\bigcup_k E_k\right) \cup \left(\bigcup_{n,m} F_{n,m}\right).$$

If $x \in G^c$, then $|f_n(x) - f_m(x)| \le \|\dot{f}_n - \dot{f}_m\|$ implies that $\{f_k(x)\}_{k\in\mathbb{N}}$ is convergent in $\mathbb{C}$ to some complex number denoted by λ_x. Thus, if $f : X \to \mathbb{C}$ is defined so that $f(x) = \lambda_x$ for $x \in G^c$ and $f(x) = 0$ for $x \in G$, then f is bounded and measurable, and $\|\dot{f} - \dot{f}_k\| \to 0$. $\square$

As with L^p-spaces, one can consider sequence spaces:

$$\ell^\infty(\mathbb{N}) = \left\{ \mathfrak{a} = \{\alpha_k\}_{k\in\mathbb{N}} \,|\, \alpha_k \in \mathbb{C}, \text{ for all } k \in \mathbb{N}, \text{ and } \sup_{k\in\mathbb{N}} |\alpha_k| < \infty \right\}.$$

Corollary 5.50. *The sequence space $\ell^\infty(\mathbb{N})$ is a Banach space with respect to the norm*

$$\|\mathfrak{a}\| = \sup_{k\in\mathbb{N}} |\alpha_k|.$$

For finite-length sequences, we use the notation $\ell^\infty(n)$, as we have already done with finite-dimensional ℓ^p-spaces.

Every function on the measurable space $(\mathbb{N}, \mathscr{P}(\mathbb{N}))$ is continuous; therefore, using counting measure μ, we may consider the sequential analogue of $C_0(\mathbb{N})$, which is a Banach space denoted by $c_0(\mathbb{N})$.

Definition 5.51. The set $c_0(\mathbb{N})$ is defined to be

$$c_0(\mathbb{N}) = \left\{ \mathfrak{a} = \{\alpha_k\}_{k \in \mathbb{N}} \,|\, \alpha_k \in \mathbb{C}, \text{ for all } k \in \mathbb{N}, \text{ and } \lim_{k \to \infty} |\alpha_k| = 0 \right\}.$$

The next result is the L^∞-version of Proposition 5.41.

Proposition 5.52. *The linear manifold $\{\varphi \,|\, \varphi \text{ is simple and essentially bounded}\}$ is dense in $L^\infty(X, \Sigma, \mu)$.*

Proof. As in the proof of Proposition 5.41, it is sufficient to prove that if f is a nonnegative essentially bounded function and if $\varepsilon > 0$, then there is a simple function φ such that ess-sup$|f - \varphi| < \varepsilon$.

The proof of Theorem 3.14 shows that if f is a nonnegative essentially bounded function, then there exists a monotone-increasing sequence of simple functions φ_n such that $\varphi_n(x) \geq n$ if $f(x) \geq n$ and $\varphi_n(x) = \frac{j-1}{2^n}$ if $\frac{j-1}{2^n} \leq f(x) < \frac{j}{2^n}$, for $j = 1, \ldots, 2^n n$. Now let $C = \text{ess-sup} f$ and choose $n_0 \in \mathbb{N}$ such that $n_0 > C$ and $n_0 > -\log_2 \varepsilon$. Thus, the set $E = \{x \in X \,|\, f(x) \geq n_0\}$ has measure zero and

$$0 \leq f(x) - \varphi_{n_0}(x) < \frac{1}{2^{n_0}} < \varepsilon, \ \forall x \notin E.$$

Hence, ess-sup$|f - \varphi_{n_0}| < \varepsilon$. □

5.6 Banach Spaces of Complex Measures

The examples of Banach spaces to this point have involved vectors with finitely many or countably infinitely many coordinates (entries), or have concerned functions or equivalence classes of functions on a topological or measurable space. The purpose of this section is to give an example of a Banach space that arises from measures theory; in this case, the functions involved are defined on a σ-algebra.

Definition 5.53. If (X, Σ) is a measurable space, then the set

$$M(X, \Sigma) = \{\nu \,|\, \nu \text{ is a complex measure on } (X, \Sigma)\}$$

is called the *space of complex measures* on (X, Σ).

As noted earlier, the set $M(X, \Sigma)$ carries the structure of a complex vector space in that, if $\alpha_j \in \mathbb{C}$ and $\nu_j \in M(X, \Sigma)$ for $j = 1, 2$, then

$$(\alpha_1 \nu_1 + \alpha_2 \nu_2)(E) = \alpha_1 \nu_1(E) + \alpha_2 \nu_2(E),$$

for every $E \in \Sigma$. Recall also from Proposition 3.69 that, if $\nu \in M(X, \Sigma)$, then the function $|\nu| : \Sigma \to [0, \infty]$ defined by

$$|\nu|(A) = \sup\left\{ \sum_{E \in \mathscr{P}_A} |\nu(E)| \,\Big|\, \mathscr{P}_A \text{ is a measurable partition of } A \right\}$$

is a measure on (X, Σ) with $|\nu|(X) < \infty$.

Theorem 5.54. *The function $\|\cdot\|$ on $M(X, \Sigma)$ defined by $\|\nu\| = |\nu|(X)$, for $\nu \in M(X, \Sigma)$, is a norm under which $M(X, \Sigma)$ is a Banach space.*

Proof. The verification that $\|\nu\| = |\nu|(X)$ defines a norm on $M(X, \Sigma)$ is left as an exercise (Exercise 5.117).

Assume that $\{\nu_k\}_{k \in \mathbb{N}}$ is a Cauchy sequence in $M(X, \Sigma)$. Because $|\nu(E)| \le |\nu|(E) \le |\nu|(X) = \|\nu\|$ for every $E \in \Sigma$ and every complex measure ν, for each $E \in \Sigma$ the sequence $\{\nu_k(E)\}_{k \in \mathbb{N}}$ is a Cauchy sequence in $\mathbb{C}$. Because $\mathbb{C}$ is complete, for every $E \in \Sigma$ there exists a unique $\nu(E) \in \mathbb{C}$ such that $\nu(E) = \lim_k \nu_k(E)$. Hence, we aim to show that the function $E \mapsto \nu(E)$ is a complex measure on (X, Σ) and that $\lim_k \|\nu - \nu_k\| = 0$.

Obviously $\nu(\emptyset) = 0$. If $\{E_i\}_{i=1}^{\ell}$ is a finite sequence of pairwise disjoint measurable sets, then

$$\nu\left(\bigcup_{i=1}^{\ell} E_i\right) = \lim_{k \to \infty} \nu_k\left(\bigcup_{i=1}^{\ell} E_i\right) = \lim_{k \to \infty} \sum_{i=1}^{\ell} \nu_k(E_i) = \sum_{i=1}^{\ell} \lim_{k \to \infty} \nu_k(E_i) = \sum_{i=1}^{\ell} \nu(E_i).$$

That is, ν is finitely additive.

Suppose now that $\{E_i\}_{i \in \mathbb{N}}$ is a countably infinite sequence of pairwise disjoint measurable sets. For every $j \in \mathbb{N}$, define $F_j = \bigcup_{i=j}^{\infty} E_i$ to obtain a sequence $\{F_j\}_{j \in \mathbb{N}}$ of measurable sets with the properties that $F_j \supset F_{j+1}$, for every j, and $\bigcap_{j=1}^{\infty} F_j = \emptyset$. If it were true that $\lim_j \nu(F_j) = 0$, then it would follow from

$$\nu\left(\bigcup_{i=1}^{\infty} E_i\right) = \nu\left(\left[\bigcup_{i=1}^{j-1} E_i\right] \bigcup F_j\right) = \sum_{i=1}^{j-1} \nu(E_i) + \nu(F_j)$$

that $\nu\left(\bigcup_{i=1}^{\infty} E_i\right) = \sum_{i=1}^{\infty} \nu(E_i)$. To prove that $\lim_j \nu(F_j) = 0$, let $\varepsilon > 0$ be selected arbitrarily and let $N \in \mathbb{N}$ be such that $\|\nu_m - \nu_n\| < \varepsilon$ for all $m, n \ge N$. Thus, in letting $m \to \infty$ and fixing $n \ge N$, we obtain $|\nu(E) - \nu_n(E)| \le \varepsilon$ for all $E \in \Sigma$. Because the sequence $\{F_j\}_{j \in \mathbb{N}}$ is descending and has empty intersection, and because the measure ν_n is finite, Proposition 3.22 on the continuity of measure (routinely modified to apply to complex measures) yields $\lim_j \nu_n(F_j) = 0$. Thus, there exists a $j_0 \in \mathbb{N}$ such that $\nu_n(E_j) < \varepsilon$ for every $j \ge j_0$. Consequently, if $j \ge j_0$, then

$$|v(F_j)| = |v(F_j) - v_n(F_j) + v_n(F_j)| \le |v(F_j) - v_n(F_j)| + |v_n(F_j)| \le 2\varepsilon.$$

Therefore, $\lim_j v(F_j) = 0$, which completes the proof that v is countably additive. Hence, $v \in M(X, \Sigma)$.

To show that $\lim_k \|v - v_k\| = 0$, suppose that $\varepsilon > 0$ and that $N \in \mathbb{N}$ is such that $\|v_m - v_n\| < \varepsilon$ for all $m, n \ge N$. If $\{E_i\}_{i=1}^r$ is an arbitrary finite measurable partition of X and $m, n \ge N$, then

$$\sum_{i=1}^r |v_m(E_i) - v_n(E_i)| \le \|v_m - v_n\| < \varepsilon.$$

Thus,

$$\sum_{i=1}^r |v(E_i) - v_n(E_i)| = \lim_{m \to \infty} \sum_{i=1}^r |v_m(E_i) - v_n(E_i)| \le \varepsilon.$$

Because the inequality above holds regardless of the size r of the partition of X, the same inequality holds for any countable measurable partition $\mathscr{P}_X$ of X, and hence it also holds for the supremum over all countable measurable partitions X. That is, $\|v - v_n\| \le \varepsilon$ for every $n \ge N$. Therefore, the Cauchy sequence $\{v_k\}_{k \in \mathbb{N}}$ is convergent in $M(X, \Sigma)$ to v. $\qquad\square$

5.7 Separable Banach Spaces

As with topological spaces, the notion of separability is an important feature that a Banach space might possess.

Definition 5.55. A Banach space V is *separable* if V has a countable dense subset.

Not surprisingly, finite-dimensional normed vector spaces are separable.

Proposition 5.56. *Every finite-dimensional Banach space is separable.*

Proof. By Proposition 5.8, for any fixed basis if $\{v_1, \ldots, v_n\}$ of V there are positive constants c and C such that

$$c \left(\sum_{j=1}^n |\alpha_j|^2 \right)^{1/2} \le \left\| \sum_{j=1}^n \alpha_j v_j \right\| \le C \left(\sum_{j=1}^n |\alpha_j|^2 \right)^{1/2},$$

for all $\alpha_1, \ldots, \alpha_n \in \mathbb{C}$. Because the countable set $(\mathbb{Q} + i\mathbb{Q})^n$ is dense in the Euclidean space $\mathbb{C}^n$, if $v = \sum_{j=1}^n \alpha_j v_j \in V$ and if $\varepsilon > 0$, then there are $\beta_1, \ldots, \beta_n \in (\mathbb{Q} + i\mathbb{Q})$ such that

$$\left\| \sum_{j=1}^{n} \alpha_j v_j - \sum_{j=1}^{n} \beta_j v_j \right\| \leq C \left(\sum_{j=1}^{n} |\alpha_j - \beta_j|^2 \right)^{1/2} < \varepsilon.$$

That is, the countable set of all vectors of the form $\sum_{j=1}^{n} \beta_j v_j$, where $\beta_j \in (\mathbb{Q} + i\mathbb{Q})$ for each j, is dense in V. $\square$

There is a close relation between the separability of $C(X)$ and the topology of X, if X is compact and Hausdorff.

Theorem 5.57. *The following statements are equivalent for a compact Hausdorff space X:*

1. *$C(X)$ is separable;*
2. *X is second countable;*
3. *X is metrisable.*

Proof. Assume that $C(X)$ is separable and that $S = \{f_n\}_{n\in\mathbb{N}}$ is a countable dense subset of $C(X)$. For each n let $V_n = f_n^{-1}(B_{1/3}(1))$, which is an open set in X, and let $\mathscr{B} = \{V_n\}_{n\in\mathbb{N}}$. Suppose now that $x \in X$ and U is a neighbourhood of x. Because X is normal, there is a neighbourhood V of x such that $\{x\} \subseteq V \subseteq \overline{V} \subseteq U$. Hence, by Urysohn's Lemma, there is a continuous function $g : X \to [0,1]$ with $g(x) = 1$ and $g(U^c) = \{0\}$. By hypothesis, there exists $f_n \in S$ with $\|f_n - g\| < \frac{1}{3}$. In particular, $\frac{1}{3} > |f_n(x) - g(x)| = |f_n(x) - 1|$. Thus, if $z \in V_n$, then $f_n(z) \neq 0$, which implies that $z \in (U^c)^c$. Thus, by letting $B = V_n$, this shows that there exists $B \in \mathscr{B}$ such that $x \in B \subseteq U$. Hence, $\mathscr{B}$ is a basis for the topology of X and, therefore, X is second countable.

Because Theorem 2.48 asserts that a compact Hausdorff space second countable if and only if it is metrisable, we assume now that d is a metric on X that induces the topology of X and we aim to prove that $C(X)$ is separable. For each $n \in \mathbb{N}$ consider the open cover $\{B_{1/n}(x)\}_{n\in\mathbb{N}}$ of X. Because X is compact, we may extract a finite subcover: $\{B_{1/n}(x_{n,j})\}_{j=1}^{n_k}$. Let $\mathscr{U} = \bigcup_{n\in\mathbb{N}}\{B_{1/n}(x_{n,j})\}_{j=1}^{n_k}$ and consider the subset $\mathscr{D} \subseteq \mathscr{U} \times \mathscr{U}$ in which

$$\left(B_{1/n}(x_{n,j}), B_{1/m}(x_{m,\ell})\right) \in \mathscr{D} \text{ if and only if } \overline{B_{1/n}(x_{n,j})} \cap \overline{B_{1/m}(x_{m,\ell})} = \emptyset.$$

Because $\mathscr{U}$ is countable, so is $\mathscr{D}$. By Urysohn's Lemma, that there exist functions $f_{n,j,m,\ell} : X \to [0,1]$ with the property that

$$f_{n,j,m,\ell}(\overline{B_{1/n}(x_{n,j})}) = \{1\} \text{ and } f_{n,j,m,\ell}(\overline{B_{1/m}(x_{m,\ell})}) = \{0\},$$

provided $\left(B_{1/n}(x_{n,j}), B_{1/m}(x_{m,\ell})\right) \in \mathscr{D}$. Let $\mathscr{F}$ be the (countable) set of all such functions $f_{n,j,m,\ell}$ and consider the sequence $\mathscr{F}^k$ defined as follows:

$$\mathscr{F}^0 = \{1\}, \text{ where 1 denotes the constant function } x \mapsto 1, \ \forall x \in X,$$

and, for $k \in \mathbb{N}$,

$$\mathscr{F}^k = \{\prod_{j=1}^{k} f_j \,|\, f_1,\ldots,f_k \in \mathscr{F}\}.$$

Thus the set $S = \mathrm{Span}_{\mathbb{Q}+i\mathbb{Q}}\left(\bigcup_{k=0}^{\infty} \mathscr{F}^k\right)$ is countable and is closed under sums and products. Now let $A = \overline{S}$. It is clear that A is a vector space over $\mathbb{C}$, but it is also an algebra. To verify this last assertion, suppose $f, g \in A$ and let $\varepsilon > 0$ be given. Then there are $f_0, g_0 \in S$ such that $\|f - f_0\| < \varepsilon$ and $\|g - g_0\| < \varepsilon$. Further,

$$\|fg - f_0 g_0\| = \|fg - f_0 g + f_0 g - f_0 g_0\| \leq \|f - f_0\| \|g\| + \|f_0\| \|g - g_0\|$$

$$\leq \|f - f_0\| \|g\| + (\|f\| + \varepsilon)\|g - g_0\|$$

$$< (\|g\| + \|f\|)\varepsilon + \varepsilon^2.$$

Thus, because $f_0 g_0 \in S$, we deduce that $fg \in \overline{S} = A$, and hence A is an algebra. We now show that A separates the points of X.

Let $x, y \in X$ be distinct and choose $n \in \mathbb{N}$ such that $\frac{1}{n} < \frac{1}{2}d(x,y)$. Because $\{B_{1/n}(x_{n,j})\}_{j=1}^{n_k}$ is a cover of X, there are j, i such that $x \in B_{1/n}(x_{n,j})$ and $y \in B_{1/n}(x_{n,i})$. The condition $\frac{1}{n} < \frac{1}{2}d(x,y)$ implies that $B_{1/n}(x_{n,j}) \cap B_{1/n}(x_{n,i}) = \emptyset$. By the same reasoning, if $\frac{1}{m} < \frac{1}{2}(\frac{1}{n} - d(y,x_{n,i}))$, then $y \in \overline{B_{1/m}(x_{m,\ell})} \subset B_{1/n}(x_{n,i})$ for some $x_{m,\ell}$. Thus, $f_{n,j,m,\ell}(x) = 1$ and $f_{n,j,m,\ell}(y) = 0$, which implies that A separates the points of X.

Because every element of $\mathscr{F}$ is a real-valued function, A is closed under complex conjugation $f \mapsto \overline{f}$. Moreover, A contains all the constant functions, since $\mathscr{F}_0 \subset A$. Finally, given that A separates the points of X, the Stone–Weierstrass Theorem yields $A = C(X)$, and so S is a countable dense subset of $C(X)$. Hence, $C(X)$ is separable. $\square$

A similar theme prevails for certain L^p-spaces.

Proposition 5.58. *If X a compact metrisable space and if μ is a regular Borel measure on a σ-algebra Σ that contains the Borel sets of X, then $L^p(X, \Sigma, \mu)$ is separable, for all $p \in \mathbb{R}$ such that $p \geq 1$.*

Proof. Let $f \in \mathscr{L}^p(X, \Sigma, \mu)$ and suppose that $\varepsilon > 0$. Theorem 5.43 asserts that $\|\dot{f} - \dot{g}\| < \varepsilon/2$ for some $g \in C(X)$. By Theorem 5.57, there exists a countable subset $\mathscr{C} \subset C(X)$ such that $\mathscr{C}$ is dense in $C(X)$. In particular, there exists $h \in \mathscr{C}$ such that $\max_{x \in X} |g(x) - h(x)| < \varepsilon/2\mu(X)^{1/p}$. Thus,

$$\|\dot{g} - \dot{h}\|^p = \int_X |g - h|^p \, d\mu < \frac{\varepsilon^p}{2^p \mu(X)}\mu(X) = (\varepsilon/2)^p.$$

Thus, $\|\dot{f} - \dot{h}\| < \varepsilon$, and so the countable set $\{\dot{h} \,|\, h \in \mathscr{C}\}$ is dense in $L^p(X, \Sigma, \mu)$. $\square$

The results established to this point can be used to prove that $L^p(\mathbb{R}^n, \mathfrak{M}(\mathbb{R}^n), m_n)$ is separable, where m_n denotes Lebesgue measure on $\mathbb{R}^n$.

Theorem 5.59. $L^p(\mathbb{R}^n, \mathfrak{M}(\mathbb{R}^n), m_n)$ *is separable for all* $n \in \mathbb{N}$, *and all* $p \in \mathbb{R}$ *with* $p \geq 1$.

Proof. Exercise 5.118.   □

In contrast to the results above, the Banach space $L^\infty(X, \Sigma, \mu)$ is generally non-separable. Indeed, the sequence space $\ell^\infty(\mathbb{N})$ is not a separable space (Exercise 5.119).

To conclude the discussion on separable spaces, we consider a Banach space of continuous complex-valued functions which at first glance is not of the form $C(X)$. Specifically, let $C(\mathbb{R}/2\pi\mathbb{Z})$ denote the set of all continuous 2π-periodic functions $f : \mathbb{R} \to \mathbb{C}$. It is straightforward to verify that under the norm

$$\|f\| = \max_{t \in \mathbb{R}} |f(t)|,$$

$C(\mathbb{R}/2\pi\mathbb{Z})$ is a Banach algebra with identity 1, the constant function. An important subset of $C(\mathbb{R}/2\pi\mathbb{Z})$ is the set $\mathfrak{T}$ of trigonometric polynomials.

Definition 5.60. A *trigonometric polynomial* is a 2π-periodic function $p : \mathbb{R} \to \mathbb{C}$ of the form

$$p(t) = \sum_{k=n}^{m} \alpha_k e^{ikt}, \quad t \in \mathbb{R}, \tag{5.8}$$

where $n \leq m$ in $\mathbb{Z}$ and $\alpha_n, \ldots, \alpha_m \in \mathbb{C}$.

The set $\mathfrak{T}$ of all trigonometric polynomials is a complex vector space, closed under multiplication and complex conjugation, and contains the constant functions. Therefore, one anticipates that the Stone-Weierstrass Theorem may have a formulation in this context, and indeed it does, yielding the classical theorem of Weierstrass on the uniform approximation of periodic continuous functions by trigonometric polynomials.

Proposition 5.61 (Trigonometric Weierstrass Approximation Theorem). *If* $f :$ $\mathbb{R} \to \mathbb{C}$ *is a continuous* 2π-*periodic function, then for every* $\varepsilon > 0$ *there is a trigonometric polynomial* $p \in \mathfrak{T}$ *such that*

$$|f(t) - p(t)| < \varepsilon, \text{ for every } t \in \mathbb{R}.$$

Proof. Recall that S^1 is the unit circle in $\mathbb{R}^2$, which in the present context we view as the unit circle $\mathbb{T} = \{z \in \mathbb{C} \mid |z| = 1\}$ in the complex plane. Every $f \in C(\mathbb{R}/2\pi\mathbb{Z})$ determines a unique function $F \in C(\mathbb{T})$ whereby

$$F(e^{it}) = f(t), \text{ for every } t \in \mathbb{R}.$$

Conversely, each $F \in C(\mathbb{T})$ determines a unique $f \in C(\mathbb{R}/2\pi\mathbb{Z})$ such that $f(t) = F(e^{it})$ for every $t \in \mathbb{R}$. In particular, let $\mathscr{T} \subset C(\mathbb{T})$ be the set of all the functions P of the form $P(e^{it}) = p(t)$, for trigonometric polynomials $p \in \mathfrak{T}$. Observe $\mathscr{T} \subset C(\mathbb{T})$ contains the constant functions and is closed under complex conjugation. Furthermore, because $e^{it} = e^{it'}$ if and only if $t' - t$ is an integer multiple of 2π, the function $P(e^{it}) = e^{it}$ separates the points of $\mathbb{T}$. Therefore, the closure of $\mathscr{T}$ in $C(\mathbb{T})$ is selfadjoint uniform algebra on $\mathbb{T}$, which by the Stone-Weierstrass Theorem can only be $C(\mathbb{T})$. Hence, if $f \in C(\mathbb{R}/2\pi\mathbb{Z})$ and $\varepsilon > 0$, then there is a $P \in \mathscr{T}$ such that

$$|f(t) - p(t)| = |F(e^{it}) - P(e^{it})| < \varepsilon, \text{ for every } t \in \mathbb{R},$$

where $F \in C(\mathbb{T})$ is the unique function determined by f and where $p \in \mathfrak{T}$ is the unique trigonometric polynomial determined by P.   □

Corollary 5.62. $C(\mathbb{R}/2\pi\mathbb{Z})$ *is a separable Banach space.*

Proof. Exercise 5.120.   □

5.8 Hilbert Space

If p and q are conjugate real numbers, then Hölder's inequality asserts that the product of a p-integrable function f with a q-integrable function g is integrable. In one special case, namely the case in which $p = q = 2$, the functions f and g come from the same function space, $\mathscr{L}^2(X, \Sigma, \mu)$. For any function $f \in \mathscr{L}^2(X, \Sigma, \mu)$, we may write $|f|^2$ as $f\bar{f}$ to obtain $\|\dot{f}\|^2 = \int_X f\bar{f} \, d\mu$. More generally, if $f, g \in \mathscr{L}^2(X, \Sigma, \mu)$, then $f\bar{g}$ is integrable, by Hölder's inequality, and so we may consider the function on the Cartesian product $\mathscr{L}^2(X, \Sigma, \mu) \times \mathscr{L}^2(X, \Sigma, \mu)$ that sends an ordered pair (f, g) to the complex number $\int_X f\bar{g} \, d\mu$. This function is linear in f and conjugate-linear in g, and has the property that (f, f) is mapped to $\|\dot{f}\|^2$. Such a function on $\mathscr{L}^2(X, \Sigma, \mu) \times \mathscr{L}^2(X, \Sigma, \mu)$ is called a *sesquilinear form*, and these properties of L^2-spaces motivate the definition of an abstract Hilbert space.

Definition 5.63. An *inner product* on a complex vector space H is a complex-valued function $\langle \cdot, \cdot \rangle$ on the Cartesian product $H \times H$ satisfying the following properties for all vectors $\xi, \xi_1, \xi_2, \eta, \eta_1, \eta_2 \in H$ and scalars $\alpha \in \mathbb{C}$:

1. $\langle \xi, \xi \rangle \geq 0$, with $\langle \xi, \xi \rangle = 0$ if and only if $\xi = 0$;
2. $\langle \xi, \eta \rangle = \overline{\langle \eta, \xi \rangle}$;
3. $\langle \xi_1 + \xi_2, \eta \rangle = \langle \xi_1, \eta \rangle + \langle \xi_2, \eta \rangle$ and $\langle \xi, \eta_1 + \eta_2 \rangle = \langle \xi, \eta_1 \rangle + \langle \xi, \eta_2 \rangle$;
4. $\langle \alpha \xi, \eta \rangle = \alpha \langle \xi, \eta \rangle$ and $\langle \xi, \alpha \eta \rangle = \bar{\alpha} \langle \xi, \eta \rangle$.

The vector space H, when considered with the inner product $\langle \cdot, \cdot \rangle$, is called an *inner product space*.

The simplest example of an inner product space is $\mathbb{C}^n$.

Example 5.64. *The vector space $\mathbb{C}^d$ is an inner product space, where $\langle\cdot,\cdot\rangle$ is defined by*

$$\langle\xi,\eta\rangle = \sum_{k=1}^{d}\xi_k\overline{\eta_k}$$

for $\xi = \begin{bmatrix}\xi_1 \\ \vdots \\ \xi_d\end{bmatrix}, \eta = \begin{bmatrix}\eta_1 \\ \vdots \\ \eta_d\end{bmatrix} \in \mathbb{C}^d.$

The Hölder Inequality in the case of $L^2(X,\Sigma,\mu)$ is what is called the Cauchy-Schwarz Inequality in abstract Hilbert space.

Proposition 5.65 (Cauchy-Schwarz Inequality). *If $\langle\cdot,\cdot\rangle$ is an inner product on a vector space H, then*

$$|\langle\xi,\eta\rangle| \le \langle\xi,\xi\rangle^{1/2}\langle\eta,\eta\rangle^{1/2}, \tag{5.9}$$

for all $\xi,\eta \in H$. Furthermore, if $\eta \ne 0$, then $|\langle\xi,\eta\rangle| = \langle\xi,\xi\rangle^{1/2}\langle\eta,\eta\rangle^{1/2}$ if and only if $\xi = \lambda\eta$ for some $\lambda \in \mathbb{C}$.

Proof. The result is trivially true if $\langle\xi,\eta\rangle = 0$. Assume, therefore, that $\langle\xi,\eta\rangle \ne 0$. For any $\lambda \in \mathbb{C}$,

$$0 \le \langle\xi-\lambda\eta,\xi-\lambda\eta\rangle = \langle\xi,\xi\rangle - \overline{\lambda}\langle\xi,\eta\rangle - \lambda\langle\eta,\xi\rangle + |\lambda|^2\langle\eta,\eta\rangle$$

$$= \langle\xi,\xi\rangle - 2\Re(\lambda\langle\eta,\xi\rangle) + |\lambda|^2\langle\eta,\eta\rangle.$$

For

$$\lambda = \frac{\langle\xi,\xi\rangle}{\langle\eta,\xi\rangle},$$

the inequality above becomes

$$0 \le -\langle\xi,\xi\rangle + \frac{\langle\xi,\xi\rangle^2\langle\eta,\eta\rangle}{|\langle\xi,\eta\rangle|^2},$$

which yields inequality (5.9). Further, note that if $|\langle\xi,\eta\rangle| = \langle\xi,\xi\rangle^{1/2}\langle\eta,\eta\rangle^{1/2}$, then $\langle\xi-\lambda\eta,\xi-\lambda\eta\rangle = 0$ for $\lambda = \frac{\langle\xi,\xi\rangle}{\langle\eta,\xi\rangle}$, which yields $\xi = \lambda\eta$.   □

With our experience with L^2-spaces, the following proposition is a natural consequence of the Cauchy-Schwarz inequality.

Proposition 5.66. *If $\langle\cdot,\cdot\rangle$ is an inner product on a vector space H, then*

$$\|\xi\| = \langle\xi,\xi\rangle^{1/2} \tag{5.10}$$

defines a norm $\|\cdot\|$ on H.

Proof. If $\xi, \eta \in H$, then $\langle \xi, \eta \rangle + \langle \eta, \xi \rangle = 2\,\mathfrak{R}\,(\langle \xi, \eta \rangle)$, and so

$$\|\xi + \eta\|^2 = \langle \xi + \eta, \xi + \eta \rangle = \|\xi\|^2 + 2\mathfrak{R}(\langle \xi, \eta \rangle) + \|\eta\|^2$$

$$\leq \|\xi\|^2 + 2\,|\langle \xi, \eta \rangle| + \|\eta\|^2 \leq \|\xi\|^2 + 2\,\|\xi\|\,\|\eta\| + \|\eta\|^2$$

$$= (\|\xi\| + \|\eta\|)^2.$$

Hence, equation (5.10) defines a norm on H. □

Observe that the Cauchy-Schwarz Inequality (5.9) now takes the form

$$|\langle \xi, \eta \rangle| \leq \|\xi\|\,\|\eta\|$$

and leads to the following useful proposition.

Proposition 5.67 (Continuity of the Inner Product). *If $\xi_0 \in H$ and $\varepsilon > 0$, then for each $\eta \in H$ there is a $\delta_\eta > 0$ such that*

$$|\langle \xi, \eta \rangle - \langle \xi_0, \eta \rangle| < \varepsilon$$

for all $\xi \in H$ with $\|\xi - \xi_0\| < \delta_\eta$.

Proof. The assertion is clear if $\eta = 0$. If $\eta \neq 0$, then let $\delta = \varepsilon \|\eta\|^{-1}$. Thus, by the Cauchy-Schwarz inequality,

$$|\langle \xi, \eta \rangle - \langle \xi_0, \eta \rangle| = |\langle \xi - \xi_0, \eta \rangle| \leq \|\xi - \xi_0\|\,\|\eta\| < \varepsilon,$$

for all $\xi \in H$ with $\|\xi - \xi_0\| < \delta_\eta$. □

Definition 5.68. A *Hilbert space* is an inner product space H such that H is a Banach space with respect to the norm $\|\xi\| = \sqrt{\langle \xi, \xi \rangle}$.

The following example is the most generic example of a Hilbert space.

Example 5.69. *The Banach space $L^2(X, \Sigma, \mu)$ is a Hilbert space with respect to the inner product*

$$\langle \dot{f}, \dot{g} \rangle = \int_X f\overline{g}\,d\mu,$$

where $f, g \in \mathscr{L}^2(X, \Sigma, \mu)$.

Proof. We need only note that $\langle \dot{f}, \dot{f} \rangle = \int |f|^2\,d\mu$, which is the square of the norm of $\dot{f}$ in the Banach space $L^2(X, \Sigma, \mu)$. Hence, the norm on the Banach space $L^2(X, \Sigma, \mu)$ is induced by the inner product given above. □

The concepts of Euclidean geometry greatly influence Hilbert space theory, starting with the idea of perpendicular vectors.

Definition 5.70. In an inner product space H, two vectors $\xi, \eta \in H$ are *orthogonal* if $\langle \xi, \eta \rangle = 0$.

Proposition 5.71 (Pythagorean Theorem). $\|\xi + \eta\|^2 = \|\xi\|^2 + \|\eta\|^2$, *for all pairs of orthogonal vectors $\xi, \eta \in H$.*

Proof. Use $\|\xi + \eta\|^2 = \langle \xi + \eta, \xi + \eta \rangle = \|\xi\|^2 + 2\Re(\langle \xi, \eta \rangle) + \|\eta\|^2$ and the fact that $\langle \xi, \eta \rangle = 0$ to obtain $\|\xi + \eta\|^2 = \|\xi\|^2 + \|\eta\|^2$.   □

A distinguished geometric property of Hilbert space is the parallelogram law below.

Proposition 5.72 (Jordan-von Neumann Theorem). *If H is an inner product space H and if $\xi, \eta \in H$, then*

$$\|\xi + \eta\|^2 + \|\xi - \eta\|^2 = 2\|\xi\|^2 + 2\|\eta\|^2. \tag{5.11}$$

Furthermore, if V is a normed vector space in which the parallelogram law (5.11) holds for all $\xi, \eta \in V$, then there is an inner product that induces the norm on V,

Proof. The parallelogram law is verified by expanding the appropriate inner products. Therefore, suppose now that V is a normed vector space and that equation (5.11) holds for all $\xi, \eta \in V$. Define $\langle \cdot, \cdot \rangle : V \times V \to \mathbb{C}$ by

$$\langle \xi, \eta \rangle = \|\xi + \eta\|^2 - \|\xi - \eta\|^2 + i\|\xi + i\eta\|^2 - i\|\xi - i\eta\|^2,$$

for $\xi, \eta \in V$. Thus, $\langle \cdot, \cdot \rangle$ is an inner product and $\|\xi\| = \langle \xi, \xi \rangle^{1/2}$ for all $\xi \in V$.   □

The notion of convexity appeared earlier in the context of Jensen's inequality. This fundamental geometric idea is especially important in Hilbert space theory, as demonstrated by Theorem 5.75 below, which has many important consequences for Hilbert spaces that are not necessarily true for arbitrary Banach spaces.

Definition 5.73. A subset K of a vector space V is a *convex set* if

$$\lambda v + (1 - \lambda)w \in K$$

for all $\lambda \in [0, 1]$ and for all $v, w \in K$.

Definition 5.74. If K is a nonempty subset of a normed vector space V and if $v_0 \in V$, then the *distance* from v_0 to K is denoted by $\text{dist}(v_0, K)$ and is defined by

$$\text{dist}(v_0, K) = \inf\{\|v_0 - v\| \mid v \in K\}. \tag{5.12}$$

The main "convexity theorem" in Hilbert space is the following result.

Theorem 5.75. *If K is a nonempty closed convex subset of a Hilbert space H, and if $\xi_0 \in H$, then there is a unique $\eta \in K$ such that*

$$\text{dist}(\xi_0, K) = \|\xi_0 - \eta\|.$$

Proof. The convexity of K will be used repeatedly in the following guise: if $\eta_1, \eta_2 \in K$, then $\frac{1}{2}(\eta_1 + \eta_2) \in K$.

By definition of distance, for each $k \in \mathbb{N}$ there is a vector $\eta_k \in K$ such that

$$\|\xi_0 - \eta_k\|^2 < (\operatorname{dist}(\xi_0, K))^2 + \frac{1}{k}.$$

Thus,

$$\begin{aligned}
\|2\xi_0 - (\eta_n + \eta_m)\|^2 + \|\eta_m - \eta_n\|^2 &= \|(\xi_0 - \eta_n) + (\xi_0 - \eta_m)\|^2 \\
&\quad + \|(\xi_0 - \eta_n) - (\xi_0 - \eta_m)\|^2 \\
&= 2\|\xi_0 - \eta_n\|^2 + 2\|\xi_0 - \eta_m\|^2 \\
&< 4(\operatorname{dist}(\xi_0, K))^2 + \tfrac{1}{n} + \tfrac{1}{m},
\end{aligned}$$

where the second equality above follows from the parallelogram law. On the other hand,

$$4(\operatorname{dist}(\xi_0, K))^2 \le 4\left\|\xi_0 - \frac{1}{2}(\eta_n + \eta_m)\right\|^2 = \|2\xi_0 - (\eta_n + \eta_m)\|^2.$$

Hence, $\|\eta_m - \eta_n\|^2 < \tfrac{1}{n} + \tfrac{1}{m}$, which proves that $\{\eta_k\}_{k \in \mathbb{N}}$ is a Cauchy sequence. Let $\eta \in H$ denote the limit of this sequence. Because K is closed, η lies in K and

$$\operatorname{dist}(\xi_0, K) \le \|\xi_0 - \eta\| = \|\xi_0 - \eta_k + \eta_k - \eta\| \le \|\xi_0 - \eta_k\| + \|\eta_k - \eta\|$$

$$< \sqrt{(\operatorname{dist}(\xi_0, K))^2 + \tfrac{1}{k}} + \|\eta_k - \eta\|.$$

In letting $k \to \infty$, the inequalities above sandwich to the equation

$$\operatorname{dist}(\xi_0, K) = \|\xi_0 - \eta\|.$$

To prove the uniqueness of the best approximant η, let $\eta' \in K$ satisfy $\operatorname{dist}(\xi_0, K) = \|\xi_0 - \eta'\|$. Thus, $\|\xi_0 - \eta\| = \|\xi_0 - \eta'\|$ and, by the parallelogram law,

$$\|(\xi_0 - \eta') + (\xi_0 - \eta)\|^2 + \|(\xi_0 - \eta') - (\xi_0 - \eta)\|^2 = 2\left(\|\xi_0 - \eta'\|^2 + \|\xi_0 - \eta\|^2\right).$$

Therefore,

$$\begin{aligned}
\|\eta - \eta'\|^2 &= 4\|\xi_0 - \eta\|^2 - 4\|\xi_0 - \tfrac{1}{2}(\eta + \eta')\|^2 \\
&= 4(\operatorname{dist}(\xi_0, K))^2 - 4\|\xi_0 - \tfrac{1}{2}(\eta + \eta')\|^2 \\
&\le 4(\operatorname{dist}(\xi_0, K))^2 - 4(\operatorname{dist}(\xi_0, K))^2 \\
&= 0,
\end{aligned}$$

and so $\eta' = \eta$. $\qquad\square$

Definition 5.76. If S is a nonempty subset of a Hilbert space H, then the *orthogonal complement* of S is the subset of H denoted by $S^{\perp}$ and defined by

$$S^{\perp} = \{\eta \in H \,|\, \langle \eta, \xi \rangle = 0, \ \forall \xi \in S\}.$$

By linearity of the inner product in the first variable, it is not difficult to see that $S^{\perp}$ is a vector subspace of H. By the Cauchy-Schwarz inequality (that is, by continuity of the inner product in the first variable), $S^{\perp}$ is closed. Thus, $S^{\perp}$ is a subspace of H for every nonempty subset $S \subseteq H$.

Proposition 5.77. *Let $M \subseteq H$ be a subspace of a Hilbert space H and let $\xi \in H$ and $\eta \in M$. The following statements are equivalent:*

1. $\mathrm{dist}(\xi, M) = \|\xi - \eta\|$;
2. $\xi - \eta \in M^{\perp}$.

Proof. Assume that $\mathrm{dist}(\xi, M) = \|\xi - \eta\|$. Let $\eta' \in M$. We aim to prove that $\langle \xi - \eta, \eta' \rangle = 0$. To this end, consider any $\alpha \in \mathbb{C}$. Because $\eta + \alpha \eta' \in M$, $\|\xi - (\eta + \alpha \eta')\| \geq \mathrm{dist}(\xi, M) = \|\xi - \eta\|$. Thus,

$$\|\xi - \eta\|^2 \leq \|(\xi - \eta) - \alpha \eta'\|^2 = \|\xi - \eta\|^2 - 2\Re\left(\alpha \langle \eta', \xi - \eta \rangle\right) + |\alpha|^2 \|\eta'\|^2,$$

which yields

$$2\Re\left(\alpha \langle \eta', \xi - \eta \rangle\right) \leq |\alpha|^2 \|\eta'\|^2. \tag{5.13}$$

The complex number $\langle \eta', \xi - \eta \rangle$ is either zero or it is not. If it is zero, then $\xi - \eta$ is orthogonal to η', as desired. Thus, assume that $\langle \eta', \xi - \eta \rangle \neq 0$ and let $\alpha = t \langle \eta', \xi - \eta \rangle$, for some $t > 0$. Inequality (5.13) becomes

$$2t \,|\langle \eta', \xi - \eta \rangle|^2 \leq t^2 \,|\langle \eta', \xi - \eta \rangle|^2 \|\eta'\|^2.$$

Because $\langle \eta', \xi - \eta \rangle \neq 0$ and $t > 0$, the inequality above implies that

$$2 \leq t \|\eta\|^2,$$

which is clearly impossible if $t \to 0^+$. Hence, it must be that $\langle \eta', \xi - \eta \rangle = 0$, which proves that $\xi - \eta$ is orthogonal to every vector $\eta' \in M$. That is, $\xi - \eta \in M^{\perp}$.

Conversely, assume that $\xi - \eta \in M^{\perp}$. If $\eta' \in M$, then $\xi - \eta$ is orthogonal to $\eta - \eta'$ because $\eta - \eta' \in M$. Invoking the Pythagorean theorem yields

$$\|\xi - \eta'\|^2 = \|(\xi - \eta) + (\eta - \eta')\|^2 = \|\xi - \eta\|^2 + \|\eta - \eta'\|^2 \geq \|\xi - \eta\|^2.$$

Thus,

$$\|\xi - \eta\| \leq \inf\{\|\xi - \eta'\| \,|\, \eta' \in M\} = \mathrm{dist}(\xi, M),$$

which proves that $\|\xi - \eta\| = \mathrm{dist}(\xi, M)$. □

Recall from linear algebra that if M and N are linear submanifolds of H, then $M + N$ denotes the set

$$M + N = \{\omega + \delta \mid \omega \in M \text{ and } \delta \in N\}$$

and $M + N$ is itself a linear submanifold of H.

Proposition 5.78. *If M and N are subspaces of a Hilbert space H such that $M \subseteq N^{\perp}$, then $M \cap N = \{0\}$ and the linear submanifold $M + N$ is a subspace.*

Proof. If $\xi \in M \cap N$, then the hypothesis $M \subseteq N^{\perp}$ implies that $\xi \in N \cap N^{\perp}$, whence $0 = \langle \xi, \xi \rangle$ and so $\xi = 0$.

To show that the linear submanifold $M + N$ is closed in the topology of H, select a vector ξ in the closure of $M + N$ and suppose that $\{\xi_k\}_{k \in \mathbb{N}}$ is a sequence in $M + N$ converging to ξ. Thus, $\{\xi_k\}_k$ is necessarily a Cauchy sequence.

For each $k \in \mathbb{N}$, there are $\omega_k \in M$ and $\delta_k \in N$ such that $\xi_k = \omega_k + \delta_k$. Because the vectors of M are orthogonal to the vectors of N, the Pythagorean Theorem applies:

$$\|\xi_m - \xi_n\|^2 = \|(\omega_m - \omega_n) + (\delta_m - \delta_n)\|^2 = \|\omega_m - \omega_n\|^2 + \|\delta_m - \delta_n\|^2.$$

Thus, the sequences $\{\omega_k\}_{k \in \mathbb{N}} \subset M$ and $\{\delta_k\}_{k \in \mathbb{N}} \subset N$ are Cauchy sequences. Let $\omega \in M$ and $\delta \in N$ be the limits, respectively, of these sequences. Then $\xi = \omega + \delta$ and so $M + N$ is closed.   □

This situation described in Proposition 5.78 is formalised by the following definition.

Definition 5.79. *If $M, N \subseteq H$ are subspaces of a Hilbert space H such that $M \subseteq N^{\perp}$, then the* orthogonal direct sum *of M and N is the subspace $M + N$ and is denoted by $M \oplus N$.*

Proposition 5.80. *If $M \subseteq H$ is a subspace, then $H = M \oplus M^{\perp}$.*

Proof. Obviously $M \oplus M^{\perp} \subseteq H$. Conversely, suppose that $\xi \in H$. Because M is a subspace, M is closed and convex. Thus, by Theorem 5.75, there is a unique $\eta \in M$ for which $\|\xi - \eta\| = \text{dist}(\xi, M)$. Therefore, $\xi - \eta \in M^{\perp}$, by Theorem 5.77. Let $\omega = \eta$ and $\delta = \xi - \eta$ to get $\omega \in M$, $\delta \in M^{\perp}$, and $\xi = \omega + \delta \in M \oplus M^{\perp}$. This proves that $H \subseteq M \oplus M^{\perp}$.   □

The direct sum decomposition in Proposition 5.80 above is internal in the sense that one decomposes an existing space H as an orthogonal direct sum of two subspaces. One could repeat this process finitely or countable infinitely many times inductively. Alternatively, one frequently has a finite or countable family of Hilbert spaces and aims to construct their (external) direct sum to create a new Hilbert space.

Proposition 5.81. *If Λ is a finite or countable set, and if $\{H_k\}_{k \in \Lambda}$ is a family of Hilbert spaces in which $\langle \cdot, \cdot \rangle_k$ denotes the inner product of H_k, then the vector space*

$$\bigoplus_{k \in \Lambda} H_k = \{(\xi_k)_{k \in \Lambda} \mid \sum_{k \in \Lambda} \|\xi_k\|^2 < \infty\}$$

is a Hilbert space with respect to the inner product

$$\langle (\xi_k)_k, (\eta_k)_k \rangle = \sum_{k \in \Lambda} \langle \xi_k, \eta_k \rangle_k.$$

Proof. Exercise 5.123.   □

If $H_k = H$ for every k and for some fixed Hilbert space H, and if $\Lambda = \mathbb{N}$, for example, then $\ell_H^2(\mathbb{N})$ denotes the Hilbert space in Proposition 5.81.

5.9 Orthonormal Bases of Hilbert Spaces

Definition 5.82. The elements of a collection $\mathcal{O}$ of vectors in an inner-product space H are said to be *orthonormal* if each $\phi \in \mathcal{O}$ has norm $\|\phi\| = 1$ and $\langle \phi, \phi \rangle = 0$ for all $\phi, \phi' \in \mathcal{O}$ in which $\phi' \neq \phi$.

If $\phi_1, \ldots, \phi_k$ are orthonormal and if $\xi \in \text{Span}\{\phi_1, \ldots, \phi_k\}$, then there are complex numbers $\alpha_1, \ldots, \alpha_k$ such that

$$\xi = \sum_{j=1}^{k} \alpha_j \phi_j.$$

Because the inner product is linear in its first variable, for each $\ell \in \{1, \ldots, k\}$ we have that

$$\langle \xi, \phi_\ell \rangle = \left\langle \sum_{j=1}^{k} \alpha_j \phi_j, \phi_\ell \right\rangle = \sum_{j=1}^{d} \alpha_j \langle \phi_j, \phi_\ell \rangle = \alpha_\ell.$$

Hence, every $\xi \in \text{Span}\{\phi_1, \ldots, \phi_k\}$ is expressed by uniquely by its *Fourier series*:

$$\xi = \sum_{j=1}^{k} \langle \xi, \phi_j \rangle \phi_j. \tag{5.14}$$

The Fourier series (5.14) also demonstrates that orthonormal vectors are necessarily linear independent.

Proposition 5.83 (Gram-Schmidt Process). *If $v_1, \ldots, v_k \in H$ are linearly independent vectors in an inner-product space H, then there are orthonormal vectors $\phi_1, \ldots, \phi_k \in H$ such that*

$$\text{Span}\{\phi_1, \ldots, \phi_k\} = \text{Span}\{v_1, \ldots, v_k\}.$$

Proof. Let $\phi_1 = \|v_1\|^{-1}v_1$ and, inductively, for each j let $\phi_j = \|v_j - w_j\|^{-1}(v_j - w_j)$, where $w_j = \sum_{i=1}^{j-1}\langle v_j, \phi_i \rangle \phi_i$. The vectors $\phi_1, \ldots, \phi_k$ are orthonormal and are contained in the k-dimensional subspace $\mathrm{Span}\{v_1, \ldots, v_k\}$. Thus, by the linear independence of orthonormal vectors, $\mathrm{Span}\{\phi_1, \ldots, \phi_k\}$ and $\mathrm{Span}\{v_1, \ldots, v_k\}$ coincide. □

As noted earlier in Theorem 5.16, the concept of a linear basis is not really suited to analysis.

Definition 5.84. A collection $\mathscr{B}$ of orthonormal vectors in a Hilbert space H is called an *orthonormal basis* if $\mathscr{B}' = \mathscr{B}$ for every set $\mathscr{B}'$ of orthonormal vectors for which $\mathscr{B}' \supseteq \mathscr{B}$.

In other words, an orthonormal basis is a maximal set of orthonormal vectors.

Proposition 5.85. *If $\mathscr{B}$ is a set of orthonormal vectors in Hilbert space H, then $\mathscr{B}$ is an orthonormal basis of H if and only if $\mathrm{Span}\,\mathscr{B}$ is dense in H.*

Proof. Exercise 5.124. □

The question of existence of orthonormal bases is not unlike that of linear bases in that it requires Zorn's Lemma to prove this fact.

Theorem 5.86. *Every nonzero Hilbert space has an orthonormal basis.*

Proof. Mimic the proof of Theorem 5.16. □

The cardinality of the basis is related to the topology of the Hilbert space by way of the following proposition.

Proposition 5.87. *A Hilbert space is separable if and only if it has a countable orthonormal basis. Moreover, all orthonormal bases of a separable Hilbert space are in bijective correspondence.*

Proof. Let H be a Hilbert space. If H has a countable orthonormal basis $\{\phi_k\}_{k\in\mathbb{N}}$, then the countable set

$$W = \mathrm{Span}_{\mathbb{Q}+i\mathbb{Q}}\{\phi_k \mid k \in \mathbb{N}\}$$

is dense in H by Proposition 5.85 and by the fact that $\mathbb{Q} + i\mathbb{Q}$ is dense in $\mathbb{C}$.

Conversely, assume that H is separable. If S is any countable, dense subset of H, then the linear submanifold

$$W = \mathrm{Span}_{\mathbb{C}}\{\omega \mid \omega \in S\}$$

is dense in H. The spanning set S must contain an algebraic basis for the vector space W. Thus, W has a countable basis and to this basis one can apply the Gram-Schmidt process to obtain a countable set $\{\phi_k\}_{k\in\mathbb{N}}$ of orthonormal vectors whose linear span $\mathscr{O}$ is dense in the closure $\overline{W}$ of W. But $\overline{W} = H$, which indicates that $\mathscr{O}^{\perp} = \{0\}$. In other words, there are no nonzero vectors orthogonal to $\{\phi_k\}_{k\in\mathbb{N}}$, which proves that $\{\phi_k\}_{k\in\mathbb{N}}$ is a maximal set of orthonormal vectors. That is, $\{\phi_k\}_{k\in\mathbb{N}}$ is an orthonormal basis of H.

If H has finite dimension n, then n is an algebraic invariant of the space H: all linear bases of H must have the same cardinality, whence any two orthonormal bases of H are in bijective correspondence. If H has infinite dimension and is separable, then all orthonormal bases of H are countably infinite and are, therefore, in bijective correspondence with one another.   □

It is also true that if H is a nonseparable Hilbert space, then all orthonormal bases of H are in bijective correspondence. The proof of this fact requires some cardinal arithmetic; however, we will not need to use such a theorem in what follows, we will not pursue this result here.

Hilbert spaces with a countable orthonormal basis are especially easy to analyse. The following theorem, which is an abstraction of classical Fourier series, illustrates this fact.

Theorem 5.88. *If $\{\phi_k\}_{k \in \mathbb{N}}$ is an orthonormal basis of a Hilbert space H, then, for every $\xi \in H$,*

$$\lim_{n \to \infty} \left\| \xi - \sum_{k=1}^{n} \langle \xi, \phi_k \rangle \, \phi_k \right\| = 0. \tag{5.15}$$

Proof. For each $n \in \mathbb{N}$ let

$$\xi_n = \sum_{k=1}^{n} \langle \xi, \phi_k \rangle \, \phi_k.$$

Observe that

$$0 \leq \| \xi - \xi_n \|^2 = \langle \xi - \xi_n, \xi - \xi_n \rangle = \| \xi \|^2 - \sum_{k=1}^{n} |\langle \xi, \phi_k \rangle|^2.$$

Hence,

$$\lim_{n \to \infty} \sum_{k=1}^{n} |\langle \xi, \phi_k \rangle|^2 \leq \| \xi \|^2,$$

which implies that the sequence $\{\xi_n\}_{n \in \mathbb{N}}$ is a Cauchy sequence. Because H is a Hilbert space, this sequence has a limit $\xi' \in H$. We shall prove that $\xi' = \xi$.

Choose $k \in \mathbb{N}$. Direct computation shows that $\langle \xi - \xi_n, \phi_k \rangle = 0$ for every $n \geq k$. Hence, if $n \geq k$, then

$$|\langle (\xi - \xi'), \phi_k \rangle| = |\langle (\xi - \xi_n + \xi_n - \xi'), \phi_k \rangle|$$

$$= |\langle (\xi - \xi_n), \phi_k \rangle + \langle (\xi_n - \xi'), \phi_k \rangle|$$

$$= |\langle (\xi_n - \xi'), \phi_k \rangle|$$

$$\leq \| \xi_n - \xi' \|.$$

(The final inequality is the Cauchy-Schwarz inequality.) Thus,

$$0 \leq \left| \langle (\xi - \xi'), \phi_k \rangle \right| \leq \lim_{n \to \infty} \| \xi_n - \xi' \| = 0.$$

Therefore, $\xi - \xi'$ is orthogonal to every vector ϕ_k in the orthonormal basis. Thus, $\xi - \xi' = 0$.   □

Equation (5.15) asserts that if $\xi \in H$ and $\{\phi_k\}_{k \in \mathbb{N}}$ is an orthonormal basis for H, then

$$\lim_{n \to \infty} \left\| \xi - \sum_{k=1}^{n} \langle \xi, \phi_k \rangle \phi_k \right\| = 0.$$

This will be expressed as

$$\xi = \sum_{k \in \mathbb{N}} \langle \xi, \phi_k \rangle \phi_k. \tag{5.16}$$

Definition 5.89. If $\{\phi_k\}_{k \in \mathbb{N}}$ is an orthonormal basis of a separable Hilbert space H, then the series

$$\xi = \sum_{k \in \mathbb{N}} \langle \xi, \phi_k \rangle \phi_k$$

is called a *Fourier series decomposition* of $\xi \in H$, and the complex numbers $\langle \xi, \phi_k \rangle$ are called the *Fourier coefficients* of ξ.

Proposition 5.90. *Assume that $\{\phi_k\}_{k \in \mathbb{N}}$ is an orthonormal basis for a separable Hilbert space H.*

1. (Parseval's Equation) *For every $\xi, \eta \in H$,*

$$\langle \xi, \eta \rangle = \sum_{k \in \mathbb{N}} \langle \xi, \phi_k \rangle \overline{\langle \eta, \phi_k \rangle}. \tag{5.17}$$

2. *For every $\xi \in H$,*

$$\| \xi \|^2 = \sum_{k \in \mathbb{N}} | \langle \xi, \phi_k \rangle |^2.$$

Proof. Express ξ in its Fourier series decomposition

$$\xi = \sum_{k \in \mathbb{N}} \langle \xi, \phi_k \rangle \phi_k,$$

and consider

$$\xi_n = \sum_{k=1}^{n} \langle \xi, \phi_k \rangle \phi_k.$$

Because $\lim_{n \to \infty} \|\xi_n - \xi\| = 0$, Proposition 5.67 shows that

$$\langle \xi, \eta \rangle = \lim_{n \to \infty} \langle \xi_n, \eta \rangle$$

$$= \lim_{n \to \infty} \left(\sum_{k=1}^{n} \langle \xi, \phi_k \rangle \langle \phi_k, \eta \rangle \right)$$

$$= \sum_{k \in \mathbb{N}} \langle \xi, \phi_k \rangle \overline{\langle \eta, \phi_k \rangle}.$$

This proves Parseval's Equation. The equation

$$\|\xi\|^2 = \sum_{k \in \mathbb{N}} |\langle \xi, \phi_k \rangle|^2$$

follows from Parseval's Equation because $\|\xi\|^2 = \langle \xi, \xi \rangle$. $\square$

We conclude with some examples of orthonormal bases.

Example 5.91 (Legendre Polynomials). *If $\phi_0(t) = \sqrt{\frac{1}{2}}$ and*

$$\phi_k(t) = \sqrt{\frac{2k+1}{k}} \frac{1}{2^k k!} \frac{d^k}{dt^k} \left[(t^2 - 1)^k \right], \quad k \in \mathbb{N},$$

then $\{\phi_k\}_{k=0}^{\infty}$ is an orthonormal basis of $L^2([-1, 1], \mathfrak{M}, m)$.

Proof. The functions ϕ_k are obtained from the linearly independent functions $1, t, t^2, \ldots$ by the Gram-Schmidt process. Thus,

$$\text{Span}\, \mathfrak{L} = \text{Span}\, \{1, t, t^2, t^3, \ldots\},$$

where $\mathfrak{L}$ denotes the set of Legendre polynomials. By the Weierstrass Approximation Theorem, if $\vartheta \in C([-1, 1])$ and $\varepsilon > 0$, then there is a element $f \in$ Span $\mathfrak{L}$ such that $|\vartheta(t) - f(t)| < \varepsilon$ for all $t \in [-1, 1]$. Furthermore, Theorem 5.43 states that $C([-1, 1])$ is dense in $L^2([-1, 1], \mathfrak{M}, m)$. Hence, Span $\mathfrak{L}$ is dense in $L^2([-1, 1], \mathfrak{M}, m)$, and so $\mathfrak{L}$ is an orthonormal basis of $L^2([-1, 1], \mathfrak{M}, m)$. $\square$

The next example is drawn from classical Fourier series.

Example 5.92 (Classical Fourier Series). *If the function $\phi_n : [-\pi, \pi] \to \mathbb{C}$ given by*

$$\phi_n(t) = \frac{e^{int}}{\sqrt{2\pi}}, \tag{5.18}$$

for every $n \in \mathbb{Z}$, then $\{\phi_n\}_{n \in \mathbb{Z}}$ is an orthonormal basis of $L^2([-\pi, \pi], \mathfrak{M}, m)$.

Proof. Because

$$\langle \phi_n, \phi_m \rangle = \int_{-\pi}^{\pi} \phi_n(t)\overline{\phi_m(t)}\,dt = \frac{1}{2\pi} \int_{-\pi}^{\pi} e^{i(n-m)t}\,dt,$$

we have $\langle \phi_n, \phi_m \rangle = 1$ if $m = n$ and $\langle \phi_n, \phi_m \rangle = 0$ otherwise. Thus, $\{\phi_n\}_{n\in\mathbb{Z}}$ is a set of orthonormal vectors in $L^2([-\pi,\pi], \mathfrak{M}, m)$.

Theorem 5.61 shows that $\mathrm{Span}\{\phi_n\}_{n\in\mathbb{Z}}$ is uniformly dense in $C(\mathbb{R}/2\pi\mathbb{Z})$, the space of all continuous 2π-periodic functions $\mathbb{R} \to \mathbb{C}$. Furthermore, $C([-\pi,\pi])$ is a dense linear submanifold of $L^2([-\pi,\pi], \mathfrak{M}, m)$ (by Theorem 5.43). Therefore, it is sufficient to show that every $f \in C([-\pi,\pi])$ can be approximated to within ε in the norm of $L^2([-\pi,\pi], \mathfrak{M}, m)$ by a 2π-periodic continuous function h. To this end, choose $f \in C([-\pi,\pi])$ and let $\varepsilon > 0$. Let $M = \max\{|f(t)|\,|\,t \in [-\pi,\pi]\}$ and choose $\delta > 0$ such that $\delta < \frac{\varepsilon^2}{8M^2}$. Let $h \in C([-\pi,\pi])$ be the function that agrees with f on $[-\pi + \delta, \pi - \delta]$, is a straight line from the point $(-\pi, 0)$ to the point $(-\pi + \delta, f(-\pi + \delta))$, and is a straight line from the point $(\pi - \delta, f(\pi - \delta))$ to the point $(\pi, 0)$. Thus, $|f(t) - h(t)| = 0$ for $t \in [-\pi + \delta, \pi - \delta]$ and $|f(t) - h(t)| \leq 2M$ for all $t \notin [-\pi + \delta, \pi - \delta]$. Hence,

$$\|f - h\|^2 = \int_{[-\pi, -\pi+\delta]} |f - h|^2\,dm + \int_{[\pi - \delta, \pi]} |f - h|^2\,dm \leq 8M^2\delta.$$

That is, $\|f - h\| < \varepsilon$.   □

The *classical Fourier coefficients* of $f \in \mathscr{L}^2([-\pi,\pi], \mathfrak{M}, m)$ are the complex numbers $\hat{f}(k)$ defined by

$$\hat{f}(k) = \langle \dot{f}, \dot{\phi}_k \rangle.$$

In particular, if $f : [-\pi,\pi] \to \mathbb{C}$ is a continuous 2π-periodic function, then the classical Fourier series

$$\sum_{k\in\mathbb{Z}} \hat{f}(k)\, e^{ikt}$$

of f converges to f in the sense that

$$\lim_{n\to\infty} \left(\int_{[-\pi,\pi]} |f(t) - \sum_{k=-n}^{n} \hat{f}(k)\, e^{ikt}|^2\,dt \right) = 0.$$

Very old books on analysis refer to this as "convergence in the mean" to f. But from our perspective, convergence in the mean is more plainly understood to be convergence in Hilbert space, as explained by Theorem 5.88.

5.10   Sums of Vectors and Hilbert Spaces

The definition of Banach space involves Cauchy sequences, but it is sometimes useful to extend this idea to nets.

Definition 5.93. If $(\Lambda, \preceq)$ is a directed set, then a net $\{v_\alpha\}_{\alpha \in \Lambda}$ in a normed vector space V is said to be a *Cauchy net* if for every $\varepsilon > 0$ there exists $\alpha_0 \in \Lambda$ such that $\|v_\alpha - v_\beta\| < \varepsilon$, for all $\alpha, \beta \in \Lambda$ such that $\alpha_0 \preceq \alpha$ and $\alpha_0 \preceq \beta$.

Proposition 5.94. *In a normed vector space V, every convergent net is a Cauchy net. Conversely, if V is a Banach space, then every Cauchy net is a convergent net.*

Proof. Exercise 5.128 $\square$

If Λ is a set, not necessarily ordered, then $\mathscr{F}(\Lambda) = \{F \subseteq \Lambda \mid F$ is a finite set$\}$ is a directed set under inclusion.

Definition 5.95. Assume that V is a normed vector space and that $\{v_\alpha\}_{\alpha \in \Lambda}$ is a collection of vectors in V.

1. A *partial sum* of the collection $\{v_\alpha\}_{\alpha \in \Lambda}$ is a vector v_F of the form $v_F = \sum_{\alpha \in F} v_\alpha$, for some finite subset $F \subseteq \Lambda$.
2. The set $\{v_F\}_{F \in \mathscr{F}(\Lambda)}$ of partial sums is *convergent* if the net $\{v_F\}_{F \in \mathscr{F}(\Lambda)}$ is convergent in V; the limit of this convergent net is denoted by $\sum_{\alpha \in \Lambda} v_\alpha$.

Proposition 5.96. *If $\{v_\alpha\}_{\alpha \in \Lambda}$ is a family of vectors in a Banach space V, and if $\sum_{\alpha \in \Lambda} \|v_\alpha\|$ is convergent in $\mathbb{R}$, then $\sum_{\alpha \in \Lambda} v_\alpha$ is convergent in V.*

Proof. Let $\varepsilon > 0$. Because $\sum_{\alpha \in \Lambda} \|v_\alpha\|$ is convergent, the net of partial sums is Cauchy. Thus, there is an $F_0 \in \mathscr{F}(\Lambda)$ such that the partial sums $\sum_{\alpha \in F_1} \|v_\alpha\|$ and $\sum_{\alpha \in F_2} \|v_\alpha\|$ differ by less than ε of $F_0 \subseteq F_1$ and $F_0 \subseteq F_2$. Indeed, the same is true of the partial sums $\sum_{\alpha \in (F_1 \cup F_2)} \|v_\alpha\|$ and $\sum_{\alpha \in F_0} \|v_\alpha\|$. Therefore, by the triangle inequality,

$$\left\| \sum_{\alpha \in F_1} v_\alpha - \sum_{\alpha \in F_2} v_\alpha \right\| \leq \left\| \sum_{\alpha \in (F_1 \cup F_2) \setminus F_0} v_\alpha \right\|$$

$$\leq \left| \sum_{\alpha \in (F_1 \cup F_2)} \|v_\alpha\| - \sum_{\alpha \in F_0} \|v_\alpha\| \right|$$

$$< \varepsilon.$$

Hence, the net $\{v_F\}_{F\in\mathscr{F}(\Lambda)}$ of partial sums of the series $\sum\limits_{\alpha\in\Lambda} v_\alpha$ is Cauchy, and so the net is convergent by Proposition 5.94. □

Proposition 5.81 indicates how the direct sum $\bigoplus\limits_{n\in\mathbb{N}} H_n$ of a countable family $\{H_n\}_{n\in\mathbb{N}}$ of Hilbert spaces H_n is itself a Hilbert space. It useful to be able to carry out a direct sum construction for arbitrary families $\{H_\alpha\}_{\alpha\in\Lambda}$ of Hilbert spaces H_α.

In the Cartesian product $\prod\limits_{\alpha\in\Lambda} H_\alpha$, denote by $\bigoplus\limits_{\alpha\in\Lambda} H_\alpha$ the set of all $\xi = (\xi_\alpha)_{\alpha\in\Lambda}$ for which $\sum\limits_{\alpha\in\Lambda} \|\xi_\alpha\|^2$ is convergent. By the usual Hilbert space inequalities, we see that $\bigoplus\limits_{\alpha\in\Lambda} H_\alpha$ is a complex vector space and that, if $\xi = (\xi_\alpha)_\alpha, \eta = (\eta_\alpha)_\alpha \in \bigoplus\limits_{\alpha\in\Lambda} H_\alpha$, then

$$\langle \xi, \eta \rangle = \sum_\alpha \langle \xi_\alpha, \eta_\alpha \rangle$$

defines an inner product on $\bigoplus\limits_{\alpha\in\Lambda} H_\alpha$.

Definition 5.97. The inner product space $\bigoplus\limits_{\alpha\in\Lambda} H_\alpha$ is called the *direct sum* of the family $\{H_\alpha\}_{\alpha\in\Lambda}$ of Hilbert spaces H_α.

Proposition 5.98. *The direct sum of a family of Hilbert spaces is a Hilbert space.*

Proof. Suppose that $\{\xi^{[n]}\}_{n\in\mathbb{N}}$ is a Cauchy sequence of vectors $\xi^{[n]} = (\xi^{[n]}_\alpha)_\alpha$ in $\bigoplus\limits_{\alpha\in\Lambda} H_\alpha$. Thus, for each α, the sequence $\{\xi^{[n]}_\alpha\}_{n\in\mathbb{N}}$ is a Cauchy sequence in H_α; let $\xi_\alpha \in H_\alpha$ denote the limit of this sequence and consider $\xi = (\xi_\alpha)_\alpha \in \prod\limits_\alpha H_\alpha$.

Let $\varepsilon > 0$. Because $\{\xi^{[n]}\}_{n\in\mathbb{N}}$ is a Cauchy sequence, there exists $n_0 \in \mathbb{N}$ such that $\|\xi^{[m]} - \xi^{[n]}\| < \varepsilon$ for all $m, n \geq n_0$. Thus, for any $F \in \mathscr{F}(\Lambda)$ and $m \geq n_0$,

$$\sum_{\alpha\in F} \|\xi^{[m]}_\alpha - \xi^{[n_0]}_\alpha\|^2 \leq \|\xi^{[m]} - \xi^{[n_0]}\|^2 < \varepsilon^2.$$

Thus, in letting $m \to \infty$,

$$\sum_{\alpha\in F} \|\xi_\alpha - \xi^{[n_0]}_\alpha\|^2 \leq \varepsilon^2,$$

for any finite subset $F \subseteq \Lambda$. Therefore, $\sum_{\alpha \in \Lambda} \|\xi_\alpha - \xi_\alpha^{[n_0]}\|^2$ converges, which implies

that $\xi - \xi^{[n_0]} \in \bigoplus_{\alpha \in \Lambda} H_\alpha$ and, thus, $\xi \in \bigoplus_{\alpha \in \Lambda} H_\alpha$. The inequality $\|\xi - \xi^{[n_0]}\| \leq \varepsilon$ shows

that ξ is the limit of the Cauchy sequence $\{\xi^{[n]}\}_{n \in \mathbb{N}}$. Hence, $\bigoplus_{\alpha \in \Lambda} H_\alpha$ is a Hilbert

space.   □

Problems

5.99. Prove that if V is a normed vector space, then the maps $a : V \times V \to V$ and $m : \mathbb{C} \times V \to V$ defined by

$$a(v_1, v_2) = v_1 + v_2 \quad \text{and} \quad m(\alpha, v) = \alpha v, \; \forall \alpha \in \mathbb{C}, \, v, v_1, v_2 \in V,$$

are continuous.

5.100. Let V be a normed vector space.

1. Prove that the open ball $B_r(v_0)$ is a convex set, for every $v_0 \in V$ and $r > 0$.
2. Prove that the closure $\overline{C}$ of a convex subset $C \subset V$ is convex.

5.101. Suppose that V and W are Banach spaces. On the Cartesian product $V \times W$ define

$$\|(v, w)\| = \max \{\|v\|, \|w\|\}.$$

1. Show that $\| \cdot \|$ is a norm on the vector space $V \times W$ under which $V \times W$ is a Banach space.
2. Prove that the norm topology on $V \times W$ coincides with the product topology on $V \times W$.

5.102. If $\mathcal{T}$ and $\mathcal{T}'$ are the norm topologies on a vector space V induced, respectively, by equivalent norms $\| \cdot \|$ and $\| \cdot \|'$ on V, then prove that $\mathcal{T} = \mathcal{T}'$.

5.103. Suppose that X is a compact topological space and that μ is a finite measure on the Borel sets of X.

1. Prove that there exists $C > 0$ such that $\|f\|_1 \leq C\|f\|_\infty$ for every $f \in C(X)$.
2. Give an example of a compact space X and a finite measure μ on the Borel sets of X in which $\| \cdot \|_1$ and $\| \cdot \|_\infty$ are not equivalent norms on $C(X)$.

5.104. In a normed vector space V, prove that

$$| \|v_1\| - \|v_2\| | \leq \|v_1 - v_2\|, \quad \forall v_1, v_2 \in V.$$

5.105. Let ρ be a seminorm on a vector space V.

1. Prove that the function $d : V \times V \to [0, \infty)$ defined by

$$d(v, w) = \rho(v - w), \quad v, w \in V,$$

is a pseudo-metric on V.
2. With respect to the topology on V induced by the seminorm ρ, prove that the vector space operations (scalar multiplication and vector addition) are continuous. That is, prove that V is a topological vector space.

5.106. If ρ is a seminorm on a topological vector space, and if $\sim$ is the relation on V defined by

$$v \sim w \quad \text{if} \quad \rho(v - w) = 0,$$

then prove that $\sim$ is an equivalence relation. Moreover, if the equivalence classes of elements of V are denoted by

$$\dot{v} = \{w \in V \,|\, w \sim v\},$$

then prove the following assertions:

1. the set $V/\sim$ of equivalence classes is a vector space under the operations

$$\dot{v} + \dot{w} = (v \dot{+} w), \, v, w \in V,$$
$$\alpha \dot{v} = (\dot{\alpha v}), \quad \alpha \in \mathbb{C}, v \in V;$$

2. the function

$$\|\dot{v}\| = \rho(v), \quad v \in V,$$

is a norm on $V/\sim$.

5.107. Let (X, Σ, μ) be a measure space and $p, q \in (1, \infty)$ be conjugate. Assume that $f, g : X \to \mathbb{R}$ are nonnegative measurable functions. Prove that if f is p-integrable and g is q-integrable, then

$$\int_X fg \, d\mu = \left(\int_X f^p \, d\mu \right)^{1/p} \left(\int_X g^q \, d\mu \right)^{1/q}$$

if and only if there is a complex number λ such that $f^p = \lambda g^q$ or $g^q = \lambda f^p$ almost everywhere.

5.108. Consider the function $f(t) = \sqrt{t}$ on a closed interval $[0, b]$, $b > 0$. Prove that there is a sequence of polynomials p_n such that

1. $p_n(0) = 0$, for every $n \in \mathbb{N}$, and
2. $\lim\limits_{n \to \infty} \left(\max\limits_{t \in [0,b]} |\sqrt{t} - p_n(t)| \right) = 0$.

5.109. Prove that if $X \subset \mathbb{R}$ is compact and if $f : X \to \mathbb{C}$ is a continuous function, then for every $\varepsilon > 0$ there is a polynomial p with complex coefficients such that $|f(t) - p(t)| < \varepsilon$ for all $t \in X$.

5.110. Prove that if X is a locally compact space, then $C_c(X)$ is a subspace of $C_b(X)$.

5.111. If X is a locally compact space, then prove that

1. $C_0(X)$ is a subspace of $C_b(X)$, and
2. $fg \in C_0(X)$, for all $f \in C_0(X)$ and $g \in C_b(X)$.

5.112. Prove that if φ is a simple function on a measure space (X, Σ, μ), then φ is p-integrable if and only if $\mu\left(\varphi^{-1}(\mathbb{C} \setminus \{0\})\right) < \infty$.

5.113. Suppose that f is an essentially bounded function on a measure space (X, Σ, μ). Prove that there exists $E \in \Sigma$ such that $\mu(E) = 0$ and $\sup |f(t)| < \infty$ for all $t \in X \setminus E$.

5.114. Prove that if f is an essentially bounded function on a measure space (X, Σ, μ), then

$$\text{ess-sup} f = \inf\left\{\alpha \in \mathbb{R} \,|\, |f|^{-1}(\alpha, \infty) \text{ has measure zero}\right\}.$$

5.115. Assume that $f \in \mathscr{L}^\infty(X, \Sigma, \mu)$ is nonnegative and that $\{f_n\}_{n \in \mathbb{N}}$ is a monotone-increasing sequence of measurable functions for which $0 \le f_n(x) \le f(x)$ and $\lim_{n \to \infty} f_n(x) = f(x)$ for all $x \in X$. Prove that each $f_n \in \mathscr{L}^\infty(X, \Sigma, \mu)$ and that $\lim_{n \to \infty} \|\dot{f} - \dot{f}_n\| = 0$.

5.116. Prove that $L^\infty(X, \Sigma, \mu)$ is an abelian Banach algebra.

5.117. Prove that the function $\|\cdot\|$ on $M(X, \Sigma)$ defined by $\|\nu\| = |\nu|(X)$ is a norm.

5.118. Prove that $L^p(\mathbb{R}^d, \mathfrak{M}(\mathbb{R}^d), m_d)$ is a separable Banach space, for every $p \ge 1$. (Suggestion: Note that $\mathbb{R}^d = \bigcup_{n \in \mathbb{N}} [-n, n]^d$.)

5.119. Prove that the Banach space $\ell^\infty(\mathbb{N})$ is not separable.

5.120. Prove that the Banach space $C(\mathbb{R}/2\pi\mathbb{Z})$ of continuous 2π-periodic complex-valued functions is separable.

5.121. Prove that in a Hilbert space H, $S^\perp$ is a subspace, for every $S \subseteq H$.

5.122. In a Hilbert space H, let M_1 and M_2 be closed subspaces.

1. Prove that $(M_1 + M_2)^\perp = M_1^\perp \cap M_2^\perp$.
2. Prove or find a counterexample to $(M_1 \cap M_2)^\perp = M_1^\perp + M_2^\perp$.

5.123. If Λ is a finite or countable set, and if $\{H_k\}_{k \in \Lambda}$ is a family of Hilbert spaces in which $\langle \cdot, \cdot \rangle_k$ denotes the inner product of H_k, then prove that the vector space

$$\bigoplus_{k \in \Lambda} H_k = \{(\xi_k)_{k \in \Lambda} \,|\, \sum_{k \in \Lambda} \|\xi_k\|^2 < \infty\}$$

is a Hilbert space with respect to the inner product

$$\langle (\xi_k)_k, (\eta_k)_k \rangle = \sum_{k \in \Lambda} \langle \xi_k, \eta_k \rangle_k.$$

5.124. Prove that if $\mathscr{B} \subset H$ is a set of orthonormal vectors in a Hilbert space H, then $\mathscr{B} \subset H$ is an orthonormal basis of H if and only if Span $\mathscr{B}$ is dense in H.

5.125. Prove that every nonzero Hilbert space has an orthonormal basis. In particular, if $\mathscr{O} \subset H$ is a set of orthonormal vectors, then prove that H has an orthonormal basis $\mathscr{B}$ that contains $\mathscr{O}$.

5.126. Suppose that in a Hilbert space H, $\phi_1, \ldots, \phi_n$ are orthonormal vectors. Prove that if $\xi_1, \ldots, \xi_n \in H$ satisfy $\|\xi_j - \phi_j\| < n^{-1/2}$ for each j, then $\xi_1, \ldots, \xi_n$ are linearly independent.

5.127. Consider the Fourier series expansions of $f_1(t) = t$ and $f_2(t) = e^{\alpha t}$ in $\mathscr{L}^2([-\pi, \pi])$. Calculate the Fourier series of each f_j and use Parseval's identity to establish each of the following formulae:

$$\sum_{n=1}^{\infty} \frac{1}{n^2} = \frac{\pi^2}{6};$$

$$\sum_{n=1}^{\infty} \frac{1}{n^2 + \alpha^2} = \frac{\pi}{\alpha} \coth(\alpha \pi).$$

5.128. Assume that V is a normed vector space.

1. Prove that every convergent net in V is a Cauchy net.
2. If V is a Banach space, then prove that every Cauchy net in V is a convergent net.

Chapter 6
Dual Spaces

In considering the elements of the vector space $\mathbb{R}^d$ as column vectors, the vector space $(\mathbb{R}^d)^t$ of row vectors is obviously related to $\mathbb{R}^d$, but is not necessarily identical to it. What, therefore, is one to make of row vectors and of the transpose $\xi \mapsto \xi^t$ operation applied to column vectors ξ? The usual product of matrices and vectors indicates that each ξ^t is a linear transformation $\mathbb{R}^d \to \mathbb{R}$ via $\xi^t(\eta) = \xi^t\eta$, for (column) vectors $\eta \in \mathbb{R}^d$. This is perhaps the simplest instance of duality, which is an association of a closely related vector space V^* of linear maps $V \to \mathbb{R}$ to a given real vector space V. This idea is at the heart of functional analysis (indeed, this is where the "functional" part of "functional analysis" enters the picture), and this notion is developed and explored in the present chapter.

6.1 Operators

Recall that a linear transformation from a vector space V to a vector space W (both over some field $\mathbb{F}$) is a function $T : V \to W$ in which T is additive and homogenous; that is, $T(\alpha_1 v_1 + \alpha_2 v_2) = \alpha_1 T(v_1) + \alpha_2 T(v_2)$ for all $v_1, v_2 \in V$ and $\alpha_1, \alpha_2 \in \mathbb{F}$. Unless the context leads to ambiguity, the notation Tv shall be used in place of $T(v)$ for linear transformations $T : V \to W$ and $v \in V$.

Our interest is with vector spaces that are Banach spaces; thus, we shall require that linear transformations between Banach space be continuous. In this regard, the essential concept is that of "boundedness of a linear transformation".

Definition 6.1. If V and W are normed vector spaces, then a linear transformation $T : V \to W$ is *bounded* if there is a constant $M > 0$ such that $\|Tv\| \leq M\|v\|$, for every $v \in V$; otherwise, T is *unbounded*.

© Springer International Publishing Switzerland 2016
D. Farenick, *Fundamentals of Functional Analysis*, Universitext,
DOI 10.1007/978-3-319-45633-1_6

A bounded linear transformation is called a *bounded operator*, or simply an *operator*, while an unbounded linear transformation is called an *unbounded operator*.

In principle, a given normed vector space can admit both bounded and unbounded operators, as the following example shows.

Example 6.2. *Let $\mathbb{C}[t]$ be the normed vector space of complex polynomials with norm $\|f\| = \max_{t \in [0,1]} |f(t)|$, and consider the linear transformations $D: \mathbb{C}[t] \to \mathbb{C}[t]$ and $J: \mathbb{C}[t] \to \mathbb{C}$ defined by $Df = \frac{df}{dt}$ and $Jf = \int_0^1 f(t)dt$, respectively. Then J is bounded and D is unbounded.*

Proof. If $f_n \in \mathbb{C}[t]$ is $f_n(t) = t^n$, then $Df_n = nf_{n-1}$ for $n \geq 2$. Thus, $\|f_n\| = 1$ for every n, but $\|Df_n\| = n - 1 = (n-1)\|f_n\|$; that is, no constant $M > 0$ exists for which $\|Df\| \leq M\|f\|$ for every $f \in V$. Hence, D is an unbounded operator.

On the other hand, for every $f \in V$,

$$\|Jf\| = \left| \int_0^1 f(t)\,dt \right| \leq \int_0^1 |f(t)|\,dt \leq \int_0^1 \|f\|\,dt = \|f\|,$$

which shows that J is a bounded operator. $\square$

As is often the case, finite-dimensional spaces do not exhibit any exotic features: all linear maps are necessarily bounded.

Proposition 6.3. *Assume that V and W are normed vector spaces.*

1. *If V has finite dimension, then every linear transformation $T: V \to W$ is bounded.*
2. *If V has infinite dimension, and if $W \neq \{0\}$, then there is a linear transformation $T: V \to W$ such that T is unbounded.*

Proof. Suppose that V has finite dimension, and fix a basis $\mathscr{B} = \{v_1, \ldots, v_n\}$ of V. Set

$$\tau = \left(\sum_{j=1}^n \|Tv_j\|^2 \right)^{1/2}.$$

By the triangle inequality and the Cauchy-Schwarz inequality,

$$\left\| T\left(\sum_{j=1}^n \alpha_j v_j \right) \right\| = \left\| \sum_{j=1}^n \alpha_j Tv_j \right\| = \sum_{j=1}^n |\alpha_j|\,\|Tv_j\| \leq \left(\sum_{j=1}^n |\alpha_j|^2 \right)^{1/2} \tau.$$

Proposition 5.8 and its proof show that there is a constant $C > 0$ such that

$$\left(\sum_{j=1}^n |\alpha_j|^2 \right)^{1/2} \leq C \left\| \sum_{j=1}^n \alpha_j v_j \right\|.$$

Hence, with $M = C\tau$, we obtain $\|Tv\| \leq M\|v\|$, for every $v \in V$.

Next, suppose that V has infinite dimension and let $\mathscr{B}$ be a linear basis of V. From $\mathscr{B}$ select a countably infinite subset $\{y_k \mid k \in \mathbb{N}\} \subseteq \mathscr{B}$. Without loss of generality, each vector of $\mathscr{B}$ may be assumed to be of norm 1. Because W is a nonzero vector space there is at least one vector $w \in W$ of norm $\|w\| = 1$. Define a linear transformation $T : V \to W$ by the following action on the vectors of the linear basis $\mathscr{B}$:

$$T(y_k) = kw, \text{ for each } k \in \mathbb{N};$$
$$T(y) = 0, \text{ for each } y \in \mathscr{B}\backslash\{y_k \mid k \in \mathbb{N}\}.$$

Note that $\|T(y_k)\| = k\|w\| = k$. Because $\|y_k\| = 1$, for every $k \in \mathbb{N}$, the linear transformation T is unbounded. $\square$

The importance of boundedness for linear maps is that boundedness is a synonym for continuity.

Proposition 6.4. *The following statements are equivalent for a linear transformation $T : V \to W$ between normed vector spaces V and W:*

1. *T is bounded;*
2. *T is continuous.*

Proof. Assume that $M > 0$ is such that $\|T(v)\| \leq M\|v\|$, for all $v \in V$. Fix $v_0 \in V$. By the linearity of T, $\|T(v) - T(v_0)\| = \|T(v - v_0)\| \leq M\|v - v_0\|$. Thus, if $\varepsilon > 0$ and if $\delta = \varepsilon/M$, then $\|v - v_0\| < \delta$ implies that $\|T(v) - T(v_0)\| < \varepsilon$. That is, T is continuous at v_0. Hence, as the choice of $v_0 \in V$ is arbitrary, T is continuous on V.

Conversely, assume that T is continuous. In particular, T is continuous at 0. Thus, for $\varepsilon = 1$ there is a $\delta > 0$ such that $\|w\| < \delta$ implies $\|T(w)\| < 1$. Let $M = 2/\delta$. If $v \in V$ is nonzero, then let $w = \frac{\delta}{2\|v\|}v$. Thus, $\|w\| < \delta$ and so $\|T(w)\| < 1$. That is, $\|Tv\| \leq M\|v\|$. $\square$

The next proposition is the first step toward the study of operators in the context of analysis.

Definition 6.5. If V and W are normed vector spaces, then the set of all operators $T : V \to W$ is denoted by $B(V, W)$. In the case where $W = V$, the notation $B(V)$ is used in place of $B(V, V)$.

Proposition 6.6. *Assume that V and W are normed vector spaces.*

1. *The set $B(V, W)$ is a vector space under the operations $(T_1 + T_2)(v) = T_1(v) + T_2(v)$ and $(\alpha T)(v) = \alpha T(v)$ for $T, T_1, T_2 \in B(V, W)$ and $v \in V$, $\alpha \in \mathbb{C}$.*
2. *The function $\|\cdot\| : B(V, W) \to \mathbb{R}$ defined by*

$$\|T\| = \sup_{0 \neq v \in V} \frac{\|T(v)\|}{\|v\|} \tag{6.1}$$

   is a norm on $B(V, W)$.
3. *If W is a Banach space, then so is $B(V, W)$.*

Proof. It is clear that $B(V, W)$ has the indicated structure of a vector space, and so we turn to the proof that (6.1) defines a norm on $B(V, W)$. To this end, note that (using the fact that every linear transformation sends zero to zero) equation (6.1) is equivalent to

$$\|T(v)\| \leq \|T\| \|v\|, \quad \forall v \in V. \tag{6.2}$$

Thus, $\|T\| = 0$ only if $\|T(v)\| = 0$ for every $v \in V$; hence, $T(v) = 0$ for all $v \in V$. This proves that T is the zero transformation: $T = 0$. It is also clear that $\|\alpha T\| = |\alpha| \|T\|$, and so we now establish the triangle inequality. Let $T_1, T_2 \in B(V, W)$. For every $v \in V$,

$$\|(T_1 + T_2)v\| = \|T_1(v) + T_2(v)\|$$

$$\leq \|T_1(v)\| + \|T_2(v)\|$$

$$\leq (\|T_1\| + \|T_2\|) \|(v)\|,$$

and so

$$\|T_1 + T_2\| = \sup_{0 \neq v \in V} \frac{\|T_1 + T_2(v)\|}{\|v\|} \leq \|T_1\| + \|T_2\|.$$

This completes the proof that equation (6.1) defines a norm on $B(V, W)$.

Suppose now that W is a Banach space and select any Cauchy sequence $\{T_k\}_{k \in \mathbb{N}}$ in $B(V, W)$. Inequality (6.2) implies that $\{T_k(v)\}_{k \in \mathbb{N}}$ is a Cauchy sequence in W for every $v \in V$. Because W is a Banach space, each sequence $\{T_k(v)\}_{k \in \mathbb{N}}$ is convergent in W; denote the limit by $T(v)$. Note that the map $v \mapsto T(v)$ is indeed a linear transformation $T : V \to W$. It remains to show that T is bounded and that $\lim_k \|T_k - T\| = 0$.

Let $\varepsilon > 0$. Because $\{T_k\}_{k \in \mathbb{N}}$ is a Cauchy sequence, there exists $N_\varepsilon \in \mathbb{N}$ such that $\|T_n - T_m\| < \varepsilon$ for all $n, m \geq N_\varepsilon$. If $v \in V$ and if $n \geq N_\varepsilon$, then

$$\|T(v) - T_n(v)\| \leq \|T(v) - T_m(v)\| + \|T_m(v) - T_n(v)\|$$

$$\leq \|T(v) - T_m(v)\| + \|T_m - T_n\| \|v\|.$$

As the inequalities above are true for all $m \in \mathbb{N}$,

$$\|T(v) - T_n(v)\| \leq \inf_{m \in \mathbb{N}} (\|T(v) - T_m(v)\| + \|T_m - T_n\| \|v\|)$$

$$\leq 0 + \varepsilon \|v\|.$$

Hence, if $n \geq N_\varepsilon$ is fixed, then $T - T_n$ is bounded and $\|T - T_n\| \leq \varepsilon$. Therefore, $T_n + (T - T_n) = T$ is bounded and $\|T - T_n\| < \varepsilon$ for all $n \geq N_\varepsilon$. This proves that the Cauchy sequence $\{T_k\}_{k \in \mathbb{N}}$ converges in $B(V, W)$ to $T \in B(V, W)$. $\quad\square$

Two linear structures affiliated with an operator $T \in B(V, W)$ are recalled below from linear algebra.

Definition 6.7. If $T : V \to W$ is a linear transformation between vector spaces V and W, then

1. the *kernel* of T is the set $\ker T = \{v \in V \mid Tv = 0\}$, and
2. the *range* of T is the set $\operatorname{ran} T = \{Tv \in W \mid v \in V\}$.

It is clear that both $\ker T$ and $\operatorname{ran} T$ are linear submanifolds of V and W, respectively. If T is continuous (that is, bounded), then $\ker T$ is normed closed and, hence, is a subspace; however, the range of a bounded operator need not be closed in general. There is one important instance in which $\operatorname{ran} T$ is always a subspace, and that occurs with operators T that preserve the norms of vectors.

Definition 6.8. An operator $T : V \to W$ acting on normed vector spaces V and W is an *isometry* if $\|Tv\| = \|v\|$ for every $v \in V$.

Of course, every isometry T is of norm $\|T\| = 1$. Moreover, $\|Tv\| = \|v\|$ implies that $Tv = 0$ if and only if $v = 0$, and so every isometry is an injection. On the other hand, by scaling an operator T by $\alpha = \|T\|^{-1}$, every nonzero operator is a scalar multiple of an operator of norm 1, and so not every operator of norm 1 need be an isometry.

Proposition 6.9. *If V is a Banach space and if $T \in B(V, W)$ is an isometry, then $\operatorname{ran} T$ is a subspace of W.*

Proof. Let $w \in W$ be in the closure of the range of T. Thus, $\|w - T(v_n)\| \to 0$, for some sequence of vectors $v_n \in V$, and so $\{T(v_n)\}_{n \in \mathbb{N}}$ is a Cauchy sequence in W. Because $\|Tv_n - Tv_m\| = \|T(v_n - v_m)\| = \|v_n - v_m\|$, the sequence $\{v_n\}_{n \in \mathbb{N}}$ is Cauchy in V. As V is a Banach space, there is a limit $v \in V$ to this sequence. Moreover,

$$\|Tv - w\| \leq \|T(v - v_n)\| + \|Tv_n - w\| = \|v - v_n\| + \|Tv_n - w\|, \quad \forall n \in \mathbb{N}.$$

Hence, $w = Tv$. That is, the range of T contains all of its limit points, implying that the range of T is closed. $\quad\square$

Definition 6.10. Two normed vector spaces V and W are said to be *isometrically isomorphic* if there is a linear isometry $T : V \to W$ such that T is a surjection.

Thus, if V and W are Banach spaces, and if $T : V \to W$ is an isometry, then V is isometrically isomorphic to the range of T (which by, Proposition 6.9, is a subspace of W). Hence, T embeds V into W and in so doing preserves the Banach space structure of V. In such cases, we say that "W contains a copy of V" because inside W that copy of V appears (as a Banach space) no different from V itself.

Example 6.11. *If* $\varphi : \mathbb{R} \to \mathbb{C}$ *is a continuous* 2π-*periodic function such that* $|\varphi(t)| = 1$ *for all* $t \in \mathbb{R}$, *and if* $1 \leq p < \infty$, *then the linear map* $M_\varphi : L^p(\mathbb{T}) \to L^p(\mathbb{T})$ *given by* $T\dot{f} = (\dot{\varphi f})$ *is an isometric isomorphism.*

Further study of operators on Banach and Hilbert spaces will be taken up in later chapters; the remainder of the present chapter is devoted to the quite special case of operators that map complex normed vector spaces V to the 1-dimensional vector space $\mathbb{C}$.

6.2   Linear Functionals

Definition 6.12. If V is a normed vector space, then a *linear functional* on V is an operator $\varphi : V \to \mathbb{C}$.

The familiar operation of integration is a basic example of a linear functional.

Example 6.13. *If* (X, Σ, μ) *is a measure space, then the map* $\varphi_\mu : L^1(X, \Sigma, \mu) \to \mathbb{C}$ *defined by*

$$\varphi_\mu(\dot{f}) = \int_X f \, d\mu,$$

for $f \in \mathscr{L}^1(X, \Sigma, \mu)$, *is a linear functional on* $L^1(X, \Sigma, \mu)$.

Proof. The map φ_μ is obviously linear. To show it is bounded, use the triangle inequality:

$$|\varphi_\mu(\dot{f})| = \left| \int_X f \, d\mu \right| \leq \int_X |f| \, d\mu = \|\dot{f}\|.$$

Hence, φ_μ is bounded of norm $\|\varphi_\mu\| \leq 1$. □

Another familiar example of a linear functional is drawn from linear algebra.

Example 6.14. *If* $\eta \in \mathbb{C}^n$, *then the map* $\varphi_\eta : \mathbb{C}^n \to \mathbb{C}$ *defined by*

$$\varphi_\eta(\xi) = \sum_{j=1}^{n} \xi_j \eta_j,$$

for $\xi \in \mathbb{C}^n$, *is a linear functional on* $\mathbb{C}^n$. *Conversely, if* φ *is a linear functional on* $\mathbb{C}^n$, *then there exists a unique* $\eta \in \mathbb{C}^n$ *such that* $\varphi = \varphi_\eta$.

Proof. It is clear that φ_η is a linear functional. Conversely, any linear map of $\mathbb{C}^n$ onto $\mathbb{C}$ will be given by a $1 \times n$ matrix whose action on $\mathbb{C}^n$ is precisely that of φ_η, for some $\eta \in \mathbb{C}^n$. □

The norm of $\|\varphi_\eta\|$ depends upon the choice of norm for $\mathbb{C}^n$.

Definition 6.15. The set $B(V, \mathbb{C})$ of all linear functionals on V is called the *dual space* of V and is denoted by V^*.

By Proposition 6.6, the dual space V^* is a Banach space for every normed vector space V. The determination of V^* from V (or *vice versa*) is an important and nontrivial problem. The next result is an example of an instance in which V^* is determined precisely from V.

Proposition 6.16 below is the first of many results that have the moniker *Riesz Representation Theorem*. Other "Riesz Representation Theorems" in the present book appear as Theorems 6.40, 6.51, and 10.1.

Proposition 6.16 (Riesz). *Suppose that p and q are conjugate real numbers.*

1. If $g = (g_k)_{k \in \mathbb{N}} \in \ell^q(\mathbb{N})$, then the function $\varphi : \ell^p(\mathbb{N}) \to \mathbb{C}$ defined by

$$\varphi(f) = \sum_{k=1}^{\infty} f_k g_k,$$

for $f = (f_k)_{k \in \mathbb{N}} \in \ell^p(\mathbb{N})$, is a linear functional on $\ell^p(\mathbb{N})$ of norm $\|\varphi\| = \|g\|$.
2. For each $\varphi \in (\ell^p(\mathbb{N}))^*$ there is a unique $g \in \ell^q(\mathbb{N})$ of norm $\|g\| = \|\varphi\|$ such that

$$\varphi(f) = \sum_{k=1}^{\infty} f_k g_k,$$

for every $f = (f_k)_{k \in \mathbb{N}} \in \ell^p(\mathbb{N})$.

Proof. The proof of the first assertion is left to the reader as Exercise 6.54.

To prove the second assertion, select $\varphi \in (\ell^p(\mathbb{N}))^*$. If $e_k \in \ell^p(\mathbb{N})$ is the vector with 1 in position k and 0 elsewhere, then let $g_k = \varphi(e_k)$ for each $k \in \mathbb{N}$. Select and fix $n \in \mathbb{N}$, and consider the element $f = (f_k)_{k \in \mathbb{N}} \in \ell^p(\mathbb{N})$ in which $f_k = 0$ for every $k > n$ and $f_k = |g_k|^{q-1} \mathrm{sgn}(\overline{g}_k)$ for each $k = 1, \dots, n$, where, for any $\alpha \in \mathbb{C}$, $\mathrm{sgn}\,\alpha$ is given by 0 if $\alpha = 0$ and by $\frac{\alpha}{|\alpha|}$ otherwise. Thus,

$$f_k g_k = |g_k|^{q-1} (g_k \mathrm{sgn}(\overline{g}_k)) = |g_k|^q \quad \text{and} \quad |f_k|^p = |g_k|^{p+q-p} = |g_k|^q.$$

Thus,

$$\varphi(f) = \varphi\left(\sum_{k=1}^{n} f_k e_k\right) = \sum_{k=1}^{n} f_k \varphi(e_k) = \sum_{k=1}^{n} |g_k|^q,$$

and so

$$\sum_{k=1}^{n} |g_k|^q = |\varphi(f)| \leq \|\varphi\| \, \|f\| = \|\varphi\| \left(\sum_{k=1}^{n} |f_k|^p\right)^{1/p} = \|\varphi\| \left(\sum_{k=1}^{n} |g_k|^q\right)^{1/p}.$$

Thus,

$$\|\varphi\| \geq \left(\sum_{k=1}^{n} |g_k|^q\right)^{1-\frac{1}{p}} = \left(\sum_{k=1}^{n} |g_k|^q\right)^{1/q}.$$

Because the inequality above is true for arbitrary $n \in \mathbb{N}$, we deduce that $g \in \ell^q(\mathbb{N})$ and that $\|g\| \leq \|\varphi\|$.

Next consider $h_k = |g_k|^{q-1} \text{sgn}(\overline{g}_k)$ for every $k \in \mathbb{N}$. Because

$$\left(\sum_{k=1}^{\infty} |h_k|^p\right)^{1/p} = \left(\sum_{k=1}^{\infty} |g_k|^q\right)^{1/p} = \|g\|^{q/p},$$

we have that $h = (h_k)_{k\in\mathbb{N}} \in \ell^p(\mathbb{N})$. Thus, by the continuity of φ, the triangle inequality, and Hölder's Inequality, we have

$$|\varphi(h)| = \left|\varphi\left(\sum_{k=1}^{\infty} h_k e_k\right)\right| = \left|\sum_{k=1}^{\infty} h_k g_k\right| \leq \sum_{k=1}^{\infty} |h_k| |g_k| \leq \|h\| \|g\|,$$

which implies that $\|\varphi\| \leq \|g\|$. This proves that $\|\varphi\| = \|g\|$.

Lastly, because each $f \in \ell^p(\mathbb{N})$ is expressed uniquely in terms of the vectors $\{e_k\}_{k\in\mathbb{N}} \subset \ell^p(\mathbb{N})$, the choice of $g \in \ell^q(\mathbb{N})$ is also uniquely determined from the equations $g_k = \varphi(e_k)$, $k \in \mathbb{N}$. $\square$

Corollary 6.17. *If p and q are conjugate real numbers, then $(\ell^p(\mathbb{N}))^*$ and $\ell^q(\mathbb{N})$ are isometrically isomorphic.*

Proof. Let $\Theta : \ell^q(\mathbb{N}) \to (\ell^p(\mathbb{N}))^*$ be given by $\Theta g = \varphi_g$, where $\varphi_g \in (\ell^p(\mathbb{N}))^*$ satisfies

$$\varphi_g(f) = \sum_{k=1}^{\infty} f_k g_k,$$

for every $f = (f_k)_{k\in\mathbb{N}} \in \ell^p(\mathbb{N})$. The map Θ is plainly linear and is also, by Proposition 6.16, isometric and surjective. Hence, Θ is a linear isometric isomorphism of $(\ell^p(\mathbb{N}))^*$ and $\ell^q(\mathbb{N})$. $\square$

Although Proposition 6.16 and Corollary 6.17 address the case of ℓ^p spaces over $\mathbb{N}$, precisely the same results hold true for the finite-dimensional Banach spaces $\ell^p(n)$, for every $n \in \mathbb{N}$.

Some other accessible examples of interesting dual spaces (such as the dual of $\ell^1(\mathbb{N})$) are addressed in Exercises 6.57 and 6.56. There is an integral version of Proposition 6.16, but establishing it requires somewhat more effort, and so further discussion of it is deferred to Section 6.6.

6.3   Hahn-Banach Extension Theorem

Notwithstanding the explicit determination of the duals of ℓ^p-spaces in the previous section, the analysis of the dual space V^* of abstract infinite-dimensional normed vector spaces V is somewhat delicate. For example, to begin with, it is not at all obvious that V^* has any nonzero elements whatsoever. Proving that V^* has elements in abundance is the principal objective of the present section, and the main result developed here, the Hahn-Banach Extension Theorem, is surely the most important foundational theorem in functional analysis. At its heart is the following linear-algebraic theorem.

Theorem 6.18 (Hahn-Banach Extension Theorem). *Assume that V is a real vector space and that $p : V \to \mathbb{R}$ is a function such that, for all $v, v_1, v_2 \in V$ and all $\alpha \geq 0$,*

1. *$p(v_1 + v_2) \leq p(v_1) + p(v_2)$, and*
2. *$p(\alpha v) = \alpha p(v)$.*

If L is a linear submanifold of V and if $\varphi : L \to \mathbb{R}$ is a linear transformation for which $|\varphi(v)| \leq p(v)$ for every $v \in L$, then there is a linear transformation $\Phi : V \to \mathbb{R}$ such that $\Phi_{|L} = \varphi$ and $-p(-v) \leq \Phi(v) \leq p(v)$ for every $v \in V$.

Proof. The linearity of φ implies that $-p(-v) \leq \varphi(v) \leq p(v)$ for every $v \in L$. If $\varphi = 0$, then we may take Φ to be $\Phi = 0$; therefore, assume that $\varphi \neq 0$.

Define a set $\mathfrak{S}$ consisting of all ordered pairs (M, ϑ) such that M is a linear submanifold of V containing L and $\vartheta : M \to \mathbb{R}$ is a linear transformation satisfying $-p(-v) \leq \vartheta(v) \leq p(v)$ for every $v \in M$. The set $\mathfrak{S}$ is nonempty since $(L, \varphi) \in \mathfrak{S}$. Define a partial order $\preceq$ on $\mathfrak{S}$ by

$$(M_1, \vartheta_1) \preceq (M_2, \vartheta_2) \quad \text{if and only if} \quad M_1 \subseteq M_2 \text{ and } \vartheta_{2|M_1} = \vartheta_1.$$

With respect to this partial order, let $\mathfrak{F}$ be any linearly ordered subset of $\mathfrak{S}$. Hence, there is a linearly ordered set Λ such that

$$\mathfrak{F} = \{(M_\lambda, \vartheta_\lambda) \,|\, \lambda \in \Lambda\}.$$

Define $M \subseteq V$ by

$$M = \bigcup_{\lambda \in \Lambda} M_\lambda,$$

and note that M is a linear submanifold of V containing L. Furthermore, the function $\vartheta : M \to \mathbb{R}$ given by $\vartheta(v) = \vartheta_\lambda(v)$, if $v \in M_\lambda$, is well defined, linear, and satisfies $-p(-v) \leq \vartheta(v) \leq p(v)$ for all $v \in M$. Thus, $(M, \vartheta) \in \mathfrak{S}$ and (M, ϑ) is an upper bound for $\mathfrak{F}$. Hence, by Zorn's Lemma, $\mathfrak{S}$ has a maximal element.

Let (M, ϑ) denote one such maximal element of $\mathfrak{G}$. Suppose $M \neq V$; thus, there exists (nonzero) $v_0 \in V \setminus M$. If $x, y \in M$, then

$$\vartheta(x) + \vartheta(y) = \vartheta(x+y) \leq p(x+y)$$

$$= p((x - v_0) + (v_0 + y))$$

$$\leq p(x - v_0) + p(x_0 + y).$$

Thus,

$$\vartheta(x) - p(x - v_0) \leq p(v_0 + y) - \vartheta(y).$$

Let $\delta_0 \in \mathbb{R}$ satisfy

$$\sup_{x \in M} (\vartheta(x) - p(x - v_0)) \leq \delta_0 \leq \inf_{y \in M} (p(v_0 + y) - \vartheta(y)).$$

Thus,

$$\vartheta(x) - \delta_0 \leq p(x - v_0) \quad \text{and} \quad \vartheta(y) + \delta_0 \leq p(v_0 + y)$$

for all $x, y \in M$.

Consider now the linear submanifold $M_1 = \{x + \alpha v_0 \mid x \in M, \ \alpha \in \mathbb{R}\}$ of V, and note that M_1 contains M and $M \neq M_1$. Define a function $\vartheta_1 : M_1 \to \mathbb{R}$ by

$$\vartheta_1(x + \alpha v_0) = \vartheta(x) + \alpha \delta_0,$$

and observe that ϑ_1 is linear. If $\alpha > 0$, then

$$\vartheta_1(x + \alpha v_0) = \alpha \left(\vartheta(\alpha^{-1} x) + \delta_0\right) \leq \alpha p(\alpha^{-1} x + v_0) = p(x + \alpha v_0),$$

while if $\alpha < 0$, then

$$\vartheta_1(x + \alpha v_0) = |\alpha| \left(\vartheta(-\alpha^{-1} x) - \delta_0\right) \leq |\alpha| p(-\alpha^{-1} x - v_0) = p(x + \alpha v_0).$$

Therefore, by the linearity of ϑ_1, $-p(-v) \leq \vartheta_1(v) \leq p(v)$ for every $v \in V$, which shows that $(M_1, \vartheta_1) \in \mathfrak{G}$. But the relation $(M, \vartheta) \preceq (M_1, \vartheta_1)$ with $M_1 \neq M$ contradicts the maximality of (M, ϑ) in $\mathfrak{G}$. Therefore, it must be that $M = V$ and so $\Phi = \vartheta$ is one desired extension of φ. $\qquad \square$

The function p in the statement of the Hahn-Banach Extension Theorem is called a sublinear functional.

Definition 6.19. A function $p : V \to \mathbb{R}$ for which $p(v_1 + v_2) \leq p(v_1) + p(v_2)$ and $p(\alpha v) = \alpha p(v)$, for all $v, v_1, v_2 \in V$ and all $\alpha \geq 0$, is called a *sublinear functional*.

A more specialised version of the Hahn-Banach Extension Theorem, in which the sublinear functional p is given by $p(v) = \|v\|$ for some norm $\|\cdot\|$, is very frequently used. Before stating this form of the result, we first indicate how to pass from a complex normed vector space to a real normed vector space and then back again.

Thus, suppose, as usual, that V is a complex vector space and let $\varphi : V \to \mathbb{C}$ be an arbitrary linear transformation. The functions $\psi_1, \psi_2 : V \to \mathbb{R}$ defined by

$$\psi_1(v) = \frac{1}{2}\left(\varphi(v) + \overline{\varphi(v)}\right) \quad \text{and} \quad \psi_2(v) = \frac{1}{2i}\left(\varphi(v) - \overline{\varphi(v)}\right)$$

satisfy $\psi_j(v_1 + v_2) = \psi_j(v_1) + \psi_j(v_2)$ and $\psi_j(rv) = r\psi_j(v)$ for $v, v_1, v_2 \in V$, $r \in \mathbb{R}$, and $j = 1, 2$. Thus, considering V as a real vector space, ψ_1 and ψ_2 are $\mathbb{R}$-linear transformations such that $\varphi(v) = \psi_1(v) + i\psi_2(v)$ for every $v \in V$. Therefore, we denote ψ_1 and ψ_2 by $\Re\varphi$ and $\Im\varphi$, respectively.

Now if V is a normed vector space and $\varphi \in V^*$, then for every unit vector $v \in V$ we have

$$|\Re\varphi(v)| \leq \frac{1}{2}\left(|\varphi(v)| + |\overline{\varphi(v)}|\right) = |\varphi(v)|,$$

which shows that $\Re\varphi$ is a bounded linear transformation and that $\|\Re\varphi\| \leq \|\varphi\|$. On the other hand, given a unit vector $v \in V$, there is a real number θ such that $|\varphi(v)| = e^{i\theta}\varphi(v)$. Thus, with $w = e^{i\theta}v \in V$, we have

$$|\Re\varphi(w)| = \left|\Re\left(\varphi(e^{i\theta}v)\right)\right| = \left|\Re\left(e^{i\theta}\varphi(v)\right)\right| = |\varphi(v)|,$$

which shows that

$$\|\varphi\| = \sup_{v \in V, \|v\|=1} |\varphi(v)| \leq \sup_{w \in V, \|w\|=1} |\varphi(w)| = \|\Re\varphi\|.$$

Hence, $\|\Re\varphi\| = \|\varphi\|$. A similar argument shows that $\|\Im\varphi\| = \|\varphi\|$.

The discussion above is summarised by the following lemma.

Lemma 6.20. *If V is a normed vector space and if $\varphi \in V^*$, then $\Re\varphi$ and $\Im\varphi$ are $\mathbb{R}$-linear maps $V \to \mathbb{R}$ such that $\|\Re\varphi\| = \|\Im\varphi\| = \|\varphi\|$.*

Conversely, suppose now that $\psi : V \to \mathbb{R}$ is a (real) linear transformation, where V is considered as a vector space over $\mathbb{R}$. Motivated by the fact that $\zeta = \Re(\zeta) - i\Re(i\zeta)$ for every $\zeta \in \mathbb{C}$, define $\psi_1 : V \to \mathbb{R}$ by $\psi_1(v) = -\psi(iv)$, for $v \in V$, and $\varphi : V \to \mathbb{C}$ by $\varphi = \psi + i\psi_1$. Observe that ψ_1 is a $\mathbb{R}$-linear transformation, φ is $\mathbb{C}$-linear transformation, and $\Re\varphi = \psi$ and $\Im\varphi = \psi_1$. If, in addition, V is a normed vector space and ψ is bounded, then the discussion preceding Lemma 6.20 shows that φ is bounded and $\|\varphi\| = \|\psi\|$. Thus:

Lemma 6.21. *If V is a normed vector space and if $\psi : V \to \mathbb{R}$ is $\mathbb{R}$-linear and bounded, then the function $\varphi : V \to \mathbb{C}$ defined by $\varphi(v) = \psi(v) - i\psi(iv)$, for $v \in V$, is $\mathbb{C}$-linear, bounded, and satisfies $\|\varphi\| = \|\psi\|$.*

The specialised form of the Hahn-Banach Extension Theorem is now ready to be formulated and proved.

Theorem 6.22 (Hahn-Banach Extension Theorem for Normed Spaces). *If L is a linear submanifold of a normed vector space V, and if $\varphi \in L^*$, then there exists $\Phi \in V^*$ such that $\Phi_{|L} = \varphi$ and $\|\Phi\| = \|\varphi\|$.*

Proof. Write $\varphi = \Re\varphi + i\Im\varphi$. Because $\Re\varphi$ and $\Im\varphi$ are $\mathbb{R}$-linear functionals on L of norm $\|\Re\varphi\| = \|\Im\varphi\| = \|\varphi\|$ (by Lemma 6.20), it is sufficient to find $\mathbb{R}$-linear extensions Ψ_1 and Ψ_2 of $\Re\varphi$ and $\Im\varphi$, respectively, such that $\|\Psi_j\| = \|\varphi\|$ for $j = 1, 2$, and to then define Φ by $\Phi = \Psi_1 + i\Psi_2$, using Lemma 6.21.

Therefore, we assume without loss of generality that $\psi : L \to \mathbb{R}$ is $\mathbb{R}$-linear, nonzero, and bounded of norm $\|\psi\|$. Thus, $\tilde{\psi} = \|\psi\|^{-1}\psi$ is $\mathbb{R}$-linear of norm $\|\tilde{\psi}\| = 1$.

Let $p : V \to \mathbb{R}$ be given by $p(v) = \|v\|$, for $v \in V$. Thus, for every $v, v_1, v_2 \in V$ and $\alpha \geq 0$, we have $p(v_1 + v_2) \leq p(v_1) + p(v_2)$, $p(\alpha v) = \alpha p(v)$, and $|\tilde{\psi}(v)| \leq p(v)$, for every $v \in L$. Hence, by Theorem 6.18, there is an $\mathbb{R}$-linear map $\tilde{\Psi} : V \to \mathbb{R}$ extending $\tilde{\psi}$ and such that $-p(-v) \leq \tilde{\Psi}(v) \leq p(v)$ for every $v \in V$; that is, such that $-\|v\| \leq \tilde{\Psi}(v) \leq \|v\|$ for all $v \in V$. Hence, $\|\tilde{\Psi}\| \leq 1 = \|\tilde{\psi}\| \leq \|\tilde{\Psi}\|$ implies that $\|\Psi\| = \|\psi\|$, where $\Psi = \|\psi\|\tilde{\Psi}$ is the desired $\mathbb{R}$-linear extension of ψ from L to V. $\square$

The Hahn-Banach Extension Theorem leads to the following result which shows that the dual space V^* of an infinite-dimensional normed vector space V contains (many) elements other than 0.

Corollary 6.23. *If v is a nonzero element of a normed vector space V, then there is a $\varphi \in V^*$ such that $\|\varphi\| = 1$ and $\varphi(v) = \|v\|$.*

Proof. On the 1-dimensional subspace $L = \{\alpha v \,|\, \alpha \in \mathbb{C}\}$, let $\varphi_0 : L \to \mathbb{C}$ be given by $\varphi_0(\alpha v) = \alpha \|v\|$. Then φ_0 is a linear transformation and $\varphi_0(v) = \|v\|$. The norm of φ_0 is 1, since $|\varphi_0(\alpha v)| = |\alpha| \|v\| = \|\alpha v\|$. By the Hahn-Banach Theorem, there is an extension of φ_0 to a linear functional $\varphi \in V^*$ such that $\|\varphi\| = \|\varphi_0\| = 1$. $\square$

6.4   The Second Dual

By Corollary 6.23, for every v in a normed vector space V, there is a $\varphi \in V^*$ such that $\|\varphi\| = 1$ and $\varphi(v) = \|v\|$. Therefore, a normed vector space admits a rich family of linear functionals. Treating the dual V^* as a normed vector space itself, then its dual $(V^*)^*$ is likewise large. This "second dual" of V is important in many regards, not the least of which is because the second dual of V contains V in a natural way, which leads to a fruitful conceptual perspective in which the elements of V act on the elements of V^* (rather than *vice versa*, according to the definition of V^*).

Definition 6.24. The *second dual* of a normed vector space V is the dual space $(V^*)^*$ of V^*, and is denoted by V^{**}.

The following proposition shows that V^{**} contains an isometric copy of V.

Proposition 6.25. *If V is normed vector space, then there exists a linear isometry $\Delta : V \to V^{**}$ such that $\Delta v(\varphi) = \varphi(v)$ for every $v \in V$ and $\varphi \in V^*$.*

Proof. For each $v \in V$, let $\omega_v : V^* \to \mathbb{C}$ be the linear transformation defined by $\omega_v(\varphi) = \varphi(v)$, for $\varphi \in V^*$. Because $|\varphi(v)| \leq \|\varphi\| \|v\|$, for every $\varphi \in V^*$, the linear map ω_v is bounded. Hence, $\omega_v \in V^{**}$, for every $v \in V$.

It is straightforward to verify that the map $\Delta : V \to V^{**}$ that sends each $v \in V$ to the function ω_v is linear (*i.e.*, $\omega_{v_1} + \omega_{v_2} = \omega_{v_1+v_2}$ and $\lambda \omega_v = \omega \lambda v$, for all $v_1, v_2, v \in V$ and $\lambda \in \mathbb{C}$). Thus, Δ is a linear transformation. Moreover, for each $v \in V$,

$$\|\Delta v\| = \sup_{\varphi \in V^*, \|\varphi\|=1} |\omega_v(\varphi)| = \sup_{\varphi \in V^*, \|\varphi\|=1} |\varphi(v)| = \|v\|.$$

Thus, Δ is a linear isometry. □

Definition 6.26. A Banach space V is said to be *reflexive* if the operator Δ in Proposition 6.25 is a surjection.

The most immediate example of a reflexive Banach space is afforded by finite-dimensional spaces.

Proposition 6.27. *Every finite-dimensional Banach space is reflexive.*

Proof. To prove this, suppose that $\mathscr{B} = \{v_1, \ldots, v_n\}$ is a basis of V, then each $v \in V$ has unique representation as linear combination of $v_1, \ldots, v_n$: $v = \sum_j \alpha_j v_j$. For each k, let $\varphi_k : V \to \mathbb{C}$ be defined by $\varphi_k \left(\sum_j \alpha_j v_j \right) = \alpha_k$. Clearly $\varphi_k \in V^*$ and $\varphi_k(v_j) = 0$, if $j \neq k$ and $\varphi_k(v_k) = 1$. To show that $\varphi_1, \ldots, \varphi_n$ are linearly independent, suppose that $\sum_j \alpha_j \varphi_j = 0$. Then, for any $k \in \{1, 2, \ldots, n\}$, $0 = \left(\sum_j \alpha_j \varphi_j \right) v_k = \sum_j \alpha_j \varphi_j(v_k) = \alpha_k$. Hence, $\varphi_1, \ldots, \varphi_n$ are linearly independent.

Using the basis $\{\varphi_1, \ldots, \varphi_n\}$ of V^*, repeat the argument above to produce a basis $\{\Phi_1, \ldots, \Phi_n\}$ of V^{**} that has the property that $\Phi_k(\varphi_k) = 1$ and $\Phi_k(\varphi_j) = 0$ if $j \neq k$. Define a function $\Delta : V \to V^{**}$ on the basis of V by $\Delta(v_k) = \Phi_k$ and extend this by linearity to all of V. The operator Δ plainly satisfies $\Delta v(\varphi) = \varphi(v)$ for every $v \in V$ and $\varphi \in V^*$. □

Proposition 6.16 provides another set of examples of reflexive spaces:

Example 6.28. *If $p > 1$, then $\ell^p(\mathbb{N})$ is a reflexive Banach space.*

6.5 Weak Topologies

Suppose that X and Y are topological spaces and that $\mathscr{F}$ is a family of functions $f : X \to Y$. Recall that Proposition 1.88 shows that the collection

$$\mathscr{B} = \left\{ f_1^{-1}(U_1) \cap \cdots \cap f_n^{-1}(U_n) \mid n \in \mathbb{N}, \ U_j \subseteq Y \text{ is an open set}, f_j \in \mathscr{F} \right\}$$

is a basis for a topology (called the weak topology induced by $\mathcal{F}$) on X with respect to which each function $f \in \mathcal{F}$ is continuous. In this section we shall consider two particular choices of $\mathcal{F}$ when X and Y are certain normed vector spaces.

Definition 6.29. If V is a normed vector space, then the *weak topology* on V is the the weak topology on V induced by the family $\mathcal{F} = V^*$, and the elements of this topology are called *weakly open sets*.

Note the use of the term "weak topology" in the setting of Banach space differs slightly from the use of the term in topology in that reference to the choice of family $\mathcal{F}$ is dropped. That is, when saying that V has the weak topology it is understood implicitly that the family of functions inducing the topology is the family V^* of all bounded linear functionals on V.

Suppose now that a normed vector space has the weak topology. If $v_0 \in V$ and $U \subset V$ is a weakly open set, then there is a basic weakly open set B such that $v_0 \in B \subseteq U$. That is, there are $\varphi_1, \ldots, \varphi_n \in V^*$ and open sets $W_1, \ldots, W_n \subseteq \mathbb{C}$ such that

$$v_0 \in \bigcap_{j=1}^{n} \varphi_j^{-1}(W_j) \subseteq U.$$

As $\varphi(v_0) \in W_j \subseteq \mathbb{C}$ for each j, there are positive real numbers $\varepsilon_1, \ldots, \varepsilon_n$ such that, for each j,

$$\{\zeta \in \mathbb{C} \,|\, |\zeta - \varphi_j(v_0)| < \varepsilon_j\} \subseteq W_j.$$

Hence,

$$v_0 \in \{v \in V \,|\, |\varphi_j(v) - \varphi_j(v_0)| < \varepsilon_j, \ \forall j = 1, \ldots, n\} \subseteq U.$$

Proposition 6.30. *If V is a finite-dimension normed vector space, then the weak topology and the norm topology on V coincide.*

Proof. Exercise 6.65. $\square$

In contrast to Proposition 6.30, the weak topology and the norm topology are strikingly different in the case of infinite-dimensional spaces. For example, in the norm topology of an infinite-dimensional Banach space V there are bounded open sets $U \subset V$ that contain $0 \in V$ (the open unit ball, for example); however, this is not at all true for the weak topology.

Proposition 6.31. *If V is an infinite-dimensional Banach space, and if $U \subset V$ is a weakly open set such that $0 \in U$, then U is unbounded. In fact, there is an infinite-dimensional subspace $L \subset V$ such that $L \subset U$.*

Proof. Choose a basic weakly open set B such that $0 \in B \subseteq U$. Thus, there are $\varphi_1, \ldots, \varphi_n \in V^*$ and open sets $W_1, \ldots, W_n \subseteq \mathbb{C}$ such that $0 \in \bigcap_{j=1}^{n} \varphi_j^{-1}(W_j)$. Let $L = \bigcap_{j=1}^{n} \ker \varphi_j$, which is a subspace of V contained in U. We need only verify that L has infinite dimension. To this end, let $\Phi : V \to \mathbb{C}^n$ be the linear transformation

$$\Phi(v) = \begin{bmatrix} \varphi_1(v) \\ \vdots \\ \varphi_n(v) \end{bmatrix}, \quad v \in V.$$

Note that $\ker \Phi = L$. The First Isomorphism Theorem in linear algebra asserts that the quotient space V/L is isomorphic to the range of Φ, which is a subspace of $\mathbb{C}^n$. Because the quotient of an infinite-dimensional vector space by a finite-dimensional subspace cannot have finite dimension, it must be that L has infinite dimension.   □

When it comes to the dual space V^* of V, it is of less interest to endow V^* with the weak topology induced by the family V^{**} than it is to endow V^* with the weak topology induced by the subfamily $\Delta(V) \subseteq V^{**}$ indicated in Proposition 6.25— namely, those functions $f : V^* \to \mathbb{C}$ for which there exist a $v \in V$ such that $f(\varphi) = \varphi(v)$ for every $\varphi \in V^*$.

Definition 6.32. If V is a normed vector space, then the *weak* topology* on V^* is the weak topology on V^* induced by the subfamily $\Delta(V) \subseteq V^{**}$ indicated in Proposition 6.25.

To be clear, if $\varphi_0 \in V^*$, then a basic weak* open subset $B \subset V^*$ that contains φ_0 has the form

$$B = \{\varphi \in V^* \,|\, |\varphi(v_j) - \varphi_0(v_j)| < \varepsilon_j, \text{ for all } j = 1, \ldots, n\},$$

for some $n \in \mathbb{N}$, $v_1, \ldots, v_n \in V$, and positive real numbers $\varepsilon_1, \ldots \varepsilon_n$.

The most important fundamental property of the weak* topology is established by the following theorem.

Theorem 6.33 (Banach-Alaoglu). *If V is a normed vector space and if $X \subset V^*$ is the closed unit ball of V^*, then X is compact in the weak* topology of V^*.*

Proof. For each $v \in V$, let $K_v = \{\lambda \in \mathbb{C} \,|\, |\lambda| \leq \|v\|\}$. Consider the space $K = \prod_{v \in V} K_v$, endowed with the product topology. By Tychonoff's Theorem (Theorem 2.14), K is a compact set. Furthermore, K is Hausdorff because each K_v is Hausdorff.

Define $f : X \to K$ by $f(\varphi) = (\varphi(v))_{v \in V}$, and note that f is an injective function. Select $\varphi \in X$ and consider an open set $W \subset K$ that contains $f(\varphi)$. Thus, there are open subsets $W_v \subseteq K_v$ such that $\varphi(v) \in W_v$, for every $v \in V$, and $W_v = K_v$ for all but at most a finite number of vectors in V, say $v_1, \ldots, v_n$, and $W = \prod_{v \in V} W_v$. Because each $\Delta(v_j)$ is continuous on V^*, the set $\Delta(v_j)^{-1}(W_{v_j})$ is open in V^*, which implies that $U_j = \{\psi \in X \mid \psi(v_j) \in W_{v_j}\}$ is open in X for each j. Thus, if $U = \bigcap_{j=1}^{n} U_j$, then U is an open set containing φ for which $f(\psi) \in W$ for every $\psi \in U$. Hence, f is continuous at φ, and so f is a continuous function on X.

On the other hand, if $U \subseteq X$ is an arbitrary open set and if $\varphi \in U$, then there is a basic open set B such that $\varphi \in B \subseteq U$. By definition, there are $v_1, \ldots, v_n \in V$ and $\varepsilon_1, \ldots, \varepsilon_n > 0$ such that, for $\psi \in X$, we have $\psi \in B$ if and only if $|\psi(v_j) - \varphi(v_j)| < \varepsilon_j$ for each $j = 1, \ldots, n$. Hence, if $W_{v_j} = \{\lambda \in K_{v_j} \mid |\lambda - \varphi(v_j)| < \varepsilon_j\}$ and if $W_v = K_v$ for every $v \in V \setminus \{v_1, \ldots, v_n\}$, then $W_\varphi = \prod_{v \in V} W_v$ is open in K and $f(\varphi) \in W_\varphi \subseteq f(U)$. Thus, $f(U) = \bigcup_{\varphi \in U} W_\varphi$, which shows that $f(U)$ is open. Hence, f^{-1} is continuous, and therefore f is a homeomorphism between X and $f(X)$.

We now show that $f(X)$ is a closed subset of K. Let $\lambda = (\lambda_v)_{v \in V} \in K$ be in the closure of $f(X)$. Suppose that $v_1, v_2 \in V$ and $\alpha_1, \alpha_2 \in \mathbb{C}$, and let $\varepsilon > 0$. Define

$$U = \{\psi \in X \mid |\psi(v) - \lambda_v| < \varepsilon, \ v \in \{v_1, v_2, \alpha_1 v_1 + \alpha_2 v_2\}\},$$

which is an open subset of X. Thus, $W = f(U)$ is open in K (because f is a homeomorphism). The open set W contains λ, and λ is in the closure of $f(X)$. Hence, there exists $\varphi \in X$ such that $f(\varphi) \in W$, and so

$$|\varphi(v_1) - \lambda_{v_1}| < \varepsilon, \quad |\varphi(v_2) - \lambda_{v_2}| < \varepsilon,$$

and

$$|\varphi(\alpha_1 v_1 + \alpha_2 v_2) - \lambda_{\alpha_1 v_1 + \alpha_2 v_2}| < \varepsilon.$$

Therefore,

$$|\lambda_{\alpha_1 v_1 + \alpha_2 v_2} - \alpha_1 \lambda_{v_1} - \alpha_2 \lambda_{v_2}| < (1 + |\alpha_1| + |\alpha_2|)\varepsilon.$$

As $\varepsilon > 0$ is arbitrary, we deduce that $\lambda_{\alpha_1 v_1 + \alpha_2 v_2} = \alpha_1 \lambda_{v_1} + \alpha_2 \lambda_{v_2}$. Therefore, the map $v \mapsto \lambda_v$ is linear and satisfies $|\lambda_v| \le \|v\|$ for all $v \in V$, implying that this map is an element of X and that $\lambda \in f(X)$. This proves that $f(X)$ is closed in K.

Because K is compact, Hausdorff, and $f(X)$ is closed in K, we deduce that $f(X)$ is compact and Hausdorff; hence, X is compact and Hausdorff. $\square$

The following striking theorem shows that all Banach spaces arise as subspaces of $C(X)$ for various choices of compact Hausdorff spaces X.

Proposition 6.34. *For every Banach space V there is a compact Hausdorff space X such that V is isometrically isomorphic to a subspace of $C(X)$.*

Proof. Let X be the closed unit ball of V^*, which is compact and Hausdorff when endowed with the weak* topology (Theorem 6.33). By Proposition 6.25, there is an isometric embedding $\Delta : V \to V^{**}$ whereby $\Delta v(\varphi) = \varphi(v)$, for all $\varphi \in X$ and $v \in V$. Thus, we need only show that $\Delta v \in C(X)$ for every $v \in V$. To this end, select $v \in V$ and consider the function $\Delta v : X \to \mathbb{C}$. To show that Δv is continuous at $\varphi \in X$, assume that W is an open set containing $\Delta v(\varphi) = \varphi(v)$. Thus, there exists $\varepsilon > 0$ such that $W_0 = \{\lambda \in \mathbb{C} \,|\, |\lambda - \varphi(v)| < \varepsilon\} \subseteq W$. If $U = \Delta v^{-1}(W_0) = \{\psi \in X \,|\, |\psi(v) - \varphi(v)| < \varepsilon\}$, which is a basic open set in X, then $\Delta v(U) \subseteq W$, which proves that Δv is continuous at φ. Hence, Δv is continuous on X. $\qquad \square$

If V is a separable Banach space, then one would hope that the topological space X that arises in Proposition 6.34 above is a compact metric space, for then the enveloping Banach space $C(X)$ that contains V as a subspace would also be separable (Theorem 5.57). This is indeed the case by the following result.

Proposition 6.35. *If V is a separable Banach space and X is the closed unit ball of V^*, then X is metrisable.*

Proof. By hypothesis, there is a countable set that is dense in V; hence, there is a countable set $\{v_n\}_{n \in \mathbb{N}} \subset X$ that is dense in the closed unit ball of V.

The compact set $\overline{\mathbb{D}} = \{z \in \mathbb{C} \,|\, |z| \leq 1\}$ is a subset of the metric space $\mathbb{C}$, and so the product space $D = \prod_{n \in \mathbb{N}} \overline{\mathbb{D}}$ of countably many copies of $\overline{\mathbb{D}}$ is compact and metrisable in the product topology (by Tychonoff's Theorem and Proposition 1.57).

Let $f : X \to D$ be given by $f(\varphi) = (\varphi(v_n))_{n \in \mathbb{N}}$. To show that f is continuous at each point of X, select $\varphi_0 \in X$ and let $W \subset D$ be an open set that contains $f(\varphi_0)$. Thus, $W = \prod_{n \in \mathbb{N}} W_n$ for some open sets $W_n \subseteq \overline{\mathbb{D}}$ for which $W_n = \overline{\mathbb{D}}$ for all n with the exception of at most finitely many $n_1, \ldots, n_m \in \mathbb{N}$. Thus, there are positive real numbers $\varepsilon_1, \ldots, \varepsilon_m$ such that $\{\lambda \in \overline{\mathbb{D}} \,|\, |\lambda - \varphi_0(v_{n_j})| < \varepsilon_j\}$, for $j = 1, \ldots, m$. The set $U = \{\varphi \in X \,|\, |\varphi(v_{n_j}) - \varphi_0(v_{n_j})| < \varepsilon_j, j = 1, \ldots, n\}$ is an open subset of X that contains φ_0 and satisfies $f(U) \subseteq W$. Hence, f is continuous at φ_0, which proves that f is a continuous map $X \to D$.

The continuous image of a compact set is compact (Proposition 2.9), which implies that $f(X)$ is compact. Furthermore, because D is metrisable, so is $f(X)$. Therefore, in particular, $f(X)$ is Hausdorff. Because $\{v_n\}_{n \in \mathbb{N}}$ is dense in the closed unit ball of V and because each $\varphi \in X$ is continuous, the map f is injective. Hence, f is a bijective continuous map from a compact space X onto a Hausdorf space $f(X)$. By Proposition 2.9, f is necessarily a homeomorphism, which implies that X is metrisable. $\qquad \square$

Corollary 6.36. *For every separable Banach space V there exists a compact metric space X and a subspace $L \subseteq C(X)$ such that V and L are isometrically isomorphic.*

Proof. Let X be the topological space given by the closed unit ball of V^* in the weak*-topology. By Proposition 6.35, X is metrisable; and, by Proposition 6.34 V is isometrically isomorphic to a subspace L of $C(X)$. $\qquad \square$

6.6   Linear Functionals on L^p and L^∞

For every $p \geq 1$, the elements of $L^p(X, \Sigma, \mu)$ are determined by p-integrable functions $f : X \to \mathbb{C}$. We shall write $f \geq 0$ for $f \in \mathscr{L}^p(X, \Sigma, \mu)$ if $f(x) \geq 0$ for every $x \in X$, and write $\dot{f} \geq 0$ if there is a function $g \in \mathscr{L}^p(X, \Sigma, \mu)$ with $g \geq 0$ and $\dot{g} = \dot{f}$. The notation $\dot{g} \leq \dot{f}$ is used to denote $\dot{f} - \dot{g} \geq 0$.

Definition 6.37. A linear functional $\varphi : L^p(X, \Sigma, \mu) \to \mathbb{C}$ is said to be *positive* if $\varphi(\dot{f}) \geq 0$ for every $\dot{f} \in L^p(X, \Sigma, \mu)$ with $\dot{f} \geq 0$.

Lemma 6.38. *Every linear functional on* $L^p(X, \Sigma, \mu)$*, where* $p \in \mathbb{R}$ *satisfies* $p \geq 1$*, is a linear combination of four positive linear functionals.*

Proof. By Lemmas 6.20 and 6.21, every $\varphi \in L^p(X, \Sigma, \mu)^*$ is a linear combination $\varphi = \Re\varphi + i\Im\varphi$ of two continuous $\mathbb{R}$-linear maps $\Re\varphi, \Im\varphi : L^p(X, \Sigma, \mu) \to \mathbb{R}$. Thus, consider $L^p(X, \Sigma, \mu)$ as a Banach space over $\mathbb{R}$ and let $\psi = \Re\varphi$, which is continuous and $\mathbb{R}$-linear. Define a real-valued function ψ_+ on the positive elements of $L^p(X, \Sigma, \mu)$ by

$$\psi_+(\dot{f}) = \sup\{\psi(\dot{g}) \mid g \in \mathscr{L}^p(X, \Sigma, \mu),\ 0 \leq g \leq f\}.$$

To confirm that the supremum above exists, note that $0 \leq g \leq f$ and $p \geq 1$ imply that $|g|^p \leq |f|^p$, and so $\|\dot{g}\| \leq \|\dot{f}\|$; thus, $|\psi(\dot{g})| \leq \|\psi\| \|\dot{g}\| \leq \|\psi\| \|\dot{f}\|$ and, therefore, $\psi_+(\dot{f})$ exists and is such that $\psi_+(\dot{f}) \leq \|\psi\| \|\dot{f}\|$.

Suppose now that $g_1, g_2 \in \mathscr{L}^p(X, \Sigma, \mu)$ satisfy $0 \leq g_j \leq f$. By the linearity of ψ, we have $\psi(\dot{g}_1) + \psi(\dot{g}_2) = \psi(\dot{g}_1 + \dot{g}_2) \leq \psi_+(\dot{f}_1 + \dot{f}_2)$; hence, $\psi_+(\dot{f}_1) + \psi_+(\dot{f}_2) \leq \psi_+(\dot{f}_1 + \dot{f}_2)$. On the other hand, if $h \in \mathscr{L}^p(X, \Sigma, \mu)$ satisfies $0 \leq h \leq f_1 + f_2$, then let $g_1(x) = \max\{h(x) - f_2(x), 0\}$ and $g_2(x) = \min\{h(x), f_2(x)\}$ to obtain p-integrable functions g_1 and g_2 with $0 \leq g_j \leq f_j$, for $j = 1, 2$, and $g_1 + g_2 = h$. Thus, $\psi(\dot{h}) = \psi(\dot{g}_1) + \psi(\dot{g}_2) \leq \psi_+(\dot{f}_1) + \psi_+(\dot{f}_2)$ yields $\psi_+(\dot{f}_1 + \dot{f}_2) \leq \psi_+(\dot{f}_1) + \psi_+(\dot{f}_2)$. Hence, ψ_+ is an additive function on the positive elements of $L^p(X, \Sigma, \mu)$. The function ψ_+ is also plainly positive-homogeneous in the sense that $\psi_+(\alpha\dot{f}) = \alpha\psi_+(\dot{f})$ for every $\alpha \geq 0$ in $\mathbb{R}$ and $\dot{f} \geq 0$ in $L^p(X, \Sigma, \mu)$.

Let $\mathscr{R} = \{\dot{f} \mid f \in \mathscr{L}^p(X, \Sigma, \mu) \text{ such that } f(x) \in \mathbb{R} \text{ for all } x \in X\}$, which is a real subspace of $L^p(X, \Sigma, \mu)$. Express each real-valued function $f \in \mathscr{L}^p(X, \Sigma, \mu)$ as the difference $f = f^+ - f^-$ of positive p-integrable functions f^+ and f^- as prescribed in equation (3.3), and define ψ_+ on $\mathscr{R}$ by

$$\psi_+(\dot{f}) = \psi_+(\dot{f}^+) - \psi_+(\dot{f}^-).$$

In expressing each $\alpha \in \mathbb{R}$ as a difference $\alpha = \alpha^+ - \alpha^-$ of positives, we see that $\psi_+(\alpha\dot{f}) = \alpha\psi_+(\dot{f})$ for every real-valued $f \in \mathscr{L}^p(X, \Sigma, \mu)$. As ψ_+ is plainly additive on $\mathscr{R}$, we deduce that ψ_+ is a continuous $\mathbb{R}$-linear map on $\mathscr{R}$. Likewise, by defining $\psi_- = \psi_+ - \psi$ on $\mathscr{R}$, the map ψ_- is $\mathbb{R}$-linear, continuous, and satisfies $\psi_-(\dot{f}) \geq 0$ for every $\dot{f} \geq 0$.

Because every $f \in \mathscr{L}^p(X, \Sigma, \mu)$ has the form $f = \Re f + i\Im f$, the maps ψ_+ and ψ_- extend from $\mathscr{R}$ to all of $L^p(X, \Sigma, \mu)$ via $\psi_+(\dot{f}) = \psi_+(\Re\dot{f}) + i\psi_+(\Im\dot{f})$ and $\psi_-(\dot{f}) = \psi_-(\Re\dot{f}) + i\psi_-(\Im\dot{f})$, in each case yielding a continuous, positive, $\mathbb{C}$-linear map $L^p(X, \Sigma, \mu) \to \mathbb{C}$ with the property that $\psi = \psi_+ - \psi_-$. Hence, $\Re\varphi$ is the difference of two positive linear functionals. A similar argument shows that $\Im\varphi$ is a difference of two positive linear functionals, which implies that φ is a linear combination of four positive linear functionals. $\square$

The main result that we aim to prove in this section is Theorem 6.40, which is based in part on the density of simple functions in L^p. A proof technique related to the density of simple functions is encompassed by the following lemma.

Lemma 6.39. *Assume that $p \geq 1$ and that φ is a positive linear functional on $L^p(X, \Sigma, \mu)$. Suppose that $g : X \to \mathbb{R}$ is a nonnegative measurable function and that $f, f_k \in \mathscr{L}^p(X, \Sigma, \mu)$, for $k \in \mathbb{N}$, have the following properties:*

1. *$0 \leq f_k(x) \leq f_{k+1}(x) \leq f(x)$, for all $x \in X$ and $k \in \mathbb{N}$;*
2. *$\lim_{k\to\infty} f_k(x) = f(x)$, for all $x \in X$;*
3. *$\varphi(\dot{f}_k) = \int_X f_k g \, d\mu$ for every $k \in \mathbb{N}$.*

Then $\varphi(\dot{f}) = \int_X fg \, d\mu$.

Proof. By Proposition 5.42, the first two of the three conditions above imply that $\lim_{k\to\infty} \|\dot{f} - \dot{f}_k\| = 0$. Because $0 \leq f_k(x)g(x) \leq f(x)g(x)$ and $\lim_k f_k(x)g(x) = f(x)g(x)$ for every $x \in X$, the Dominated Convergence Theorem yields

$$\lim_{k\to\infty} \int_X |fg - f_k g| \, d\mu = 0.$$

Hence,

$$\left| \varphi(\dot{f}) - \int_X fg \, d\mu \right| = \left| \varphi(\dot{f}) - \varphi(\dot{f}_k) + \varphi(\dot{f}_k) - \int_X fg \, d\mu \right|$$

$$\leq \|\dot{f} - \dot{f}_k\| \, \|\varphi\| + \left| \int_X (f_k - f)g \, d\mu \right|$$

$$\leq \|\dot{f} - \dot{f}_k\| \, \|\varphi\| + \int_X |f_k g - fg| \, d\mu.$$

Thus, by letting $k \to \infty$, we obtain $\varphi(\dot{f}) = \int_X fg \, d\mu$. $\square$

The following result is our second instance of a *Riesz Representation Theorem*.

Theorem 6.40 (Riesz). *Suppose that p and q are conjugate real numbers. If (X, Σ, μ) is a σ-finite measure space, and if $\Omega : L^q(X, \Sigma, \mu) \to L^p(X, \Sigma, \mu)^*$ is defined by*

$$\Omega(\dot{g})[\dot{f}] = \int_X fg \, d\mu, \tag{6.3}$$

for all $f \in \mathscr{L}^p(X, \Sigma, \mu)$ and $g \in \mathscr{L}^q(X, \Sigma, \mu)$, then Ω is a linear isometric isomorphism.

Proof. Let $g \in \mathscr{L}^q(X, \Sigma, \mu)$ be fixed. By Hölder's inequality (Proposition 4.53),

$$\int_X |fg| \, d\mu \le \left(\int_X |f|^p \, d\mu \right)^{1/p} \left(\int_X |g|^q \, d\mu \right)^{1/q} \quad \forall f \in \mathscr{L}^p(X, \Sigma, \mu).$$

Hence, the function $\varphi_g : L^p(X, \Sigma, \mu) \to \mathbb{C}$ defined by $\varphi_g(\dot{f}) = \int_X fg \, d\mu$ is a linear functional on $L^p(X, \Sigma, \mu)$ for which $\|\varphi_g(\dot{f})\| \le \|\dot{f}\| \, \|\dot{g}\|$ for all $\dot{f} \in L^p(X, \Sigma, \mu)$. Therefore, the function Ω indeed takes values in the dual of $L^p(X, \Sigma, \mu)$ and is given unambiguously by $\Omega(\dot{g}) = \varphi_g$. The map Ω is plainly linear, and the inequality $\|\Omega(\dot{g})\| = \|\varphi_g\| \le \|\dot{g}\|$ implies that Ω is continuous.

To show that Ω is isometric, we need to show that $\|\dot{g}\| \le \|\varphi_g\|$. Let $\zeta_g : X \to \mathbb{C}$ be the function $\zeta_g = \operatorname{sgn} g$—namely, the measurable function whose value at $x \in X$ is 0, if $g(x) = 0$, and is $\frac{g(x)}{|g(x)|}$, if $g(x) \ne 0$. Thus, $|g(x)| = \zeta_g(x)g(x)$ for all $x \in X$. Let $f : X \to \mathbb{C}$ be defined by $f = |g|^{q/p}\zeta_g$. Observe that f is p-integrable and that $fg = |g|^{q/p}|g| = |g|^{1+\frac{q}{p}} = |g|^q$. Thus,

$$\|\dot{g}\|^q = \int_X |g|^q \, d\mu = \int_X fg \, d\mu = |\varphi_g(\dot{f})| \le \|\varphi_g\| \, \|\dot{f}\|$$

$$= \|\varphi_g\| \left(\int_X |f|^p \, d\mu \right)^{1/p} = \|\varphi_g\| \left(\int_X |g|^q \, d\mu \right)^{1/p} = \|\varphi_g\| \, \|\dot{g}\|^{q/p}.$$

Hence, because $q - \frac{q}{p} = 1$, we deduce that $\|\dot{g}\| \le \|\varphi_g\|$, which proves that Ω is a linear isometry.

What remains, therefore, is to prove is that Ω is surjective. By Lemma 6.38, it is sufficient to show that every positive linear functional on $L^p(X, \Sigma, \mu)$ is in the range of Ω. To this end, suppose that φ is a nonzero positive linear functional.

Assume, in the first instance, that $\mu(X) < \infty$. Define a function $\nu : \Sigma \to \mathbb{R}$ by $\nu(E) = \varphi(\dot{\chi}_E)$, and note that $\nu(E) \ge 0$ (because $\dot{\chi}_E \ge 0$). Suppose that $\{E_k\}_{k \in \mathbb{N}}$ is a sequence of pairwise disjoint measurable sets and let $E = \bigcup_{k \in \mathbb{N}} E_k$. If $G_n = \bigcup_{k=1}^{n} E_k$, then $\dot{\chi}_{G_n} = \sum_{k=1}^{n} \dot{\chi}_{E_k}$ and, thus, $\nu(G_n) = \sum_{k=1}^{n} \nu(E_k)$. Therefore, if $F_n = E \setminus G_n$, then G_n and F_n are disjoint with union E, and so

$$\nu(F_n) = \varphi(\dot{\chi}_E) - \left(\sum_{k=1}^n \varphi(\dot{\chi}_{E_k})\right) = \nu(E) - \left(\sum_{k=1}^n \nu(E_k)\right).$$

The sequence $\{\chi_{G_n}^p\}_{n\in\mathbb{N}}$ is monotone increasing to χ_E^p, and therefore the Monotone Convergence Theorem yields $\lim_{n\to\infty}\|\dot{\chi}_{G_n} - \dot{\chi}_E\| = 0$. Hence, by the continuity of φ,

$$\nu(E) = \lim_{n\to\infty}\sum_{k=1}^n \nu(E_k) = \sum_{k=1}^\infty \nu(E_k).$$

This proves that ν is countably additive. That is, ν is a measure on (X, Σ).

Now suppose that $E \in \Sigma$ satisfies $\mu(\chi_E) = 0$. Thus, $\|\dot{\chi}_E\| = 0$ and so $\nu(E) = \varphi(\dot{\chi}_E) = 0$. Therefore, ν is absolutely continuous with respect to μ. By the Radon-Nikodým Theorem (Theorem 4.36), there exists $g \in \mathscr{L}^1(X, \Sigma, \mu)$ such that $g(x) \geq 0$, for every $x \in X$, and

$$\nu(E) = \int_E g\,d\mu, \quad \forall E \in \Sigma.$$

Hence, if h is a simple function in canonical form (see equation (4.1)), then

$$\varphi(\dot{h}) = \int_X hg\,d\mu.$$

Suppose that $f \in \mathscr{L}^p(X, \Sigma, \mu)$ satisfies $f(x) \geq 0$ for every $x \in X$. Proposition 5.41 shows that there is a monotone-increasing sequence $\{h_k\}_{k\in\mathbb{N}}$ of simple functions h_k such that $0 \leq h_k(x) \leq f(x)$ for all $x \in X$ and $\lim_k h_k(x) = f(x)$ for each $x \in X$. Therefore, the equations $\varphi(\dot{h}_k) = \int_X h_k g\,d\mu$ for every $k \in \mathbb{N}$ and Lemma 6.39 yield the desired formula $\varphi(\dot{f}) = \int_X fg\,d\mu$. Because every element of $L^p(X, \Sigma, \mu)$ is a linear combination of positive elements, the formula $\varphi(\dot{f}) = \int_X fg\,d\mu$ holds for every $f \in \mathscr{L}^p(X, \Sigma, \mu)$.

To show that g is q-integrable, let $E_k = \{x \in X \,|\, g(x) \leq k\}$ for each $k \in \mathbb{N}$ and define $f_k = \chi_{E_k}\zeta_g g^{q-1}$. Because $\zeta_g(x) \in \{0, 1\}$ for all $x \in X$, the function f_k is bounded. Furthermore, $\mu(E_k) \leq \mu(X) < \infty$ implies that f_k is p-integrable, and so $\varphi(\dot{f}_k) = \int_X f_k g\,d\mu$. Therefore, using $f_k g = g^q \chi_{E_k}$, we have

$$\int_{E_k} g^q\,d\mu = \int_X f_k g\,d\mu = \varphi(\dot{f}_k) \leq \|\varphi\|\,\|\dot{f}_k\|$$

$$= \|\varphi\| \left(\int_X |f_k|^p\,d\mu\right)^{1/p} = \|\varphi\| \left(\int_X g^q\,d\mu\right)^{1/p}.$$

This proves that

$$\left(\int_{E_k} g^q\,d\mu\right)^{(1-\frac{1}{p})} = \left(\int_{E_k} g^q\,d\mu\right)^{1/q} \leq \|\varphi\|.$$

Because $\int_X g_k^q \, d\mu = \int_{E_k} g^q \, d\mu$, where $g_k = \chi_{E_k} g$, and because $g(x)^q = \lim_k g_k(x)^q$ for every $x \in X$, Fatou's Lemma (Corollary 4.9) yields the first of the inequalities below:

$$\int_X g^q \, d\mu \le \liminf_k \int_X g_k^q \, d\mu \le \|\varphi\|^q.$$

Hence, $g \in \mathscr{L}^q(X, \Sigma, \mu)$, which thereby completes the proof (under the assumption that $\mu(X) < \infty$) that Ω is surjective.

Assume now that $\mu(X) = \infty$ and that $\varphi \in L^p(X, \Sigma, \mu)^*$ is a positive linear functional. Because (X, Σ, μ) is σ-finite, there is an increasing sequence $\{E_k\}_{k \in \mathbb{N}}$ of measurable sets E_k of finite measure such that $X = \bigcup_{k \in \mathbb{N}} E_k$. If, for each $k \in \mathbb{N}$, $\Sigma_k = \{E_k \cap A \mid A \in \Sigma\}$ and $\mu_k = \mu_{|\Sigma_k}$, then the linear map $f \mapsto \tilde{f}$ whereby a function $f : E_k \to \mathbb{C}$ is sent to a function $\tilde{f} : X \to \mathbb{C}$ in which $\tilde{f}(x) = f(x)$, for $x \in E_k$, and $\tilde{f}(x) = 0$, for $x \in E_k^c$, induces a natural linear isometry $T_{k,p} : L^p(E_k, \Sigma_k, \mu_k) \to L^p(X, \Sigma, \mu)$. Therefore, we view $L^p(E_k, \Sigma_k, \mu_k)$ as a subspace of $L^p(X, \Sigma, \mu)$ consisting of equivalence classes of p-integrable functions on X that vanish on E_k^c. This is also true for the conjugate q, via $T_{k,q} : L^q(E_k, \Sigma_k, \mu_k) \to L^q(X, \Sigma, \mu)$.

Setting $\varphi_k = \varphi_{|L^p(E_k, \Sigma_k, \mu_k)}$ yields an element of the dual space of $L^p(E_k, \Sigma_k, \mu_k)$. Note that if $m > n$, then $\varphi_{m|L^p(E_n, \Sigma_n, \mu_n)} = \varphi_n$. Furthermore, because $\mu(E_k) < \infty$, the linear isometry $\Omega_k : L^q(E_k, \Sigma_k, \mu_k) \to L^p(E_k, \Sigma_k, \mu_k)^*$ given by equation (6.3) yields an element $\Omega_k^{-1}(\varphi_k)$ of $L^q(E_k, \Sigma_k, \mu_k)$. For each $k \in \mathbb{N}$ select a representative $g_k \in \mathscr{L}^q(E_k, \Sigma_k, \mu_k)$ such that $\dot{g}_k = \Omega_k^{-1}(\varphi_k)$. Hence,

$$\varphi_k(\dot{f}_k) = \int_{E_k} f_k g_k \, d\mu_k = \int_X \tilde{f}_k \tilde{g}_k \, d\mu,$$

for every $f_k \in \mathscr{L}^p(E_k, \Sigma_k, \mu_k)$. Because any two representatives in $\mathscr{L}^q(E_k, \Sigma_k, \mu_k)$ for a single equivalence class in $L^q(E_k, \Sigma_k, \mu_k)$ will differ only on a set of measure zero, we see that for $m > n$ the set $\{x \in E_n \mid g_m(x) \ne g_n(x)\}$ has measure zero. Let

$$F = \bigcup_{n=1}^\infty \bigcup_{m=n}^\infty \{x \in E_n \mid g_m(x) \ne g_n(x)\},$$

which is a measurable set with $\mu(F) = 0$. Define $g : X \to \mathbb{C}$ by $g(x) = 0$ if $x \in F$, and for $x \in F^c$ by $g(x) = g_k(x)$ for any $k \in \mathbb{N}$ that satisfies $x \in E_k$. Therefore g is a measurable function and $g(x) = \lim_k \tilde{g}_k(x)$ for all $x \in X \setminus F$. Because

$$\int_{E_k} g_k^q \, d\mu_k = \|\varphi_k\|^q \le \|\varphi\|^q$$

for every k, Fatou's Lemma again yields

$$\int_X g^q \, d\mu \le \liminf_k \int_X \tilde{g}_k^q \, d\mu = \int_{E_k} g_k^q \, d\mu_k \le \|\varphi\|^q,$$

which proves that g is q-integrable.

To complete the proof, select any nonnegative $f \in \mathscr{L}^p(X, \Sigma, \mu)$ and, for each $k \in \mathbb{N}$, let $f_k = f \chi_{E_k}$. Thus, $\dot{f}_k$ is an element of the subspace $L^p(E_k, \Sigma_k, \mu_k)$ and

$$\varphi(\dot{f}_k) = \varphi_k(\dot{f}_k) = \int_{E_k} f_k g_k \, d\mu = \int_{E_k \cap F^c} f_k g \, d\mu = \int_X f_k g \, d\mu.$$

Furthermore, $\{f_k\}_{k \in \mathbb{N}}$ is a monotone increasing sequence of nonnegative p-integrable functions in which $f_k(x) \le f(x)$ and $\lim_k f_k(x) = f(x)$ for all $x \in X$. Hence, by Lemma 6.39, we have $\varphi(\dot{f}) = \int_X fg \, d\mu$. Hence, because the positive elements span $L^p(X, \Sigma, \mu)$, the formula $\varphi(\dot{f}) = \int_X fg \, d\mu$ holds for every $f \in \mathscr{L}^p(X, \Sigma, \mu)$. □

Corollary 6.41. *If positive p and q are conjugate real numbers, and if (X, Σ, μ) is a σ-finite measure space, then for every $\varphi \in L^p(X, \Sigma, \mu)^*$ there exists a unique $\dot{g} \in L^q(X, \Sigma, \mu)$ such that $\|\dot{g}\| = \|\varphi\|$ and $\varphi(\dot{f}) = \int_X fg \, d\mu$, for every $\dot{f} \in L^p(X, \Sigma, \mu)$.*

The hypothesis that (X, Σ, μ) be a σ-finite measure space, in Theorem 6.40 above, can be removed; doing so, however, is rather subtle. The monograph of Bartle [6] details how such an extension of Theorem 6.40 is achieved.

The case of $L^\infty(X, \Sigma, \mu)$ is similar to that of L^p, but there are some key differences, which we make note of below.

Definition 6.42. An element $\dot{f} \in L^\infty(X, \Sigma, \mu)$ is said to be *positive* if $\dot{g} = \dot{f}$ for some nonnegative function $g \in \mathscr{L}^\infty(X, \Sigma, \mu)$.

Note that if $f \in \mathscr{L}^\infty(X, \Sigma, \mu)$ is such that ess-ran $f \subseteq [0, \infty)$, then $\dot{f}$ is a positive element of $L^\infty(X, \Sigma, \mu)$. As with L^p-spaces, we shall write write $\dot{f} \ge 0$ for positive elements, and use the notation $\dot{g} \le \dot{f}$ to denote $\dot{f} - \dot{g} \ge 0$.

Definition 6.43. A linear functional $\varphi : L^\infty(X, \Sigma, \mu) \to \mathbb{C}$ is said to be *positive* if $\varphi(\dot{f}) \ge 0$ for every $\dot{f} \in L^\infty(X, \Sigma, \mu)$ with $\dot{f} \ge 0$.

Lemma 6.44. *Every linear functional on $L^\infty(X, \Sigma, \mu)$ is a linear combination of four positive linear functionals.*

Proof. Exercise 6.68. □

Theorem 6.45 (Riesz). *If (X, Σ, μ) is a σ-finite measure space, then the function $\Omega : L^\infty(X, \Sigma, \mu) \to L^1(X, \Sigma, \mu)^*$ defined by*

$$\Omega(\dot{g})[\dot{f}] = \int_X fg \, d\mu, \tag{6.4}$$

for all $f \in \mathscr{L}^1(X, \Sigma, \mu)$ and $g \in \mathscr{L}^\infty(X, \Sigma, \mu)$, is a linear isometric isomorphism.

Proof. Let $g \in \mathscr{L}^\infty(X, \Sigma, \mu)$ be fixed. Because

$$\int_X |fg| \, d\mu \leq \left(\int_X (\text{ess-sup} \, |g|) |f| \, d\mu \right) = \|g\| \, \|f\|,$$

the function $\varphi_g : L^1(X, \Sigma, \mu) \to \mathbb{C}$ defined by $\varphi_g(\dot{f}) = \int_X fg \, d\mu$ is a linear functional on $L^1(X, \Sigma, \mu)$ and $\|\varphi_g(\dot{f})\| \leq \|\dot{f}\| \, \|\dot{g}\|$ for all $\dot{f} \in L^1(X, \Sigma, \mu)$. Therefore, the function Ω takes values in the dual of $L^1(X, \Sigma, \mu)$ and is given unambiguously by $\Omega(\dot{g}) = \varphi_g$. The map Ω is linear, and the inequality $\|\Omega(\dot{g})\| = \|\varphi_g\| \leq \|\dot{g}\|$ implies that Ω is bounded.

Assume, to begin with, that $\mu(X) < \infty$. To show that Ω is isometric, let $\varepsilon > 0$ and $A_\varepsilon = \{x \in X \mid \|\varphi\| + \varepsilon < |g(x)|\}$. For each $n \in \mathbb{N}$, let $E_n = \{x \in X \mid |g(x)| \leq n\}$. Thus, $A = \bigcup_{n \in \mathbb{N}} E_n \cap A$ for every $A \in \Sigma$. Set $f_n = \chi_{E_n \cap A_\varepsilon} \frac{\overline{g}}{|g|}$ and note that

$$\int_X |f_n| \, d\mu = \int_{E_n \cap A_\varepsilon} \left| \frac{\overline{g}}{|g|} \right| d\mu = \mu(E_n \cap A_\varepsilon) \leq \mu(X) < \infty.$$

Therefore, $f_n \in \mathscr{L}^1(X, \Sigma, \mu)$ and

$$\|\varphi\| \mu(E_n \cap A_\varepsilon) = \|\varphi\| \, \|\dot{f}_n\| \geq \int_X |f_n g| \, d\mu = \int_{E_n \cap A_\varepsilon} |g| \, d\mu \geq (\|\varphi\| + \varepsilon) \mu(E_n \cap A_\varepsilon).$$

By continuity of measure,

$$\|\varphi\| \mu(A_\varepsilon) \geq (\|\varphi\| + \varepsilon) \|\mu(A_\varepsilon).$$

Hence, $\mu(A_\varepsilon) = 0$, which implies that ess-sup $|g| \leq \|\varphi\| + \varepsilon$. However, as the choice of $\varepsilon > 0$ is arbitrary, ess-sup $|g| \leq \|\varphi\|$, whence $\|\dot{g}\| \leq \|\varphi_g\|$. That is, Ω is an isometry.

To prove is that Ω is surjective, suppose that $\varphi \in L^1(X, \Sigma, \mu)^*$ is a nonzero positive linear functional. Define a function $\nu : \Sigma \to \mathbb{R}$ by $\nu(E) = \varphi(\dot{\chi}_E)$. As shown in the proof of Theorem 6.40, ν is a measure on (X, Σ, μ), absolutely continuous with respect to μ. Therefore, by the Radon-Nikodým Theorem (Theorem 4.36), there exists measurable function g such that $g(x) \geq 0$, for every $x \in X$, and $\nu(E) = \int_E g \, d\mu$, for all $E \in \Sigma$. Hence, $\varphi(\dot{h}) = \int_X hg \, d\mu$ for every simple function h.

Suppose that $f \in \mathscr{L}^1(X, \Sigma, \mu)$ satisfies $f(x) \geq 0$ for every $x \in X$. Using Proposition 5.41 we find a monotone-increasing sequence $\{h_k\}_{k \in \mathbb{N}}$ of simple functions h_k such that $0 \leq h_k(x) \leq f(x)$ for all $x \in X$ and $\lim_k h_k(x) = f(x)$ for each $x \in X$. Therefore, the equations $\varphi(\dot{h}_k) = \int_X h_k g \, d\mu$ for every $k \in \mathbb{N}$ and Lemma 6.39 yield the desired formula $\varphi(\dot{f}) = \int_X fg \, d\mu$. Because every element of $L^1(X, \Sigma, \mu)$ is a linear combination of positive elements, the formula $\varphi(\dot{f}) = \int_X fg \, d\mu$ holds for every $f \in \mathscr{L}^1(X, \Sigma, \mu)$. The proof above, where it is shown that Ω is isometric, also shows that ess-sup $|g| \leq \|\varphi\|$, and therefore $g \in \mathcal{L}^\infty(X, \Sigma, \mu)$. Hence, Ω is a surjective, thereby completing the proof of the theorem in the case where $\mu(X)$ is finite.

The proof of the remainder of the theorem is similar to the corresponding part of the proof of Theorem 6.40 and is, therefore, left as an exercise (Exercise 6.69). □

Corollary 6.46. *If (X, Σ, μ) is σ-finite, then for each $\varphi \in L^1(X, \Sigma, \mu)^*$ there is a unique $\dot{g} \in L^\infty(X, \Sigma, \mu)$ such that $\varphi(\dot{f}) = \int_X fg\, d\mu$, for every $\dot{f} \in L^1(X, \Sigma, \mu)$, and $\|\dot{g}\| = \|\varphi\|$.*

6.7 Linear Functionals on $C(X)$

As with L^p and L^∞ spaces, the notion of positive elements has a key role in determining features of linear functionals on $C(X)$, where X is a compact Hausdorff space.

Definition 6.47. *If X is a compact Hausdorff space, an element $f \in C(X)$ is said to be* positive *if $f(x) \geq 0$ for every $x \in X$.*

As in the previous section, if $f, g \in C(X)$ are real-valued functions, then the notation $f \leq g$ is used to denote that $g - f \geq 0$.

Definition 6.48. *A linear functional $\varphi : C(X) \to \mathbb{C}$ is said to be* positive *if $\varphi(f) \geq 0$ for every $f \in C(X)$ with $f \geq 0$.*

The next result illustrates a somewhat surprising fact: if a linear transformation on $C(X)$ preserves positivity, then the linear transformation is necessarily continuous.

Proposition 6.49. *If X is a compact space and if $\varphi : C(X) \to \mathbb{C}$ is a linear transformation for which $\varphi(f) \geq 0$ for every positive $f \in C(X)$, then φ is continuous.*

Proof. Let $1 \in C(X)$ denote the constant function $x \mapsto 1 \in \mathbb{C}$. For any $f \in C(X)$ satisfying $f(x) \geq 0$ for all $x \in X$, we have that $0 \leq f \leq \|f\|1$, which implies that $0 \leq \varphi(f) \leq \|f\|\varphi(1)$ in $\mathbb{R}$.

Suppose now that $g \in C(X)$ is real valued and write $g = g^+ - g^-$, where $g^+, g^- \in C(X)$ are given by $g^+ = \frac{1}{2}(|g| + g)$ and $g^- = \frac{1}{2}(|g| - g)$ and satisfy $0 \leq g^+$ and $0 \leq g^-$. Thus, $\|g^+\| \leq \|g\|$ and $\|g^-\| \leq \|g\|$, and

$$|\varphi(g)| \leq |\varphi(g^+)| + |\varphi(g^-)| \leq \left(\|g^+\| + \|g^-\|\right)\varphi(1) \leq 2\|g\|\varphi(1).$$

Now let $h \in C(X)$ be arbitrary and write $h = \Re h + i\Im h$. Because the real-valued functions $\Re h$ and $\Im h$ are given by $\Re h = \frac{1}{2}(h + \overline{h})$ and $\Im h = \frac{1}{2i}(h - \overline{h})$, we have that $\|\Re h\| \leq \|h\|$ and $\|\Im h\| \leq \|h\|$. Thus,

$$|\varphi(h)| \leq |\varphi(\Re h)| + |\varphi(\Im h)| \leq 2\|\Re h\|\varphi(1) + 2\|\Im h\|\varphi(1) \leq 4\|h\|\varphi(1).$$

Hence, φ is bounded. □

A straightforward adaption of the proof of Lemma 6.38 yields:

Lemma 6.50. *Every linear functional on $C(X)$ is a linear combination of four positive linear functionals.*

Examples of positive linear functionals are point evaluations $f \mapsto f(x_0)$, and integration $f \mapsto \int_X f \, d\mu$. However, the two examples are related, for if X is compact and Hausdorff and if $x_0 \in X$, then $\int_X f \, d\mu = f(x_0)$, if μ is the point-mass measure $\mu = \delta_{\{x_0\}}$.

The following theorem, Theorem 6.51, is another *Riesz Representation Theorem*, and is without doubt one of the major achievements of analysis. The version of the theorem that is proved here requires the compact Hausdorff space X to be second countable; under this assumption, the topology on X is metrisable, which allows one to invoke Proposition 3.35 to show that the σ-algebra constructed from a particular outer measure includes the Borel sets of X. Even so, Theorem 6.51 is true for arbitrary compact Hausdorff spaces, but showing that the measurable space that is constructed in the proof actually contains the Borel sets of X is a much more delicate task without the assumption that X be second countable (see [10, 50]). When appealing to Theorem 6.51 at later points of the present book, it will always be the case that the topological space under consideration is a second countable compact Hausdorff space, and in many ways the specific version of theorem proved here is in fact the most important of all cases.

Theorem 6.51 (Riesz). *If X is a second countable compact Hausdorff space, and if φ is a positive linear functional on $C(X)$, then there exists a unique regular Borel measure μ on the Borel sets of X such that*

$$\varphi(f) = \int_X f \, d\mu,$$

for every $f \in C(X)$.

Proof. An important topological feature of (locally) compact Hausdorff spaces was noted in the second version of Urysohn's Lemma (Corollary 2.44) in Chapter 2: if K and U are nonempty subsets of X such that K is compact, U is open, and $K \subseteq U$, then there exists a continuous function $f : X \to [0, 1]$ such that $f(K) = \{1\}$ and $\mathrm{supp} f$ is a compact subset of U. This result is key to linking the topology of X to elements of $C(X)$ and, ultimately, to the functional φ.

We begin with the construction of μ from the positive linear functional φ. Let $\mathscr{I} = \{f \in C(X) \mid 0 \leq f(x) \leq 1, \ \forall x \in X\}$ and let $S : \mathscr{T} \to \mathbb{R}$ (where $\mathscr{T}$ is the topology of X) be defined by

$$s(U) = \sup\{\varphi(f) \mid f \in \mathscr{I} \text{ and } \mathrm{supp} f \subseteq U\}.$$

Note that the compactness of X implies that the support of any $f \in C(X)$ is compact. Now define a function $\mu^* : \mathscr{P}(X) \to \mathbb{R}$ on the power set $\mathscr{P}(X)$ of X by

$$\mu^*(S) = \inf\{s(U) \mid U \in \mathscr{T} \text{ and } S \subseteq U\}.$$

Our claim is that μ^* is an outer measure.

Because $\varphi(f) \geq 0$ for every $f \in \mathscr{I}$, the function s has nonnegative values. Furthermore, if U_1 and U_2 are open sets such that $U_1 \subseteq U_2$, then clearly $s(U_1) \leq s(U_2)$; thus, $\mu^*(U) = s(U)$, for every $U \in \mathscr{T}$. Suppose the $\{U_k\}_{k \in \mathbb{N}}$ is a countable collection of open sets and let $U = \bigcup_{k \in \mathbb{N}} U_k$. If $f \in \mathscr{I}$ satisfies supp $f \subseteq U$, then the compactness of supp f implies that there are finite many U_k that cover supp f, say $U_{k_1}, \ldots, U_{k_n}$. Let $\{h_1, \ldots, h_n\}$ be a partition of unity of supp f subordinate to $\{U_{k_j}\}_{j=1}^n$ (Proposition 2.41). Thus, $fh_j \in \mathscr{I}$ and supp $fh_j \subseteq U_{k_j}$, and so

$$\varphi(f) = \varphi\left(f \sum_{j=1}^n h_j\right) = \sum_{j=1}^n \varphi(fh_j) \leq \sum_{j=1}^n \mu^*(U_{k_j}) \leq \sum_{k=1}^\infty \mu^*(U_k).$$

Therefore,

$$\mu^*\left(\bigcup_{k \in \mathbb{N}} U_k\right) = \sup\left\{\varphi(f) \mid f \in \mathscr{I} \text{ and supp } f \subseteq \bigcup_{k \in \mathbb{N}} U_k\right\} \leq \sum_{k=1}^\infty \mu^*(U_k),$$

which shows that μ^* is countably subadditive on open sets. To handle the case of arbitrary subsets of X, assume that $\{S_k\}_{k \in \mathbb{N}}$ is a countable collection of subsets $S_k \subseteq X$, and let $\varepsilon > 0$. By definition, for each $k \in \mathbb{N}$ there is an open set U_k containing S_k and such that $\mu^*(U_k) < \mu^*(E_k) + \frac{\varepsilon}{2^k}$. Therefore,

$$\mu^*\left(\bigcup_{k \in \mathbb{N}} E_k\right) \leq \mu^*\left(\bigcup_{k \in \mathbb{N}} U_k\right) \leq \varepsilon + \sum_{k=1}^\infty \mu^*(E_k).$$

As $\varepsilon > 0$ is arbitrary, the inequality above implies that μ^* is countably subadditive, which completes the proof that μ^* is an outer measure on X.

One other feature of μ^* to mention before proceeding further is: if U_1 and U_2 are disjoint open sets, then $\mu^*(U_1 \cup U_2) = \mu^*(U_1) + \mu^*(U_2)$. To see this, let $\varepsilon > 0$ and select $f_1, f_2 \in \mathscr{I}$ such that supp $f_j \subseteq U_j$ and $\mu^*(U_j) < \varphi(f_j) + \varepsilon/2$. Because $U_1 \cap U_2 = \emptyset$, the function $f_1 + f_2$ is an element of $\mathscr{I}$ and supp $(f_1 + f_2) \subseteq (U_1 \cup U_2)$. Thus,

$$\mu^*(U_1) + \mu^*(U_2) \leq \varphi(f_1) + \varphi(f_2) + \varepsilon = \varphi(f_1 + f_2) + \varepsilon \leq \mu^*(U_1 \cup U_2) + \varepsilon.$$

Hence, $\mu^*(U_1) + \mu^*(U_2) \leq \mu^*(U_1 \cup U_2) \leq \mu^*(U_1) + \mu^*(U_2)$ implies that μ^* is additive on the union of two disjoint open sets.

To show that every Borel set of X belongs to the σ-algebra $\mathfrak{M}_{\mu^*}(X)$ induced by μ^*, recall that, because the compact Hausdorff space X is second countable, the topology on X is induced by a metric d on X (Theorem 2.48). Suppose that A_1 and A_2 are subsets of X such that dist $(A_1, A_2) > 0$. By continuity of the metric d, it will also be true that dist $(\overline{A}_1, \overline{A}_2) > 0$. Because X is normal, there are open disjoint sets V_1 and V_2 such that $\overline{A}_j \subseteq V_j$. Let $\varepsilon > 0$ be given. If W is an open set containing $A_1 \cup A_2$, then let $U_j = W \cap V_j$ so that $U_1 \cap U_2 = \emptyset$ and $A_j \subset U_j$. Thus,

$$\mu^*(A_1) + \mu^*(A_2) \leq \mu^*(U_1) + \mu^*(U_2) = \mu^*(U_1 \cup U_2) \leq \mu^*(W).$$

The infimum of the right-hand side over all open sets W that contain $A_1 \cup A_2$ yields $\mu^*(A_1 \cup A_2)$, which shows that $\mu^*(A_1 \cup A_2) = \mu^*(A_1) + \mu^*(A_2)$. Therefore, by Proposition 3.35, every Borel set of X belongs to the σ-algebra $\mathfrak{M}_{\mu^*}(X)$ induced by μ^*.

Thus, let Σ denote the σ-subalgebra of $\mathfrak{M}_{\mu^*}(X)$ generated by the topology of X, and let $\mu : \Sigma \to \mathbb{R}$ be the measure defined by $\mu(E) = \mu^*(E)$, for every $E \in \Sigma$. We shall now show that μ is a regular measure. By definition of μ^*, we already have that

$$\mu(E) = \inf\{\mu(U) \mid U \text{ is open and } U \supseteq E\}, \tag{6.5}$$

for every $E \in \Sigma$. To complete the proof of the regularity of μ, we need to show (by Definition 3.58) that $\mu(U) = \sup\{\mu(K) \mid K \text{ is compact and } K \subseteq U\}$ for every open set $U \subseteq X$. Now if $U = X$, then there is nothing to show because U is compact. Likewise, if $\mu(U) = 0$, then $\mu(K)\mu(U) = 0$ for every compact $K \subseteq U$. Thus, assume that $U \neq X$ and that $\mu(U) > 0$. Clearly, $\sup\{\mu(K) \mid K \text{ is compact and } K \subseteq U\} \leq \mu(U)$. Conversely, suppose that $\varepsilon > 0$ satisfies $0 < \mu(U) - \varepsilon$; by definition, there exists a $f \in \mathscr{I}$ with support supp$f \subseteq U$ and $\varphi(f) > \mu(U) - \varepsilon$. Let $K = \text{supp}f$; we shall compute $\mu(K)$ using equation (6.5). To this end, let W be an open set containing K. Because $K = \text{supp}f \subseteq W$, the definition of μ yields $\mu(W) \geq \varphi(f) \geq \mu(U) - \varepsilon$. Hence,

$$\mu(U) \geq \mu(K) = \inf\{\mu(W) \mid W \text{ is open and } W \supseteq K\} \geq \mu(U) - \varepsilon,$$

which implies that $\mu(U) = \sup\{\mu(K) \mid K \text{ is compact and } K \subseteq U\}$. Hence, μ is a regular Borel measure.

We now show that if $K \subseteq X$ is compact, then

$$\mu(K) = \inf\{\varphi(f) \mid f \in \mathscr{I} \text{ and } f(K) = \{1\}\}. \tag{6.6}$$

To this end, select $f \in \mathscr{I}$ such that $f(K) = \{1\}$, and let $U_\alpha = f^{-1}((\alpha, \infty))$, for each $\alpha \in (0,1)$. Note that $K \subseteq U_\alpha$, for each α. Suppose that $g \in \mathscr{I}$ has support supp$g \subseteq U_\alpha$. If $x \in \text{supp}g$, then $0 \leq \alpha g(x) \leq \alpha \leq f(x)$; and if $x \notin \text{supp}g$, then $0 = \alpha g(x) = \leq f(x)$. Thus, $\alpha g \leq f$ in $C(X)$, and so

$$\alpha\varphi(g) = \varphi(\alpha g) \leq \varphi(f).$$

Thus,

$$\mu(U_\alpha) = \sup\{\varphi(g) \mid g \in \mathscr{I} \text{ and } \text{supp}g \subseteq U_\alpha\} \leq \frac{\varphi(f)}{\alpha}.$$

As the inequality above holds for every $\alpha \in (0, 1)$, we have

$$\mu(K) \leq \inf_\alpha \mu(U_\alpha) \leq \inf_\alpha \frac{1}{\alpha} \varphi(f) = \varphi(f).$$

Hence,

$$\mu(K) \leq \inf\{\varphi(f) \mid f \in \mathscr{I} \text{ and } f(K) = \{1\}\}.$$

To show that the inequality above can be reversed, select $\varepsilon > 0$. By definition, there is an open set W containing K and such that $\mu(W) < \mu(K) + \varepsilon$. By Corollary 2.44, there exists $f \in \mathscr{I}$ with supp$f \subseteq W$ and $f(K) = \{1\}$. Therefore, $\varphi(f) \leq \mu(W)$, by definition of μ. Hence,

$$\mu(K) \leq \inf\{\varphi(f) \mid f \in \mathscr{I} \text{ and } f(K) = \{1\}\} \leq \mu(W) < \mu(K) + \varepsilon,$$

which implies equation (6.6).

The integral representation of φ may now be established. Select any $f \in \mathscr{I}$. Fix $n \in \mathbb{N}$ and for $k = 1, \ldots, n$ define $U_k = f^{-1}\left(\left(\frac{k-1}{n}, \infty\right)\right)$, and let $U_0 = X$ and $U_{n+1} = \emptyset$. These sets form a descending sequence

$$X = \overline{U}_0 = U_0 \supseteq \overline{U}_1 \supseteq U_1 \supseteq \overline{U}_2 \supseteq U_2 \supseteq \cdots \supseteq \overline{U}_n \supseteq U_n \supseteq \overline{U}_{n+1} = U_{n+1} = \emptyset.$$

Define $f_k = \frac{1}{n}\chi_{U_{k+1}} + (f - \frac{k-1}{n})\chi_{U_k \setminus U_{k+1}}$. Because the sets U_j are open, each f_k is continuous. If $x \in U_k \setminus U_{k+1}$, then $\frac{k-1}{n} < f(x) \leq \frac{k}{n}$, and so

$$0 < \left(f(x) - \frac{k-1}{n}\right) \leq \frac{k}{n} - \frac{k-1}{n} = \frac{1}{n},$$

for all $x \in U_k \setminus U_{k+1}$. Thus,

$$\int_X f_k \, d\mu = \frac{1}{n}\mu(U_{k+1}) + \int_{U_k \setminus U_{k+1}} (f - \frac{k-1}{n}) \, d\mu$$

$$\leq \frac{1}{n}\mu(U_{k+1}) + \frac{1}{n}[\mu(U_k) - \mu(U_{k+1})]$$

$$= \frac{1}{n}\mu(U_k).$$

The fact that $f_k(x) = 1/n$ for every $x \in U_{k+1}$ yields $\frac{1}{n}\mu(U_{k+1}) \leq \int_X f_k \, d\mu$. Hence, by summing over all k we obtain the inequalities

$$\frac{1}{n}\sum_{k=1}^n \mu(U_{k+1}) \leq \int_X f \, d\mu \leq \frac{1}{n}\sum_{k=1}^n \mu(U_k). \tag{6.7}$$

The function f_k is zero on U_k^c; thus, $\{x \in X \,|\, f_k(x) \neq 0\} \subseteq U_k$, which implies that the support of f_k satisfies $\mathrm{supp} f_k \subseteq \overline{U}_k \subseteq U_{k-1}$. Therefore, by definition of μ, and noting that nf_k and f_k have the same support, we have that $\varphi(nf_k) \leq \mu(U_{k-1})$. On the other hand, by continuity, $f_k(x) = \frac{1}{n}$ for all $x \in \overline{U}_{k+1}$. Therefore, by the compactness of $\overline{U}_{k+1}$ and because $nf_k \in \mathscr{I}$ satisfies $nf_k\left(\overline{U}_{k+1}\right) = \{1\}$, equation (6.6) shows that $\mu(U_{k+1}) \leq \varphi(nf_k)$. Hence, by summing over all k and dividing by n we obtain the inequalities

$$\frac{1}{n} \sum_{k=1}^{n} \mu(U_{k+1}) \leq \varphi(f) \leq \frac{1}{n} \sum_{k=1}^{n} \mu(U_{k-1}). \tag{6.8}$$

Therefore, using the fact that $|\gamma - \delta| \leq (b+c) - 2a$ if $\gamma \in [a,b]$ and $\delta \in [a,c]$, inequalities (6.7) and (6.8) yield

$$\left| \int_X f \, d\mu - \varphi(f) \right| \leq \frac{1}{n} \left(\mu(U_0) + \mu(U_1) - \mu(U_n) \right) \leq \frac{2\mu(X)}{n}.$$

Because the choice of $n \in \mathbb{N}$ is arbitrary, we obtain $\varphi(f) = \int_X f \, d\mu$. The integral formula for arbitrary $f \in C(X)$ follows from the fact that the positive functions span $C(X)$ and the fact both φ and the integral are linear maps.

To prove the uniqueness of μ, suppose that $\tilde{\mu}$ is another regular Borel measure for which $\varphi(f) = \int_X f \, d\tilde{\mu}$, for every $f \in C(X)$. Choose any open set U and let $f \in \mathscr{I}$ have support contained in U. Thus,

$$\varphi(f) = \int_X f \, d\tilde{\mu} = \int_U f \, d\tilde{\mu} \leq \int_U d\tilde{\mu} = \tilde{\mu}(U).$$

By definition of μ, the inequality above yields $\mu(U) \leq \tilde{\mu}(U)$. To prove the reverse inequality, let $\varepsilon > 0$ and select a compact set K such that $K \subseteq U$ and $\tilde{\mu}(U) < \tilde{\mu}(K) + \varepsilon$. Select $f \in \mathscr{I}$ with $\mathrm{supp} f \subseteq U$ and $f(K) = \{1\}$. Therefore, equation (6.6) and the definition of μ yield

$$\tilde{\mu}(U) < \tilde{\mu}(K) + \varepsilon \leq \varphi(f) + \varepsilon \leq \mu(U) + \varepsilon.$$

Hence, $\tilde{\mu}(U) = \mu(U)$ for all open sets U. By regularity of the measures, we deduce that $\tilde{\mu}(E) = \mu(E)$ for all Borel sets E. $\square$

Corollary 6.52. *If X is a second countable compact Hausdorff space and if φ is a linear functional on $C(X)$, then there exist a regular Borel measures $\mu_1, \ldots, \mu_4$ on the Borel sets of X, such that, for every $f \in C(X)$,*

$$\varphi(f) = \int_X f \, d\nu,$$

where ν is the complex measure $\nu = (\mu_1 - \mu_2) + i(\mu_3 - \mu_4)$.

Proof. Lemma 6.50 and its proof show that φ has the form $\varphi = (\varphi_1 - \varphi_2) + i(\varphi_3 - \varphi_4)$, for some positive linear functionals φ_j on $C(X)$. Theorem 6.51 gives a representing regular Borel measure μ_j for each φ_j. $\square$

For each complex measure ν on the Borel sets of X, the linear map $f \mapsto \int_X f \, d\nu$ is bounded. Thus, the dual space of $C(X)$ may be identified with the Banach space $M(X, \Sigma)$ of complex measures. One would like such an identification to be isometric, and such is the content of the next theorem which characterises the dual space of $C(X)$.

Theorem 6.53. *If X is a second countable compact Hausdorff space, then the map $\Phi : M(X, \Sigma) \to C(X)^*$ defined by*

$$\Phi(\nu)[f] = \int_X f \, d\nu,$$

for every $f \in C(X)$, is a linear isometric surjection.

Proof. Exercise 6.70. $\square$

Problems

6.54. Suppose that p and q are conjugate real numbers. Prove that if $g = (g_k)_{k \in \mathbb{N}} \in \ell^q(\mathbb{N})$, then the function $\varphi : \ell^p(\mathbb{N}) \to \mathbb{C}$ defined by

$$\varphi(f) = \sum_{k=1}^{\infty} f_k g_k,$$

for $f = (f_k)_{k \in \mathbb{N}} \in \ell^p(\mathbb{N})$, is a linear functional on $\ell^p(\mathbb{N})$ of norm $\|\varphi\| = \|g\|$.

6.55. Consider the Banach space $\ell^1(\mathbb{N})$.

1. Prove that if $g = (g_k)_{k \in \mathbb{N}} \in \ell^\infty(\mathbb{N})$, then the function $\varphi_g : \ell^1(\mathbb{N}) \to \mathbb{C}$ defined by

$$\varphi_g(f) = \sum_{k=1}^{\infty} f_k g_k,$$

   for $f = (f_k)_{k \in \mathbb{N}} \in \ell^1(\mathbb{N})$, is a linear functional on $\ell^1(\mathbb{N})$ of norm $\|\varphi\| = \|g\|$.
2. Prove that the function $\Theta : \ell^\infty(\mathbb{N}) \to \left(\ell^1(\mathbb{N})\right)^*$ defined by $\Theta(g) = \varphi_g$ (as above) is a linear isometric isomorphism of $\ell^\infty(\mathbb{N})$ and $\left(\ell^1(\mathbb{N})\right)^*$.

6.56. Recall that $c_0(\mathbb{N})$ is the subspace of $\ell^\infty(\mathbb{N})$ given by

$$c_0(\mathbb{N}) = \{(f_k)_{k \in \mathbb{N}} \mid \lim_{k \to \infty} f_k = 0\}.$$

1. Prove that if $g = (g_k)_{k \in \mathbb{N}} \in \ell^1(\mathbb{N})$, then the function $\varphi_g : c_0(\mathbb{N}) \to \mathbb{C}$ defined by

$$\varphi_g(f) = \sum_{k=1}^{\infty} f_k g_k,$$

for $f = (f_k)_{k \in \mathbb{N}} \in c_0(\mathbb{N})$, is a linear functional on $c_0(\mathbb{N})$ of norm $\|\varphi\| = \|g\|$.

2. Prove that the function $\Theta : \ell^1(\mathbb{N}) \to (c_0(\mathbb{N}))^*$ defined by $\Theta(g) = \varphi_g$ (as above) is a linear isometric isomorphism of $\ell^1(\mathbb{N})$ and $(c_0(\mathbb{N}))^*$.

6.57. Prove that if $g = (g_k)_{k \in \mathbb{N}} \in \ell^{\infty}(\mathbb{N})$, then the function $\varphi_g : \ell^1(\mathbb{N}) \to \mathbb{C}$ defined by

$$\varphi_g(f) = \sum_{k=1}^{\infty} f_k g_k,$$

for $f = (f_k)_{k \in \mathbb{N}} \in \ell^1(\mathbb{N})$, is a linear functional on $\ell^1(\mathbb{N})$ of norm $\|\varphi\| = \|g\|$. Furthermore, prove that the function $\Theta : \ell^{\infty}(\mathbb{N}) \to (\ell^1(\mathbb{N}))^*$ defined by $\Theta(g) = \varphi_g$ (as above) is a linear isometric isomorphism of $\ell^{\infty}(\mathbb{N})$ and $(\ell^1(\mathbb{N}))^*$.

6.58. Prove that if V and W are Banach spaces, then the Banach space $B(V, W)$ is nonzero.

6.59. Suppose that M is a proper subspace of a Banach space V and that $v \in V$ is nonzero and $v \notin M$. Prove that there exists $\varphi \in V^*$ such that $\varphi(v) = 1$ and $\varphi(w) = 0$ for every $w \in M$. (Suggestion: consider the linear submanifold $L = \{w + \alpha v \mid w \in M, \ \alpha \in \mathbb{C}\}$ and the linear map $\varphi_0 : L \to \mathbb{C}$ defined by $\varphi_0(w + \alpha v) = \alpha$.)

6.60. Assume that V is a real vector space and that $p : V \to \mathbb{R}$ is a sublinear functional. Modify the proof of Theorem 6.18 to show that if L is a linear submanifold of V and if $\varphi : L \to \mathbb{R}$ is a linear transformation for which $\varphi(v) \leq p(v)$ for every $v \in L$, then there is a linear transformation $\Phi : V \to \mathbb{R}$ such that $\Phi_{|L} = \varphi$ and $-p(-v) \leq \Phi(v) \leq p(v)$ for every $v \in V$.

6.61. Suppose that V is a Banach space.

1. (a) Prove that if V^* is a separable, then V is separable.
2. (b) Show by example that there are separable Banach spaces V for which V^* is nonseparable.

6.62. Prove that $\ell^p(\mathbb{N})$ is a reflexive Banach space, for all $p \in \mathbb{R}$ such that $p > 1$.

6.63. Prove that neither $c_0(\mathbb{N})$ nor $\ell^1(\mathbb{N})$ is a reflexive Banach space.

6.64. Let V be a normed vector space and suppose that $\varphi_1, \ldots, \varphi_n \in V^*$. Let

$$L = \bigcap_{j=1}^{n} \ker \varphi_j.$$

Assume that $\varphi \in V^*$ satisfies $\varphi(\xi) = 0$, for every $\xi \in L$.

1. Let $\Phi : V \to \mathbb{C}^n$ be given by $\Phi(v) = \begin{bmatrix} \varphi_1(v) \\ \vdots \\ \varphi_n(v) \end{bmatrix}$, $v \in V$. Show that there is a linear

functional $\psi : \mathbb{C}^n \to \mathbb{C}$ such that $\varphi(v) = \psi(\Phi(v))$, for every $v \in V$.

2. Show that there are $\lambda_1, \ldots, \lambda_n \in \mathbb{C}$ such that $\varphi = \sum_{j=1}^{n} \lambda_j \varphi_j$.

6.65. Prove that the weak topology and the norm topology on a finite-dimensional normed vector space coincide.

6.66. Suppose that H is an infinite-dimensional separable Hilbert space.

1. Prove that the closed unit ball of H is weakly compact.
2. Prove that the zero vector is in the weak closure of the unit sphere of H.

6.67. Prove that the vector space operations on V^* are continuous in the weak* topology, for every normed vector space V.

6.68. Prove that every linear functional on $L^\infty(X, \Sigma, \mu)$ is a linear combination of four positive linear functionals.

6.69. Prove that if Theorem 6.45 is true for finite measure spaces, then it is also true for σ-finite measure spaces.

6.70. Assume that X is a second countable compact Hausdorff space, and define a linear map $\Phi : M(X, \Sigma) \to C(X)^*$ by $\Phi(\nu) = \varphi_\nu$, where

$$\varphi_\nu(f) = \int_X f \, d\nu,$$

for every $f \in C(X)$. Prove that Φ is an isometric surjection.

6.71. Suppose that μ is a finite regular Borel measure on $\mathbb{R}$, and consider the functions $g_n : \mathbb{R} \to \mathbb{C}$, for $n \in \mathbb{Z}$, defined by $g_n(t) = e^{int}$. Prove that if $\int_{\mathbb{R}} g_n \, d\mu = 0$ for every $n \in \mathbb{Z}$, then $\mu = 0$.

Chapter 7
Convexity

The linear character of functional analysis underscores the entire subject. In addition to geometric structures such as subspaces, it can be important to consider subsets C of vector spaces V that are locally linear in the sense that C contains the line segment in V between every pair of points of C. That is, if $u, v \in C$, then so is $tu + (1-t)v$ for every $t \in [0, 1]$. Such sets are said to be convex and they are crucial structures in functional analysis, useful for both the geometrical and topological information that they reveal about a space V.

7.1 Convex Sets

Recall that a subset C of a vector space V is *convex* if $tu + (1-t)v \in C$ for all $u, v \in C$ and every $t \in [0, 1]$. More generally, a *convex combination* of elements $v_1, \dots, v_n$ in a vector space V is a sum of the form

$$\sum_{j=1}^{n} t_j v_j, \tag{7.1}$$

where each $t_j \in [0, 1]$ and $\sum_{j=1}^{n} t_j = 1$.

Proposition 7.1. *A subset C is convex if and only if C contains every convex combination of its elements.*

Proof. Exercise 7.27. □

The scalars that arise in (7.1) are real, and so convexity can be studied completely within the realm of real vector spaces. However, functional analysis is mostly carried

© Springer International Publishing Switzerland 2016
D. Farenick, *Fundamentals of Functional Analysis*, Universitext,
DOI 10.1007/978-3-319-45633-1_7

out over the complex field. Thus, in this chapter, we shall use real vector spaces when it is helpful to do so and then make use of the fact that every complex vector space is also a real vector space.

Of course any subspace of a vector space is a convex set. Some basic convex sets of real and complex numbers are:

1. any interval J of real numbers;
2. the open unit disc $\mathbb{D}$ of complex numbers;
3. the closed unit disc $\overline{\mathbb{D}}$ of complex numbers;
4. the right halfplane $\{z \in \mathbb{C} \mid \Re z \geq 0\}$ of $\mathbb{C}$.

For every subset $S \subseteq V$ there is a smallest convex set C that contains S.

Definition 7.2. If $S \subseteq V$, then the set of all convex combinations of elements of S is called the *convex hull* of S and is denoted by $\mathrm{Conv}\, S$.

The following theorem is very useful in the convexity theory of finite-dimensional space.

Theorem 7.3 (Caratheódory). *If V is an n-dimensional vector space over $\mathbb{R}$ and if $S \subset V$ is nonempty, then for each $v \in \mathrm{Conv}\, S$ there are $v_1, \ldots, v_m \in S$ such that v is a convex combination of $v_1, \ldots, v_m$ and $m \leq n + 1$.*

Proof. Let $v \in \mathrm{Conv}\, S$. Thus, v is a convex combination of $v_1, \ldots, v_m \in S$, say

$$v = \sum_{j=1}^{m} t_j v_j,$$

where each $t_j \neq 0$. If $m \leq (n+1)$, then the desired conclusion is reached. Therefore, suppose that $m > (n+1)$. Let $\tilde{V} = \mathbb{R} \times V$ and let $\tilde{v}_j = (1, v_j) \in \tilde{V}$, for $1 \leq j \leq m$. Since $m > n + 1 = \dim \tilde{V}$, the vectors $\tilde{v}_1, \ldots, \tilde{v}_m \in \tilde{W}$ are linearly dependent. Thus, there are $\alpha_1, \ldots, \alpha_m \in \mathbb{R}$, not all zero, such that $\sum_j \alpha_j \tilde{v}_j = 0$; hence, $\sum_j \alpha_j = 0$ in $\mathbb{R}$ and $\sum_j \alpha_j v_j = 0$ in V.

Let i be such that

$$\left| \frac{\alpha_j}{t_j} \right| \leq \left| \frac{\alpha_i}{t_i} \right|, \quad \forall j \neq i,$$

and set $s_j = t_j - \dfrac{\alpha_j t_i}{\alpha_i}$ for every j. Then

$$s_j \in [0,1], \ s_i = 0, \ \sum_{j=1}^{m} s_j = \sum_{j=1}^{m} t_j = 1, \text{ and } v = \sum_{j=1}^{m} s_j v_j.$$

But the number of nonzero summands in $\displaystyle\sum_{j=1}^{m} s_j v_j$ is less than m since $s_i = 0$. Hence, if v can be expressed as a convex combination of $m > n + 1$ elements of S, then v

can be expressed as a convex combination of $m - 1$ of these same elements. This shows, by iteration of the argument, that the number of summands can be reduced to $n + 1$. □

The upper bound of $n + 1$ in Caratheódory's Theorem is sharp, as one sees by considering a triangle C in $\mathbb{R}^2$ with vertex set S.

Corollary 7.4. *If V is an n-dimensional vector space over $\mathbb{C}$ and if $S \subset V$ is nonempty, then for each $v \in \text{Conv}\, S$ there are $u_1, \ldots, u_m \in S$ such that v is a convex combination of $u_1, \ldots, u_m$ and $m \leq 2n + 1$.*

7.2 Separation Theorems

This section establishes a cornerstone result called the Hahn-Banach Separation Theorem that provides analytic information from the knowledge that two convex sets are disjoint. This theorem is applied to subsets of Banach spaces in a variety of topologies (norm, weak, weak*, etc.), and so the development below will not immediately make use of Banach spaces and linear functionals, but rather topological vector spaces V and linear transformations $\varphi : V \to \mathbb{C}$ that are continuous in the topology of V.

Recall that a vector space V is a *topological vector space* if V is a topological space for which addition and scalar multiplication are continuous functions $V \times V \to V$ and $\mathbb{C} \times V \to V$, respectively.

Proposition 7.5. *If V is a topological vector space, then*

1. *W is an open subset of V if and only if there exists $w_0 \in W$ and open neighbourhood U of 0 such that $W = \{w_0\} + U$, and*
2. *for each open neighbourhood U of $0 \in V$ and $v \in V$ there exists a $\varepsilon > 0$ such that $\lambda v \in U$ for all $\lambda \in \mathbb{C}$ with $|\lambda| < \varepsilon$.*

Proof. For (1), select $w_0 \in W$ and let $U = \{w_0 - w \,|\, w \in W\}$. Because scalar multiplication and vector addition are continuous, U is an open set and $W = \{w_0\} + U$. For (2), because scalar multiplication $m : \mathbb{C} \times V \to V$ is continuous, the map $m_v : \mathbb{C} \to V$ defined by $m_v(\alpha) = m(\alpha, v) = \alpha v$ is continuous for each fixed $v \in V$. Thus if $U \subset V$ is an open neighbourhood of $0 \in V$, then $m_v^{-1}(U)$ is an open neighbourhood of $0 \in \mathbb{C}$. Hence, there exists a $\varepsilon > 0$ such that $\{\lambda \in \mathbb{C} \,|\, |\lambda| < \varepsilon\} \subseteq m_v^{-1}(U)$. □

The next analytical tool is a sublinear functional (Definition 6.19) known as the *Minkowski functional*.

Proposition 7.6. *If C is an open convex subset of a topological vector space V such that $0 \in C$, then the map $p : V \to \mathbb{R}$ defined by*

$$p(v) = \inf\{t > 0 \,|\, t^{-1} v \in C\}, \tag{7.2}$$

for $v \in V$, is a sublinear functional and $C = \{v \in V \,|\, p(v) < 1\}$.

Proof. The sublinearity of p is evident. Suppose that $v \in V$ satisfies $p(v) < 1$. Then there exists $t \in (0, 1)$ and $u \in C$ such that $\frac{1}{t}v = u$. That is, $v = tu = tu + (1-t)0 \in C$. Conversely, if $v \in C$, then there is an open set W such that $v \in W \subset C$. By Proposition 7.5, W may be taken to be $W = \{v\} + U$ for some open neighbourhood U of 0 for which $0 \in U \subseteq C$. Proposition 7.5 shows that there is a $\varepsilon > 0$ such that $\varepsilon v \in U$. Thus, $(1+\varepsilon)v = v + \varepsilon v \in W \subset C$ implies that $p(v) \leq (1+\varepsilon)^{-1} < 1$.   □

Theorem 7.7 (Hahn-Banach Separation Theorem). *Assume that C_1 and C_2 are nonempty, disjoint convex subsets of a topological vector space V such that C_1 is open. Then there are a linear transformation $\varphi : V \to \mathbb{C}$ and a $\gamma \in \mathbb{R}$ such that φ is continuous and*

$$\Re\varphi(v_1) < \gamma \leq \Re\varphi(v_2), \quad \forall\, v_1 \in C_1,\, v_2 \in C_2. \tag{7.3}$$

Proof. Let $C = \bigcup_{v_1 \in C_1} \{v_1 - v_2 \,|\, v_2 \in C_2\}$, and note that C is open and convex. Further, $0 \notin C$, as $C_1 \cap C_2 = \emptyset$. Select $v_0 \in C$ and let $C_0 = \{v_0 - v \,|\, v \in C\}$. Thus, C_0 is open, convex, and $0 \in C_0$. Let p be the Minkowski functional (7.2) associated with C_0. Because $0 \notin C$ implies $v_0 \notin C_0$, Proposition 7.6 yields $p(v_0) \geq 1$.

Now let $L = \mathrm{Span}_{\mathbb{R}}\{v_0\}$ and define $\mu_0 : L \to \mathbb{R}$ by $\mu_0(\lambda v_0) = \lambda p(v_0)$, for all $\lambda \in \mathbb{R}$. The function μ is linear over $\mathbb{R}$ and satisfies $\mu_0(\lambda v_0) = p(\lambda v_0)$ if $\lambda \geq 0$. If $\lambda < 0$, then $\mu_0(\lambda v_0) = \lambda p(v_0) < 0 \leq p(\lambda_0)$. Thus, $\mu_0(w) \leq p(w)$ for every $w \in L$. By the version of the Hahn-Banach Extension Theorem in Exercise 6.60, μ_0 extends to a linear transformation $\mu : V \to \mathbb{R}$ such that $-p(-v) \leq \mu(v) \leq p(v)$ for every $v \in V$.

Because C_0 is open and contains 0, for each $v \in V$ there exists $\delta > 0$ such that $\delta v \in C_0$; thus,

$$-1 > -p(-\delta v) = \delta(-p(-v)) \leq \delta\mu(v) = \mu(\delta v) \leq p(\delta v) < 1.$$

Therefore, $|\mu(u)| < 1$ for every $u \in U = C_0 \cap (-C_0)$, an open neighbourhood of 0. Hence, for every $\varepsilon > 0$ such that $\varepsilon < 1$, $W_\varepsilon = (-\varepsilon, \varepsilon)$ is a neighbourhood of $\mu(0) = 0$ in $\mathbb{R}$ and the set $U_\varepsilon = \varepsilon U \subset U$ is a neighbourhood of $0 \in V$ such that $\mu(U_\varepsilon) \subseteq W_\varepsilon$. Thus, μ is continuous at $0 \in V$. By Proposition 7.5, μ is therefore continuous at every $v \in V$, which implies that $\mu : V \to \mathbb{R}$ is a continuous linear transformation. Because scalar multiplication and vector addition are continuous, the linear transformation $\varphi : V \to \mathbb{C}$ defined by $\varphi(v) = \mu(v) - i\mu(iv)$, for $v \in V$, is continuous and $\mu = \Re\varphi$.

Let $N = \ker\mu = \mu^{-1}(\{0\})$, which is closed by the continuity of μ. If $v \in C$, then $v_0 - v \in C_0$ and so

$$1 < p(v_0 - v) \geq \mu(v_0 - v) = \mu(v_0) - \mu(v) = p(v_0) - \mu(v);$$

that is, $\mu(v) > p(v_0) - 1 \geq 0$, which implies $\mu(v) \neq 0$. Thus, $\ker\mu \cap C = \emptyset$. Therefore, because C is convex and μ is continuous, $\mu(C)$ is a connected subset of $\mathbb{R}$. But $\ker\mu \cap C = \emptyset$ implies that $\mu(C) \subset (-\infty, 0)$ or $\mu(C) \subset (0, \infty)$. Without loss of generality, assume that $\mu(C) \subset (-\infty, 0)$. Hence, $\mu(v_1) < \mu(v_2)$, for all $v_1 \in C_1$, $v_2 \in C_2$. Therefore, there is a $\gamma \in \mathbb{R}$ such that

$$\sup_{v_1 \in C_1} \mu(v_1) \leq \gamma \leq \inf_{v_2 \in C_2} \mu(v_2).$$

Because C_1 is an open convex set and because μ is a continuous linear transformation, the interval $\mu(C_1)$ is open in $\mathbb{R}$. Hence, $\gamma \notin \mu(C_1)$ and so the inequalities (7.3) hold.   □

In the case of Banach spaces, the following "Hahn-Banach Separation Theorem" is of particular importance.

Corollary 7.8. *If C_1 and C_2 are convex sets in a Banach space V such that $C_1 \cap C_2 = \emptyset$, C_1 is compact, and C_2 is closed, then there exist $\varphi \in V^*$ and $\gamma_1, \gamma_2 \in \mathbb{R}$ such that*

$$\Re\varphi(v_1) < \gamma_1 < \gamma_2 < \Re\varphi(v_2), \qquad \forall\, v_1 \in C_1,\ v_2 \in C_2. \tag{7.4}$$

Proof. Assume that $C_1, C_2 \subset V$ are disjoint convex sets and that C_1 is compact and C_2 is closed. By Proposition 5.14, there exists $\varepsilon > 0$ such that $(C_1 + B_\varepsilon(0)) \cap C_2 = \emptyset$. Note that $C_1 + B_\varepsilon(0)$ is convex and open. Thus, by Theorem 7.7, there exist $\varphi \in V^*$ and $\gamma \in \mathbb{R}$ such that

$$\Re\varphi(v_1) < \gamma \leq \Re\varphi(v_2), \qquad \forall\, v_1 \in C_1 + B_\varepsilon(0),\ v_2 \in C_2.$$

However, as C_1 is compact, $\Re\varphi(C_1)$ is compact and $\Re\varphi(C_1 + B_\varepsilon(0))$ has compact closure, disjoint from C_2. Therefore, there are $\gamma_1, \gamma_2 \in \mathbb{R}$ such that

$$\Re\varphi(v_1) < \gamma_1 < \gamma_2 < \Re\varphi(v_2),$$

for all $v_1 \in C_1$ and $v_2 \in C_2$.   □

A somewhat surprising consequence of Hahn-Banach Separation Theorem is the following result about topology, which demonstrates the essential role convexity theory plays in understanding Banach spaces.

Proposition 7.9. *Let C be a nonempty convex subset of a Banach space V. If $\overline{C}$ and $\overline{C}^{wk}$ denote the closures of C in the norm and weak topologies of V, then $\overline{C} = \overline{C}^{wk}$.*

Proof. Because $\overline{C}$ is convex and closed, if $v_0 \notin \overline{C}$, then by the Hahn-Banach Separation Theorem (Corollary 7.8) there exist $\varphi \in V^*$ and $\gamma \in \mathbb{R}$ such that, for all $v \in \overline{C}$, $\mu(v) < \gamma < \mu(v_0)$, where $\mu = \Re\varphi$. Therefore, there exists $\varepsilon > 0$ such that $|\mu(v) - \mu(v_0)| \geq \varepsilon$ for all $v \in \overline{C}$. In particular, for $v \in \overline{C}$,

$$|\varphi(v) - \varphi(v_0)|^2 = (\mu(v) - \mu(v_0))^2 + (\mu(iv_0) - \mu(iv))^2 \geq (\mu(v) - \mu(v_0))^2 \geq \varepsilon^2.$$

Hence, $v_0 \notin \overline{C}^{wk}$. This proves that $\overline{C}^{wk} \subseteq \overline{C}$.

Conversely, for any set C, it is always the case that $\overline{C} \subseteq \overline{C}^{\text{wk}}$.   □

To make additional concrete applications of the Hahn-Banach Separation Theorem, one needs to have a sufficient amount of information concerning continuous linear maps $V \to \mathbb{C}$ on topological vector spaces V. The following proposition is one such example.

Proposition 7.10. *If V^* is the dual of a normed vector space V, then*

1. *V^* is a Hausdorff topological vector space in the weak*-topology, and*
2. *a linear transformation $\Phi : V^* \to \mathbb{C}$ is continuous with respect to the weak*-topology if and only if there exists $v \in V$ such that $\Phi(\varphi) = \varphi(v)$ for every $\varphi \in V^*$.*

Proof. The proof of the first statement is left as an exercise (Exercise 7.30).

Linear maps of the form $\varphi \mapsto \varphi(v)$, for fixed $v \in V$, are continuous by the definition of weak*-topology. Suppose, conversely, that $\Phi : V^* \to \mathbb{C}$ is an arbitrary linear transformation and is continuous with respect to the weak*-topology. Thus, if $\mathbb{D} \subset \mathbb{C}$ is the open unit disc in $\mathbb{C}$, then there is a basic weak*-open set $U \subset V^*$ such that $0 \in U \subseteq \Phi^{-1}(\mathbb{D})$. By definition, such a set U has the form

$$U = \bigcap_{j=1}^{n} \{ \psi \in V^* \mid |\psi(v_j)| < \varepsilon \},$$

for some $v_1, \ldots, v_n \in V$ and $\varepsilon > 0$. Therefore, $|\Phi(\psi)| < 1$ for all $\psi \in U$. In particular, if $\varphi \in V^*$ is arbitrary and is nonzero on at least one v_j, then

$$\frac{\varepsilon}{2 \max_\ell |\varphi(v_\ell)|} \varphi \in U,$$

and so $|\Phi(\varphi)| < \varepsilon^{-1} \max_\ell |\varphi(v_\ell)|$. Hence

$$|\Phi(\varphi)| \leq \frac{2 \max_\ell |\varphi(v_\ell)|}{\varepsilon}, \tag{7.5}$$

for every $\varphi \in V^*$.

Consider the linear transformation $T : V^* \to \mathbb{C}^n$ defined by

$$T(\varphi) = \begin{bmatrix} \varphi(v_1) \\ \vdots \\ \varphi(v_n) \end{bmatrix}.$$

On the range of T, define a function $\Lambda_0 : \operatorname{ran} T \to \mathbb{C}$ by $\Lambda_0(T\varphi) = \Phi(\varphi)$. This function Λ_0 is well defined because, if $\varphi_1, \varphi_2 \in V^*$ satisfy $T\varphi_1 = T\varphi_2$, then $(\varphi_1 - \varphi_2)(v_j) = 0$ for each $j = 1, \ldots, n$, and so inequality (7.5) gives $\Phi(\varphi_1 - \varphi_2) = 0$. Because Λ_0 is also linear, Λ_0 extends to a linear functional Λ on $\mathbb{C}^n$. Hence, Λ is given by an $n \times 1$ matrix $\Lambda = [\alpha_1 \, \alpha_2 \, \cdots \, \alpha_n]$ and, for every $\varphi \in V^*$,

$$\Phi(\varphi) = \langle T\varphi, \gamma \rangle = \sum_{j=1}^{n} \alpha_j \varphi(v_j) = \varphi \left(\sum_{j-1}^{n} \alpha_j v_j \right) = \varphi(v),$$

where $v = \sum_{j=1}^{n} \alpha_j v_j$. □

With Proposition 7.10 in hand, the following Hahn-Banach Separation Theorem holds for dual spaces.

Proposition 7.11. *Assume that V^* is the dual of a normed vector space V. If $K \subset V^*$ is weak*-compact and convex, and if $\varphi_0 \notin K$, then there exist $\gamma \in \mathbb{R}$ and $v_0 \in V$ such that*

$$\Re \varphi_0(v_0) < \gamma \le \Re \varphi(v_0)$$

for every $\varphi \in K$.

Proof. The dual space V^* is a Hausdorff topological vector space in the weak*-topology. By the Hausdorff property, for each $\varphi \in K$ there are weak*-open sets U_φ and W_φ such that $\varphi_0 \in U_\varphi$, $\varphi \in W_\varphi$, and $U_\varphi \cap W_\varphi = \emptyset$. Because K is compact, there are finitely members $W_{\varphi_1} \dots, W_{\varphi_n}$ of the open cover $\{W_\varphi\}_{\varphi \in K}$ that cover K. Hence, if

$$U = \bigcap_{j=1}^{n} U_{\varphi_j} \quad \text{and} \quad W = \bigcup_{j=1}^{n} W_{\varphi_j},$$

then U and V are disjoint open sets with $\varphi_0 \in U$ and $K \subset V$. The proof of Theorem 7.7 shows that there is a convex open set U_0 that contains the origin and Proposition 7.5 shows that there is a $\varepsilon > 0$ such that $C_1 = \{\varphi_0\} + \varepsilon U_0 \subseteq U$. Hence, C_1 is an open convex set disjoint from K. By Theorem 7.7 there exist a weak*-continuous linear transformation $\Phi : V^* \to \mathbb{C}$ and $\gamma \in \mathbb{R}$ such that $\Re \Phi(\varphi_0) < \gamma \le \Re \Phi(\varphi)$ for all $\varphi \in K$. However, by Proposition 7.10, the map Φ is given by evaluation at some point $v_0 \in V$. Hence, $\Re \varphi_0(v_0) < \gamma \le \Re \varphi(v_0)$ for every $\varphi \in K$.

□

7.3 Extreme Points

Triangles and squares are determined by their vertices, a disc by its boundary circle, and a Euclidean ball in $\mathbb{R}^3$ by its boundary surface (sphere). The general concept in convexity theory that captures these phenomena is that of an extreme point.

Definition 7.12. An element v in a convex subset $C \subseteq V$ is an *extreme point* of C if the equation

$$v = \sum_{j=1}^{n} t_j v_j, \text{ where each } v_j \in C, \sum_{j=1}^{n} t_j = 1, \text{ and each } t_j \in (0, 1), \tag{7.6}$$

holds only for $v_1 = \cdots = v_n = v$. The set of all extreme points of C is denoted by $\text{ext}\, C$.

Thus, if C is convex, then $v \in C$ is an extreme point if v is not a proper convex combination of elements other than itself. Geometrically this means:

Proposition 7.13. *Let C be a convex set and $v \in C$. The following statements are equivalent:*

1. *v is an extreme point of C;*
2. *v is not interior to any line segment contained in C (i.e., if there are $v_1, v_2 \in C$ and $t \in (0,1)$ such that $v = tv_1 + (1-t)v_2$, then $v_1 = v_2 = v$).*

Proof. Exercise 7.32. $\square$

Proposition 7.13 implies that the extreme points of a convex set C must lie on the topological boundary of C. If not all points of the boundary are extreme points, then the boundary contains line segments. On the other hand, if every point on the boundary is an extreme point, then we may think of the convex set as having a curved or rounded boundary.

Example 7.14. *The extreme points of the closed unit disc in the complex plane are precisely the points on the unit circle.*

Proof. To verify this assertion, note that if $\zeta \in \text{ext}\,\overline{\mathbb{D}}$, then necessarily $|\zeta|$ has modulus 1 by Proposition 7.13. Conversely, suppose that $\zeta \in \overline{\mathbb{D}}$ is such that $|\zeta| = 1$. Thus, there exists $\theta \in \mathbb{R}$ such that $\zeta = \cos\theta + i\sin\theta$. Assume that $\zeta = t\lambda + (1-t)\mu$, for some $t \in (0,1)$ and $\lambda, \mu \in \overline{\mathbb{D}}$. The triangle inequality yields $|\lambda| = |\mu| = 1$ and so $\lambda = \cos\alpha + i\sin\alpha$ and $\mu = \cos\beta + i\sin\beta$ for some $\alpha, \beta \in \mathbb{R}$. Hence,

$$1 = \cos^2\theta + \sin^2\theta = 1 + 2t^2 - 2t + 2t(1-t)\left(\cos(\alpha - \beta)\right),$$

which implies that $\cos(\alpha - \beta) = -1$; that, is $\alpha = \beta + (2k+1)\pi$ for some $k \in \mathbb{Z}$. Thus, $\lambda = \mu = \zeta$. $\square$

Example 7.15. *The set of extreme points of the closed unit ball of a Hilbert space H is the set of unit vectors of H.*

Proof. Let H_1 denote the closed unit ball of H. Choose any unit vector $\xi \in H$ and assume that $\xi = t\xi_1 + (1-t)\xi_2$ for some $\xi_1, \xi_2 \in H_1$ and $t \in (0,1)$. Thus,

$$1 = \langle \xi, \xi \rangle = t\langle \xi_1, \xi \rangle + (1-t)\langle \xi_2, \xi \rangle$$

expresses the number 1 as a proper convex combination of the complex numbers $\langle \xi_1, \xi \rangle$ and $\langle \xi_2, \xi \rangle$ in the closed unit disc $\overline{\mathbb{D}} \subset \mathbb{C}$ (as $|\langle \xi_j, \xi \rangle| \leq \|\xi_j\| \|\xi\| \leq 1$). Example 7.14 shows that 1 is an extreme point of $\overline{\mathbb{D}}$, and therefore $1 = \langle \xi_j, \xi \rangle$ for each $j = 1, 2$, which give cases of equality in the Cauchy-Schwarz Inequality. Hence, $\xi_j = \lambda_j \xi$ for some $\lambda_j \in \mathbb{C}$; but $1 = \langle \xi_j, \xi \rangle = \lambda_j \langle \xi, \xi \rangle = \lambda_j$ yields $\xi_1 = \xi_2 = \xi$, and so $\xi \in \text{ext}\,H_1$. $\square$

Convex sets that are not topologically closed may fail to have extreme points. For example, the open unit ball in a Hilbert space does not have any extreme points. But even closedness of a convex set is an insufficient condition for extreme points to exist.

Example 7.16. *The closed unit ball of $L^1([0,1], \mathfrak{M}, m)$ has no extreme points.*

Proof. Extreme points of the closed unit ball in any Banach space, if they exist, necessarily lie on the unit sphere. Thus, suppose that $f \in \mathscr{L}^1([0,1], \mathfrak{M}, m)$ is such that $\|\dot{f}\| = 1$ and select a Borel subset $E \subset [0,1]$ such that $\int_E |f| \, dm = 1/2$. Define $h, g \in \mathscr{L}^1([0,1], \mathfrak{M}, m)$ by $h = 2f\chi_E$ and $g = 2f\chi_{E^c}$. Then $\dot{h}, \dot{g} \in L^1([0,1], \mathfrak{M}, m)$ are unit vectors and $\dot{f} = \frac{1}{2}(\dot{h} + \dot{g})$. Because $\dot{g} \neq \dot{f}, \dot{f}$ is not an extreme point of the closed unit ball of $L^1([0,1], \mathfrak{M}, m)$.   □

The main value in knowing the extreme points in a convex set (if they exist) is that they can be used to recover the convex set itself by way of the convex hull. For example, in the closed unit ball of a Hilbert space, every point is either an extreme point or an average of two extreme points. Indeed, the zero vector is evidently an average of ξ and $-\xi$, for any unit vector $\xi \in H$. Suppose that $\eta \in H$ is of norm $\|\eta\| < 1$. Let $\xi_1 = -\|\eta\|^{-1}\eta$ and $\xi_2 = \|\eta\|^{-1}\eta$, so that ξ_1 and ξ_2 are extreme points of H_1. With $t = \frac{1}{2}(1 - \|\eta\|)$ we obtain $\eta = t\xi_1 + (1-t)\xi_2$, and so the closed unit ball of H is the convex hull of the unit sphere—that is, $H_1 = \text{Conv}(\text{ext}\,H_1)$.

In contrast to Example 7.16, a compact convex set will possess extreme points and, moreover, knowledge of the extreme points of a compact convex set K is sufficient to recover the entire set K.

Theorem 7.17 (Kreĭn-Milman). *If K is a nonempty compact convex subset of a Banach space space V, then*

1. the set $\text{ext}\,K$ of extreme points of K is nonempty, and
2. K is the closure of the convex hull of the set $\text{ext}\,K$,

Proof. Consider a convex subset $F \subseteq K$ with the property that if $v \in F$ and $tv_1 + (1-t)v_2 \in F$, for some $t \in (0,1)$ and $v_1, v_2 \in K$, then $v_1, v_2 \in F$. (Such sets F are called *faces*.) Assume further that F is compact and define

$$\mathfrak{S}_F = \{G \subseteq F \,|\, G \text{ is a nonempty compact face of } F\}.$$

Use reverse inclusion to partially order $\mathfrak{S}_F$: $G_1 \preceq G_2$ if and only if $G_2 \subseteq G_1$. Let $\mathfrak{L}$ be a linearly ordered subset of $\mathfrak{S}_F$. Hence, if $G_1, \ldots, G_n \in \mathfrak{L}$, then there is a j_0 such that $G_i \preceq G_{j_0}$ for all $1 \leq i \leq n$. Hence, $\emptyset \neq G_{j_0} \subseteq G_1 \cap \cdots \cap G_n$. Therefore, the family $\mathfrak{L}$ of compact sets has the finite intersection property, and so $F_0 \neq \emptyset$, where

$$F_0 = \bigcap_{G \in \mathfrak{L}} G.$$

As F_0 is a nonempty compact face of F, F_0 is an upperbound in $\mathfrak{S}_F$ for $\mathfrak{L}$. Zorn's Lemma implies that $\mathfrak{S}_F$ has a maximal element, say E.

Let $\varphi \in V^*$, $\gamma = \max\{\Re\varphi(v) \,|\, v \in E\}$, and $E_\varphi = \{v \in E \,|\, \varphi(v) = \gamma\}$. If $t \in (0,1)$ and $v_1, v_2 \in E$ are such that $tv_1 + (1-t)v_2 \in E_\varphi$, then

$$\gamma = t\Re\varphi(v_1) + (1-t)\Re\varphi(v_2) \leq t\gamma + (1-t)\gamma = \gamma,$$

which implies that $v_1, v_2 \in E_\varphi$. Hence, E_φ is a face of E and, as well, a face of F. Furthermore, E_φ is closed, and so it is compact (since it is a closed subset of a compact set). Thus, $E_\varphi \in \mathfrak{S}_F$ and $E \preceq E_\varphi$. By the maximality of E in $\mathfrak{S}_F$, $E = E_\varphi$. We conclude, therefore, that for every $\varphi \in V^*$ and all $v_1, v_2 \in E$, $\Re\varphi(v_1) = \Re\varphi(v_2)$. The formula

$$\varphi(v) = \Re\varphi(v) - i\Re\varphi(iv)$$

yields $\varphi(v_1 - v_2) = 0$ for every $\varphi \in V^*$, which proves that $v_1 = v_2$. Hence, E is a singleton set $\{v_0\}$. But $\{v_0\}$ is a face of K if and only if v_0 is an extreme point of K, and so the set $\mathrm{ext}\,K$ of extreme points of K is nonempty.

Next let $C = \mathrm{Conv}\,(\mathrm{ext}\,K)$ and consider the compact convex subset $\overline{C}$ of K. If $\overline{C} \neq K$, then let $w_0 \in K \setminus \overline{C}$. By the Hahn-Banach Separation Theorem (Theorem 7.7), there are $\varphi \in V^*$ and $\gamma_1, \gamma_2 \in \mathbb{R}$ such that

$$\Re\varphi(v) < \gamma_1 < \gamma_2 < \Re\varphi(v_0), \quad \forall v \in \overline{C}.$$

If $\delta = \max\{\Re\varphi(w) \,|\, w \in K\}$, then $\gamma_2 < \Re\varphi(w_0) \leq \delta$. The set

$$K_\varphi = \{w \in K \,|\, \Re\varphi(w) = \delta\}$$

is a compact face of K. Therefore, by the proof of the first statement, there is an extreme point v_0 of K in K_φ. Thus, $\Re\varphi(v_0) = \delta$ and $\Re\varphi(v_0) < \delta$ (since $v_0 \in \mathrm{ext}K \subset \overline{C}$), which is a contradiction. Hence, it must be that $\overline{C} = K$. $\qquad\square$

By changing the topology on a Banach space, one may produce different versions of the Kreĭn-Milman Theorem. One of the most useful versions occurs with the dual space in its weak*-topology.

Theorem 7.18 (Kreĭn-Milman Theorem: Weak*-Topology Version). *If K is a nonempty weak*-compact convex subset of the dual space V^* of a normed vector space V, then*

1. the set $\mathrm{ext}\,K$ of extreme points of K is nonempty, and
2. K is the weak-closure of the convex hull of the set $\mathrm{ext}\,K$,*

Proof. The proof proceeds as in the proof of Theorem 7.17 to produce, using Zorn's Lemma, a minimal face E of K.

Fix $v \in V$ and define $\gamma = \max\{\Re\varphi(v) \,|\, \varphi \in E\}$ and let $C_v = \{\varphi \in E \,|\, \Re\varphi(v) = \gamma\}$. The set C_v is a weak*-closed convex subset of K; thus, C_v is weak*-compact. As in the proof of Theorem 7.17, C_v is a face of K such that $C_v \subseteq E$. Therefore, $C_v = E$ and, hence, E is a singleton set, whence K has an extreme point.

As in the proof of Theorem 7.17, the Hahn-Banach Separation Theorem is required for the second statement. In the case of the weak*-topology, it is the version of the Hahn-Banach Separation Theorem given in Proposition 7.11 that yields the result. $\qquad\square$

One consequence of Theorem 7.18 is that it tells us something about the geometry of the closed unit ball of dual spaces. Thus, Banach spaces that fail to have this property are not the duals of other Banach spaces.

Example 7.19. $L^1([0, 1], \mathfrak{M}, m)$ *is not isometrically isomorphic to the dual space* V^* *of any normed vector space* V.

Proof. Theorem 7.18 says that the closed unit ball of V^* has an extreme point. However, Example 7.16 shows that the closed unit ball of $L^1([0, 1], \mathfrak{M}, m)$ has no extreme points. □

7.4 Extremal Regular Probability Measures

The purpose of this section is to give an interesting example in which the extreme points of a compact convex set may be determined completely.

Assume that (X, Σ) is a measurable space in which X is a compact Hausdorff space and Σ is the σ-algebra of Borel sets of X. Consider the set

$$P(X, \Sigma) = \{\mu \in M(X, \Sigma) \mid \mu(E) \geq 0, \ \forall E \in \Sigma, \ \mu(X) = 1\},$$

which is a subset of the closed unit sphere of the Banach space $M(X, \Sigma)$ of regular complex measures on (X, Σ). Because $M(X, \Sigma)$ is the dual space of $C(X)$, $M(X, \Sigma)$ carries a weak*-topology. Furthermore, because $P(X, \Sigma)$ is weak*-compact (Exercise 7.36), the Kreĭn-Milman Theorem implies that $P(X, \Sigma)$ is the weak*-closure of the closed convex hull of the extreme points of $P(X, \Sigma)$.

Recall that if $K \subset X$ is a closed subset, then the σ-algebra Σ_K of Borel sets of K is given by $\Sigma_K = \{K \cap E \mid E \in \Sigma\}$.

Definition 7.20. The *support* of a regular measure μ on a measurable space (X, Σ), where X is a compact Hausdorff space and Σ is the σ-algebra of Borel sets of X, is the smallest closed subset $K \subseteq X$ for which $\mu(X \setminus K) = 0$.

Lemma 7.21. *Assume* $K_\mu \subset X$ *is the support of* $\mu \in P(X, \Sigma)$. *Then* μ *is an extreme point of* $P(X, \Sigma)$ *if and only if the restriction* $\mu_{|\Sigma_{K_\mu}}$ *of* μ *to* Σ_{K_μ} *is an extreme point of* $P(K_\mu, \Sigma_{K_\mu})$.

Proof. Assume that μ is an extreme point of $P(X, \Sigma)$. Let $\mu_0, \mu_1, \mu_2 \in P(K_\mu, \Sigma_{K_\mu})$ and such that $\mu_{|\Sigma_{K_\mu}} = \mu_0 = \frac{1}{2}(\mu_1 + \mu_2)$. Define $\tilde{\mu}_j : \Sigma \to \mathbb{R}$ by $\tilde{\mu}_j(E) = \mu_j(E \cap K_\mu)$ for all $E \in \Sigma$, to obtain $\tilde{\mu}_j \in P(X, \Sigma)$. Because K_μ is the support of μ, $\mu(E) = \mu(E \cap K_\nu)$ for all $E \in \Sigma$; thus, $\mu = \frac{1}{2}(\tilde{\mu}_1 + \tilde{\mu}_2)$, and so $\mu = \tilde{\mu}_1 = \tilde{\mu}_2$, and so $\mu_0 = \mu_1 = \mu_2$.

Conversely, assume that $\mu_0 = \mu_{|\Sigma_{K_\mu}}$ is an extreme point of $P(K_\mu, \Sigma_{K_\mu})$. Let $\mu = \frac{1}{2}(\mu_1 + \mu_2)$ for $\mu_1, \mu_2 \in P(X, \Sigma)$. If $E \in \Sigma$ satisfies $\mu(E) = 0$, then $0 = \mu(E) \geq \frac{1}{2}\mu_j \geq 0$ implies that $\mu_j = 0$. Thus, $\mu_j \ll \mu$. If we show that the support of each μ_j is contained in the support of μ, then we conclude that $\mu_1 = \mu_2 = \mu$.

Thus, it remains to prove that if $\omega, \mu \in P(X, \Sigma)$ is such that $\omega \ll \mu$, then $K_\omega \subset K_\mu$. To this end, let $U = (X \setminus K_\mu) \cap (X \setminus K_\omega)$, which is open, and let $K = (X \setminus U) \cap K_\omega$, which is closed. Thus,

$$\omega(X \setminus K) = \omega(U \cup (X \setminus K_\omega)) \leq \omega(U) + \omega(X \setminus K_\omega) = \omega(U).$$

Now since $U \subset X \setminus K_\mu$, we have $\mu(U) \leq \mu(X \setminus K_\mu) = 0$. Thus, $\omega \ll \mu$ implies that $\omega(U) = 0$ and so $\omega(X \setminus K) = 0$. Hence, $K \subset K_\omega$ and $\omega(X \setminus K) = 0$ which implies that $K = K_\omega$ by definition of support and by the above arguments. Hence, $K_\omega = K = (X \setminus U) \cap K_\omega = (K_\mu \cup K_\omega) \cap K_\omega$ implies that $K_\omega \subset K_\mu$.   □

The following theorem is a cornerstone of probability theory.

Theorem 7.22. *The following statements are equivalent for* $\mu \in P(X, \Sigma)$:

1. μ *is an extreme point of* $P(X, \Sigma)$;
2. $\mu = \delta_{\{x_0\}}$ *for some* $x_0 \in X$.

Proof. Assume that μ is an extreme point of $P(X, \Sigma)$. By Lemma 7.21, we may replace X with the support of μ, and so we assume without loss of generality that $X = K_\mu$. Suppose, contrary to what we aim to prove, that the support X of μ contains at least two points, x and y. Because X is a normal topological space (Proposition 2.34), there are open subsets U_x and V_y containing x and y, respectively, and such that $U_x \cap U_y = \emptyset$. Because X is the support of μ, both U_x and V_y have positive measure. Therefore, if $f = \chi_{U_x}$ and $g = \chi_{U_y}$, then $\dot{f}$ and $\dot{g}$ are linearly independent elements of $L^1(X, \Sigma, \mu)$. Moreover, $\dot{f}$ is clearly linearly independent of $\dot{1}$, where $1 = \chi_X$. Hence, by Exercise 6.59, there is a linear functional φ on $L^1(X, \Sigma, \mu)$ such that $\varphi(\dot{f}) = 1$ and $\varphi(\dot{1}) = 0$.

Theorem 6.45 asserts that linear functionals on $L^1(X, \Sigma, \mu)$ are determined by elements of $L^\infty(X, \Sigma, \mu)$. Hence, there exists $\phi \in \mathscr{L}^\infty(X, \Sigma, \mu)$ such that $\|\dot{\phi}\| = 1$ in $L^\infty(X, \Sigma, \mu)$ and $\int_X \phi \, d\mu = 0$. Define $\tilde{\mu} : \Sigma \to \mathbb{R}$ by $\tilde{\mu}(E) = \int_E \phi \, d\mu$. Let $\mu_1 = \mu + \tilde{\mu}$ and $\mu_2 = \mu - \tilde{\mu}$. Note that

$$\mu_1(E) = \int_E d(\mu + \tilde{\mu}) = \int_E d\mu + \int_E \phi \, d\mu = \int_E (1 + \phi) \, d\mu.$$

Since $1 + \phi$ is nonnegative for almost all $x \in X$, this final integral above is nonnegative. Likewise $\mu_2(E)$ is nonnegative. Further,

$$\mu_1(X) = \int_X d(\mu + \tilde{\mu}) = \mu(X) + \int_X \phi \, d\mu = \mu(X) + 0 = 1.$$

Hence, $\mu_1, \mu_2 \in P(X, \Sigma)$ and $\mu = \frac{1}{2}\mu_1 + \frac{1}{2}\mu_2$. Because $\mu_1 \neq \mu$ (as $\dot{\phi} \neq 0$), the measure μ is not an extreme point of $P(X, \Sigma)$. This contradiction implies that X must be a singleton set, which is to say that μ is a point-mass measure.

The proof that point-mass measures $\delta_{\{x_0\}}$ are extremal is left as an exercise (Exercise 7.36).   □

7.5 Integral Representations of Compact Convex Sets

The Kreĭn-Milman Theorem asserts that every element x_0 of a compact convex set K in a Banach space V is a limit point of the set of all convex combinations of the extreme points of K. However, suppose that x_0 is already a convex combination of extreme points of K, say $x_0 = \sum_{j=1}^{m} \tau_j x_j$, where $x_1, \ldots, x_m \in \text{ext}\, K$ and where $\tau_1, \ldots, \tau_m \in [0, 1]$ satisfy $\sum_j \tau_j = 1$. Then, for every linear functional $\varphi : V \to \mathbb{C}$,

$$\varphi(x_0) = \sum_{j=1}^{m} \tau_j \varphi(x_j).$$

That is,

$$\varphi(x_0) = \int_K \varphi \, d\mu, \qquad (7.7)$$

for every $\varphi \in V^*$, where μ is the probability measure $\mu = \sum_{j=1}^{m} \tau_j \delta_{\{x_j\}}$ on the Borel sets of K. Observe that not only is μ a measure on the Borel sets of K, but that it is in fact supported on the Borel set $E = \{x_1, \ldots, x_m\}$ (in the sense that $\mu(K \setminus E) = 0$.

If it were possible, for each $x_0 \in K$, that the integral equation (7.7) held for some regular Borel probability measure μ supported on the extreme points of K, then this would represent a sharpening of the Kreĭn-Milman Theorem. The goal of this section is to prove such a result (Theorem 7.25), due to Choquet, in the case where the topology of K is metrisable.

Definition 7.23. If C is a convex set, then a function $h : C \to \mathbb{C}$ is an *affine function* if

$$h(\lambda x + (1 - \lambda)y) = \lambda h(x) + (1 - \lambda)h(y),$$

for every $x, y \in C$ and $\lambda \in [0, 1]$.

In cases where the convex set C is a subset of a vector space V, one of the most immediate ways to produce an affine function h on C is to take any linear map $\varphi : V \to \mathbb{C}$ on V and then consider $h = \varphi|_C$.

Recall that $f : C \to \mathbb{R}$ is convex if $f(\lambda x + (1 - \lambda)y) \leq \lambda f(x) + (1 - \lambda)f(y)$ for all $x, y \in C$ and $\lambda \in [0, 1]$, and that $f : C \to \mathbb{R}$ is concave if the function $-f$ is convex. Thus, every affine function h on a convex set C is both convex and concave.

Lemma 7.24. *If $f : K \to \mathbb{R}$ is a function on a compact, convex topological space K such that f is bounded above by $q \in \mathbb{R}$, then the function $\bar{f}$ defined by*

$$\bar{f}(x) = \inf\{h(x) \,|\, h \in \text{Aff}\, K, f \leq h\}, \; x \in X,$$

is concave, upper-semicontinuous, and bounded above by q. Furthermore, if f itself is concave and upper-semicontinuous, then $\bar{f} = f$.

Proof. By hypothesis, there is a real number $q \in \mathbb{R}$ with $f(x) \leq q$ for every $x \in K$. Therefore, if $\mathscr{F} = \{h \in \mathrm{Aff}\,K \,|\, f \leq h\}$, then the constant function $r \in \mathscr{F}$ and, hence, $\bar{f}(x) \leq r$ for all $x \in K$, thereby establishing that $\bar{f}$ is bounded above by r.

To prove that $\bar{f}$ is concave, select $x, y \in K$ and $\lambda \in [0, 1]$. Thus,

$$\bar{f}(\lambda x + (1-\lambda)y) = \inf\{h(\lambda x + (1-\lambda)y) \,|\, h \in \mathscr{F}\}$$

$$= \inf\{\lambda h(x) + (1-\lambda)h(y) \,|\, h \in \mathscr{F}\}$$

$$\geq \lambda \inf\{h(x) \,|\, h \in \mathscr{F}\} + (1-\lambda)\inf\{h(y) \,|\, h \in \mathscr{F}\}$$

$$= \lambda \bar{f}(x) + (1-\lambda)\bar{f}(y).$$

Thus, $\bar{f}$ is a concave function.

Now select $\alpha \in \mathbb{R}$ and consider $\bar{f}^{-1}((-\infty, \alpha))$. If $x \in \bar{f}^{-1}((-\infty, \alpha))$, then $\bar{f}(x) < \alpha$ implies that there exists $h \in \mathscr{F}$ such that $\bar{f}(x) \leq h(x) < \alpha$, which implies that $x \in h^{-1}((-\infty, \alpha))$. If, on the other hand, $x \in h^{-1}((-\infty, \alpha))$ for some $h \in \mathscr{F}$, then $x \in \bar{f}^{-1}((-\infty, \alpha))$. Hence,

$$\bar{f}^{-1}((-\infty, \alpha)) = \bigcup_{h \in \mathscr{F}} h^{-1}((-\infty, \alpha)),$$

which is an open set since each $h \in \mathscr{F}$ is continuous.

Now assume that $f : K \to \mathbb{R}$ is upper-semicontinuous and concave. Because f is upper-semicontinuous, the graph $G(f) = \{(x, f(x)) \in V \times \mathbb{R} \,|\, x \in K\}$ of f is a closed set. Further, the concavity of f implies that $G(f)$ is convex.

Suppose, contrary to what we aim to prove, that $\bar{f} \neq f$. Thus, $f(x_1) < \bar{f}(x_1)$ for at least one $x_1 \in K$, which implies that $(x_1, \bar{f}(x_1))$ is separated from the closed convex set $G(f)$. Therefore, by the Separation Theorem (Theorem 7.7), there is a (real) linear functional $\varphi : V \times \mathbb{R} \to \mathbb{R}$ and a $\delta \in \mathbb{R}$ such that $\varphi((x, f(x))) < \delta < \varphi((x_1, \bar{f}(x_1)))$ for every $x \in K$. In particular, using that $V \times \mathbb{R}$ is a vector space,

$$0 < \varphi[(x_1, \bar{f}(x_1)) - (x_1, f(x_1))] = \varphi((0, f(x_1) - \bar{f}(x_1)))$$

$$= (f(x_1) - \bar{f}(x_1))\varphi((0, 1)),$$

and therefore $\varphi((0, 1)) > 0$.

For each $x \in K$, let $\ell_x : \mathbb{R} \to \mathbb{R}$ denote the function given by $\ell_x(s) = \varphi((x, s))$. If $s < s'$, then $\ell_x(s') - \ell_x(s) = \varphi((0, s' - s)) = (s' - s)\varphi((0, 1))$, and so ℓ_x is a strictly increasing function with $\lim_{s \to \infty} \ell_x(s) = \infty$. Thus, because $\ell_x(f(x)) < \delta$, there is a unique $r \in \mathbb{R}$ for which $\ell_x(r) = \delta$. Therefore, let $h : K \to \mathbb{R}$ denote the function $h(x) = r$, where $r \in \mathbb{R}$ is the unique real number such that $\varphi((x, r)) = \delta$. Note that the continuity of φ implies that h is continuous. Further, if $x, y \in K$ and $\lambda \in [0, 1]$, and if $r = h(x)$ and $s = h(y)$, then

$$\delta = \lambda\varphi\left((x,r)\right) + (1-\lambda)\varphi\left((y,s)\right)$$

$$= \varphi\left((\lambda x, \lambda r) + ((1-\lambda)y, (1-\lambda)s)\right)$$

$$= \varphi\left((\lambda x + (1-\lambda)y, \lambda r + (1-\lambda)s)\right).$$

Hence, $h(\lambda x + (1-\lambda)y) = \lambda h(x) + (1-\lambda)h(y)$, which shows that h is an affine function on K, and so $h \in \mathrm{Aff}\,K$. Because, for every $x \in K$, ℓ_x is strictly increasing and $\ell_x(f(x)) = \varphi((x,f(x))) < \delta$, we have that $f(x) < h(x)$ for all $x \in K$. Hence, $\overline{f}(x) \le h(x)$ for each $x \in K$. However, $\varphi((x_1, h(x_1))) = \delta$ and $\varphi((x_1, \overline{f}(x_1))) > \delta$—that is, $\ell_{x_1}(h_1(x_1)) < \ell_{x_1}(\overline{f}(x_1))$—implies that $h_1(x_1) < \overline{f}(x_1)$, by the strict monotonicity of ℓ_{x_1}. But this contradicts $\overline{f} \le h$. Therefore, this contradiction demonstrates that it must be that $\overline{f} = f$. $\square$

Theorem 7.25 (Choquet). *Assume that K is a convex subset of a real topological vector space V. If K is compact, Hausdorff, and second countable, then the set $\mathrm{ext}\,K$ of extreme points of K is a G_δ-set and for every $x_0 \in K$ there exists a regular Borel probability measure μ supported on $\mathrm{ext}\,K$ such that*

$$\varphi(x_0) = \int_{\mathrm{ext}\,K} \varphi\,d\mu$$

for every $\varphi \in V^*$.

Proof. The topological conditions on K imply, by Theorem 2.48, that there is a metric d on K that induces the topology of K. For each $n \in \mathbb{N}$, consider the subset

$$K_n = \left\{ x \in K \,|\, x = \frac{1}{2}(y+z),\ y,z \in K,\ d(y,z) \ge \frac{1}{n} \right\}.$$

Each K_n is a closed subset of K and $x \notin \mathrm{ext}\,K$ if and only if there is some $n \in \mathbb{N}$ for which $x \in K_n$. Hence,

$$\mathrm{ext}\,K = \bigcap_{n \in \mathbb{N}} K_n^c.$$

That is, $\mathrm{ext}\,K$ is a G_δ set; in particular, $\mathrm{ext}\,K$ is Borel measurable.

The topological conditions on K also imply that $C(K)$ is a separable Banach space (Theorem 5.57); thus, $\mathrm{Aff}\,K$ is also separable and, therefore, there is a countable subset $\{h_n\}_{n\in\mathbb{N}} \subset \mathrm{Aff}\,K$ such that $\{h_n\}_{n\in\mathbb{N}}$ is dense in the unit ball of $\mathrm{Aff}\,K$. Because $\|h_n\| \le 1$ for all n, the series $\sum_{n=1}^{\infty} 2^{-n}(h_n)^2$ converges uniformly to some $f \in C(K)$. As the function $\psi(t) = t^2$ has strictly positive second derivative on $\mathbb{R}$, the function ψ is strictly convex (Proposition 4.46); therefore, because h_n is affine, the function $\psi \circ h_n = (h_n)^2$ is strictly convex. Thus, f is a convex function.

In fact, f itself is strictly convex. Indeed, if $x, y \in K$ are distinct, then there is an $n \in \mathbb{N}$ such that $h_n(x) \neq h_n(y)$ (because $\mathrm{Aff}\,K$ separates the points of K). Hence, for all $\lambda \in (0, 1)$,

$$\psi \circ h_n(\lambda x + (1 - \lambda)y) < \lambda h_n(x)^2 + (1 - \lambda)h_n(y)^2,$$

which implies that $f(\lambda x + (1 - \lambda)y) < \lambda f(x) + (1 - \lambda)f(y)$.

Fix $x_0 \in K$ and define $p : C_\mathbb{R}(K) \to \mathbb{R}$ by $p(g) = \overline{g}(x_0)$, where $\overline{g}$ is the upper envelope of g. Observe that $p(g_1 + g_2) \leq p(g_1) + p(g_2)$ and $p(rg) = rp(g)$, for all $g_1, g_2, g \in C_\mathbb{R}(K)$ and all $r \in [0, \infty)$. Hence p is a sublinear functional on $C_\mathbb{R}(K)$.

Assume that $h \in \mathrm{Aff}\,K$. Because h is affine, continuous, and bounded, Lemma 7.24 asserts that $\overline{h} = h$. And if $r \geq 0$, then $\overline{rf} = r\overline{f}$ and, therefore, $\overline{h + rf} = h + r\overline{f}$. On the other hand, if $r < 0$, then $h + rf$ is concave and so $\overline{h + rf} = h + rf$, again by Lemma 7.24. But since $f \leq \overline{f}$ and $r < 0$, we have that $\overline{h + rf} = h + rf \geq h + r\overline{f}$. Hence, if $L = \{h + rf \mid h \in \mathrm{Aff}\,K, r \in \mathbb{R}\}$, then L is a linear submanifold in $C_\mathbb{R}(K)$ and the function $\omega : L \to \mathbb{R}$ defined by $\omega(h + rf) = h(x_0) + r\overline{f}(x_0)$ is linear and satisfies $\omega(h + rf) \leq p(h + rf)$ for every $h + rf \in L$. Therefore, by the Hahn–Banach Extension Theorem, there is a linear map $\Omega : C_\mathbb{R}(K) \to \mathbb{R}$ such that $\Omega_{|L} = \omega$ and $\Omega(g) \leq p(g) = \overline{g}(x_0)$ for every $g \in C_\mathbb{R}(K)$.

Suppose that $g \in C_\mathbb{R}(K)$ is such that $g \geq 0$ and let $q = -g$. Then $q \leq 0$ implies that $\overline{q} \leq 0$ and so $\Omega(q) \leq \overline{q}(x_0) \leq 0$. That is, $\Omega(g) \geq 0$, and so Ω is continuous, by Proposition 6.49. The Riesz Representation Theorem (Theorem 6.51) yields a regular Borel measure μ such that $\Omega(g) = \int_K g \, d\mu$ for every $g \in C_\mathbb{R}(K)$. The constant function $x \mapsto 1$ is affine and so $\Omega(1) = 1 = \mu(K)$, which shows that μ is a probability measure.

Integration preserves order; thus, if $h \in \mathrm{Aff}\,K$ satisfies $f \leq h$, then $\overline{f} \leq h$ and

$$\overline{f}(x_0) = \omega(f) = \int_K f \, d\mu \leq \int_K \overline{f} \, d\mu \leq \int_K h \, d\mu = \omega(h) = \overline{h}(x_0) = h(x_0).$$

Hence

$$\int_K \overline{f} \, d\mu \leq \inf\{h(x_0) \mid h \in \mathrm{Aff}\,K, f \leq h\} = \overline{f}(x_0),$$

which proves that $\int_K \overline{f} \, d\mu = \int_K f \, d\mu$.

Let $E = \{x \in K \mid \overline{f}(x) = f(x)\}$. Because $f \leq \overline{f}$ and $\int_K \overline{f} \, d\mu = \int_K f \, d\mu$, we have that $\int_{E^c} (\overline{f} - f) \, d\mu = 0$. Hence, as $\overline{f} - f$ is strictly positive on E^c, $\mu(E^c) = 0$. Therefore, μ is supported on E.

If $x \in K \setminus \mathrm{ext}\,K$, then there are distinct $y, z \in K$ such that $x = \frac{1}{2}(y + z)$. Because f is strictly convex,

$$f(x) < \frac{1}{2}(f(x) + f(y)) \leq \frac{1}{2}(\overline{f}(x) + \overline{f}(y)) \leq \overline{f}\left(\frac{1}{2}(y + z)\right) = \overline{f}(x),$$

where the final inequality above holds because $\overline{f}$ is concave. Hence, $E \subseteq \mathrm{ext}\,K$, which proves that μ is supported on the Borel set $\mathrm{ext}\,K$.

Lastly, if $\varphi \in V^*$, then $\varphi_{|K}$ is a continuous affine function $K \to \mathbb{R}$, and so

$$\varphi(x_0) = \overline{\varphi_{|K}}(x_0) = \omega(\varphi_{|K}) = \int_{\text{ext}\,K} \varphi \, d\mu,$$

which completes the proof. $\square$

7.6   The Range of Non-Atomic Measures

If one considers the counting measure μ on a finite set X, then the range $R_\mu = \{\mu(E) \,|\, E \in \Sigma\}$ of μ is also a finite set. On the other hand, if m is Lebesgue measure on $[a, b]$, then the range of m is a continuum, namely the closed interval $[0, b - a]$. From the point of view of convexity, R_m is convex, while R_ν is not convex (assuming X has at least two elements). The explanation for the convexity of R_m goes beyond the particulars of Lebesgue measure—rather, it is the property of Lebesgue measure being non-atomic that is at play here.

Recall from Definition 3.55 that a measure μ on a measurable space (X, Σ) is non-atomic if there are no atoms for μ; that is, there are no sets $E \in \Sigma$ with the property that $\mu(E) > 0$ and one of $\mu(E \cap F)$ or $\mu(E \cap F^c)$ is 0 for every $F \in \Sigma$.

Theorem 7.26 (Lyapunov). *If μ is a finite non-atomic measure on a measurable space (X, Σ), then the range of μ is the closed interval $[0, \mu(X)]$.*

Proof. The Banach space $L^\infty(X, \Sigma, \mu)$ is the dual of $L^1(X, \Sigma, \mu)$ (Theorem 6.45), and so $L^\infty(X, \Sigma, \mu)$ carries a weak*-topology. If

$$\mathscr{I} = \{\psi \in \mathscr{L}^\infty(X, \Sigma, \mu) \,|\, \text{ess-ran}\, \psi \subseteq [0, 1]\} \text{ and } I = \{\dot{\psi} \in L^\infty(X, \Sigma, \mu) \,|\, \psi \in \mathscr{I}\},$$

then I is a convex subset of the closed unit ball of $L^\infty(X, \Sigma, \mu)$, which is compact with respect to the weak*-topology (Theorem 6.33). Therefore, the weak*-compactness of I will follow from showing that I is weak*-closed. To this end, suppose that $\psi \in \mathscr{L}^\infty(X, \Sigma, \mu)$ is such that $\dot{\psi} \notin I$. Thus, there is a $\lambda \in \text{ess-ran}\, \psi$ such that $\lambda \notin [0, 1]$. Select an open set $V \subset \mathbb{C}$ that contains λ but does not intersect $[0, 1]$; by definition of essential range, the measurable set $E = \psi^{-1}(V)$ has positive measure. The function $g = \frac{1}{\mu(X)} \chi_E \in \mathscr{L}^1(X, \Sigma, \mu)$ induces a weak*-open neighbourhood $U \subset L^\infty(X, \Sigma, \mu)$ of $\dot{\psi}$ given by

$$U = \left\{ \dot{\varphi} \in L^\infty(X, \Sigma, \mu) \,|\, \int_X \varphi g \, d\mu = \frac{1}{\mu(X)} \int_E \varphi \, d\mu \in V \right\}.$$

Note that ess-ran $\varphi \nsubseteq [0, 1]$ for every $\dot{\varphi} \in U$; on the other hand, if $\vartheta \in \mathscr{I}$, then $\int_E \vartheta \, d\mu \in [0, \mu(X)]$ and so $\dot{\vartheta} \notin U$. Hence, $U \cap I = \emptyset$, which proves that I^c is a weak*-open set.

Next, consider the map $\mathscr{E} : I \to \mathbb{R}$ defined by $\mathscr{E}(\dot{\psi}) = \int_X \psi \, d\mu$. Observe that $\mathscr{E}$ is an affine function and that it is continuous with respect to the weak*-topology on I. Hence, the range K of $\mathscr{E}$ is a compact convex subset of $\mathbb{R}$. Indeed, because $\int_X \psi \, d\mu \in [0, \mu(X)]$ for all $\psi \in \mathscr{I}$, and because 0 and $\mu(X)$ are plainly elements of the set K, we deduce from the convexity of K that $K = \mathscr{E}(I) = [0, \mu(X)]$.

Select $a \in K$ and consider the set $I_a = \{\dot{\psi} \in I \mid \mathscr{E}(\dot{\psi}) = a\}$. Because $I_a = I \cap \mathscr{E}^{-1}(\{a\})$ is the intersection of two weak*-closed sets, the set I_a is a weak*-closed subset of the closed unit ball of $L^\infty(X, \Sigma, \mu)$; hence, I_a is weak*-compact. The set I_a is also convex. Therefore, by the Kreĭn-Milman Theorem, I_a has an extreme point $\dot{\varphi}$. Assume, for the moment, that there is a measurable set $E \subseteq X$ for which $\dot{\varphi} = \dot{\chi}_E$; thus, $a = \mathscr{E}(\dot{\varphi}) = \int_X \chi_E \, d\mu = \int_E d\mu = \mu(E)$. The choice of $a \in K = [0, \mu(X)]$ being arbitrary would yield $[0, \mu(X)] \subseteq \{\mu(E) \mid E \in \Sigma\} \subseteq [0, \mu(X)]$, thereby completing the proof.

Therefore, it remains to show that there is a measurable set $E \subseteq X$ for which $\dot{\varphi} = \dot{\chi}_E$. Assume, on the contrary, that $\dot{\varphi} \neq \dot{\chi}_E$ for every measurable set E. Thus, the essential range of φ contains at least one point λ different from 0 and 1. Therefore, there exists $\varepsilon > 0$ such that $(\varepsilon, 1 - \varepsilon)$ is an open neighbourhood in $\mathbb{R}$ of λ. If V is any open set in $\mathbb{C}$ for which $V \cap \mathbb{R} = (\varepsilon, 1 - \varepsilon)$, then, by definition of essential range, the measurable set $E = \varphi^{-1}(V)$ has positive measure. Because μ is a non-atomic measure, there is a measurable proper subset F of E such that both F and $E \setminus F$ have positive measure. Likewise, there are measurable proper subsets $G_1 \subset F$ and $G_2 \subset (E \setminus F)$ such that $0 < \mu(G_1) < \mu(F)$ and $0 < \mu(G_2) < \mu(E \setminus F)$. In the 1-dimensional real vector space $\mathbb{R}$ any two real numbers are linearly dependent. Thus, there are $\alpha, \beta \in \mathbb{R}$ not both zero such that

$$\alpha \left(\mu(G_1) - \mu(F)\right) + \beta \left(\mu(E \setminus F) - \mu(G_2)\right) = 0.$$

By multiplying the equation above by an appropriate constant, we may assume that α and β have been scaled so as to satisfy $|\alpha| < \varepsilon$ and $|\beta| < \varepsilon$. Consider now the measurable function

$$\vartheta = \alpha \left(\chi_{G_1} - \chi_F\right) + \beta \left(\chi_{E \setminus F} - \chi_{G_2}\right),$$

which has the properties that $\int_X \vartheta \, d\mu = 0$ and $\varphi \pm \vartheta \in \mathscr{I}$. If $\varphi_1 = \varphi + \vartheta$ and $\varphi_2 = \varphi - \vartheta$, then $\int_X \varphi_1 \, d\mu = \int_X \varphi_2 \, d\mu = a$, which is to say that $\dot{\varphi}_1, \dot{\varphi}_2 \in I_a$. However, $\dot{\varphi} \neq \dot{\varphi}_j$ for each j and $\dot{\varphi} = \frac{1}{2}(\dot{\varphi}_1 + \dot{\varphi}_2)$ contradict the fact that $\dot{\varphi}$ is an extreme point of I_a. Therefore, it must be that $\dot{\varphi} = \dot{\chi}_E$ for some $E \in \Sigma$. $\qquad\square$

Problems

7.27. Prove that a subset C of a vector space V is convex if and only if C contains every convex combination of its elements.

7.28. Determine the extreme points of a closed interval $[a, b]$ in $\mathbb{R}$.

7.29. Let V be a topological vector space and assume that $C \subset V$ is an open convex set. If $\mu : V \to \mathbb{R}$ is a continuous linear transformation for which $\mu(v) \neq 0$ for every $v \in C$, then prove that $\mu(C)$ is an open interval of $\mathbb{R}$.

7.30. Prove that the dual space V^* of a normed vector space V is a Hausdorff topological vector space in the weak*-topology.

7.31. A cone in a finite-dimensional normed vector space V is a convex subset $C \subset V$ such that $\tau v \in C$, for every $\tau \in \mathbb{R}^+$ and $v \in C$. Let $C^\dagger = \{\varphi \in V^* \,|\, \varphi(v) \geq 0, \, \forall v \in C\}$.

1. Prove that $C^\dagger$ is a cone in the dual space V^*.
2. Prove that $C^{\dagger\dagger} = C$.

7.32. Let C be a convex set and $v \in C$. Prove that the following statements are equivalent:

1. v is an extreme point of C;
2. if there are $v_1, v_2 \in C$ and $\tau \in (0, 1)$ such that $v = \tau v_1 + (1 - \tau)v_2$, then $v_1 = v_2 = v$.

7.33. Let $C = [0, 1] \times [0, 1] \times [0, 1] \subset \mathbb{R}^3$.

1. Determine all the faces of C.
2. Of the faces found, identify those that correspond to extreme points of C.

7.34. Let C be a convex set in a vector space V. Show that if $F_1 \subseteq C$ is a face of C and $F_2 \subseteq F_1$ is a face of F_1, then F_2 is a face of C.

7.35. Let $\{e_n\}_{n \in \mathbb{N}}$ denote the canonical coordinate vectors of $\ell^1(\mathbb{N})$. Prove that the extreme points of the closed unit ball of $\ell^1(\mathbb{N})$ are precisely the vectors of the form $e^{i\theta} e_n$, for some $\theta \in \mathbb{R}$ and $n \in \mathbb{N}$.

7.36. Assume that X is a compact Hausdorff space and Σ is the σ-algebra of Borel sets of X, and consider the set

$$P(X, \Sigma) = \{\mu \in M(X, \Sigma) \,|\, \mu(E) \geq 0, \, \forall E \in \Sigma, \, \mu(X) = 1\},$$

which is a subset of the closed unit sphere of the Banach space $M(X, \Sigma)$ of regular complex measures on (X, Σ).

1. Prove that $P(X, \Sigma)$ is weak*-compact.
2. Prove that if $x_0 \in X$, then the point-mass measure $\delta_{\{x_0\}}$ is an extreme point of $P(X, \Sigma)$.

7.37. Prove that the vector space operations on V^* are continuous in the weak* topology, for every normed vector space V.

7.38. Suppose that V is a locally convex topological vector space and that $K \subset V$ is a convex, compact Hausdorff space. Prove that if $x_0 \in K$ is an extreme point of K and μ is a regular Borel probability measure such that $\varphi(x_0) = \int_K \varphi \, d\mu$ for every $\varphi \in V^*$, then $\mu = \delta_{\{x_0\}}$, a point-mass measure concentrated on $\{x_0\}$.

7.39. If C is a convex set in a Banach space V, then an element $v_0 \in C$ is *exposed* if there is a $\varphi \in V^*$ such that $\Re\varphi(v) < \Re\varphi(v_0)$, for all $v \in C \setminus \{v_0\}$. Furthermore, an exposed point v_0 of C is *strongly exposed* if, for any sequence $\{v_k\}_{k \in \mathbb{N}} \subset C$, the sequence $\{\Re\varphi(v_k)\}_{k \in \mathbb{N}}$ converges to $\Re\varphi(v_0)$ only if $\{v_k\}_{k \in \mathbb{N}}$ converges to v_0.

1. Prove that every exposed point of C is an extreme point of C.
2. Determine the strongly exposed points of $\ell^1(\mathbb{N})$.
3. Determine the strongly exposed points of $\ell^p(\mathbb{N})$, for $p > 1$.

Part IV
Operator Theory

Chapter 8
Banach Space Operators

If Banach spaces are viewed as the metric analogue of the notion of vector space, then the concept of an operator is correspondingly viewed as the continuous analogue of the notion of a linear transformation of a vector space.

Recall that one can compose linear transformation $T : V \to W$ and $S : W \to Z$ to produce linear transformation $ST : V \to W$ defined by $ST(v) = S(Tv)$, for $v \in V$. Similarly, one has the obvious notions of sum $T_1 + T_2$ and scalar multiplication αT (where $\alpha \in \mathbb{C}$) for linear transformations $T, T_1, T_2 : V \to W$. Of particular importance are cases in which $W = V$, for in these cases the set of all linear transformations $V \to V$ has the structure of an associative algebra. In considering only those linear transformations $T : V \to V$ of a Banach space V that are continuous, it turns out that the resulting set is also an associative algebra, known as a Banach algebra.

8.1 Examples of Operators

To this point we have encountered operators in the form of linear functionals and as surjective isometries between certain Banach spaces. This section is a brief sampling of operators of a more general type.

8.1.1 Matrices

The most accessible and most familiar examples of operators are to be found with matrices. An $m \times n$ matrix $T = [t_{ij}]_{1 \leq i \leq m, 1 \leq j \leq n}$, where each $t_{ij} \in \mathbb{C}$, is a linear transformation $\mathbb{C}^n \to \mathbb{C}^m$. The entries of the i-th column of T represent the entries in

© Springer International Publishing Switzerland 2016
D. Farenick, *Fundamentals of Functional Analysis*, Universitext,
DOI 10.1007/978-3-319-45633-1_8

the vector $Te_i \in \mathbb{C}^m$, where $e_1,\ldots,e_n \in \mathbb{C}^n$ are the canonical basis vectors (canonical in the sense that e_i is a column vector with 1 in the i-th entry and 0 in every other entry). If $\mathbb{C}^n$ and $\mathbb{C}^m$ are endowed with norms, then T is an operator by Proposition 6.3.

Another convenient way to analyse matrices is to consider $\mathbb{C}^n$ as a space of functions. Specifically, let $X = \{1,2,\ldots,n\}$ and let Σ be the power set of X. Consider $L^\infty(X,\Sigma,\mu)$, where μ is counting measure. Thus, each $f \in L^\infty(X,\Sigma,\mu)$ is a function $f : X \to \mathbb{C}$ in which

$$\|f\| = \max\{f(k) \mid 1 \le k \le n\}.$$

Now assume $Y = \{1,\ldots,m\}$, Ω is the power set of Y, and ν is counting measure on Y. As vector spaces, there are the obvious isomorphisms

$$\mathbb{C}^n \cong L^\infty(X,\Sigma,\mu) \quad \text{and} \quad \mathbb{C}^m \cong L^\infty(Y,\Omega,\nu).$$

Hence, an $m \times n$ complex matrix T is an operator

$$T : L^\infty(X,\Sigma,\mu) \to L^\infty(Y,\Omega,\nu).$$

There is nothing special about the choice of the L^∞-norm, and one could just as easily consider a matrix T as an operator

$$T : L^p(X,\Sigma,\mu) \to L^{p'}(Y,\Omega,\nu)$$

for any choice of $p,p' \in [1,\infty)$.

8.1.2   Operators on Finite-Dimensional Banach Spaces

If V and W are finite-dimensional vector spaces, then by choosing bases for V and W each operator T will have a matrix representation with respect to these bases. Specifically, if $\mathscr{B}_V = \{v_1,\ldots,v_n\}$ and $\mathscr{B}_W = \{w_1,\ldots,w_m\}$ are bases for V and W, respectively, then, for each $i = 1,\ldots,n$,

$$Tv_i = \sum_{j=1}^{n} t_{ij}w_j$$

for some unique choice of $t_{i1},\ldots,t_{im} \in \mathbb{C}$. The $m \times n$ matrix $\tilde{T} = [t_{ij}]_{1 \le i \le m, 1 \le j \le n}$ is a representation of T with respect to the bases $\mathscr{B}_V$ and $\mathscr{B}_W$.

8.1.3 Finite-Rank Operators

An operator $T \in B(V)$ is said to have *finite rank* if the range of T has finite dimension. In such cases, the rank of T is defined to be the dimension of the range of T. If $\psi \in V^*$ and $w \in V$ are nonzero, then the operator $F_{\psi,w}$ on V defined by $F_{\psi,w}(v) = \psi(w)v$ is of rank 1. Conversely, if F is a rank-1 operator, then there exist nonzero $\psi \in V^*$ and $w \in V$ such that $F = F_{\psi,w}$. Furthermore, if T has finite rank $n \in \mathbb{N}$, then there are rank-1 operators $F_1, \ldots, F_n \in B(V)$ such that $T = \sum_{j=1}^{n} F_j$. The proofs of these facts are left to the reader (Exercise 8.60).

8.1.4 Integral Operators

Let $p \in [1, \infty)$ and consider Banach spaces $L^p(X, \Sigma, \mu)$ and $L^p(Y, \Omega, \nu)$ for some finite measure spaces (X, Σ, μ) and (Y, Ω, ν). Assume that $\kappa : X \times Y \to \mathbb{C}$ is a bounded measurable function and that $M = \sup\{|\kappa(s,t)| \,|\, (s,t) \in X \times Y\}$. For each $s \in Y$, let $\kappa_s : X \to \mathbb{C}$ be given by $\kappa_s(t) = \kappa(s,t)$. Thus, for each $f \in \mathscr{L}^p(X, \Sigma, \mu)$ we obtain a function $(\mathscr{K}f) : Y \to \mathbb{C}$ defined by

$$(\mathscr{K}f)(s) = \int_X \kappa_s f \, d\mu.$$

Because $|\kappa_s(t)| \leq M$ for every $(s,t) \in X \times Y$, we deduce that

$$\int_X |\mathscr{K}f|^p \, d\nu = \int_Y \int_X |\kappa_s|^p |f|^p \, d\mu \, d\nu \leq M^p \nu(Y)^p \int_X |f|^p \, d\mu.$$

Hence, the function $(\mathscr{K}f)$ is p-integrable. Moreover, the map $f \mapsto \mathscr{K}f$ is plainly linear, and so we obtain an operator $K : L^p(X, \Sigma, \mu) \to L^p(Y, \Omega, \nu)$ defined by $K(\dot{f}) = \dot{g}$, where $g = \mathscr{K}f \in \mathscr{L}^p(Y, \Omega, \nu)$, of norm $\|K\| \leq M$.

8.1.5 Multiplication Operators on $C(X)$

Let X be a compact Hausdorff space and fix a function $\psi \in C(X)$. The map $M_\psi : C(X) \to C(X)$ defined by $M_\psi f = \psi f$ is a bounded operator and it is clear that $\|M_\psi\| \leq \|\psi\| = \max\{|\psi(x)| \,|\, x \in X\}$. By taking $f(x) = 1$ for all $x \in X$, we see that $\|M_\psi f\| = \|\psi\| \|f\|$. Hence, $\|M_\psi\| = \|\psi\|$.

8.1.6 Integral Operators on $C([0, 1])$

If $\kappa : [0, 1] \times [0, 1] \to \mathbb{C}$ is continuous, then the linear transformation $K : C([0, 1]) \to C([0, 1])$ defined by

$$Kf(s) = \int_0^1 \kappa(s, t)f(t)\,dt,$$

for $f \in C([0, 1])$ is a linear transformation in which

$$\|Kf\| = \max_{s \in [0,1]} \left| \int_0^1 \kappa(s, t)f(t)\,dt \right| \leq \max_{s,t \in [0,1]} |\kappa(s, t)| \int_0^1 |f(t)|\,dt \leq \|\kappa\|\,\|f\|,$$

where $\|\kappa\|$ is the norm of κ as an element of $C([0, 1] \times [0, 1])$. Operators of the form K are called *integral operators*.

Another type of integral operator occurs with functions $\kappa : [0, 1] \times [0, 1] \to \mathbb{C}$ that are continuous on $0 \leq t \leq s \leq 1$. In particular, if $\kappa(s, t) = 1$ for $0 \leq t \leq s \leq 1$ and $\kappa(s, t) = 0$ otherwise, then the corresponding integral operator K is given by

$$Kf(s) = \int_0^s f(t)\,dt,$$

for $f \in C([0, 1])$. This particular operator is known as the *Volterra integral operator* and is frequently denoted by V rather than K.

8.1.7 Multiplication Operators on $L^p(X, \Sigma, \mu)$, $1 \leq p < \infty$

Assume that (X, Σ, μ) μ is a σ-finite measure space. If $\psi : X \to \mathbb{C}$ is an essentially bounded Borel measurable function and if $\dot{f} \in L^p(X, \Sigma, \mu)$, where $p \geq 1$, then

$$\int_X |\psi \cdot f|^p\,d\mu \leq \int_X (\text{ess-sup }\psi)^p\,|f|^p\,d\mu = (\text{ess-sup }\psi)^p \int_X |f|^p\,d\mu < \infty$$

implies that ψ induces a linear transformation $M_\psi : L^p(X, \Sigma, \mu) \to L^p(X, \Sigma, \mu)$ via $M_\psi(\dot{f}) = (\dot{\psi f})$ for all $\dot{f} \in L^p(X, \Sigma, \mu)$. The inequality above shows that M_ψ is bounded and that

$$\|M_\psi\| \leq \|\dot{\psi}\|_\infty = \text{ess-sup }\psi.$$

To show, conversely, that $\|\dot{\psi}\|_\infty \leq \|M_\psi\|$, assume first that ψ is a simple function such that $\dot{\psi} \neq 0$. Thus, if $\alpha \in \mathbb{C}$ satisfies $|\alpha| = \text{ess-sup }\psi$ and if $E = \psi^{-1}(\{\alpha\})$, then $\mu(E) > 0$. Because X is σ-finite, there is a sequence of measurable sets $F_n \in \Sigma$, each with finite measure, such that $X = \bigcup_n F_n$. Hence, there is at least one $n \in \mathbb{N}$ for which $F = E \cap F_n$ is a nonempty set of finite measure. Let $f = \mu(F)^{-1/p}\chi_F$ and note that $\|\dot{f}\| = 1$ and $\|M_\psi \dot{f}\| = |\alpha| = \|\dot{\psi}\|_\infty$. Hence, $\|M_\psi\| = \|\dot{\psi}\|_\infty$ if ψ is a simple function with $\dot{\psi} \neq 0$, and $\|M_\psi\| = \|\dot{\psi}\|_\infty$ is trivially true for $\dot{\psi} = 0$.

Assume now that ψ is an arbitrary essential bounded function. By Proposition 5.52, there is a sequence $\{\varphi\}_{n \in \mathbb{N}}$ of simple functions such that $\lim_n \|\dot{\psi} - \dot{\varphi}_m\|_\infty = 0$. Because $\|M_\psi \dot{f} - M_{\varphi_n} \dot{f}\| = \|M_{\psi - \varphi_n} \dot{f}\| \le \|\dot{\psi} - \dot{\varphi}_n\|_\infty \|\dot{f}\|$ for every $\dot{f} \in L^p(X, \Sigma, \mu)$, we deduce that $\|M_\psi - M_{\varphi_n}\| \le \|\dot{\psi} - \dot{\varphi}_n\|_\infty$. Hence, $\lim_n \|M_\psi - M_{\varphi_n}\| = 0$ implies that $\|M_\psi\| = \lim_n \|M_{\varphi_n}\| = \lim_n \|\dot{\varphi}_n\|_\infty = \|\dot{\psi}\|_\infty$.

8.1.8 Weighted Unilateral Shift Operators

Consider an element $\alpha = (\alpha)_{k \in \mathbb{N}} \in \ell^\infty(\mathbb{N})$ and the linear transformation $S_\alpha : \ell^p(\mathbb{N}) \to \ell^p(\mathbb{N})$, for $p \in [1, \infty]$, defined by

$$S_\alpha v = S_\alpha \left(\begin{bmatrix} v_1 \\ v_2 \\ v_3 \\ \vdots \end{bmatrix} \right) = \begin{bmatrix} 0 \\ \alpha_1 v_1 \\ \alpha_2 v_2 \\ \vdots \end{bmatrix}, \quad v \in \ell^p(\mathbb{N}).$$

The linear transformation S_α is clearly bounded of norm $\|S_\alpha\| = \sup_k |\alpha_k|$, and S_α is called a *weighted unilateral shift operator*.

If $\alpha_k = 1$ for every $k \in \mathbb{N}$, then the resulting weighted unilateral shift operator S_α is denoted simply by S and is called the *unilateral shift operator* on $\ell^p(\mathbb{N})$. Note that S is an isometry, but is not surjective.

8.1.9 Adjoint Operators

Theorem 8.1. *If V and W are Banach spaces, and if $T \in B(V, W)$, then there is a unique operator $T^* : W^* \to V^*$ with the property that*

$$T^* \psi(v) = \psi(Tv), \qquad \forall \psi \in W^*, v \in V. \tag{8.1}$$

Furthermore, if $T_1, T_2, T \in B(V, W)$ and $\alpha_1, \alpha_2 \in \mathbb{C}$, then

1. $\|T^*\| = \|T\|$,
2. $(\alpha_1 T_1 + \alpha_2 T_2)^* = \alpha_1 T_1^* + \alpha_2 T_2^*$, *and*
3. *if $W = V$, then $(T_1 T_2)^* = T_2^* T_1^*$.*

Proof. To prove the first assertion, observe that equation (8.1) above defines a linear transformation of W^* into V^* such that

$$\|T^* \psi\| = \sup_{v \in V, \|v\|=1} |\psi(Tv)| \le \|\psi\| \|T\|.$$

Thus, T^* is a bounded linear transformation of norm $\|T^*\| \le \|T\|$.

Conversely, for every $\varepsilon > 0$ there is a unit vector $v \in V$ such that $\|T\| < \|Tv\| + \varepsilon$. Further, Corollary 6.23 of the Hahn-Banach Extension Theorem shows that there is a $\psi \in W^*$ of norm $\|\psi\| = 1$ such that $\psi(Tv) = \|Tv\|$. Thus,

$$\|T\| < \|Tv\| + \varepsilon = |\psi(Tv)| + \varepsilon = \|T^*\psi(v)\| + \varepsilon \leq \|T^*\| \|\psi\| \|v\| + \varepsilon = \|T^*\| + \varepsilon.$$

Hence, $\|T\| \leq \|T^*\|$.

To prove the uniqueness of T, suppose that $T' \in B(W^*, V^*)$ satisfies equation (8.1) for all $\psi \in W^*$ and $v \in V$. Thus, for a fixed $v \in V$, $0 = \psi(Tv) - \psi(T'v) = \psi(Tv - T'v)$ for all $\psi \in W^*$. By Corollary 6.23, this means that $Tv - T'v = 0$. Because the choice of $v \in V$ is arbitrary, we deduce that $T' = T$.

The proofs of the remaining statements are left as an exercise (Exercise 8.67).

$\square$

Definition 8.2. The operator T^* is called the *adjoint* of T.

The explicit determination of the adjoint of a given operator $T \in B(V, W)$ depends to a certain extent on how well one understands the dual spaces of V and W. For example, if $p, q > 1$ satisfy $p^{-1} + q^{-1} = 1$, then by identifying the dual of ℓ^p as ℓ^q one sees that the adjoint of the weighted unilateral shift operator $S_\alpha : \ell^p(\mathbb{N}) \to \ell^p(\mathbb{N})$ is the operator $S_\alpha^* : \ell^q(\mathbb{N}) \to \ell^q(\mathbb{N})$ defined by

$$(S_\alpha)^*\varphi = S_\alpha^* \left(\begin{bmatrix} \varphi_1 \\ \varphi_2 \\ \varphi_3 \\ \vdots \end{bmatrix} \right) = \begin{bmatrix} \alpha_1\varphi_2 \\ \alpha_2\varphi_3 \\ \vdots \end{bmatrix}, \quad \varphi \in \ell^q(\mathbb{N}). \tag{8.2}$$

Indeed, if $v \in \ell^p(\mathbb{N})$ and $\varphi \in \ell^q(\mathbb{N}) = \ell^p(\mathbb{N})^*$, then

$$\varphi(S_\alpha v) = \sum_{k=1}^{\infty} \alpha_k v_k \varphi_{k+1} = \sum_{k=1}^{\infty} \alpha_k \varphi_{k+1} v_k = \left(S_\alpha^*\varphi \right)(v),$$

which shows that S_α^* has the form given by (8.2).

8.2 Mapping Properties of Operators

Because operators are continuous, the kernel $\ker T = \{v \in V \mid Tv = 0\}$ of an operator $T : V \to W$ is necessarily closed, as $\ker T$ is the pre-image of the closed set $\{0\}$ in W. It is natural, therefore, to ask about the range $\operatorname{ran} T = \{Tv \mid v \in V\}$ of an operator. A simple sufficient condition for an operator to have closed range is that it be bounded below in the sense of the following definition.

Definition 8.3. An operator $T \in B(V, W)$ is *bounded below* if there exists $\delta > 0$ such that

$$\delta \|v\| \leq \|Tv\|, \quad \forall v \in V.$$

The positive real number δ is called a *lower bound* for T.

Proposition 8.4. *If V and W are Banach spaces, then every operator $T : V \to W$ that is bounded below will have closed range.*

Proof. The proof is a variant of the proof of Proposition 6.9 and is, therefore, left as an exercise (Exercise 8.61). □

If one has a collection Λ of operators $T : V \to W$ in which V is a Banach space, it may or may not be true that Λ is a bounded set in the sense that there is a $K > 0$ such that $\|T\| \leq K$ for all $T \in \Lambda$. The following theorem indicates that if Λ fails to be bounded, then there is a dense G_δ set $G \subseteq V$ for which the norms of Tv, for $v \in G$ and $T \in \Lambda$, are arbitrarily large.

Theorem 8.5 (Principle of Uniform Boundedness). *If $\Lambda \subseteq B(V, W)$ is a nonempty set of operators, where V is a Banach space and W is a normed vector space, then exactly one of the following two statements holds:*

1. there is a $K > 0$ such that $\|T\| \leq K$ for every $T \in \Lambda$; or
2. there is a dense G_δ-set $G \subseteq V$ such that

$$\sup_{T \in \Lambda} \|Tv\| = \infty \quad \forall v \in G.$$

Proof. For each $T \in \Lambda$ let $f_T : V \to \mathbb{R}$ be the (nonlinear) continuous function $f_T(v) = \|Tv\|$, for $v \in V$. By continuity, $f_T^{-1}([0, n]) = \{v \in V \mid \|Tv\| \leq n\}$ is a closed set and, therefore,

$$K_n = \bigcap_{T \in \Lambda} f_T^{-1}([0, n])$$

is a closed subset of V for every $n \in \mathbb{N}$. Let $U_n = V \setminus K_n$, which is open. Either every U_n is dense in V or there is at least one $n \in \mathbb{N}$ for which U_n is not dense.

Case #1: U_n is not dense for some n. In this case, fix such an n and choose $v_0 \in K_n = V \setminus U_n$ so that v_0 lies outside the closure of U_n. Thus, there is a $\rho > 0$ with $B_\rho(v_0) \cap U_n = \emptyset$. Hence, if $\varepsilon = \rho/2$, then $v_0 + v \in K_n$ for all $v \in V$ that satisfy $\|v\| \leq \varepsilon$. That is, if $\|v\| \leq \varepsilon$ and $T \in \Lambda$, then

$$\|Tv\| = \|(Tv_0 + T_v) - Tv_0\| \leq \|T(v_0 + v)\| + \|Tv\| \leq 2n.$$

Hence,

$$\|T\| \leq \frac{2n}{\varepsilon}, \quad \forall T \in \Lambda,$$

which proves that Λ is a bounded set.

Case #2: U_n is dense in V for every $n \in \mathbb{N}$. In this case, let $G \subseteq V$ be the G_δ-set defined by

$$G = \bigcap_{n \in \mathbb{N}} U_n .$$

By the Baire Category Theorem, G is a dense in V. By definition, if $v \in U_n$, then there is a $T \in \Lambda$ with $\|Tv\| > n$. That is, if $v \in G$, then $\sup_{T \in \Lambda} \|Tv\| = \infty$.   □

The next mapping property of operators to be considered originates in topology.

Definition 8.6. If X and Y are topological spaces and if $f : X \to Y$ is a function, then f is called an *open map* if $f(U)$ is open in Y for every open set U in X.

Using the terminology of open maps, one can say that if a continuous bijection $f : X \to Y$ is an open map, then f is a homeomorphism. Our aim is to achieve a similar statement for linear operators, and key result in achieving this aim is the following fundamental theorem.

Theorem 8.7 (Open Mapping Theorem). *Every surjective operator between Banach spaces is an open map.*

Proof. Assume that V and W are Banach spaces and that $T : V \to W$ is a surjective operator. Our first objective is to prove that there exists a $\delta > 0$ such that

$$\{w \in W \,|\, \|w\| < \delta\} \subseteq \{Tv \,|\, v \in V \text{ and } \|v\| < 1\}. \tag{8.3}$$

Assuming that inclusion (8.3) holds, there is an open ball Q_0 of radius δ in W such that $0 \in Q_0 \subset T(U)$, where U is the open unit ball in V. Because in a normed vector space every open ball is obtained by translation and scaling of an open ball about the origin, we deduce that for every $w \in T(U)$ there is an open ball Q_w of radius δ_w about w such that $Q_w \subset T(U)$, thereby establishing that $T(U)$ is open. Similarly, an open ball P in V may be translated and scaled to the open unit ball U, which shows that $T(P)$ is open. Thus, the proof of the theorem hinges on establishing the inclusion (8.3).

Set $U_k = \{v \in V \,|\, \|v\| < k\}$ for each $k \in \mathbb{N}$ and note that, because T is surjective, $W = \bigcup_k T(U_k)$ Thus, if K_k is the closure of each $T(U_k)$, then $W = \cup_k K_k$. By the Baire Category Theorem, there is at least one $n \in \mathbb{N}$ for which the interior of K_n is nonempty. Select $w_0 \in \text{int} K_n$; thus, there is a $\gamma > 0$ such that $w_0 + \tilde{w} \in \text{int} K_n \subset K_n = \overline{T(U_n)}$ for every $\tilde{w} \in W$ with $\|\tilde{w}\| < \gamma$. Let $\delta = \gamma/(4n)$.

Choose w from the open ball $B_\delta(0)$ in W and set $\tilde{w} = \frac{1}{2}\gamma \|w\|^{-1} w$. Because w_0 and $w_0 + \tilde{w}$ are in the closure of $T(U_n)$, there are sequences $\{u_j\}_j$ and $\{z_j\}_j$ in U_n with $w_0 = \lim_j Tu_j$ and $w_0 + \tilde{w} = \lim_j Tz_j$. Thus, with $v_j = z_j - u_j$ we have that $\{v_j\}_j$ is a sequence in U_{2n} with $\tilde{w} = \lim_j Tv_j$. Therefore, there is some $v_0 \in U_{2n}$ with $\|Tv_0 - \tilde{w}\| < \frac{1}{2}\gamma \|w\|^{-1} \varepsilon$. Hence, if $v = (2\|w\|\gamma^{-1})v_0$, then $\|v\| < \|w\|/\delta < 1$ and $\|Tv - w\| < \varepsilon$.

The previous paragraph shows that for each nonzero $w \in B_\delta(0)$ there exists $v \in V$ of norm $\|v\| < \|w\|/\delta < 1$ such that $\|Tv - w\| < \varepsilon$. We aim to replace $\|Tv - w\| < \varepsilon$ with an equality. The key point is that δ is independent of ε. Thus, assume that $\varepsilon > 0$ also satisfies $\varepsilon < 1$. Fix $w \in B_\delta(0)$ and set $\varepsilon_1 = \frac{1}{2}\delta\varepsilon$. By the previous paragraph, there exists $v_1 \in V$ with $\|v_1\| < \delta^{-1}\|w\|$ and $\|Tv_1 - w\| < \varepsilon_1$. Because $w - Tv_1 \in B_\delta(0)$, we apply the previous paragraph again with $\varepsilon_2 = \frac{1}{2}\varepsilon_1$ to obtain a vector $v_2 \in V$ with $\|v_2\| < \delta^{-1}\|w - Tv_1\|$ and $\|Tv_2 - (w - Tv_1)\| < \varepsilon_2$. Repetition of the argument yields, inductively, a sequence $\{v_j\}_{j \in \mathbb{N}}$ in V such that, for every $j \in \mathbb{N}$,

$$\|v_j\| < \frac{\varepsilon}{2^{j-1}} \quad \text{and} \quad \left\|w - \sum_{j=1}^{k} Tv_j\right\| < \frac{\delta\varepsilon}{2^k}.$$

Hence, $\{\sum_{j=1}^{k} v_j\}_k$ is a Cauchy sequence in V with limit $v = \sum_{j=1}^{\infty} v_j$ of norm

$$\|v\| \leq \sum_{j=1}^{\infty} \|v_j\| \leq \|v_1\| + \varepsilon \sum_{j=2}^{\infty} \frac{1}{2^{j-1}} < 1 + \varepsilon.$$

By continuity of T, $Tv = w$.

The arguments of the paragraph above demonstrate that for every $w \in B_\delta(0)$ and $0 < \varepsilon < 1$ there is a vector $v \in V$ of norm $\|v\| < 1 + \varepsilon$ such that $Tv = w$. Suppose that $w \in B_\delta(0)$ is of norm $\|w\| < \frac{\delta}{1+\varepsilon}$ and let $v \in V$ be a vector of norm $\|v\| < 1 + \varepsilon$ that satisfies $Tv = (1 + \varepsilon)w$. Therefore, $\tilde{v} = (1 + \varepsilon)^{-1}v$ is in the open unit ball of V and $T\tilde{v} = w$. Because ε is an arbitrary positive number in the interval $(0, 1)$,

$$\{w \in W \mid \|w\| < \delta\} = \bigcup_{0 < \varepsilon < 1} \left\{w \in W \mid \|w\| < \frac{\delta}{1+\varepsilon}\right\} \subseteq \{Tv \mid v \in V, \|v\| < 1\},$$

which completes the proof. $\qquad\qquad\qquad\qquad\qquad\qquad\qquad\qquad\qquad\qquad\qquad\qquad\qquad\Box$

8.3 Inversion of Operators

Our first goal in this section is to prove that if $T : V \to W$ is a bijective operator, then the inverse linear transformation $T^{-1} : W \to V$ is also an operator.

Proposition 8.8. *If V and W are Banach spaces, then the following statements are equivalent for $T \in B(V, W)$.*

1. T is a bijection;
2. T is bounded below and has dense range.

Proof. (1) $\Rightarrow$ (2). Assume that T is a bijection. Because T is surjective, it is trivial that T has dense range; further, the Open Mapping Theorem asserts that there is a $\delta > 0$ for which

$$\{w \in W \mid \|w\| < \delta\} \subseteq \{Tv \mid v \in V \text{ and } \|v\| < 1\}.$$

In other words, using that T is also injective, for every $w \in W$ such that $\|w\| < \delta$ there is a unique $v \in V$ with $\|v\| < 1$ and $Tv = w$. The contrapositive of this statement is: $\|Tv\| \geq \delta$ for every $v \in V$ such that $\|v\| \geq 1$. Hence, if $v \in V$ is nonzero, then $\|v\|^{-1}v$ is a unit vector and so $\|T(\|v\|^{-1}v)\| \geq \delta$; that is, $\|Tv\| \geq \delta\|v\|$, which proves (2).

(2) $\Rightarrow$ (1). Assume that T is bounded below by $\delta > 0$ and that T has dense range. Because T is bounded below, T has closed range (Proposition 8.4) and is obviously injective. But to have both dense range and closed range is to say that T is surjective. Hence, T is a bijection. $\square$

Proposition 8.8 has an important consequence: if an operator is bijective, then its inverse is also an operator.

Corollary 8.9. *If V and W are Banach spaces and if $T \in B(V,W)$ is a bijection, then the linear transformation $T^{-1} : W \to V$ is bounded.*

Proof. Ordinary linear algebra demonstrates that the inverse function $T^{-1} : W \to V$ of T is a linear transformation. By Proposition 8.8, T is bounded below by some $\delta > 0$ (and $\operatorname{ran} T$ is dense). Thus, T^{-1} is bounded and $\|T^{-1}\| \leq \delta^{-1}$ by the following computation:

$$\|T^{-1}w\| = \|T^{-1}(Tv)\| = \|v\| \leq \delta^{-1}\|w\|,$$

where $w = Tv$. $\square$

Two criteria for the singularity of an operator are:

Corollary 8.10. *If V and W are Banach spaces, then an operator $T \in B(V,W)$ fails to be invertible if*

1. *there is a sequence of unit vectors $v_k \in V$ with $\inf_k \|Tv_k\| = 0$, or*
2. *if the range of T is not dense in W.*

Another consequence of the inversion theorem for operators is a useful result called the Closed Graph Theorem. Recall from Exercise 5.101 that the Cartesian product $V \times W$ of Banach spaces V and W is a Banach space in its product topology.

Definition 8.11. *If X and Y are topological spaces and if $f : X \to Y$ is a function, then the graph of f is the set $G(f) = \{(x,f(x)) \in X \times Y \mid x \in X\}$.*

Theorem 8.12 (Closed Graph Theorem). *If the graph of a linear transformation $T : V \to W$ of Banach spaces V and W is closed, then T is continuous.*

Proof. Because T is linear, the graph $G(T)$ of T is a linear submanifold of $V \times W$. The hypothesis that $G(T)$ is closed implies that $G(T)$ is itself a Banach space.

Let p_1 and p_2 denote the projection maps of $V \times W$ onto V and W (the first and second coordinates), respectively. Recall (or observe) that p_1 and p_2 are continuous functions. In considering p_1 restricted to the graph $G(T)$ of T we obtain a continuous linear bijection $s = p_{1|G(T)} : G(T) \to V$. By Corollary 8.9, the linear inverse $s^{-1} : V \to G(T)$ of s is continuous. Hence, so is the linear map $T = p_2 \circ s^{-1}$. $\quad\square$

8.4 Idempotents and Complemented Subspaces

Definition 8.13. A linear transformation $E : V \to V$ of a vector space V is *idempotent* if $E^2 = E$.

If E is an idempotent, then so is $1 - E$ and their product satisfies $E(1 - E) = (1 - E)E = 0$. If an idempotent E is continuous, then both $\operatorname{ran} E$ and $\ker E$ are subspaces. The latter is clear; and to show the former, note that $\operatorname{ran} E$ is the kernel of the continuous idempotent $1 - E$.

The most important of the algebraic properties of idempotent operators are described in the next proposition.

Proposition 8.14. *The following properties hold for idempotents E and F acting on a Banach space V:*

1. *$E + F$ is an idempotent if and only if $EF = FE = 0$;*
2. *$E - F$ is an idempotent if and only if $EF = FE = F$;*
3. *if $EF = FE$, then EF is idempotent with range $\operatorname{ran} E \cap \operatorname{ran} F$ and kernel $\ker E + \ker F$.*

Proof. The proof of (1) is left as an exercise (Exercise 8.64).

To prove (2), assume that $EF = FE = F$. Thus, $(E - F)^2 = E^2 - EF - FE + F^2 = E - F$, which shows that $E - F$ is an idempotent. Conversely, if $E - F$ is an idempotent, then $EF + FE = 2F$, and so $2F - EF - FE = (1 - E)F + F(1 - E) = 0$. Hence, $((1 - E) + F)^2 = (1 - E)^2 + (1 - E)F + F(1 - E) + F^2 = (1 - E) + F$. Thus, $(1 - E) + F$ is idempotent. Therefore, assertion (1) implies that $1 - E$ and F are each idempotent and satisfy $(1 - E)F = F(1 - E) = 0$, Hence, $EF = FE = F$.

To prove (3), note that $EF = FE$ implies that $(EF)^2 = EF^2E = E^2F = FE$, which implies that EF is idempotent. Because $\operatorname{ran} EF = \operatorname{ran} FE$, $\operatorname{ran} EF \subseteq \operatorname{ran} E$, and $\operatorname{ran} FE \subseteq \operatorname{ran} F$, we deduce that $\operatorname{ran} EF \subseteq \operatorname{ran} E \cap \operatorname{ran} F$. Conversely, if $v \in \operatorname{ran} E \cap \operatorname{ran} F$, then $Ev = Fv = v$ and so $v \in \operatorname{ran} EF$. This proves that $\operatorname{ran} EF = \operatorname{ran} E \cap \operatorname{ran} F$. If $v \in \ker E$ and $w \in \ker F$, then $EF(v + w) = FEv + EFw = 0$, which implies that $\ker E + \ker F \subseteq \ker EF$. Conversely, if $v \in \ker EF$, then $v = Fv + (1 - F)v$. Because $EFv = 0$ only if $Fv \in \ker E$ and $F(1 - E)v = 0$ only if $(1 - F)v \in \ker F$, we see that $EFv = 0$ implies that $v \in \ker E + \ker F$. Hence, $\ker EF \subseteq \ker E + \ker F$. $\quad\square$

The range of an idempotent is a subspace (Exercise 8.65), and so it is not surprising that idempotents are used to reflect algebraically certain geometric aspects of Banach spaces.

Definition 8.15. Two subspaces M and N of a Banach space V are *complementary subspaces* if

1. $M \cap N = \{0\}$ and
2. $M + N = V$.

The most immediate example of a complementary pair of subspaces is that furnished by two mutually orthogonal subspaces M and $N = M^{\perp}$ of a Hilbert space H.

There is an intimate connection between complementary pairs of subspaces and continuous linear idempotents.

Proposition 8.16. *The following statements are equivalent for a pair of subspaces M and N of a Banach space V:*

1. *M and N are a complementary pair;*
2. *there exists an idempotent $E \in B(V)$ such that $\operatorname{ran} E = M$ and $\ker E = N$.*

Proof. Suppose that M and N are a complementary pair of subspaces of V. Thus, for each $v \in V$ there are unique $u \in M$ and $w \in N$ such that $v = u + v$. Hence, the linear transformation $E : V \to V$ defined by $E(u + w) = u$, for all $u \in M$ and $w \in N$, is an idempotent with $\operatorname{ran} E = M$ and $\ker E = N$. Therefore, all that is required to show is that E is continuous. We shall use the Closed Graph Theorem (Theorem 8.12) to do so.

Let $(v, z) \in V \times V$ be an element in the closure of the graph $G(E)$ of E. Thus, there is a sequence $\{(v_k, Ev_k)\}_{k \in \mathbb{N}}$ in $G(E)$ such that $\lim_k \|v - v_k\| = \lim_k \|z - Ev_k\| = 0$. Each v_k is expressed uniquely as $v_k = u_k + w_k$ for some $u_k \in M$ and $w_k \in N$. Therefore, $Ev_k = u_k$ and so $\lim_k \|z - u_k\| = \lim_k \|z - Ev_k\| = 0$; because M is closed, this implies that $z \in M$. Likewise, $v - z = \lim_k ((u_k + w_k) - u_k) = \lim_k w_k \in N$, as N is closed. Because $N = \ker E$ we have $0 = E(v - z) = Ev - Ez = Ev - z$, which implies that $(v, z) \in G(E)$. Therefore, the graph $G(E)$ of E is closed, and therefore E is continuous.

Conversely, if $E \in B(V)$ is an idempotent operator such that $\operatorname{ran} E = M$ and $\ker E = N$, then $1 = E + (1 - E)$ implies that $M + N = V$. If $v \in M \cap N$, then $v = Ev = (1 - E)v$ and so $v = Ev = E(1 - E)v = 0v = 0$. $\qquad\square$

Proposition 8.17. *If M and N are a complementary pair of subspaces of a Banach V, then the Banach spaces V/M and N are isomorphic.*

Proof. By Proposition 8.16, there exists an idempotent $E \in B(V)$ such that $\operatorname{ran} E = N$ and $\ker E = M$. Thus, the function $T : V/M \to N$ given by $T\dot{v} = Ev$ is a well-defined linear bijection. The inverse $S : N \to V/M$ of T is given by $Sw = \dot{w}$, for $w \in N$. Because $\|Sw\| = \|\dot{w}\| \leq \|w\|$, S is continuous; hence, so is T (by Corollary 8.9). $\qquad\square$

If one has a single subspace M of a Banach space V at hand, it is natural to ask whether M is part of a complementary pair or not. Such a subspace M is said to be complemented.

Definition 8.18. A subspace M of a Banach space V is said to be *complemented in* V if there is a subspace N of V, called the *complement* of M, such that M and N form a complementary pair of subspaces of V.

Besides subspaces of Hilbert spaces, other examples of complemented subspaces are finite-dimensional subspaces.

Proposition 8.19. *Every finite-dimensional subspace of a Banach space is complemented.*

Proof. Let $\{v_1, \ldots, v_n\}$ be a basis of a finite-dimensional space M of a Banach space V. As in the proof of Proposition 6.27, there are linear functionals $\tilde{\varphi}_j : M \to \mathbb{C}$ such that $\tilde{\varphi}_j(v_k) = 0$, for $j \neq k$, and $\tilde{\varphi}_j(v_j) = 1$, for each $1 \leq j \leq n$. By the Hahn-Banach Extension Theorem, each of these linear functionals has an extension to a continuous linear function $\varphi_j : V \to \mathbb{C}$. Let $N = \bigcap_{j=1}^{n} \ker \varphi_j$, which is a subspace of V such that $M \cap N = \{0\}$. Choose any $v \in V$ and let $w = \sum_{j=1}^n \varphi_j(v) v_j \in M$. Consider $z = v - w$. Because $\varphi_j(z) = \varphi_j(v) - \varphi_j(w) = \varphi_j(v) - \varphi_j(v) = 0$, we deduce that $z \in N$. Thus, $v = w + z \in M + N$. $\qquad\square$

However, as the next result shows, not every subspace of a Banach space need be complemented.

Proposition 8.20. *The subspace $c_0(\mathbb{N})$ has no complement in $\ell^\infty(\mathbb{N})$.*

Proof. Write c_0 for $c_0(\mathbb{N})$ and ℓ^∞ for $\ell^\infty(\mathbb{N})$. Let $X = (0,1) \cap \mathbb{Q}$ and let $\alpha : \mathbb{N} \to X$ be a bijection. By Exercise 1.101, there is an uncountable family $\{X_\lambda\}_{\lambda \in \Lambda}$ of infinite subsets $X_\lambda \subseteq X$ such that $X_\lambda \cap X_{\lambda'}$ is a finite set for all $\lambda, \lambda' \in \Lambda$ for which $\lambda' \neq \lambda$. For each $\lambda \in \Lambda$, let $f_\lambda = (f_\lambda(n))_{n \in \mathbb{N}} \in \ell^\infty$ satisfy $f_\lambda(n) = 0$, if $\alpha(n) \notin X_\lambda$, and $f_\lambda(n) = 1$, if $\alpha(n) \in X_\lambda$. Because $f_\lambda - f_{\lambda'} \notin c_0$ whenever $\lambda' \neq \lambda$, the family $\{\dot{f}_\lambda\}_{\lambda \in \Lambda} \subset \ell^\infty/c_0$ is uncountable.

Choose $\varphi \in (\ell^\infty/c_0)^*$ and, for every $n \in \mathbb{N}$, let $Z_\varphi(n) = \{\dot{f}_\lambda \mid |\varphi(\dot{f}_\lambda)| \geq \frac{1}{n}\}$. Suppose that $\dot{f}_{\lambda_1}, \ldots, \dot{f}_{\lambda_m} \in Z_\varphi(n)$. For each j, let

$$\beta_j = \operatorname{sgn}\left(\varphi(\dot{f}_{\lambda_j})\right) = \frac{\overline{\varphi(\dot{f}_{\lambda_j})}}{|\varphi(\dot{f}_{\lambda_j})|}.$$

Let $v = \sum_{j=1}^{m} \beta_j f_{\lambda_j} \in \ell^\infty$. Because $X_{\lambda_j} \cap X_{\lambda_k}$ is a finite set for $k \neq j$,

$$\|\dot{v}\| = \inf\{\|v - h\| \mid h \in c_0\} = \max_{1 \leq j \leq m} |\beta_j| = 1.$$

Furthermore,

$$\varphi(\dot{v}) = \sum_{j=1}^{m} \beta_j \varphi(\dot{f}_{\lambda_j}) = \sum_{j=1}^{m} |\varphi(\dot{f}_{\lambda_j})| \geq \sum_{j=1}^{m} \frac{1}{n} = \frac{m}{n}.$$

Thus, $\|\varphi\| = \|\dot{v}\| \, \|\varphi\| \geq |\varphi(\dot{v})| = \varphi(\dot{v}) \geq m/n$ implies that m is bounded above and, hence, that $Z_\varphi(n)$ is a finite set. Therefore, the set $Z_\varphi = \{\dot{f}_\lambda \, | \, \varphi(\dot{f}_\lambda) \neq 0\} = \bigcup_{n \in \mathbb{N}} Z_\varphi(n)$

is countable. Consequently, $\bigcup_{k \in \mathbb{N}} Z_{\varphi_k}$ is countable for every countable set $\{\varphi_k\}_{k \in \mathbb{N}}$ of linear functionals on ℓ^∞/c_0.

Assume, contrary to what we aim to prove, that there exists a complement N to c_0 in ℓ^∞. By Proposition 8.17, there is a continuous linear isomorphism $T : \ell^\infty/c_0 \to N$. For each $k \in \mathbb{N}$ let $\gamma_k \in N^*$ be the linear functional for which $\gamma_k(g) = g_k$, for all $g = (g_n)_{n \in \mathbb{N}} \in N$. Observe that if $\gamma_k(g) = 0$ for every $k \in \mathbb{N}$, then necessarily $g = 0$. Now let $\varphi_k = \gamma_k \circ T$, which is a linear functional on ℓ^∞/c_0 for every $k \in \mathbb{N}$.

On the one hand, because T is an isomorphism, if $w \in \ell^\infty/c_0$ is such that $\varphi_k(w) = 0$ for every $k \in \mathbb{N}$, then necessarily $w = 0$. On the other hand, by the argument of the previous paragraph, the set $\{\lambda \in \Lambda \, | \, \exists k \in \mathbb{N} \text{ such that } \varphi_k(\dot{f}_\lambda) \neq 0\}$ is countable. Hence, because Λ is uncountable, there is a $\lambda \in \Lambda$ such that the nonzero element $\dot{f}_\lambda$ of ℓ^∞/c_0 satisfies $\varphi_k(\dot{f}_\lambda) = 0$ for every $k \in \mathbb{N}$. Therefore, this contradiction implies that N and ℓ^∞/c_0 cannot be isomorphic; that is, c_0 is not complemented in ℓ^∞. $\square$

Corollary 8.21. $c_0(\mathbb{N})$ *is not the range of any idempotent* $E \in B(\ell^\infty(\mathbb{N}))$.

Part of the definition for a pair of subspaces M and N to be a complementary pair is that $M + N = V$—in other words, $M + N$ is also closed. This leads naturally to the question of whether $M + N$ is closed for every two subspaces M and N such that $M \cap N = \{0\}$. The answer is yes in one of the most important cases of all:

Proposition 8.22. *Suppose that M is a proper subspace of a Banach space V and that $v \in V$ is nonzero and $v \notin M$. Then $M + \mathrm{Span}\{v\}$ is closed.*

Proof. By Exercise 6.59, there is a $\varphi \in V^*$ such that $\varphi(v) = 1$ and $\varphi(w) = 0$, for every $w \in M$. Suppose that $y \in V$ is in the closure of $M + \mathrm{Span}\{v\}$; thus, there are sequences $\{w_n\}_{n \in \mathbb{N}}$ in M and $\{\alpha_n\}_{n \in \mathbb{N}}$ in $\mathbb{C}$ such that $\|(w_n + \alpha_n v) - y\| \to 0$. Hence, $\alpha_n = \varphi(w_n + \alpha_n v)$ approaches $\varphi(y)$ as $n \to \infty$, which implies that $\{w_n\}_{n \in \mathbb{N}}$ converges to $y - \varphi(y)v$. Because M is closed, $y - \varphi(y)v$ must therefore belong to M, and so $y = (y - \varphi(y)v) + \varphi(y)v$ belongs to $M + \mathrm{Span}\{v\}$. $\square$

Corollary 8.23. *If M and N are subspaces of V such that $M \cap N = \{0\}$ and N has finite dimension, then $M + N$ is a subspace.*

8.5 Compact Operators

Definition 8.24. An operator K on a Banach space V is said to be *compact* if for every sequence $\{v_n\}_{n \in \mathbb{N}}$ of unit vectors $v_n \in V$ there is a subsequence $\{Kv_{n_j}\}_{j \in \mathbb{N}}$ of $\{Kv_n\}_{n \in \mathbb{N}}$ such that $\{Kv_{n_j}\}_{j \in \mathbb{N}}$ is convergent.

An alternate definition for compactness is as follows (Exercise 8.74): an operator K on a Banach space V is compact if and only if the image of the closed unit sphere of V under the operator K has compact closure.

If V has finite dimension, then Proposition 5.22 shows that the closed unit ball of V is compact. Therefore, by Theorem 2.19, any sequence $\{v_n\}_{n \in \mathbb{N}}$ of unit vectors $v_n \in V$ admits a convergent subsequence $\{v_{n_j}\}_{j \in \mathbb{N}}$. Thus, using the continuity of operators, for each $T \in B(V)$ the sequence $\{Tv_n\}_{n \in \mathbb{N}}$ admits a convergent subsequence—namely, $\{Tv_{n_j}\}_{j \in \mathbb{N}}$. That is, if $v = \lim_j v_{n_j}$, then

$$\lim_{j \to \infty} Tv_{n_j} = T\left(\lim_{j \to \infty} v_{n_j}\right) = Tv.$$

Hence, every operator $T \in B(V)$ is compact if V has finite dimension.

A variant of the argument above is the following observation, which will be used frequently.

Proposition 8.25. *If V is a Banach space, $\lambda \in \mathbb{C}$ is nonzero, and if $T = \lambda 1$, then T is a compact operator only if V has finite dimension.*

Proof. Assume that V has infinite dimension and choose $\delta \in (0, 1)$. By Proposition 5.21, there exists a sequence $\{v_n\}_{n \in \mathbb{N}}$ of unit vectors $v_n \in V$ such that $\{v_n\}_{n \in \mathbb{N}}$ does not admit any Cauchy subsequences. Thus, the sequence $\{Tv_n\}_{n \in \mathbb{N}}$ does not admit any Cauchy subsequences and, hence, does not admit any convergent subsequences. Therefore, T is compact only if V has finite dimension. $\square$

Compact operators exhibit the following algebraic and analytic features.

Proposition 8.26. *The following operators are compact:*

1. *αK, for every $\alpha \in \mathbb{C}$ and all compact operators K;*
2. *$K_1 + K_2$, for all compact operators K_1 and K_2;*
3. *KT and TK, for all compact operators K and all $T \in B(V)$;*
4. *F, for all $F \in B(V)$ with finite-dimensional range;*
5. *K, for all operators K for which there exists a sequence $\{K_n\}_{n \in \mathbb{N}}$ of compact operators K_n such that $\lim_n \|K - K_n\| = 0$.*

Proof. The proof of the first four assertions is left as an exercise (Exercise 8.70). To prove the final statement, assume that there exists a sequence $\{K_n\}_{n \in \mathbb{N}}$ of compact operators K_n such that $\|K - K_n\| < \frac{1}{n}$ for very $n \in \mathbb{N}$. As K_1 is compact, there is a subsequence $\{v_{1,j}\}_j$ of $\{v_i\}_i$ such that $\{K_1 v_{1,j}\}_j$ is convergent. As K_2 is compact, there is a subsequence $\{v_{2,j}\}_j$ of $\{v_{1,j}\}_j$ such that $\{K_2 v_{2,j}\}_j$ and $\{K_1 v_{2,j}\}_j$ are convergent. Inductively, for each $n \in \mathbb{N}$ there is a sequence $\{v_{n,j}\}_j$ such that

1. $\{v_{n,j}\}_j$ is a subsequence of $\{v_{n-1,j}\}_j$, and
2. $\{K_\ell v_{n,j}\}_j$ is convergent for all $1 \le \ell \le n$.

Now let $\varepsilon > 0$. Choose $p \in \mathbb{N}$ such that $\|K - K_p\| < \varepsilon$. Claim: $\{Kv_{n,n}\}_n$ is a Cauchy sequence. Note that if $n \ge p$, then $\{K_p v_{n,n}\}_n$ is a subsequence of the convergent sequence $\{K_p v_{p,n}\}_n$. Hence, $\{K_p v_{n,n}\}_n$ is convergent and, therefore, Cauchy. Thus,

$$\|Kv_{n,n} - Kv_{m,m}\| \leq \|(K-K_p)v_{n,n}\| + \|K_p v_{n,n} - K_p v_{m,m}\| + \|(K-K_p)v_{m,m}\|$$

$$< 2\varepsilon + \|K_p v_{n,n} - K_p v_{m,m}\|$$

Thus, $\{Kv_{n,n}\}_n$ is a Cauchy sequence and, hence, convergent. This proves that K is compact.   □

Corollary 8.27. *The set of compact operators acting on a Banach space V is a subspace of $B(V)$.*

Proof. The conclusion is a direct consequence of assertions (1), (2), and (5) of Proposition 8.26.   □

If M is a subspace of V, then the range of the identity operator restricted to M (as a map $M \to V$) obviously has closed range. The next proposition asserts that the same is true for any perturbation $1 - K$ of the identity operator 1 by a compact operator K.

Proposition 8.28. *If K is a compact operator acting on a Banach space V, and if $M \subseteq V$ is a subspace, then*

$$\{w - Kw \mid w \in M\}$$

is a subspace of V.

Proof. Let $M \subseteq V$ be a subspace and assume that $M \cap \ker(1-K) = \{0\}$. (The case where $M \cap \ker(1-K) \neq \{0\}$ will be handled at the end of the proof.) Let

$$L = \overline{\{w - Kw \mid w \in M\}}$$

and suppose that $y \in L$. We aim to prove that $y \in \{w - Kw \mid w \in M\}$. Thus, there is a sequence of vectors $y_n \in \{w - Kw \mid w \in M\}$ with limit y; that is, there is a sequence of vectors $w_n \in M$ such that $\|y - (w_n - Kw_n)\| \to 0$. Assume that the sequence $\{w_n\}_{n \in \mathbb{N}}$ admits a subsequence of vectors w_{n_k} in which $\|w_{n_k}\| \to \infty$. If this is the case, then let $v_k \in M$ denote the unit vector $\|w_{n_k}\|^{-1} w_{n_k}$. The compactness of K implies that the sequence $\{Kv_k\}_{k \in \mathbb{N}}$ admits a convergent subsequence $\{Kv_{k_j}\}_{j \in \mathbb{N}}$ with limit, say, $z \in V$. Note that

$$\|v_{k_j} - Kv_{k_j}\| = \frac{1}{\|w_{n_{k_j}}\|} \|w_{n_{k_j}} - Kw_{n_{k_j}}\|.$$

As $j \to \infty$ we have $\|w_{n_{k_j}} - Kw_{n_{k_j}}\| \to \|y\|$ and $\|w_{n_{k_j}}\| \to \infty$. Therefore, $\{v_{k_j}\}_{j \in \mathbb{N}}$ converges to z, which implies that $z \in M$. However, we now have that $z \in M$ and $z - Kz = \lim_j(v_{k_j} - Kv_{k_j}) = 0$; that is, $z \in M \cap \ker(1-K) = \{0\}$. This is a contradiction because $0 = z = \lim_j v_{k_j}$, where each $\|v_{k_j}\| = 1$. Therefore, it cannot happen that the sequence $\{w_n\}_{n \in \mathbb{N}}$ admits a subsequence of vectors w_{n_k} in which $\|w_{n_k}\| \to \infty$.

Because there is a $\gamma > 0$ such that $\|w_n\| \leq \gamma$ for all $n \in \mathbb{N}$, and because K is a compact operator, the sequence $\{Kw_n\}_{n \in \mathbb{N}}$ admits a convergent subsequence $\{Kw_{n_k}\}_{k \in \mathbb{N}}$ with limit, say, $x \in V$. Now since

$$Kw_{n_k} \to x \quad \text{and} \quad (w_{n_k} - Kw_{n_k}) \to y,$$

we conclude that $w_{n_k} \to x + y$, whence $x + y \in M$. Thus, if $w = x + y$, then

$$(1 - K)w = (1 - K)(x + y) = \lim_{k \to \infty} (w_{n_k} - Kw_{n_k}) = y,$$

which proves that the linear submanifold $\{w - Kw \,|\, w \in M\}$ is closed.

Suppose next that $M \cap \ker(1 - K) \neq \{0\}$. Thus, if $F = M \cap \ker(1 - K)$, then F is a finite-dimensional subspace. (If not, then F is an infinite-dimensional space upon which the compact operator K acts as the identity, in contradiction to the fact that the identity operator 1 is not compact.) Proposition 8.19 asserts that F is complemented in M; hence, there is a subspace $N \subset M$ such that $N \cap F = \{0\}$ and $M = N + F$. Consequently, each $w \in M$ has the form $w_1 + w_0$, where $w_1 \in N$ and $(1 - K)w_0 = 0$. Hence,

$$\{w - Kw \,|\, w \in M\} = \{w_1 - Kw_1 \,|\, w_1 \in N\}. \tag{8.4}$$

Since $N \cap \ker(1 - K) = \{0\}$, the linear submanifold in (8.4) is closed by the arguments developed initially. $\qquad\square$

Corollary 8.29. *If K is a compact operator on a Banach space V, then the range of $1 - K$ is closed.*

For operators acting on finite-dimensional spaces, injectivity implies surjectivity. Although this fails to be true in infinite-dimensional spaces (even for compact operators), this property holds for compact perturbations of the identity operator.

Proposition 8.30. *If K is a compact operator acting on a Banach space V such that $1 - K$ is injective, then $1 - K$ is surjective.*

Proof. Assume, contrary to what we aim to prove, that $1 - K$ is not a surjection. Set $M_0 = V$ and let

$$M_n = \{(1 - K)^n v \,|\, v \in V\} = \{(1 - K)w \,|\, w \in M_{n-1}\}, \quad \forall n \in \mathbb{N}.$$

Thus, $M_0 \supset M_1 \supset M_2 \supset \cdots \supset M_n \supset M_{n+1} \supset \ldots$ is a descending sequence of subspaces (by Proposition 8.28). This sequence is in fact proper, by the following argument. Suppose for some $n \in \mathbb{N}$ we have $M_n = M_{n+1}$. Because $1 - K$ is not surjective, there is a $v_0 \in V \backslash \mathrm{ran}\,(1 - K)$. On the other hand, $(1 - K)^n v_0 \in M_n = M_{n+1}$, and so there is a vector $w_0 \in V$ such that $(1 - K)^{n+1} w_0 = (1 - K)v_0$. That is,

$$(1 - K)^n \left[(1 - K)w_0 - v_0 \right] = 0.$$

Because $1 - K$ and, hence, $(1 - K)^n$ are injective, we conclude that $v_0 = (1 - K)w_0$. But this would place v_0 in the range of $1 - K$, in contradiction to $v_0 \in V\backslash\mathrm{ran}\,(1 - K)$. Hence, it must indeed be true that the sequence $\{M_n\}_{n\in\mathbb{N}\cup\{0\}}$ is properly descending. Moreover, if $v \in M_n$, then $v = (1 - K)^n w$ for some $w \in V$; hence, $Kv = (1 - K)^n Kw$ (as K and $1 - K$ commute). In other words, each M_n is invariant under K.

Let $\delta \in (0, 1)$ be fixed. For each $n \in \mathbb{N}$ there is a vector in the quotient space M_{n-1}/M_n of norm δ. Since $\delta < 1$, this means that there is a unit vector $v_n \in M_{n-1}$ such that $\|v_n - f\| \geq \delta$ for all $f \in M_n$. If $j < k$, then $(1 - K)v_j \in (1 - K)(M_{j-1}) = M_j$ and $Kv_k \in M_k \subset M_j$. Thus, $(1 - K)v_j + Kv_k \in M_j$ and so

$$\delta \leq \|v_j - [(1 - K)v_j + Kv_k]\| = \|Kv_j - Kv_k\|.$$

Therefore, the sequence $\{Kv_n\}_{n\in\mathbb{N}}$ does not admit a Cauchy subsequence and, hence, does not admit a convergent subsequence. This contradicts the fact that K is compact. Thus, the original assumption that $1 - K$ is not surjective cannot hold. That is, $1 - K$ is necessarily surjective.   □

An important feature of compact operators is that the adjoint $K^* \in B(V^*)$ is compact if $K \in B(V)$ is compact.

Proposition 8.31. *If $K \in B(V)$ is a compact operator, then $K^* \in B(V^*)$ is a compact operator.*

Proof. Let B_{V^*} and B_V denote the closed unit balls of V^* and V. Suppose that $\{\varphi_n\}_{n\in\mathbb{N}}$ is a sequence in B_{V^*}. Because B_{V^*} is weak* compact, by the Banach-Alaoglu Theorem (Theorem 6.33), there is a $\varphi \in B_{V^*}$ such that for every weak* open neighbourhood U of φ in B_{V^*} and $j \in \mathbb{N}$ there is an $n \geq j$ such that $\varphi_n \in U$. (That is, φ is a weak* limit point of $\{\varphi_n\}_{n\in\mathbb{N}}$.)

Let $\varepsilon > 0$ and fix $j \in \mathbb{N}$. Because $\overline{K(B_V)}$ is compact, there are $w_1, \ldots, w_m \in K(B_V)$ such that

$$\overline{K(B_V)} \subset \bigcup_{k=1}^{m} \{w \in V \mid \|w - w_k\| < \varepsilon\}.$$

Consider the weak* open neighbourhood $U \subset B_{V^*}$ of φ that is given by

$$U = \{\psi \in B_{V^*} \mid |\psi(w_k) - \varphi(w_k)| < \varepsilon,\ 1 \leq k \leq m\}.$$

Because φ is a weak* limit point of $\{\varphi_n\}_{n\in\mathbb{N}}$, there is a $n_j \geq j$ such that $\varphi_{n_j} \in U$. If $v \in B_v$, then there is a $1 \leq k \leq m$ such that $\|Kv - w_k\| < \varepsilon$. Hence,

$$|K^*\varphi(v) - K^*\varphi_{n_j}(v)| = |\varphi(Kv) - \varphi_{n_j}(Kv)|$$

$$\leq |\varphi(Kv - w_k) - \varphi_{n_j}(Kv - w_k)|$$

$$+ |\varphi(w_k) - \varphi_{n_j}(w_k)|$$

$$< 3\varepsilon.$$

Hence,

$$\|K^*\varphi - K\varphi_{n_j}\| = \sup_{\|v\|\le 1} |K^*\varphi(v) - K^*\varphi_{n_j}(v)| < 3\varepsilon.$$

As the choice of $\varepsilon > 0$ and $j \in \mathbb{N}$ are arbitrary, this proves that there is a subsequence $\{\varphi_{n_j}\}_j$ of $\{\varphi_n\}_n$ such that $\{K^*\varphi_{n_j}\}_j$ is convergent (to $K^*\varphi$). Hence, K^* is a compact operator. $\square$

Proposition 8.32. *If K is a compact operator acting on a Banach space V such that $1 - K$ is surjective, then $1 - K$ is injective.*

Proof. Because K is compact, so are K^* and K^{**} (Proposition 8.31). Because $1 - K$ is surjective, the defect spectrum $\sigma_d(1 - K)$ is empty. But since $\sigma_d(1 - K) = \sigma_p(1 - K^*)$, we conclude that $1 - K^*$ is injective. As K^* is compact, Proposition 8.30 implies $1 - K^*$ is surjective. Therefore, the defect spectrum is $1 - K^*$, and so the point spectrum of $(1 - K^*)^*$ is empty. In other words, $1 - K^{**}$ is injective. The restriction of $1 - K^{**}$ to the subspace V of V^{**} is precisely $1 - K$; as $1 - K^{**}$ remains injective on any smaller domain, $1 - K$ is, therefore, an injective operator. $\square$

8.6   Operator Algebra

The notion of a Banach algebra has already been encountered in our discussion of the Stone-Weierstrass Theorem. The algebraic basis for the concept of Banach algebra is that of an associative algebra. Recall from Definition 5.24 that an *algebra* a complex vector space A that has a product (or multiplication) such that, for all $a, b, c \in A$ and all $\alpha \in \mathbb{C}$,

$$(a+b)c = (ac+bc), \quad a(b+c) = ab+ac, \quad a(bc) = (ab)c,$$

and

$$(\alpha a)(b) = a(\alpha b) = \alpha(ab).$$

Recall that, if $ab = ba$, for all $a, b \in A$, then A is called an *abelian* algebra, and if there is an element $1 \in A$ such that $a1 = 1a = a$, for every $a \in A$, then A is said to be a *unital* algebra and 1 is the *multiplicative identity* of A.

Definition 5.25 asserts that a *Banach algebra* is an algebra A together with a norm $\|\cdot\|$ on A such that

1. $\|xy\| \le \|x\|\|y\|$, for all $x, y \in A$, and
2. A is a Banach space under the norm $\|\cdot\|$.

Recall that if A is a unital algebra, then A is a *unital Banach algebra* if $\|1\| = 1$.

A norm $\|\cdot\|$ that satisfies the condition $\|xy\| \leq \|x\|\,\|y\|$, for all $x, y \in A$, is called a *submultiplicative norm*.

Definition 8.33. A subset J of a Banach algebra A is an *ideal* of A if

1. J is a subspace of A, and
2. $ax \in J$ and $xa \in J$, for all $a \in A$ and $x \in J$.

If an ideal J of A satisfies $J \neq \{0\}$ and $J \neq A$, then J is called a *proper ideal* of A.

Observe that, by definition, ideals are closed in the norm topology. We shall also have need of the purely algebraic notion of ideal, which in the context of Banach algebras are called algebraic ideals.

Definition 8.34. A subset I of a Banach algebra A is an *algebraic ideal* of A if

1. I is a linear submanifold of A, and
2. $ax \in I$ and $xa \in I$, for all $a \in A$ and $x \in I$.

If an algebraic ideal I of A satisfies $I \neq \{0\}$ and $I \neq A$, then I is called a *proper algebraic ideal* of A.

The set $B(V)$, where V is a normed vector space, is a unital associative algebra whereby the product ST of operators $S, T \in B(V)$ is simply composition—namely, $ST(v) = S(Tv)$, for all $v \in V$—and the multiplicative identity 1 of $B(V)$ is the identity operator $1v = v$, for all $v \in V$.

Theorem 8.35. *Assume that V is a Banach space.*

1. *The set $B(V)$ is a unital Banach algebra.*
2. *If V has infinite dimension, then the set $K(V)$ of all compact operators on V is a proper ideal of $B(V)$.*
3. *If V has infinite dimension, then the set $F(V)$ of all finite-rank operators on V is a proper algebraic ideal of $K(V)$.*
4. *The norm-closure $\overline{F(V)}$ of $F(V)$ is an ideal of $B(V)$ and $\overline{F(V)} \subseteq J$ for every nonzero ideal J of $B(V)$.*

Proof. The norm on $B(V)$ is plainly submultiplicative and the norm of the identity operator 1 is 1. By Proposition 6.6, $B(V)$ is a Banach space. Hence, $B(V)$ is a unital Banach algebra.

Proposition 8.26 shows that $K(V)$ is an ideal of $B(V)$, while Proposition 8.25 indicates that $1 \notin K(V)$ if V has infinite dimension. Proposition 8.26 also shows that the algebraic ideal $F(V)$ of $B(V)$ is a subset of $K(V)$. Therefore, because $F(V)$ is plainly an algebraic ideal of $K(V)$, we need only verify that there is a compact operator on V of infinite rank.

Let $M_1 = V$ and select a unit vector $v_1 \in M_1$. By Proposition 8.19, the 1-dimensional subspace $\mathrm{Span}\{v_1\}$ has a complement M_2. Select a unit vector $v_2 \in M_2$. In the Banach space M_2, the 1-dimensional subspace $\mathrm{Span}\{v_2\}$ has a complement M_3. Note that M_3 is also a complement to $\mathrm{Span}\{v_1, v_2\}$ in V. Because V has infinite dimension, proceeding by induction yields a properly descending sequence of subspaces $M_1 \supset M_2 \supset \ldots$ and unit vectors $v_k \in M_k$ such that $v_k \notin M_{k+1}$. Indeed, for each $k \in \mathbb{N}$ the subspace

$$N_k = (\mathrm{Span}\{v_1, \ldots, v_{k-1}\}) + M_{k+1}$$

is closed (by Proposition 8.23 or by noting that N_k is a complement to $\mathrm{Span}\{v_k\}$ in V) and $v_k \notin N_k$. Therefore, by Exercise 6.59, there exists $\varphi_k \in V^*$ of norm $\|\varphi_k\| = 1$ such that $\varphi_k(v_k) \neq 0$ and $\varphi(w) = 0$ for all $w \in N_k$—in particular, $\varphi_k(v_\ell) = 0$ for all $\ell \neq k$.

Define, for each $n \in \mathbb{N}$, the finite-rank operator K_n on V given by

$$K_n v = \sum_{k=1}^{n} 2^{-k} \varphi_k(v) v_k,$$

for $v \in V$. If $m > n$ and $v \in V$, then

$$\|(K_m - K_n)v\| \leq \sum_{k=n+1}^{m} 2^{-k} |\varphi_k(v)| \, \|v_k\| = \|v\| \sum_{k=n+1}^{n} 2^{-k} = \|v\|(s_m - s_n),$$

where $\{s_j\}_{j \in \mathbb{N}}$ is the sequence of partial sums $s_j = \sum_{k=1}^{j} 2^{-k}$ of the convergent series $\sum_{k=1}^{\infty} 2^{-k}$. Hence, $\{K_n\}_{n \in \mathbb{N}}$ is a Cauchy sequence of compact operators, and so this sequence converges to some compact operator K. If $k \in \mathbb{N}$ is fixed and $n \geq k$, then there is a nonzero $\alpha_k \in \mathbb{R}$ such that $K_n v_k = \alpha_k v_k$ and so $K v_k = \lim_n K_n v_k = \alpha_k v_k$. This proves that the range of the compact operator K contains the countable set $\{v_k\}_{k \in \mathbb{N}}$ of linearly independent vectors, which implies that K has infinite rank.

It is straightforward to prove that the norm closure $\overline{F(V)}$ of $F(V)$ is also an ideal of $K(V)$—and, hence, of $B(V)$. Suppose now that J is a nonzero ideal of $B(V)$. Consider an arbitrary operator $F \in F(V)$ of rank-1. By Exercise 8.60, there exist (nonzero) $w \in V$ and $\varphi \in V$ such that $F = F_{\varphi,w}$, where $F_{\varphi,w} v = \varphi(v)w$ for all $v \in V$. Because $J \neq \{0\}$, there are nonzero $T \in J$ and $u \in V$ such that $Tu \neq 0$; and, by Corollary 6.23, there exists $\psi \in V^*$ such that $\psi(Tu) = 1$. Let $S = F_{\psi,w} T F_{\varphi,u}$; because J is an ideal, S is an element of J. However, $Sv = \varphi(v)w$ for all $v \in V$ implies that $S = F$, and thus $F \in J$. This proves that J contains every rank-1 operator. Exercise 8.60 asserts that every finite rank operator is a (finite) sum of rank-1 operators, which implies that $F(V) \subseteq J$. Thus, because J is closed, $\overline{F(V)} \subseteq J$.   □

Theorem 8.35 above shows that $\overline{F(V)}$ is a minimal ideal of $B(V)$. Interestingly, for certain separable Banach spaces V the ideal $\overline{F(V)}$ is a proper ideal of $K(V)$, while for other separable Banach spaces (such as separable Hilbert spaces) $\overline{F(V)}$ coincides with $K(V)$. An example of a space V in which $\overline{F(V)}$ is a proper ideal of $K(V)$ is given by P. Enflo in [24].

8.7   The Spectrum of an Operator

Another instance whereby the theory of operators on Banach spaces takes a great deal of inspiration from the theory of linear transformations is in the notion of the spectrum of an operator, which is the algebraic analogue of the concept of eigenvalue in linear algebra. However, owing to the possible infinite dimensionality of the domain of an operator, the study of the spectrum in operator theory relies a great deal more upon analysis than it does upon algebra. At the heart of the matter is the development of a mathematical result for operator theory that essentially fulfills the role that the fundamental theorem of algebra plays in linear algebra.

Earlier we determined that the linear inverse T^{-1} of a bijective linear operator $T : V \to W$ between Banach spaces is also an operator. In the case where $W = V$, the inverse operator $T^{-1} \in B(V)$ satisfies $T^{-1}T = TT^{-1} = 1$, where $1 \in B(V)$ denotes the identity operator $1v = v$, for $v \in V$. This gives rise to the usual algebraic formulation of inverse:

Definition 8.36. If V is a Banach space, then an operator $T \in B(V)$ is *invertible in $B(V)$* if there exists an operator $S \in B(V)$ such that $ST = TS = 1$ (the identity operator $1v = v$, for all $v \in V$).

Of course, if such an operator $S \in B(V)$ in which $ST = TS = 1$ exists, then S is necessarily unique and is denoted by T^{-1}.

Recall, from linear algebra, the following theorem:

Theorem 8.37. *If T is an operator acting on a finite-dimensional Banach space V, then the following statements are equivalent for a complex number λ:*

1. *λ is an eigenvalue of T;*
2. *$T - \lambda 1$ is not invertible in $B(V)$.*

The eigenvalue problem in linear algebra is settled by the characteristic polynomial in the sense that the eigenvalues of a matrix T coincide precisely with the roots λ of the characteristic polynomial $c_T(z) = \det(z1 - T)$. Therefore, it is a consequence of the Fundamental Theorem of Algebra is that every complex matrix has an eigenvalue. Thus, the eigenvalue problem for matrices is as much a problem in algebra as it is in operator theory. The corresponding theorem in operator theory states that every operator on a Banach space has nonempty spectrum (Theorem 8.42). One proof of the Fundamental Theorem of Algebra is via complex analysis (specifically, Liouville's Theorem on bounded entire functions) and, not surprisingly, it is by a very similar route that Theorem 8.42 is proved.

Lemma 8.38. *If $T \in B(V)$ and $\lambda \in \mathbb{C}$ are such that $|\lambda| > \|T\|$, then $(T - \lambda 1)^{-1}$ exists and*

$$\lim_{|\lambda| \to \infty} \|(T - \lambda 1)^{-1}\| = 0.$$

Proof. Observe that $T - \lambda 1 = -\lambda(1 - \frac{1}{\lambda}T)$. By hypothesis, $\|\frac{1}{\lambda}T\| < 1$; thus,

$$\sum_{k=0}^{\infty} \left(\frac{\|T\|}{|\lambda|}\right)^k = \frac{1}{1 - \|\frac{1}{\lambda}T\|} = \frac{|\lambda|}{|\lambda| - \|T\|},$$

and so the series $\sum_{k=0}^{\infty} \lambda^{-k} T^k$ converges to some $S \in B(V)$ with $\|S\| \le \frac{|\lambda|}{|\lambda| - \|T\|}$. Because

$$\left(1 - \frac{1}{\lambda}T\right)\left(1 + \frac{1}{\lambda}T + \cdots + \frac{1}{\lambda^k}T^k\right) = 1 - \frac{1}{\lambda^{k+1}}T^{k+1},$$

we conclude that $S = (1 - \frac{1}{\lambda}T)^{-1}$, which implies that $T - \lambda 1$ is invertible. Moreover,

$$\|(T - \lambda 1)^{-1}\| = \frac{1}{|\lambda|}\|(1 - \frac{1}{\lambda}T)^{-1}\| \le \frac{1}{|\lambda|}\left(\frac{1}{1 - \|\frac{1}{\lambda}T\|}\right) = \frac{1}{|\lambda| - \|T\|}.$$

Hence, $\lim_{|\lambda| \to \infty} \|(T - \lambda 1)^{-1}\| = 0$. $\qquad\square$

Lemma 8.39. *If $T \in B(V)$ satisfies $\|T\| < 1$, then*

$$\|1 - (1 - T)^{-1}\| \le \frac{\|T\|}{1 - \|T\|}.$$

Proof. Lemma 8.38 shows that $(1 - T)^{-1} = \sum_{k=0}^{\infty} T^k$ and $\|(1 - T)^{-1}\| \le \frac{1}{1 - \|T\|}$. Further,

$$\|1 - (1 - T)^{-1}\| = \left\|1 - \sum_{k=0}^{\infty} T^k\right\| = \left\|\sum_{k=1}^{\infty} T^k\right\|$$

$$= \left\|T\left(-\sum_{k=0}^{\infty} T^k\right)\right\| = \|-T(1 - T)^{-1}\|$$

$$\le \|T\|\,\|(1 - T)^{-1}\|.$$

Hence, $\|1 - (1 - T)^{-1}\| \le \frac{\|T\|}{1 - \|T\|}$. $\qquad\square$

Lemma 8.40. *If $T \in B(V)$ and $\lambda_0 \notin \sigma(T)$, then $(T - \lambda 1)^{-1}$ exists for all $\lambda \in \mathbb{C}$ for which $|\lambda - \lambda_0| < \|(T - \lambda_0 1)^{-1}\|^{-1}$.*

Proof. For all $\lambda \in \mathbb{C}$ such that $|\lambda - \lambda_0| < \|(T - \lambda_0 1)^{-1}\|^{-1}$, Lemma 8.39 asserts that $\left(1 + (\lambda_0 - \lambda)(T - \lambda_0 1)^{-1}\right)$ is invertible. However,

$$T - \lambda 1 = T - \lambda_0 1 - (\lambda_0 - \lambda)1 = (T - \lambda_0 1)\left(1 + (\lambda_0 - \lambda)(T - \lambda_0 1)^{-1}\right),$$

Thus, $(T - \lambda 1)^{-1}$ exists for all $\lambda \in \mathbb{C}$ for which $|\lambda - \lambda_0| < \|(T - \lambda_0 1)^{-1}\|^{-1}$. □

Lemma 8.41. *Suppose that* $T \in B(V)$, $\varphi \in B(V)^*$, *and* $\lambda_0 \notin \sigma(T)$. *There is an* $\varepsilon > 0$ *such that if* $\Omega = \{\lambda \in \mathbb{C} \mid |\lambda - \lambda_0| < \varepsilon\}$, *then* $\Omega \cap \sigma(T) = \emptyset$ *and the function* $f : \Omega \to \mathbb{C}$ *defined by* $f(\lambda) = \varphi\left((T - \lambda 1)^{-1}\right)$ *is differentiable at* λ_0.

Proof. Let $\varepsilon = \|(T - \lambda_0 1)^{-1}\|^{-1}$ and $\Omega = \{\lambda \in \mathbb{C} \mid |\lambda - \lambda_0| < \varepsilon\}$. By Lemma 8.39, $\Omega \cap \sigma(T) = \emptyset$. Define $f : \Omega \to \mathbb{C}$ by $f(\lambda) = \varphi\left((T - \lambda 1)^{-1}\right)$. For every $\lambda \in \Omega$,

$$(T - \lambda 1)^{-1} - (T - \lambda_0)^{-1} = (\lambda - \lambda_0)\left((T - \lambda 1)^{-1}(T - \lambda_0 1)^{-1}\right).$$

This yields a difference quotient:

$$\frac{1}{\lambda - \lambda_0}\left((T - \lambda 1)^{-1} - (T - \lambda_0)^{-1}\right) = (T - \lambda 1)^{-1}(T - \lambda_0 1)^{-1}.$$

This limit, as $\lambda \to \lambda_0$, appears to be $(T - \lambda_0 1)^{-2}$. This is indeed true, since

$$\|(T - \lambda_0 1 + (\lambda_0 - \lambda)1)^{-1} - (T - \lambda_0 1)^{-1} + (\lambda_0 - \lambda)(T - \lambda_0 1)^{-2}\|$$

$$\leq 2\|(T - \lambda_0 1)^{-1}\|^3 |\lambda - \lambda_0|^2.$$

Hence,

$$\lim_{\lambda \to \lambda_0} \frac{f(\lambda) - f(\lambda_0)}{\lambda - \lambda_0} = \varphi((T - \lambda_0 1)^{-2}),$$

which implies that f is differentiable at $\lambda_0 \in \mathbb{C}$. □

We are now prepared to prove every operator T on a Banach space V has a nonempty compact spectrum.

Theorem 8.42. *If* V *is a Banach space and* $T \in B(V)$, *then* $\sigma(T)$ *is a nonempty compact subset of* $\{\zeta \in \mathbb{C} \mid |\zeta| \leq \|T\|\}$.

Proof. Lemma 8.38 shows that $\sigma(T) \subseteq \{\zeta \in \mathbb{C} \mid |\zeta| \leq \|T\|\}$; in addition, $\mathbb{C} \setminus \sigma(T)$ is an open set, by Lemma 8.40. Thus, $\sigma(T)$ is bounded and closed, and hence $\sigma(T)$ is compact. It remains to show that $\sigma(T)$ is nonempty.

Assume, contrary to what we aim to prove, that $\sigma(T) = \emptyset$. Choose any $\varphi \in B(V)^*$ and let $f : \mathbb{C} \to \mathbb{C}$ be defined by $f(\lambda) = \varphi((T - \lambda 1)^{-1})$. By Lemma 8.41, f is differentiable at each $\lambda_0 \in \mathbb{C}$. Hence, f is holomorphic on $\mathbb{C}$. Lemma 8.38 shows that $\lim_{|\lambda| \to \infty} \|(T - \lambda 1)^{-1}\| = 0$; therefore, as φ is bounded, $\lim_{|\lambda| \to \infty} f(\lambda) = 0$ as well.

On the compact set $\{\zeta \in \mathbb{C} \mid |\zeta| \leq \|T\|\}$, the continuous map f is bounded, and on the complement of this set f tends to 0 as $|\lambda| \to \infty$. Therefore, f is bounded on its entire domain $\mathbb{C}$. But the only bounded entire functions are the constant functions. Thus, there is a $\alpha \in \mathbb{C}$ such that $f(\lambda) = \alpha$, for all $\lambda \in \mathbb{C}$. Because $\lim_{\lambda \to \infty} f(\lambda) = 0$, the constant α must in fact be zero. This, therefore, proves that $\varphi((T - \lambda 1)^{-1}) = 0$ for all $\varphi \in B(V)^*$ and $\lambda \in \mathbb{C}$. By the Hahn-Banach Theorem, this implies that $(T - \lambda 1)^{-1} = 0$, for each λ, which is impossible since 0 is not invertible. Therefore, it must be that $\sigma(T) \neq \emptyset$. $\qquad \square$

The methods used to prove Theorem 8.42 also lead to the following continuity assertion: namely, that the set-valued map $T \mapsto \sigma(T)$ is upper semicontinuous.

Proposition 8.43. *If $T \in B(V)$, then for every open set $U \subseteq \mathbb{C}$ that contains $\sigma(T)$ there exists $\delta > 0$ such that $\sigma(S) \subset U$ for every $S \in B(V)$ for which $\|S - T\| < \delta$.*

Proof. Assume that U is an open subset of $\mathbb{C}$ that contains the spectrum of T. By Lemma 8.38, the function $\lambda \mapsto \|(T - \lambda 1)^{-1}\|$ tends to 0 as $|\lambda| \to \infty$. Hence, there exists $M > 0$ such that $|(T - \lambda 1)^{-1}\| < M$ for all $\lambda \notin U$. Let $\delta = M^{-1}$. If $S \in B(V)$ satisfies $\|S - T\| < \delta$, and if $\lambda \notin U$, then

$$\|(S - \lambda 1) - (T - \lambda 1)\| = \|S - T\| < \delta = \frac{1}{M} < \frac{1}{\|(T - \lambda 1)^{-1}\|}.$$

The invertibility of $T - \lambda 1$ and this inequality above show, by Lemma 8.40, that $S - \lambda 1$ is invertible. Therefore, if $\lambda \in \sigma(S)$, then λ cannot be exterior to U. $\qquad \square$

Although $\sigma(T)$ lies in the closed disc of radius $\|T\|$ and centre $0 \in \mathbb{C}$, there could be a smaller disc with the same centre that contains $\sigma(T)$. The radius of the smallest of such discs is called the spectral radius.

Definition 8.44. The *spectral radius* of $T \in B(V)$ is the quantity $\operatorname{spr} T$ defined by

$$\operatorname{spr} T = \max_{\lambda \in \sigma(T)} |\lambda|.$$

Theorem 8.45. *If $T \in B(V)$, then $\lim_n \|T^n\|^{1/n}$ exists and*

$$\operatorname{spr} T = \lim_n \|T^n\|^{1/n}. \tag{8.5}$$

Proof. If $\lambda \in \sigma(T)$, then

$$(T^n - \lambda^n 1) = (T - \lambda 1) \sum_{j=1}^{n} \lambda^{j-1} T^{n-j} = \left(\sum_{j=1}^{n} \lambda^{j-1} T^{n-j} \right) (T - \lambda 1).$$

If $T^n - \lambda^n 1$ were invertible, then by the equations above $(T - \dot{\lambda}1)$ would have a left and a right inverse and, thus, be invertible (Exercise 8.69). Therefore, $T^n - \lambda^n 1$ is not invertible. Thus, $\lambda^n \in \sigma(T^n)$ and $|\lambda|^n \leq \|T^n\|$. Hence,

$$\mathrm{spr}\, T \leq \liminf_n \|T^n\|^{1/n}.$$

Let $\Omega, \Delta \subseteq \mathbb{C}$ be the open sets

$$\Omega = \{\zeta \in \mathbb{C} \mid |\zeta|\,(\mathrm{spr}\, T) < 1\} \quad \text{and} \quad \Delta = \{\lambda \in \mathbb{C} \mid |\lambda|\,\|T\| < 1\}.$$

Note that $\Delta \subseteq \Omega$ because $\mathrm{spr}\, T \leq \|T\|$.

Now, choose any $\varphi \in B(V)^*$ and define $f : \Omega \to \mathbb{C}$ by

$$f(\lambda) = \varphi\left((1 - \lambda T)^{-1}\right).$$

If $\lambda \in \Delta$, then $(1 - \lambda T)^{-1}$ is a geometric series (Lemma 8.38); hence,

$$f(\lambda) = \sum_{n=0}^{\infty} \lambda^n \varphi(T^n), \quad \forall\, \lambda \in \Delta.$$

On the other hand, if $\zeta \in \Omega$ is nonzero, then

$$f(\zeta) = \frac{1}{\zeta} \varphi\left((\frac{1}{\zeta}1 - T)^{-1}\right).$$

By Lemma 8.41, f is differentiable at each nonzero point of Δ. Thus, since $0 \in \Delta \subseteq \Omega$, f is holomorphic on the disc Ω. By the uniqueness of the power series expansion about the origin, we obtain

$$f(\zeta) = \sum_{n=0}^{\infty} \zeta^n \varphi(T^n), \quad \forall\, \zeta \in \Omega.$$

Hence, $\lim_n |\varphi(\zeta^n T^n)| = 0$ for every $\zeta \in \Omega$. Thus, for each $\zeta \in \Omega$ there is an $M_{\zeta,\varphi} > 0$ such that $|\varphi(\zeta^n T^n)| \leq M_{\zeta,\varphi}$ for all $n \in \mathbb{N}$.

Now fix $\zeta \in \Omega$ and consider the family $\{\zeta^n T^n \mid n \in \mathbb{N}\}$. Because $B(V)$ embeds into $B(V)^{**}$ isometrically as a Banach space of operators on $B(V)^*$, we consider the family $\{\zeta^n T^n \mid n \in \mathbb{N}\}$ as acting on $B(V)^*$ in this way—namely,

$$\zeta^n T^n\,(\varphi) = \varphi(\zeta^n T^n), \quad \varphi \in B(V)^*.$$

By the Uniform Boundedness Principle (Theorem 8.5), either there is an $R_\zeta > 0$ such that $\|\zeta^n T^n\| \leq R_\zeta$ for all $n \in \mathbb{N}$ or $\sup_n \|\zeta^n T^n \varphi\| = \infty$ for a dense set of $\varphi \in B(V)^*$. However, the latter situation cannot occur, since $|\varphi(\zeta^n T^n)| \leq M_{\zeta,\varphi}$ for all $n \in \mathbb{N}$. Hence, there is an $R_\zeta > 0$ such that $\|\zeta^n T^n\| \leq R_\zeta$ for all $n \in \mathbb{N}$. Thus,

$$|\zeta|\,\|T^n\|^{1/n} \leq R_\zeta^{1/n}.$$

Now choose a nonzero $\zeta \in \Omega$. Therefore, $\operatorname{spr} T < 1/|\zeta|$ and

$$\limsup_n \|T^n\|^{1/n} \leq \frac{\limsup_n R_\zeta^{1/n}}{|\zeta|} = \frac{1}{|\zeta|}.$$

Hence,

$$\limsup_n \|T^n\|^{1/n} \leq \inf_{\zeta \in \Omega \setminus \{0\}} \frac{1}{|\zeta|} = \operatorname{spr} T.$$

This proves that

$$\limsup_n \|T^n\|^{1/n} \leq \operatorname{spr} T \leq \liminf_n \|T^n\|^{1/n}.$$

That is, $\lim_n \|T^n\|^{1/n}$ exists and equals $\operatorname{spr} T$. $\square$

The next result concerns the relationship between the spectra of ST and TS.

Proposition 8.46. $\sigma(ST) \cup \{0\} = \sigma(TS) \cup \{0\}$, *for all $S, T \in B(V)$.*

Proof. If $1 - ST$ is invertible, then $(1 - TS)(1 + TZS) = (1 + TZS)(1 - TS) = 1$, where $Z = (1 - ST)^{-1}$. Interchanging the roles of S and T leads to: $1 - ST$ is invertible if and only if $1 - TS$ is invertible. Hence, if $\lambda \neq 0$, then $ST - \lambda 1$ is invertible if and only if $TS - \lambda 1$ is invertible. $\square$

If $X \subset \mathbb{C}$, and if f is a polynomial, then $f(X)$ denotes the set $\{f(\zeta) \mid \zeta \in X\}$.

Theorem 8.47 (Polynomial Spectral Mapping Theorem). *If $T \in B(V)$, then*

$$f(\sigma(T)) = \sigma(f(T))$$

for every complex polynomial f.

Proof. Let f be a complex polynomial and suppose that $\lambda \in \sigma(T)$. Let $g(t) = f(t) - f(\lambda)$. As $g(\lambda) = 0$, there is a polynomial h such that $g(t) = (t - \lambda)h(t)$. Hence, $f(T) - f(\lambda)1 = g(T) = (T - \lambda 1)h(T) = h(T)(T - \lambda 1)$. If $f(T) - f(\lambda)1$ were invertible, then these equations imply that $T - \lambda 1$ has a left and right inverse, from which we would conclude that $T - \lambda 1$ is invertible (Exercise 8.69). Therefore, as $T - \lambda 1$ is not invertible, it must be that $f(T) - f(\lambda)1$ is not invertible. That is, $f(\lambda) \in \sigma(f(T))$. This proves that $f(\sigma(T)) \subseteq \sigma(f(T))$.

Conversely, suppose that $\omega \notin f(\sigma(T))$. Thus, $\omega \neq f(\lambda)$, for all $\lambda \in \sigma(T)$. Let $h(t) = f(t) - \omega$ and factor h: $h(t) = (t - \omega_1)^{n_1} \cdots (t - \omega_m)^{n_m}$. Since $h(\lambda) \neq 0$ for all $\lambda \in \sigma(T)$, $\lambda \neq \omega_j$ for every j and $\lambda \in \sigma(T)$. That is, $\omega_j \notin \sigma(T)$ for each j. The factorisation h leads to the following expression for $h(T)$:

$$f(T) - \omega 1 = (T - \omega_1 1)^{n_1} \cdots (T - \omega_m 1)^{n_m}.$$

Because $\omega_j \notin \sigma(N)$ for each j, $f(T) - \omega 1$ is a product of invertible operators and is, hence, invertible. Thus, $\omega \notin \sigma(f(T))$, and so $\sigma(f(T)) \subseteq f(\sigma(T))$. $\square$

8.8 Eigenvalues and Approximate Eigenvalues

By Proposition 8.8, an operator T on a Banach space V is invertible if and only if T is bounded below and has dense range. This fact leads to the following definitions for subsets of the spectrum.

Definition 8.48. Assume that T is an operator on a Banach space V.

1. The *point spectrum* of T is the set $\sigma_p(T)$ of all $\lambda \in \mathbb{C}$ for which the operator $T - \lambda 1$ on V is not injective.
2. The *approximate point spectrum* of T is the set $\sigma_{ap}(T)$ of all $\lambda \in \mathbb{C}$ for which the operator $T - \lambda 1$ on V is not bounded below.
3. The *defect spectrum* of T is the set $\sigma_d(T)$ of all $\lambda \in \mathbb{C}$ for which the operator $T - \lambda 1$ on V does not have dense range.

The elements of $\sigma_p(T)$ are precisely the *eigenvalues* of T—that is, the set of $\lambda \in \mathbb{C}$ for which there is nonzero vector v (called an *eigenvector*) such that $Tv = \lambda v$. The elements of $\sigma_{ap}(T)$ are called *approximate eigenvalues* of T. If $\lambda \in \sigma_{ap}(T)$ and if $\{v_k\}_{k \in \mathbb{N}}$ is a sequence of unit vectors for which $\lim_k \|Tv_k - \lambda v_k\| = 0$, then the vectors v_k are called *approximate eigenvectors*.

Proposition 8.49. *If V is a Banach space and $T \in B(V)$, then $\sigma_d(T) = \sigma_p(T^*)$.*

Proof. Suppose that $\lambda \in \sigma_d(T)$; thus, $\overline{\mathrm{ran}\,(T - \lambda 1)}$ is a proper subspace of V. Let $W = V/\overline{\mathrm{ran}\,(T - \lambda 1)}$ and let $q : V \to W$ be the canonical quotient map $q(v) = \dot{v}$, for $v \in V$. Since W is nonzero, there exists $\psi \in W^*$ of norm $\|\psi\| = 1$. Let $\varphi \in V^*$ be given by $\varphi = \psi \circ q$ and note that $\overline{\mathrm{ran}\,(T - \lambda 1)} \subseteq \ker \varphi$. Thus, $T^*\varphi = \lambda \varphi$, which shows that $\lambda \in \sigma_p(T^*)$.

Conversely, assume that $\lambda \in \sigma_p(T^*)$. Thus, there exists $\varphi \in V^*$ of norm 1 such that $T^*\varphi = \lambda \varphi$; that is, $\varphi(Tv - \lambda v) = 0$ for all $v \in V$. If the range of $T - \lambda 1$ were dense, then we would conclude that $\varphi(w) = 0$, for all $w \in V$, and this would imply that $\varphi = 0$, by Corollary 6.23. But since $\varphi \neq 0$, the range of $T - \lambda 1$ cannot be dense. Hence, $\lambda \in \sigma_d(T)$. $\square$

Proposition 8.50. *If V is a Banach space and $T \in B(V)$, then $\sigma_{ap}(T)$ is compact and $\partial \sigma(T) \subseteq \sigma_{ap}(T)$.*

Proof. To show that $\sigma_{ap}(T)$ is closed, let $\lambda \in \mathbb{C} \backslash \sigma_{ap}(T)$. Thus, there is a $\delta > 0$ such that $\|(T - \lambda 1)v\| \geq \delta \|v\|$, for every $v \in V$. Let $\varepsilon = \delta/2$. If $\mu \in \mathbb{C}$ satisfies $|\lambda - \mu| < \varepsilon$, then, for every $v \in V$,

$$\delta \|v\| \leq \|(T - \lambda 1)v\| = \|(T - \mu 1)v + (\mu - \lambda)v\| \leq \|(T - \mu 1)v\| + |\lambda - \mu| \|v\|,$$

which implies that $T - \mu 1$ is bounded below by $\delta/2$. Hence, the complement of $\sigma_{ap}(T)$ is open, which proves that $\sigma_{ap}(T)$ is closed. As $\sigma(T)$ is bounded, $\sigma_{ap}(T)$ is compact.

Suppose that $\lambda \in \partial \sigma(T)$. Therefore, there is a sequence $\lambda_n \in \mathbb{C} \backslash \sigma(T)$ such that $|\lambda_n - \lambda| \to 0$. If there were a $\gamma > 0$ for which $\|(T - \lambda_n 1)^{-1}\| < \gamma$ for every $n \in \mathbb{N}$, then it would be true that $T - \lambda 1$ is invertible (by Lemma 8.40). Specifically, if n_0 is such that $|\lambda_{n_0} - \lambda| < 1/\gamma$, then the distance between the invertible operator $T - \lambda_{n_0} 1$ and the operator $T - \lambda 1$ satisfies

$$\|(T - \lambda 1) - (T - \lambda_{n_0} 1)\| = |\lambda_{n_0} - \lambda| < 1/\gamma < \|(T - \lambda_{n_0} 1)^{-1}\|^{-1}.$$

This would imply that $T - \lambda 1$ is invertible, contrary to the fact that $\lambda \in \sigma(T)$. Hence, it must be that $\{\|(T - \lambda_n 1)^{-1}\|\}_{n \in \mathbb{N}}$ is an unbounded sequence; that is, $\|(T - \lambda_n 1)^{-1}\|^{-1} \to 0$.

By definition of norm, for each $n \in \mathbb{N}$ there is a unit vector $w_n \in V$ such that

$$\|(T - \lambda_n 1)^{-1}\| < \|(T - \lambda_n 1)^{-1} w_n\| + \frac{1}{n}.$$

Let

$$v_n = \frac{1}{\|(T - \lambda_n 1)^{-1} w_n\|} (T - \lambda_n 1)^{-1} w_n, \quad \forall n \in \mathbb{N}.$$

Thus, $\|v_n\| = 1$ and

$$(T - \lambda 1) v_n = \frac{1}{\|(T - \lambda_n 1)^{-1} w_n\|} w_n + (\lambda_n - \lambda) v_n.$$

Hence,

$$\|(T - \lambda 1) v_n\| \leq \frac{1}{\|(T - \lambda_n 1)^{-1}\|^{-1} - (1/n)} + |\lambda - \lambda_n|,$$

which converges to 0 as $n \to \infty$. That is, $\lambda \in \sigma_{\mathrm{ap}}(T)$. $\qquad\square$

Corollary 8.51. *Every Banach space operator has an approximate eigenvalue.*

We turn now to some sample calculations of the eigenvalue and approximate eigenvalue problem.

Example 8.52. *The eigenvalue problem* $\int_0^t \left(\int_s^1 f(u) \, dm(u) \right) dm(s) = \lambda f(t)$, *for almost all* $t \in [0, 1]$, *in the Hilbert space* $L^2([0, 1], \mathfrak{M}, m)$ *admits countably many solutions* $\{\lambda_k\}_{k \in \mathbb{N}}$ *given by*

$$\lambda_k = \frac{4}{(2k - 1)^2 \pi^2},$$

where each eigenvalue λ_k *has a corresponding eigenvector*

$$f_k(t) = 2i \sin(t / \sqrt{\lambda_k}).$$

Proof. Consider the unit square $[0,1] \times [0,1] \subset \mathbb{R}^2$ and denote by $\mathscr{L}^2$ and L^2, respectively, the spaces $\mathscr{L}^2([0,1],\mathfrak{M},m)$ and $L^2([0,1],\mathfrak{M},m)$. Define a linear transformation $\mathscr{K} : \mathscr{L}^2 \to \mathscr{L}^2$ by

$$\mathscr{K}f(t) = \int_0^t \left(\int_s^1 f(s)\,dm(s) \right) dm(t), \quad f \in \mathscr{L}^2.$$

The linear transformation $\mathscr{K}$ induces an operator $K : L^2 \to L^2$ in which $\|K\dot{f}\| \leq \|\dot{f}\|$, for all $f \in \mathscr{L}^2$. The eigenvalue problem $K\dot{f} = \lambda\dot{f}$ in L^2 corresponds to an eigenvalue-type problem in $\mathscr{L}^2$: namely,

$$\int_0^t \left(\int_s^1 f(u)\,dm(u) \right) dm(s) = \lambda f(t), \text{ for almost all } t \in [0,1].$$

As the left-hand side of the equation above is twice differentiable almost everywhere, differentiation once leads to

$$\int_t^1 f(u)\,dm(u) = \lambda \frac{df}{dt}.$$

Differentiation a second time yields

$$-f = \lambda \frac{d^2 f}{dt^2}.$$

Evaluation of the two differential equations above at the boundary values for t gives $f(0) = f'(1) = 0$.

Notice that $\langle K\dot{g},\dot{g} \rangle \geq 0$, for every $\dot{g} \in L^2$, and so $0 \leq \langle K\dot{f},\dot{f} \rangle = \lambda \|\dot{f}\|^2$ implies that $\lambda \geq 0$. Therefore, the general solution of $\lambda f'' + f = 0$ is $f(t) = \alpha_1 e^{it/\sqrt{\lambda}} + \alpha_2 e^{-it/\sqrt{\lambda}}$. But $f(0) = f'(1) = 0$ implies that $\alpha_2 = -\alpha_1$. Hence, $f(t) = 2i\sin(t/\sqrt{\lambda})$. To satisfy both $f \neq 0$ and $f'(1) = 0$, it is necessary and sufficient that $\frac{1}{\sqrt{\lambda}} = \frac{\pi}{2}(2k-1)$ for some $k \in \mathbb{N}$. Hence, the eigenvalues of K are

$$\lambda_k = \frac{4}{(2k-1)^2\pi^2}, \quad k \in \mathbb{N},$$

and the corresponding eigenvectors $\dot{f}_k \in L^2$ are determined by the functions $f_k(t) = 2i\sin(t/\sqrt{\lambda_k})$.   □

Recall that the *essential range* of an essentially bounded function $\psi : X \to \mathbb{C}$ on a measure space (X, Σ, μ) is the set ess-ran ψ of all $\lambda \in \mathbb{C}$ for which

$$\mu\left(f^{-1}(U)\right) > 0,$$

for every neighbourhood $U \subseteq \mathbb{C}$ of λ, and that Proposition 5.45 shows that

$$\text{ess-ran }\psi = \bigcap_{E \in \Sigma, \mu(E^c)=0} \overline{\psi(E)}.$$

Example 8.53. *If (X, Σ, μ) is a measure space, then the approximate point spectrum of the multiplication operator M_ψ on $L^p(X, \Sigma, \mu)$, where $1 \le p < \infty$, is given by*

$$\sigma_{\text{ap}}(M_\psi) = \text{ess-ran}\,\psi.$$

Proof. Recall from Example 8.1.7 that if $\psi \in \mathscr{L}^\infty(X, \Sigma, \mu)$, then M_ψ is defined by $M_\psi \dot{f} = \dot{\psi f}$, for $\dot{f} \in L^p(X, \Sigma, \mu)$ and has norm $\|M_\psi\| \le \text{ess-sup}\,\psi$. If, in addition, (X, Σ, μ) is a σ-finite measure space, then $\|M_\psi\| = \text{ess-sup}\,\psi$; however, this equality is not required for the spectral calculation below, and so we need not assume anything special of the measure space (X, Σ, μ).

Suppose that $\lambda \in \text{ess-ran}\,\psi$. Choose any $\varepsilon > 0$ and for each $E \in \Sigma$ for which $\mu(E^c) = 0$ let $F_E \subset E$ be the measurable subset of X defined by

$$F_E = \psi^{-1}(B_\varepsilon(\lambda)) \cap E.$$

Note that each F_E is nonempty because $\lambda \in \overline{\psi(E)}$. Now let

$$F = \bigcup_{E \subset X,\, \mu(X \backslash E) = 0} F_E$$

and consider the function $f = \chi_F$. Thus, $\mu(F) > 0$ and

$$\|(M_\psi - \lambda 1)\dot{f}\|^p = \int_F |\psi(t) - \lambda|^p |f(t)|^p \, d\mu(t) \le \varepsilon^p \|\dot{f}\|^p.$$

Hence, $M_\psi - \lambda 1$ is not bounded below, which proves that $\lambda \in \sigma_{\text{ap}}(M_\psi)$. Thus, ess-ran $\psi \subseteq \sigma_{\text{ap}}(M_\psi)$.

Conversely, suppose that $\lambda \notin \text{ess-ran}\,\psi$. Thus, there exists at least one $E \in \Sigma$ such that $\mu(E^c) = 0$ and $\lambda \notin \overline{\psi(E)}$, and so there is a $\delta > 0$ such that $|\psi(t) - \lambda| \ge \delta > 0$ for all $t \in E$. Since $0 = \mu(E^c)$, $\|\dot{f}\|^p = \int_E |f|^p \, d\mu$, for every $f \in \mathscr{L}^p(X, \Sigma, \mu)$. Further, for any $f \in \mathscr{L}^p(X, \Sigma, \mu)$,

$$\|(M_\psi - \lambda 1)\dot{f}\|^p = \int_E |\psi - \lambda|^p |f|^p \, d\mu \ge \delta^p \int_E |f|^p \, d\mu = \delta^p \|\dot{f}\|^p.$$

Thus, $M_\psi - \lambda 1$ is bounded below, and so $\lambda \notin \sigma_{\text{ap}}(M_\psi)$. Hence, $\sigma_{\text{ap}}(M_\psi) \subseteq \text{ess-ran}\,\psi$, which completes the proof that $\sigma_{\text{ap}}(M_\psi) = \text{ess-ran}\,\psi$. $\square$

It can happen that M_ψ has no eigenvalues. For example, if $X = [0, 1]$, $\Sigma = \mathfrak{M}$, $\mu = m$, and $\psi(t) = t$, then $M_\psi \dot{f} = \lambda \dot{f}$ implies that $0 = t f(t) - \lambda f(t)$ for almost all $t \in [0, 1]$, which can only happen if $f(t) = 0$ almost everywhere. But, in this case, $\dot{f} = 0$ in $L^p(X, \Sigma, \mu)$, thereby violating the requirement that an eigenvector be nonzero.

8.9 The Spectra and Invariant Subspaces of Compact Operators

The nonzero points of the spectrum of a compact operator behave rather closely to the eigenvalues of a matrix, as shown by the Fredholm Alternative below.

Theorem 8.54 (The Fredholm Aternative). *Assume that K is a compact operator on a Banach space V and that $\lambda \in \mathbb{C}$ is nonzero. Then exactly one of the following statements holds:*

1. *$Kv = \lambda v$ for some nonzero $v \in V$.*
2. *$K - \lambda 1$ is invertible.*

Proof. Propositions 8.30 and 8.32 show that $1 - K$ is injective if and only if $1 - K$ is surjective. Therefore, if $\lambda \in \mathbb{C}$ is nonzero, then replacing K by the compact operator $\frac{1}{\lambda}K$ and using the fact that $\sigma(K) = \sigma_{\mathrm{ap}}(K) \cup \sigma_{\mathrm{d}}(K)$, we obtain $\lambda \in \sigma_{\mathrm{p}}(K)$ or $\lambda \notin \sigma(K)$. $\qquad\square$

The Fredholm Alternative has implications for what properties the spectrum of a compact operator may exhibit.

Theorem 8.55. *If K is a compact operator acting on an infinite-dimensional Banach space V, then*

1. *$0 \in \sigma(K)$,*
2. *each nonzero $\lambda \in \sigma(K)$ is an eigenvalue of finite geometric multiplicity,*
3. *$\sigma(K)$ is a finite or countably infinite set, and*
4. *if $\sigma(K)$ is infinite, then 0 is the only limit point of $\sigma(K)$.*

Proof. The fact that V has infinite dimension implies that $K(V)$ is a proper ideal of $B(V)$ (Theorem 8.35), and so no compact operator can possess an inverse. Thus, $0 \in \sigma(K)$.

Assume next that $\lambda \in \sigma(K)$ is nonzero. Theorem 8.54 (Fredholm Alternative) asserts that λ is an eigenvalue of K. By Proposition 8.25, the dimension of $\ker(K - \lambda 1)$ must be finite.

To show that $\sigma(K)$ is a finite or countably infinite set, it is enough to assume that $\sigma(K)$ is infinite and to prove that 0 is the only limit point of $\sigma(K)$. Therefore, assume that $\sigma(K)$ is infinite. Assume that 0 is not a limit point of $\sigma(K)$. There exist, therefore, $1 > \delta > 0$ and a sequence $\{\lambda_n\}_n \subset \sigma(K)$ (distinct elements) with the property that $|\lambda_n| \geq \delta$ for every $n \in \mathbb{N}$. Each λ_n is an eigenvalue of K; let $v_n \in V$ be corresponding eigenvectors of length 1. Fix $m \in \mathbb{N}$ and let $f_1, \ldots, f_m \in \mathbb{C}[t]$ be polynomials such that $f_i(\lambda_j) = 0$, for $j \neq i$, and $f_i(\lambda_i) = 1$. (Such polynomials exist; for example, one could use the Lagrange interpolation to construct them.) Therefore,

$$\text{if } \alpha_1, \ldots, \alpha_m \in \mathbb{C} \text{ satisfy } \sum_{j=1}^{m} \alpha_j v_j = 0, \text{ then } 0 = f(K)\left(\sum_{j=1}^{m} \alpha_j v_j\right) = \sum_{j=1}^{m} \alpha_j f(\lambda_j) v_j, \text{ for}$$

every $f \in \mathbb{C}[t]$. In particular, if $f = f_i$, then $0 = \alpha_i v_i$, and so $\alpha_i = 0$. This proves that the sequence $\{v_n\}_n$ consists of linearly independent vectors.

For each $m \in \mathbb{N}$ let $M_m = \mathrm{Span}\{v_1, \ldots, v_m\}$, which yields an ascending sequence $M_1 \subset M_2 \subset \cdots$ of finite-dimensional subspaces of V. By Lemma 5.19, there is a sequence of unit vectors $w_n \in V$ such that $w_n \in M_{n+1}$, $\|w_n - u\| \geq \delta$ for all $u \in M_n$, and $\|w_n - w_m\| \geq \delta$ if $m \neq n$. Each subspace M_n is spanned by eigenvectors of K; thus, K maps M_n back into itself. Therefore, $K - \lambda_n 1$ maps M_n into M_{n-1}, by the following calculation:

$$(K - \lambda_n 1)\left(\sum_{j=1}^{n} \alpha_j v_j\right) = \sum_{j=1}^{n} \alpha_j (\lambda_j - \lambda_n) v_j = \sum_{j=1}^{n-1} \alpha_j (\lambda_j - \lambda_n) v_j.$$

Therefore, if $\ell < n$, then $u = K w_\ell - (K - \lambda_n 1) w_n \in M_{n-1}$. Hence,

$$\|K w_n - K w_\ell\| = \|\lambda_n w_n + (K - \lambda_n 1) w_n - K w_\ell\| = \|\lambda_n w_n - u\| = |\lambda_n| \, \|w_n - \tfrac{1}{\lambda_n} u\|$$

$$\geq |\lambda_n| \delta \geq \delta^2 > 0.$$

This means that $\{K w_n\}_n$ does not admit a convergent subsequence, in contradiction to the compactness of K. Therefore, it must be that 0 is a limit point of $\sigma(K)$. The same argument shows that no nonzero $\lambda \in \sigma(K)$ could possibly be a limit point of $\sigma(K)$. $\qquad\square$

Definition 8.56. If $\mathscr{S} \subseteq B(V)$ is a nonempty subset, then a subspace $M \subseteq V$ is *invariant for* $\mathscr{S}$ if $Sv \in M$ for all $S \in \mathscr{S}$ and $v \in M$. If an invariant subspace M for $\mathscr{S}$ is neither $\{0\}$ nor V, then M is a *nontrivial invariant subspace* for $\mathscr{S}$. The set of all subspaces $M \subseteq V$ that are invariant under $\mathscr{S}$ is denoted by $\mathrm{Lat}\,\mathscr{S}$ and is called the *invariant-subspace lattice* of $\mathscr{S}$.

If $\mathscr{S}$ is a singelton set $\mathscr{S} = \{T\}$, for some operator T on V, then the notation $\mathrm{Lat}\,T$ is used in place of $\mathrm{Lat}\,\mathscr{S}$.

One of the most basic examples of invariant subspaces comes from eigenvectors (if they exist): if for some nonzero $v \in V$ and $\lambda \in \mathbb{C}$ one has $Tv = \lambda v$, then the 1-dimensional subspace $\mathrm{Span}\{v\}$ is T-invariant. Indeed, if λ is an eigenvalue of T, then entire eigenspace $\ker(T - \lambda 1)$ is T-invariant. More generally, $\ker(T - \lambda 1)^k \in \mathrm{Lat}\,T$ for every $k \in \mathbb{N}$.

Definition 8.57. A subspace $M \subseteq V$ is *hyperinvariant* for an operator $T \in B(V)$ if M is an invariant subspace for $\{T\}'$, where

$$\{T\}' = \{S \in B(V) \,|\, ST = TS\}.$$

Because $T \in \{T\}'$, any hyperinvariant subspace for T is an invariant subspace for T.

Proposition 8.58. *If λ is an eigenvalue of T, then $\ker(T - \lambda 1)$ is hyperinvariant for T.*

Proof. Exercise 8.77. $\qquad\square$

The existence of nontrivial invariant subspaces depends on both the particular Banach space at hand and on the size of the set $\mathscr{S}$ of operators. But even if $\mathscr{S}$ is a singleton set $\mathscr{S} = \{T\}$, the operator T may fail to admit any nontrivial invariant subspaces [46]. It is still an open problem whether every operator acting on a separable Hilbert space has nontrivial subspaces. A positive result, however, is the following theorem of Lomonosov.

Theorem 8.59 (Lomonosov). *If* $K \in B(V)$ *is a nonzero compact operator on an infinite-dimensional Banach space V, then K has a nontrivial hyperinvariant subspace.*

Proof. Because $K \neq 0$, we may assume without loss of generality that $\|K\| = 1$.

If K has an eigenvalue λ, then $\ker(K - \lambda 1)$ is hyperinvariant for K. If $\lambda \neq 0$, then the compactness of K implies the finite dimensionality of $\ker(K - \lambda 1)$, which means that this hyperinvariant subspace is indeed nontrivial. If $\lambda = 0$, then $K \neq 0$ implies that $\ker K \neq V$, and so again $\ker K$ is a nontrivial hyperinvariant subspace.

Assume, therefore, that K has no eigenvalues. Let $v_0 \in V$ be any vector for which $\|Kv_0\| > 1$. Observe that, because $\|K\| = 1$, the norm of v_0 necessarily satisfies $\|v_0\| > 1$. Let U be the open set $U = \{v \in V \mid \|v - v_0\| < 1\}$ and for each $S \in \{K\}'$ let W_S be the open set $W_S = S^{-1}(U) = \{v \in V \mid \|Sv - v_0\| < 1\}$. If $v \in \overline{U}$, then $\|v - v_0\| \leq 1$ and so $\|Kv - Kv_0\| \leq 1$. Therefore, the condition $\|Kv_0\| > 1$ implies that the zero vector does not belong to $K(\overline{U})$ or its closure $C = \overline{K(\overline{U})}$, which by the compactness of the operator K is a compact subset of V. Hence, C is a compact subset of $V \setminus \{0\}$.

Because $\|v_0\| > 1$, the zero vector neither belongs to $\overline{U}$ nor to any of the open sets W_S, for $S \in \{K\}'$. Thus, $0 \notin \bigcup_{S \in \{K\}'} W_S$. Suppose, though, that $\bigcup_{S \in \{K\}'} W_S = V \setminus \{0\}$. Thus, $\{W_S\}_{S \in \{K\}'}$ is an open cover of the compact set C and so there are $S_1, \ldots, S_n \in \{K\}'$ such that $C \subset \bigcup_{j=1}^{n} W_{S_j}$. The vector Kv_0 is an element of $K(U)$ and, hence, of C; thus, $Kv_0 \in W_{S_{i_1}}$ for some $i_1 \in \{1, \ldots, n\}$. Thus, $S_{i_1}(Kv_0) \in U$ and $KS_{i_1}Kv_0 \in C$, which therefore implies that $KS_{i_1}Kv_0 \in W_{S_{i_2}}$, for some $i_2 \in \{1, \ldots, n\}$, and that $S_{i_2}KS_{i_1}Kv_0 \in U$. Continuing inductively and using the fact that $S_jK = KS_j$ for all j we deduce that for every $m \in \mathbb{N}$ there is an $i_m \in \{1, \ldots, n\}$ such that

$$S_{i_m}S_{i_{m-1}} \cdots S_{i_1}K^m v_0 = S_{i_m}KS_{i_{m-1}} \cdots S_{i_1}Kv_0 \in U.$$

Now if $\alpha = \max_j \|S_j\|$ and if $\tilde{S}_{i_k} = \frac{1}{\alpha}S_{i_k}$, then

$$\tilde{S}_{i_m}\tilde{S}_{i_{m-1}} \cdots \tilde{S}_{i_1} (\alpha K)^m v_0 \in U$$

and is of norm no greater than $\|(\alpha K)^m\| \|v_0\|$. The fact that K has no eigenvalues implies, by the Fredholm Alternative, that the spectrum of K is the singleton set $\sigma(K) = \{0\}$. Therefore, $\sigma(\alpha K) = \{0\}$ which implies, by the Spectral Radius Formula, that $0 = \lim_m \|(\alpha K)^m\|^{1/m}$. Hence, $\lim_m \|(\alpha K)^m\| = 0$ which in turn implies that $0 \in \overline{U}$ because $\tilde{S}_{i_m} \tilde{S}_{i_{m-1}} \cdots \tilde{S}_{i_1} (\alpha K)^m v_0 \in U$ for all m. However, this conclusion is in contradiction to $0 \notin \overline{U}$. Therefore, the assumption that $\bigcup_{S \in \{K\}'} W_S = V \setminus \{0\}$ cannot be true, and so there is at least one nonzero vector $v \in V$ for which $v \notin W_S$ for every $S \in \{K\}'$.

With this vector v define a subspace M of V by

$$M = \overline{\{Sv \mid S \in \{K\}'\}}.$$

Note that M is hyperinvariant for K and that $0 \neq v \in M$; thus, the only question is whether $M \neq V$. If it so happened that $M = V$, then $\{Sv \mid S \in \{K\}'\}$ would be dense in V, and in particular there would exist some operator $S \in \{K\}'$ for which $\|Sv - v_0\| < 1$. However, this can never happen because $v \notin W_S$ for every $S \in \{K\}'$. Hence, $M \neq V$. $\qquad\square$

Problems

8.60. Recall that an operator $T \in B(V)$ has finite rank if the range of T has finite dimension. In such cases, the rank of T is defined to be the dimension of the range of T.

1. Prove that if $\psi \in V^*$ and $w \in V$ are nonzero, then the operator $F_{\psi,w}$ on V defined by $F_{\psi,w}(v) = \psi(w)v$ is of rank 1.
2. Prove that if F is a rank-1 operator, then there exist nonzero $\psi \in V^*$ and $w \in V$ such that $F = F_{\psi,w}$.
3. Prove that if T has finite rank $n \in \mathbb{N}$, then there are rank-1 operators $F_1, \ldots, F_n \in B(V)$ such that $T = \sum_{j=1}^{n} F_j$.

8.61. Prove that if V and W are Banach spaces, then every operator $T : V \to W$ that is bounded below will have closed range.

8.62. If $\varphi \in \mathscr{L}^\infty(X, \Sigma, \mu)$ has the property that $|\varphi(x)| = 1$ for almost all $x \in X$, then show the multiplication $M_\varphi : L^p(X, \Sigma, \mu) \to L^p(X, \Sigma, \mu)$, for a fixed $p \in [1, \infty)$, defined by

$$M_\varphi(\dot{f}) = \dot{\varphi}\dot{f}$$

is an isometric operator.

8.63. Let H be a Hilbert space and suppose that $\xi, \xi_k \in H$, $k \in \mathbb{N}$, satisfy

$$\lim_{k \to \infty} \langle \xi_k, \eta \rangle = \langle \xi, \eta \rangle, \quad \forall \eta \in H.$$

1. Prove that the set $S = \{\xi_k \mid k \in \mathbb{N}\}$ is uniformly bounded.
2. Prove that $\|\xi\| \leq \liminf_k \|\xi_k\|$.

8.64. Prove that if E and F are idempotents acting on a Banach space V, then $E + F$ is an idempotent if and only if $EF = FE = 0$.

8.65. Prove that the range of an idempotent operator E on a Banach space V is closed.

8.66. Suppose that V is a Banach space and that $T \in B(V)$ is invertible. Prove that $\|T^{-1}\| \geq \|T\|^{-1}$.

8.67. Prove the following properties of the adjoint T of an operator $T \in B(V, W)$:

1. $\|T^*\| = \|T\|$,
2. $(T_1 + T_2)^* = T_1^* + T_2^*$,
3. if T is invertible, then T^* is invertible and $(T^*)^{-1} = (T^{-1})^*$, and
4. if $W = V$, then $(T_1 T_2)^* = T_2^* T_1^*$.

8.68. Let V be a Banach space and consider V as a subspace of V^{**}. Define T^{**} to be $(T^*)^*$. Prove that the restriction of T^{**} to V is T. That is, $T^{**}|_V = T$.

8.69. Assume that $R, S, T \in B(V)$ satisfy $ST = TR = 1$. Prove that T is invertible.

8.70. Prove that the following operators are compact:

1. αK, for every $\alpha \in \mathbb{C}$ and all compact operators K;
2. $K_1 + K_2$, for all compact operators K_1 and K_2;
3. KT and TK, for all compact operators K and all $T \in B(V)$;
4. F, for all $F \in B(V)$ with finite-dimensional range;

8.71. Let X be a compact Hausdorff space, select $\psi \in C(X)$, and let $T_\psi : C(X) \to C(X)$ be the operator of multiplication by ψ: $T_\psi f = \psi f$, for all $f \in C(X)$. For each $\lambda \in \mathbb{C}$, let $K_\lambda = \overline{\{t \in X \mid \psi(t) \neq \lambda\}}$ (the closure in X).

1. Prove that $\{t \in X \mid \psi(t) \neq \lambda\}$ is an open set, for every $\lambda \in \mathbb{C}$.
2. Prove that if $\lambda \in \sigma_p(T_\psi)$ if and only if $K_\lambda \neq X$.

8.72. Assume that $p, q \in (1, \infty)$ satisfy $1/p + 1/q = 1$. Define a linear transformation $S : \ell^p(\mathbb{N}) \to \ell^p(\mathbb{N})$ by

$$
Sv = S\left(\begin{bmatrix} v_1 \\ v_2 \\ v_3 \\ v_4 \\ \vdots \end{bmatrix} \right) = \begin{bmatrix} 0 \\ \alpha_1 v_1 \\ \alpha_2 v_2 \\ \alpha_3 v_3 \\ \vdots \end{bmatrix}, \quad \forall v \in \ell^p(\mathbb{N}),
$$

where $\{\alpha_k\}_{k \in \mathbb{N}} \subset \mathbb{C}$ is a sequence for which $\sup_k |\alpha_k| < \infty$.

1. Prove that S is an operator on $\ell^p(\mathbb{N})$.
2. Determine an explicit form for the adjoint operator S^* on $\ell^q(\mathbb{N})$.

3. If $\alpha_{2j} = 0$ and $\alpha_{2j-1} = 1$, for all $j \in \mathbb{N}$, then compute $\|S\|$.

8.73. Prove that $\|1 - (\lambda 1 - T)^{-1}(S - T)\| \geq 1$, if $S, T \in B(V)$ and $\lambda \in \sigma(S) \cap \sigma(T)^c$.

8.74. Prove that an operator K on a Banach space V is compact if and only if the set $\{Kv \mid v \in V, \|v\| = 1\}$ has compact closure.

8.75. Consider the unit square $X = [0,1] \times [0,1] \subset \mathbb{R}^2$, and let $\kappa \in C(X)$. Fix $1 \leq p < \infty$ and let L^p denote $L^p([0,1], m)$ (Lebesgue measure). Consider the integral operator $T_\kappa \in B(L^p)$.

1. Prove that if κ is a polynomial, which means κ has the form

$$\kappa(t,s) = \sum_{i=0}^{m} \sum_{j=0}^{n} \alpha_{ij} t^i s^j, \quad \text{for some } \alpha_{ij} \in \mathbb{C},$$

then T_κ is an operator of finite rank.
2. If $\kappa \in C(X)$ is arbitrary, prove that for every $\varepsilon > 0$ there is a polynomial $q \in \mathbb{C}[t,s]$ such that $|\kappa(t,s) - q(t,s)| < \varepsilon$, for all $t,s \in [0,1]$.
3. If $\kappa \in C(X)$ is arbitrary, prove that for every $\varepsilon > 0$ there is a finite rank operator $F \in B(L^p)$ such that $\|T_\kappa - F\| < \varepsilon$.

Prove that if $S, T \in B(V)$ satisfy $ST = TS$, then

8.76. Prove that if $T \in B(V)$ satisfies $TK = KT$ for every compact operator K, then $T = \lambda 1$ for some $\lambda \in \mathbb{C}$.

8.77. Prove that if λ is an eigenvalue of T, then $\ker(T - \lambda 1)$ is hyperinvariant for T.

8.78. Suppose that $T \in B(V)$. Prove that if V has finite dimension or is nonseparable, then T has a nontrivial hyperinvariant subspace.

8.79. Determine the hyperinvariant subspaces of the operator S on $\ell^2(n)$ given by

$$S = \begin{bmatrix} 0 & 1 & 0 & \cdots & 0 \\ 0 & 0 & 1 & \ddots & \vdots \\ 0 & 0 & \ddots & \ddots & 0 \\ \vdots & & \ddots & \ddots & 1 \\ 0 & \cdots & \cdots & 0 & 0 \end{bmatrix}.$$

Chapter 9
Spectral Theory in Banach Algebras

A Banach algebra is an associative algebra A with a norm $\|\cdot\|$ such that A is a Banach space in this norm, and the norm satisfies $\|xy\| \leq \|x\|\,\|y\|$ on all products xy of elements $x, y \in A$. Two examples of Banach algebras encountered to this point are: (i) $C(X)$, the algebra of continuous functions on a compact Hausdorff space X, and (ii) $B(V)$, the algebra of all bounded linear operators acting on V. A third class of Banach algebras has also been encountered, but the algebraic structure of these algebras has not yet been studied; these are the Banach algebras $L^\infty(X, \Sigma, \mu)$, whose algebraic structure is noted in Exercises 9.43 and 9.44.

In the case of the algebra $B(V)$, the notion of *spectrum* for Banach space operators was analysed in detail in the Chapter 8, motivated by the notion of eigenvalue from linear algebra. However, the initial properties of the spectrum of an operator T carried out in Chapter 8 made little or no specific reference to the action of T on the space V, but rather made reference to the behaviour of T as an element of the algebra $B(V)$. Following this mode of thinking, the present chapter develops spectral theory within the general context of Banach algebras, and concludes with a few applications.

9.1 Banach Algebras

Recall from Definition 5.25 that a *Banach algebra* is a complex associative algebra A together with a norm $\|\cdot\|$ on A such that $\|xy\| \leq \|x\|\,\|y\|$, for all $x, y \in A$, and A is a Banach space under the norm $\|\cdot\|$. Furthermore, if A is a unital algebra, then A is a *unital Banach algebra* if $\|1\| = 1$. A Banach algebra A is *abelian* if $xy = yx$ for every $x \in A$.

The submultiplicativity of the norm, which is to say that $\|xy\| \leq \|x\|\,\|y\|$, for all $x, y \in A$, ensures that multiplication is a continuous map $A \times A \to A$ (Exercise 9.45). The natural maps between Banach algebras are homomorphisms.

© Springer International Publishing Switzerland 2016
D. Farenick, *Fundamentals of Functional Analysis*, Universitext,
DOI 10.1007/978-3-319-45633-1_9

Definition 9.1. If A and B are Banach algebras, then a function $\phi : A \to B$ is

1. a *homomorphism*, if ϕ is a bounded linear operator and $\phi(xy) = \phi(x)\phi(y)$ for all $x, y \in A$, and is
2. an *isomorphism*, if ϕ is a bijective homomorphism.

If A and B are unital Banach algebras, then a homomorphism $\phi : A \to B$ for which $\phi(1_A) = 1_B$ is called a *unital homomorphism*.

Definition 9.2. Suppose that A is a unital Banach algebra.

1. An element $x \in A$ is *invertible* if there is an element $y \in A$ such that $xy = yx = 1$.
2. The *spectrum* of an element $x \in A$ is the set $\sigma(x)$ defined by

$$\sigma(x) = \{\lambda \in \mathbb{C} \,|\, x - \lambda 1 \text{ is not invertible in } A\}.$$

3. The *spectral radius* of x is the quantity $\mathrm{spr}\, x$ defined by

$$\mathrm{spr}\, x = \sup_{\lambda \in \sigma(x)} |\lambda|.$$

As with operators in the Banach algebra $B(V)$, if an element $x \in A$ is invertible, then the element $y \in A$ for which $xy = yx = 1$ is necessarily unique and is denoted by x^{-1} and is called the *inverse* of x. Furthermore, the results about invertible operators and their spectra established in Section 8.7 made use only of the fact that $B(V)$ is a unital Banach algebra—that is, at no point was the action of an operator T on the Banach space V considered. Therefore, the results of Section 8.7 carry over to unital Banach algebras *verbatim*, including the following theorem.

Theorem 9.3. *If A is a unital Banach algebra and if $x, y \in A$, then*

1. *x is invertible if $\|1 - x\| < 1$,*
2. *$\sigma(x)$ is a nonempty compact subset of $\{\lambda \in \mathbb{C} \,|\, |\lambda| \leq \|x\|\}$,*
3. *$\sigma(xy) \cup \{0\} = \sigma(yx) \cup \{0\}$, and*
4. *$\mathrm{spr}\, x = \lim_n \|x^n\|^{1/n}$*

Recall that a *division ring* is a unital ring in which every nonzero element has an inverse. Fields are of course division rings, but there exist division rings, such as the ring of quaternions, that are nonabelian.

Corollary 9.4 (Gelfand-Mazur). *If a unital Banach algebra A is a division ring, then A is isometrically isomorphic to the Banach algebra of complex numbers.*

Proof. If $x \in A$, then x has at least one spectral element $\lambda \in \sigma(x)$. Hence, $x - \lambda 1$ is not invertible. Because A is a division ring, it must be that $x - \lambda 1 = 0$, and so $x = \lambda 1$. Note that $|\lambda| = \|x\|$ and that the map $x \mapsto \lambda$ is an isometric isomorphism of A with $\mathbb{C}$. ☐

Proposition 9.6 below adds to the list of Banach algebra properties mentioned above in Theorem 9.3.

Definition 9.5. If A is a unital Banach algebra, then the *general linear group* of A is the set GL(A) consisting of all invertible elements of A.

Proposition 9.6. *If A is a unital Banach algebra, then*

1. *GL(A) is a multiplicative group,*
2. *GL(A) is an open subset of A, and*
3. *the inverse map $x \mapsto x^{-1}$ is continuous on GL(A).*

Proof. The set GL(A) is evidently a multiplicative group in the product of A, and so we now prove that GL(A) is an open set. Select $x \in$ GL(A). If $y \in A$ satisfies $\|y - x\| < \|x^{-1}\|^{-1}$, then

$$\|1 - x^{-1}y\| = \|x^{-1}(x - y)\| \leq \|x^{-1}\| \, \|x - y\| < 1,$$

and Proposition 9.3 shows that $x^{-1}y$ is invertible. Thus, $y = xg$ for some $g \in$ GL(A), implying that y is invertible. Hence, GL(A) is an open set.

Suppose now that $x \in$ GL(A) and that $\varepsilon > 0$ is given. Select $\delta > 0$ such that $\delta < \|x^{-1}\|^{-1}(1 + \varepsilon^{-1}\|x^{-1}\|)^{-1}$ and suppose that $y \in A$ satisfies $\|y - x\| < \delta$. Because $\delta < \|x^{-1}\|^{-1}$, the previous paragraph shows that $y \in$ GL(A). Furthermore, since $\|1 - x^{-1}y\| < 1$,

$$y^{-1}x = (x^{-1}y)^{-1} = \sum_{n=0}^{\infty}(1 - x^{-1}y)^{n},$$

implying that

$$y^{-1} = \left(\sum_{n=0}^{\infty}(1 - x^{-1}y)^{n}\right)x^{-1} = x^{-1} + \sum_{n=1}^{\infty}(1 - x^{-1}y)^{n}x^{-1}.$$

Therefore,

$$\|y^{-1} - x^{-1}\| \leq \|x^{-1}\| \sum_{n=1}^{\infty}\|x^{-1}\|^{n}\|x - y\|^{n}$$

$$= \|x^{-1}\|^{2}\|x - y\| \sum_{n=0}^{\infty}\|x^{-1}\|^{n}\|x - y\|^{n}$$

$$= \frac{\|x^{-1}\|^{2}\|x - y\|}{1 - \|x^{-1}\| \, \|x - y\|}.$$

Thus, $\|x - y\| < \delta < \|x^{-1}\|^{-1}(1 + \varepsilon^{-1}\|x^{-1}\|)^{-1}$ yields

$$\|x^{-1}\| + \varepsilon^{-1}\|x^{-1}\|^{2} < \|x - y\|^{-1}$$

and

$$\frac{\|x^{-1}\|^2}{\varepsilon} < \|x-y\|^{-1} - \|x^{-1}\| = \frac{1 - \|x^{-1}\|\,\|x-y\|}{\|x-y\|}.$$

Hence, if $\|y - x\| < \delta$, then $\|y^{-1} - x^{-1}\| < \varepsilon$, thereby proving that inversion is continuous.   □

Corollary 9.7. *The group of invertible operators acting on a Banach space is an open set.*

Definition 9.8. If A is a Banach algebra, then a subset $B \subseteq A$ is a *Banach subalgebra* of A if B is a Banach algebra in the norm and the algebra operations of A. If A is a unital Banach algebra and a Banach subalgebra B of A contains the multiplicative identity of A, then B is a *unital Banach subalgebra* of A.

If B is a unital Banach subalgebra of a unital Banach algebra A, and if $x \in B$, then there are two possibilities for the spectrum of x:

1. a spectrum denoted by $\sigma_A(x)$ consisting all $\lambda \in \mathbb{C}$ for which $x - \lambda 1$ has no inverse in A, and
2. a spectrum denoted by $\sigma_B(x)$ consisting all $\lambda \in \mathbb{C}$ for which $x - \lambda 1$ has no inverse in B.

Evidently, if $x - \lambda 1$ is invertible in B, then it is invertible in A. Hence, $\mathbb{C} \setminus \sigma_B(x) \subseteq \mathbb{C} \setminus \sigma_A(x)$, implying that $\sigma_A(x) \subseteq \sigma_B(x)$. On the other hand, if $x \in B$ is invertible in A, there it can happen that $x^{-1} \notin B$. A concrete case of this is provided by the following example.

Example 9.9. *Suppose that $\mathbb{D}$ is the open unit disc in $\mathbb{C}$, and let $A(\mathbb{D})$ denote the disc algebra*

$$A(\mathbb{D}) = \{f \,|\, f \in C(\overline{\mathbb{D}}) \text{ and } f_{|\mathbb{D}} \text{ is an analytic function}\},$$

which is a unital Banach subalgebra of $C(\partial \mathbb{D})$. If $f \in A(\mathbb{D})$ is given by $f(z) = z$, for $z \in \overline{\mathbb{D}}$, then f is invertible in $C(\partial \mathbb{D})$, but not in $A(\mathbb{D})$.

Proof. The function inverse of the function f in the algebra $C(\partial \mathbb{D})$ is precisely the function $f^{-1}(z) = \overline{z}$. However, the map $z \mapsto \overline{z}$ is not analytic on $\mathbb{D}$, which implies that $f^{-1} \notin A(\mathbb{D})$.   □

The most general assertion relating the spectra $\sigma_A(x)$ and $\sigma_B(x)$ is given below.

Proposition 9.10 (Spectral Permanence). *If B is a unital Banach subalgebra of a unital Banach algebra A, then*

$$\sigma_A(x) \subseteq \sigma_B(x) \text{ and } \partial \sigma_B(x) \subseteq \partial \sigma_A(x)$$

for every $x \in B$.

Proof. The inclusion $\sigma_A(x) \subseteq \sigma_B(x)$ has already been noted. By definition of boundary, if $\lambda \in \partial\sigma_B(x)$, then there is a sequence of distinct points $\lambda_k \in (\mathbb{C} \setminus \sigma_B(x))$ such that $|\lambda - \lambda_k| \to 0$. Therefore, each $x - \lambda_k 1$ is invertible in A and

$$\lim_{k \to \infty} \|(x - \lambda 1) - (x - \lambda_k 1)\| = 0.$$

Suppose that it is not true that $\lambda \in \sigma_A(x)$. Then the element $x - \lambda 1$ is invertible in A and is a limit of invertible elements $x - \lambda_k 1$. Therefore, using the continuity of inversion in $GL(A)$ (Proposition 9.6), we have that

$$\lim_{k \to \infty} \|(x - \lambda 1)^{-1} - (x - \lambda_k 1)^{-1}\| = 0.$$

However, each $(x - \lambda_k 1)^{-1}$ is an element of B, and therefore $(x - \lambda 1)^{-1} \in B$ also, in contradiction of the fact that $x - \lambda 1$ has no inverse in B. Therefore, it must be that $\lambda \in \sigma_A(x)$. Furthermore, because $\lambda = \lim_k \lambda_k$ and each λ_k lies outside $\sigma_A(x)$, we deduce that $\lambda \in \partial\sigma_A(x)$. $\qquad\square$

It sometimes happens that equality in the spectral inclusion $\sigma_A(x) \subseteq \sigma_B(x)$ is achieved.

Proposition 9.11. *If B is a unital abelian Banach subalgebra of a unital Banach algebra A, and if B has the property that*

$$\{y \in A \,|\, yz = zy, \text{ for all } z \in B\} \subseteq B,$$

then $\sigma_A(x) = \sigma_B(x)$, for every $x \in B$.

Proof. Let $x \in B$; the inclusion $\sigma_A(x) \subseteq \sigma_B(x)$ is known from Proposition 9.10. Therefore, suppose that $\lambda \notin \sigma_A(x)$. Thus, for some $y \in A$, $(x - \lambda 1)y = y(x - \lambda 1) = 1$. Thus, for any z in the abelian algebra B, we have $z(x - \lambda 1) = (x - \lambda 1)z$, and so

$$yz = yz(x - \lambda 1)y = y(x - \lambda 1)zy = zy.$$

Therefore, by the hypothesis on B, y must be an element of B, which shows that $\lambda \notin \sigma_B(x)$. Hence, $\sigma_B(x) \subseteq \sigma_A(x)$. $\qquad\square$

The algebra B that appears in Proposition 9.11 is called a maximal abelian subalgebra of A.

Definition 9.12. A unital Banach subalgebra B of a unital Banach algebra A is called a *maximal abelian subalgebra* of A if

1. B is abelian, and
2. $\{y \in A \,|\, yz = zy, \text{ for all } z \in B\} \subseteq B$.

9.2   Ideals and Quotients

As was the case with $B(V)$, it is interest to consider two-sided ideals of Banach algebras.

Definition 9.13. An *ideal* of a Banach algebra A is a subset $J \subseteq A$ such that

1. J is a subspace of A, and
2. $ax \in J$ and $xa \in J$ for all $a \in J$ and $x \in A$.

Furthermore, if an ideal J of A is satisfies $J \neq A$, then J is said to be a *proper ideal* of A.

As we had done in our study of the Banach algebra $B(V)$, if a linear submanifold J of A satisfies $ax \in J$ and $xa \in J$ for all $a \in J$ and $x \in A$, then J is called an *algebraic ideal* of A.

Example 9.14. *Suppose that X is a compact Hausdorff space, and let A be any Banach subalgebra of $C(X)$. If $Y \subseteq X$ is a closed subset of X such that $Y \neq X$, then the set*

$$J_Y = \{f \in A \,|\, f(t) = 0, \text{ for every } t \in Y\}$$

is an ideal of A. Furthermore, if A is unital, then J_Y is a proper ideal of A.

Proof. If $f \in J_Y$ and $g \in A$, and if $t \in Y$, then $fg(t) = f(t)g(t) = 0g(t) = 0$, which implies that $fg \in J_Y$. Likewise, $\alpha_1 f_1 + \alpha_2 f_2 \in J_Y$, for all $f_1, f_2 \in J_Y$ and $\alpha_1, \alpha_2 \in \mathbb{C}$. Thus, J_Y is an algebraic ideal of A. To show that J_Y is closed, observe that, if $f \in \overline{J_Y}$, and if $\{f_n\}_{n \in \mathbb{N}}$ is a sequence in J_Y converging in A to f, then for every $t \in Y$ and every $n \in \mathbb{N}$, the inequality

$$|f(t)| = |f(t) - f_n(t)| \leq \|f - f_n\|$$

implies that $f(t) = 0$. Hence, J_Y is an ideal of A.

If A is unital, then $1 \notin J_Y$ because $1(t) = 1 \neq 0$, for every $t \in Y$. Hence, $J_Y \neq A$, which implies that J_Y is proper.   □

Proposition 9.15. *If A is a Banach algebra, and if J is an algebraic ideal of A, then $\overline{J}$ is an ideal of A. Moreover, if A is unital and if the algebraic ideal J satisfies $J \neq A$, then $\|1 - x\| \geq 1$, for every $x \in \overline{J}$, and $\overline{J}$ is a proper ideal of A.*

Proof. By continuity of multiplication, scalar multiplication, and sum, the set $\overline{J}$ is also an algebraic ideal of A.

Suppose that A is unital and $J \neq A$. If there were an element $x \in J$ such that $\|1 - x\| < 1$, then x would be invertible, and so $1 = x^{-1}x$ would be an element of J, implying that $J = A$, which is in contradiction to the hypothesis on A. Hence, it must be that $\|1 - x\| \geq 1$, for every $x \in J$. Therefore, it is also true that $\|1 - y\| \geq 1$ for every $y \in \overline{J}$. Hence, $1 \notin \overline{J}$, which implies that $\overline{J}$ is a proper ideal.   □

If J is an ideal of a Banach algebra A, then A/J is a Banach space in the quotient norm. The following result shows that the quotient space is also a Banach algebra.

Proposition 9.16. *If J is an ideal of a Banach algebra A, then A/J is a Banach algebra with respect to the quotient norm and multiplication defined by $\dot{x}\dot{y} = (\dot{xy})$. Furthermore, if A is a unital Banach algebra and $J \neq A$, then A/J is a unital Banach algebra.*

Proof. The fact that $\dot{x}\dot{y} = (\dot{xy})$ is a well-defined associative product on A/J is a simple standard fact from ring theory. It is also clear that A/J is not just a ring, but an associate algebra as well. By Proposition 5.18, A/J is a Banach space. Therefore, the only issue left to address is the submultiplicativity of the quotient norm.

If $x, y \in A$, then

$$\|(\dot{xy})\| = \inf_{a \in J} \|xy - a\| \leq \inf_{a,b \in J} \|(x-a)(y-b)\|$$

$$\leq \left(\inf_{a \in J} \|x-a\| \right) \left(\inf_{b \in J} \|xy - b\| \right)$$

$$= \|\dot{x}\| \, \|\dot{y}\|.$$

Now if A is a unital Banach algebra with $J \neq A$, then $\dot{1}$ is a multiplicative identity for the ring A/J. Furthermore, Theorem 9.3 states that a is invertible for every element $a \in J$ that satisfies $\|1 - a\| < 1$. However, as $J \neq A$, the ideal J contains no invertible elements. Hence, $\|1 - a\| \geq 1$ for every $a \in J$, and so

$$1 \geq \|\dot{1}\| = \inf_{a \in J} \|1 - a\| \geq 1,$$

implying that A is a unital Banach algebra. □

Definition 9.17. An ideal N of a Banach algebra A is a *maximal ideal* if

1. N is a proper ideal of A, and
2. $M = N$, for every proper ideal M of A for which $N \subseteq M$.

Example 9.18. *If X is a compact Hausdorff space, and if $x_0 \in X$, then the set*

$$J_{\{x_0\}} = \{f \in C(X) \,|\, f(x_0) = 0\}$$

is a maximal ideal of $C(X)$.

Proof. The point set $\{x_0\}$ is closed in X; thus, Example 9.14 shows that $J_{\{x_0\}}$ is a proper ideal of $C(X)$. Suppose that M is an ideal of $C(X)$ such that $J_{\{x_0\}} \subseteq M$ and $M \neq J_{\{x_0\}}$. Select $h \in M$ such that $h \notin J_{\{x_0\}}$; therefore, $h(x_0) \neq 0$.

On the Banach algebra $C(X)/J_{\{x_0\}}$, define a function $\phi : C(X)/J_{\{x_0\}} \to \mathbb{C}$ by $\phi(\dot{g}) = g(x_0)$. If $g_1, g_2 \in C(X)$ satisfy $\dot{g}_1 = \dot{g}_2$, then $g_1(x_0) = g_2(x_0)$, and so ϕ is a well-defined function. Indeed, ϕ is a unital, surjective homomorphism. Let $\lambda = \frac{1}{\phi(\dot{h})}$; as ϕ is surjective, there is a $\dot{k} \in C(X)/J_{\{x_0\}}$ such that $\lambda = \varphi(\dot{k})$. Hence,

$$\phi(\dot{k}\dot{h}) = \phi(\dot{k})\phi(\dot{h}) = 1 = \phi(\dot{1}),$$

which implies that $\dot{1} - \dot{k}\dot{h} \in \ker\phi$; that is, $1 - kh \in J_{\{x_0\}}$. Since $h \in M$ and $J_{\{x_0\}} \subseteq M$, we deduce that $1 \in M$, and so $M = A$. Therefore, the ideal $J_{\{x_0\}}$ is maximal. □

As with rings (such as the ring of even integers), some Banach algebras may fail to possess a maximal ideal. However, in the case of unital Banach algebras, maximal ideals always exist.

Proposition 9.19. *If J is a proper ideal of a unital Banach algebra A, then there exists a maximal ideal N of A such that $J \subseteq N$.*

Proof. Let $\mathfrak{S}$ be the set of all proper ideals I of A such that $J \subseteq I$ and $I \neq A$. Define a partial order $\preceq$ on $\mathfrak{S}$ by $I \preceq K$ if $I, K \in \mathfrak{S}$ satisfy $I \subseteq K$. Suppose that $\mathfrak{E}$ is a linearly ordered subset $\mathfrak{S}$. Define $K = \bigcup_{I \in \mathfrak{E}} I$. Because $\mathfrak{E}$ is linearly ordered, the set $\bigcup_{I \in \mathfrak{E}} I$ is an algebraic ideal of A; hence, K is an ideal of A. If it were true that $K \in \mathfrak{S}$, then K would be an upper bound for $\mathfrak{E}$. Therefore, $\mathfrak{S}$ would satisfy the hypotheses of Zorn's Lemma, implying that $\mathfrak{S}$ has a maximal element N. Clearly a maximal element N of $\mathfrak{S}$ is a maximal ideal of A.

Therefore, all that is left to prove is that $K \in \mathfrak{S}$. To this end, note that $J \subseteq K$, and so it remains to show that $K \neq A$. Assume, on the contrary, that $K = A$. Since $\bigcup_{I \in \mathfrak{E}} I$ is dense in K, there exists $I \in \mathfrak{E}$ and $y \in I$ such that $\|1 - y\| < 1$. On the other, because $I \neq A$, we have that $\|1 - y\| \geq 1$ (Proposition 9.15), in contradiction to $\|1 - y\| < 1$. Hence, it must be that $K \neq A$. □

9.3 Abelian Banach Algebras

The study of unital abelian Banach algebras is very closely related to the space of maximal ideals in such an algebra, and maximal ideals arise from a special type of homomorphism which is called a character.

Definition 9.20. If A is a unital abelian Banach algebra, then a unital homomorphism $\rho : A \to \mathbb{C}$ is called a *character* of A.

Proposition 9.21. *Suppose that A is a unital abelian Banach algebra. If ρ is a character of A, then $\ker\rho$ is a maximal ideal of A. Conversely, if N is a maximal ideal of A, then there exists a unique character $\rho : A \to \mathbb{C}$ such that $N = \ker\rho$.*

Proof. The kernel of every homomorphism of A is an ideal of A and the map $\dot{\rho} : A/\ker\rho \to \mathbb{C}$ in which $\dot{\rho}(\dot{x}) = \rho(x)$ is a well-defined isomorphism of Banach algebras. Hence, $A/\ker\rho$ is a field. If $M \subset A$ is an ideal of A that properly contains $\ker\rho$, then there is an $x \in M$ such that $\dot{x}$ is nonzero in $A/\ker\rho$. Because $A/\ker\rho$ is a field, there is a $y \in A$ such that $\dot{y} = \dot{x}^{-1}$. Thus, $xy - 1 \in \ker\rho \subset M$ yields $1 \in M$ (because $xy \in M$), and so $M = A$. Hence, $\ker\rho$ is a maximal ideal of A.

Conversely, suppose that N is a maximal ideal of A. Select a nonzero $\dot{x} \in A/N$; thus, $x \notin N$. Define

$$M_0 = \{ax + y \,|\, a \in A, \ y \in N\}.$$

Evidently, M_0 is an algebraic ideal of A, and so $M = \overline{M_0}$ is an ideal of A that contains N. As $x \in M$ and $x \notin N$, the maximality of N implies that M be must A. Hence, there are $a \in A$ and $y \in N$ such that $\| 1 - (ax + y) \| < 1$. Passing to the quotient and noting that $\dot{y} = \dot{0}$, we deduce that $\| \dot{1} - \dot{a}\dot{x} \| < 1$, and so $\dot{a}\dot{x}$ is invertible. Hence, there exists $z \in A$ such that $(\dot{a}\dot{x})\dot{z} = \dot{1}$. The commutativity of A yields $\dot{x}(\dot{a}\dot{z}) = (\dot{a}\dot{z})\dot{x} = \dot{1}$, implying that $\dot{x}$ is invertible. Therefore, A/N is a division ring. By the Gelfand-Mazur Theorem, there is an isometric isomorphism $\vartheta : A/N \to \mathbb{C}$. Thus, if $\pi : A \to A/N$ is the canonical quotient homomorphism given by $\pi(x) = \dot{x}$, for all $x \in A$, then $\rho = \vartheta \circ \pi$ is a character on A with kernel $\ker\rho = N$.

To prove the uniqueness of ρ, suppose that $\tilde{\rho} : A \to \mathbb{C}$ is a character such that $\ker\tilde{\rho} = \ker\rho$. For every $x \in A$, the element $x - \tilde{\rho}(x)1$ belongs to $\ker\tilde{\rho}$. Therefore, $x - \tilde{\rho}(x)1$ also belongs to $\ker\rho$, which implies that

$$0 = \rho(x - \tilde{\rho}(x)1) = \rho(x) - \tilde{\rho}(x),$$

and so $\tilde{\rho} = \rho$. □

Proposition 9.21 identifies a bijective correspondence between maximal ideals and characters, which leads to a bijective correspondence between the sets $\mathscr{M}_A$ and $\mathscr{R}_A$, defined below.

Definition 9.22. If A is a unital abelian Banach algebra, then

1. the *maximal ideal space* of A is the set

$$\mathscr{M}_A = \{N \subseteq A \,|\, N \text{ is a maximal ideal of } A\}, and$$

2. the *character space* of A is the set

$$\mathscr{R}_A = \{\rho \in A^* \,|\, \rho \text{ is a unital homomorphism } A \to \mathbb{C}\}.$$

An important fact about the character space is that it is a compact Hausdorff space.

Proposition 9.23. *The character space $\mathscr{R}_A$ of a unital abelian Banach algebra is a weak*-compact subset of the unit sphere of the dual space A^* of A.*

Proof. If $\rho \in \mathscr{R}_A$, then $\rho(1) = 1$ and, thus, $\|\rho\| \geq 1$. Suppose it is true that $\|\rho\| > 1$; thus, there exists $x \in A$ with $\|x\| = 1$ and $\rho(x) > 1$. Let $z = \rho(x)^{-1}x$. Because

$$1 = \frac{\rho(x)}{\rho(x)} > \frac{1}{\rho(x)} = \|z\|,$$

the series $\displaystyle\sum_{n=1}^{\infty} z^n$ converges in A to an element y that satisfies $y = z + zy$. Thus, $\rho(y) = \rho(z) + \rho(z)\rho(y) = 1 + \rho(y)$, implying that $0 = 1$. Therefore, it must be that $\|\rho\| = 1$, which proves that $\mathscr{R}_A$ is a subset of the unit sphere of A^*.

To prove that $\mathscr{R}_A$ is weak*-closed, suppose that $\vartheta \in A^*$ is in the weak*-closure of A^*. Thus, if $x, y \in A$ and $\varepsilon > 0$, then the weak*-open neighbourhood

$$U = \{\phi \in A^* \,|\, |\phi(z) - \vartheta(z)| < \varepsilon, \, z \in \{1, x, y, xy\}\}$$

intersects $\mathscr{R}_A$, and so there exists $\rho \in \mathscr{R}_A$ such that $\rho \in U$. Thus,

$$|\vartheta(xy) - \vartheta(x)\vartheta(y)| \leq |\vartheta(xy) - \rho(xy)| + |\rho(y) - \vartheta(y)| \, |\rho(x)|$$

$$+ |\rho(x) - \vartheta(x)| \, |\vartheta(y)|$$

$$< \varepsilon(1 + \|x\| + \|\vartheta\| \, \|y\|).$$

Hence, as $\varepsilon > 0$ is arbitrary, $\vartheta(xy) = \vartheta(x)\vartheta(y)$, which proves that ϑ is a character on A.

The unit ball of A^* is a weak*-compact and Hausdorff; hence, the weak*-closed subset $\mathscr{R}_A$ has these same two topological properties. $\qquad\square$

Let $I : \mathscr{R}_A \to \mathscr{M}_A$ be defined by $I(\rho) = \ker \rho$. By Proposition 9.21, I is a bijection. Therefore, using Proposition 9.21, one can endow the maximal ideal space $\mathscr{M}_A$ with a topology $\mathscr{T}$ via

$$\mathscr{T} = \{U \subseteq \mathscr{M}_A \,|\, I^{-1}(U) \text{ is open in } \mathscr{R}_A\},$$

where $\mathscr{R}_A$ has the weak*-topology.

Proposition 9.24. *The maximal ideal space $\mathscr{M}_A$ of a unital abelian Banach algebra A is a compact Hausdorff space.*

Proof. The map $I : \mathscr{R}_A \to \mathscr{M}_A$ is a homeomorphism. $\qquad\square$

The importance of characters in spectral theory is demonstrated by the following result.

Proposition 9.25. *If A is a unital abelian Banach algebra, and if $x \in A$, then*

$$\sigma(x) = \{\rho(x) \,|\, \rho \in \mathscr{R}_A\}.$$

Proof. If $\rho \in \mathscr{R}_A$ and $\lambda = \rho(x)$, then $(x - \lambda 1) \in \ker \rho$. Because $\ker \rho$ is a proper ideal of A, it contains no invertible elements. Hence, $\lambda \in \sigma(x)$.

Conversely, suppose that $\lambda \in \sigma(x)$. Because $x - \lambda 1$ is not invertible, there is a maximal ideal N of A such that $x - \lambda 1 \in N$ (Exercise 9.51). By Proposition 9.21, there exists a character $\rho : A \to \mathbb{C}$ such that $\ker \rho = N$. Hence, $\rho(x) = \lambda$. □

The following result is a basic application of Proposition 9.25 that is useful for studying equations in Banach algebras. (See Exercise 9.48, for example.)

Proposition 9.26. *If A is a unital Banach algebra, not necessarily abelian, and if $x, y \in A$ satisfy $xy = yx$, then*

$$\sigma(x - y) \subseteq \{\lambda - \mu \mid \lambda \in \sigma(x), \ \mu \in \sigma(y)\}.$$

Proof. Suppose that there exists a maximal abelian subalgebra B of A that contains both x and y. In this case, Proposition 9.11 shows that $\sigma(z) = \sigma_B(z)$, for every $z \in B$. In particular, with $z = x - y$, this fact and Proposition 9.25 yield

$$\sigma(x - y) = \sigma_B(x - y) = \{\rho(x - y) \mid \rho \in \mathscr{R}_B\} = \{\rho(x) - \rho(y) \mid \rho \in \mathscr{R}_B\}$$

$$\subseteq \{\lambda - \mu \mid \lambda \in \sigma_B(x), \ \mu \in \sigma_B(y)\}$$

$$= \{\lambda - \mu \mid \lambda \in \sigma(x), \ \mu \in \sigma(y)\}.$$

Therefore, we need only establish the existence of the maximal abelian subalgebra B.

If A itself is abelian, then take $B = A$. Thus, assume that A is nonabelian and let B_0 be the closure of the unital abelian algebra of all elements of the form $\sum_{j=0}^{m} \sum_{k=0}^{n} \alpha_{kj} x^j y^k$, where each $\alpha_{jk} \in \mathbb{C}$. Thus, B_0 is a unital abelian Banach subalgebra of A. Let $\mathfrak{S}$ denote the set of all unital abelian Banach subalgebras C of A for which $B_0 \subseteq C$. The set $\mathfrak{S}$ is nonempty, as $B_0 \in \mathfrak{S}$. Let $\preceq$ be defined by $C_1 \preceq C_2$, if $C_1, C_2 \in \mathfrak{S}$ satisfy $C_1 \subseteq C_2$. Suppose that $\mathfrak{L}$ is a linearly ordered subset of $\mathfrak{S}$, and define $\tilde{C}$ to be the closure of $\bigcup_{C \in \mathfrak{L}} C$. By continuity of the algebraic operations, $\tilde{C} \in \mathfrak{S}$ and is an upper bound for $\mathfrak{L}$. Hence, by Zorn's Lemma, $\mathfrak{S}$ has a maximal element B.

To show that B has the property $\{a \in A \mid az = za, \text{ for all } z \in B\} \subseteq B$, select an element $a \in A$ for which $az = za$, for every $z \in B$. Thus, the closure $\tilde{B}$ of the set $\{f(a)z_1 + z_2 \mid z_1, z_2 \in B, f \in \mathbb{C}[t]\}$ is a unital abelian Banach subalgebra that contains B_0. Since, $B \subseteq \tilde{B}$, and because B is maximal in $\mathfrak{S}$, we deduce that $\tilde{B} \subseteq B$ and, in particular, that $a \in B$. □

We are now prepared for the major fundamental result of this section.

Theorem 9.27 (Gelfand). *If A is a unital abelian Banach algebra, then there exists a contractive unital homomorphism $\Gamma : A \to C(\mathscr{R}_A)$ such that, for every $x \in A$,*

1. $\Gamma(x)[\rho] = \rho(x)$, for all $\rho \in \mathscr{R}_A$,
2. the range of the function $\Gamma(x)$ is $\sigma(x)$, and
3. $\|\Gamma(x)\| = \mathrm{spr}\,x$.

Furthermore, $\ker \Gamma = \bigcap_{N \in \mathscr{M}_A} N$.

Proof. The map $\Gamma : A \to C(\mathscr{R}_A)$ is plainly unital, linear, and multiplicative.

For $x \in A$, the function $A^* \to \mathbb{C}$ given by $\varphi \mapsto \varphi(x)$ is weak*-continuous. Hence, the map $\rho \mapsto \rho(x)$ determines a continuous function $\Gamma(x)$ on the compact Hausdorff space $\mathscr{R}_A$. Proposition 9.25 shows that the range of $\Gamma(x)$ is $\sigma(x)$, and so the maximal modulus of $\Gamma(x)[\rho]$, as ρ varies through the character space $\mathscr{R}_A$, is the spectral radius of x. Therefore, $\|\Gamma(x)\| = \mathrm{spr}\,x \leq \|x\|$, which implies that $\|\Gamma\| \leq 1$. Note that $\Gamma(x) = 0$ if and only if $\rho(x) = 0$ for all $\rho \in \mathscr{R}_A$; that is, $\Gamma(x) = 0$ if and only if $x \in \ker \rho$ for every $\rho \in \mathscr{R}_A$. That is, by Proposition 9.21, $\Gamma(x) = 0$ if and only if $x \in N$ for every maximal ideal N of A.   □

Definition 9.28. If A is a unital abelian Banach algebra, then the *Gelfand transform* of A is the contractive unital homomorphism $\Gamma : A \to C(\mathscr{R}_A)$ described in Theorem 9.27.

We turn now to some examples of character spaces, beginning with an example that involves $C(X)$ itself.

Example 9.29. *If X is a compact Hausdorff space, then the character space of $C(X)$ is homeomorphic to X.*

Proof. Example 9.18 and Exercise 9.52 show that the maximal ideals of X are of the form

$$J_{\{x_0\}} = \{f \in C(X) \,|\, f(x_0) = 0\},$$

for $x_0 \in X$. Therefore, by Proposition 9.21, the characters of $C(X)$ are all functions $\rho_x : C(X) \to \mathbb{C}$ of the form $\rho_x(f) = f(x)$, for $f \in C(X)$, and $x \in X$. Thus, we aim to prove that the map $\psi : X \to \mathscr{R}_{C(X)}$, defined by $\psi(x) = \rho_x$, is a homeomorphism. Since $\mathscr{R}_{C(X)}$ is a compact Hausdorff space, it is sufficient, by Proposition 2.9, to prove that ψ is continuous and bijective.

The surjectivity of ψ is apparent. If $x_1, x_2 \in X$ are distinct, then by Urysohn's Lemma there is a function $f \in C(X)$ with $f(x_1) = 1$ and $f(x_2) = 0$; hence, ρ_{x_1} and ρ_{x_2} take different values at f, which shows that ψ is injective.

To prove the continuity of ψ, fix $x_0 \in X$ and consider a basic weak* open neighbourhood U of $\psi(x_0)$:

$$U = \{\rho \in \mathscr{R}_{C(X)} \,|\, |\rho(f_j) - \rho_{x_0}(f_j)| < \varepsilon_j, \text{ for all } j = 1, \dots, n\},$$

for some $n \in \mathbb{N}, f_1, \dots, f_n \in C(X)$, and positive real numbers $\varepsilon_1, \dots, \varepsilon_n$. The equations $|\rho(f_j) - \rho_{x_0}(f_j)| = |\rho(f_j) - f_j(x_0)|$ imply that

$$\psi^{-1}(U) = \{x \in X \mid |f_j(x) - f_j(x_0)| < \varepsilon_j, \text{ for all } j = 1, \ldots, n\}$$

$$= \bigcap_{j=1}^{n} f_j^{-1} \left(B_{\varepsilon_j}(f_j(x_0)) \right),$$

which is open in X by the continuity of the functions f_j. Thus, ψ is continuous, which completes the proof that ψ is a homeomorphism. $\qquad \square$

The next example makes use of the Stone-Čech compactification. Recall that $C_b(X)$ denotes the unital abelian Banach algebra of bounded, continuous complex-valued functions on a locally compact Hausdorff space X.

Example 9.30. *If X is a locally compact Hausdorff space, then the character space of $C_b(X)$ is homeomorphic to βX.*

Proof. By Theorem 2.71, for every $f \in C_b(X)$ there exists a unique $\tilde{f} \in C(\beta X)$ such that $\tilde{f} \circ \iota_X(x) = f(x)$, for every $x \in X$, where $\iota_X : X \to \beta X$ is a topological embedding of X into βX as a dense open subset, homeomorphic with X. Therefore, the function $\pi : C_b(X) \to C(\beta X)$ defined by $\pi(f) = \tilde{f}$ is a unital isometric isomorphism of abelian Banach algebras. Therefore, the character space of $C_b(X)$ is homeomorphic to the character space of $C(\beta X)$. Because βX is compact, Example 9.29 shows that βX is homeomorphic to the character space of $C(\beta X)$, and so the same is true of the character space of $C_b(X)$ (Exercise 9.53). $\qquad \square$

Example 9.31. *The maximal ideal space of $\ell^\infty(\mathbb{N})$ is homeomorphic to $\beta \mathbb{N}$.*

Proof. By endowing $\mathbb{N}$ with the discrete topology, every function $\mathbb{N} \to \mathbb{C}$ is continuous. Thus, the bounded functions are precisely those given by the elements of $\ell^{(\mathbb{N})}$. Therefore, because $\ell^\infty(\mathbb{N}) = C_b(\mathbb{N})$, Example 9.30 shows that the character space (and, hence, the maximal ideal space) of $\ell^\infty(\mathbb{N})$ is homeomorphic to $\beta \mathbb{N}$. $\qquad \square$

Another algebraic structure of interest is the kernel of the Gelfand transform

Definition 9.32. The *radical* of a unital abelian Banach algebra is the ideal $\mathrm{Rad}\, A$ of A defined by

$$\mathrm{Rad}\, A = \bigcap_{N \in \mathscr{M}_A} N.$$

If $\mathrm{Rad}\, A = \{0\}$, then A is called a *semisimple* Banach algebra.

With this terminology:

Proposition 9.33. *The Gelfand transform of a unital abelian Banach algebra A is an injection if and only if A is semisimple.*

If X is a locally compact Hausdorff space, then $C_b(X)$ is a semisimple Banach algebra (Exercise 9.56). In contrast, the next example illustrates a situation where the radical is so large as to be of co-dimension 1.

Example 9.34. *Fix $n \in \mathbb{N}$ such that $n \geq 2$, and for each n-tuple $(z_0, z_1, \ldots, z_{n-1})$ of complex numbers, let $T(z_0, z_1, \ldots, z_{n-1})$ denote the $n \times n$ upper triangular Toeplitz matrix*

$$
T(z_0, z_1, \ldots, z_{n-1}) = \begin{bmatrix} z_0 & z_1 & z_2 & \cdots & z_{n-1} \\ 0 & z_0 & z_1 & \ddots & \vdots \\ 0 & 0 & \ddots & \ddots & z_2 \\ \vdots & & \ddots & \ddots & z_1 \\ 0 & \ldots & \ldots & 0 & z_0 \end{bmatrix}.
$$

If $\mathscr{T}_n = \{T(z_0, z_1, \ldots, z_{n-1}) \mid z_0, z_1, \ldots, z_{n-1} \in \mathbb{C}\}$, then $\mathscr{T}_n$ is a unital abelian Banach algebra of operators acting on the Hilbert space $\ell^2(n)$ such that:

1. the character space of $\mathscr{T}_n$ is given by $\mathscr{R}_{\mathscr{T}_n} = \{\rho\}$, where

$$
\rho\left(T(z_0, z_1, \ldots, z_{n-1})\right) = z_0,
$$

for all $T(z_0, z_1, \ldots, z_{n-1}) \in \mathscr{T}_n$;
2. Rad $\mathscr{T}_n = \{T(0, z_1, \ldots, z_{n-1}) \mid z_1, \ldots, z_{n-1} \in \mathbb{C}\}$; and
3. $\mathscr{T}_n / \text{Rad } \mathscr{T}_n$ is isometrically isomorphic to $\mathbb{C}$.

Proof. If $T(z_0, z_1, \ldots, z_{n-1})$ and $T(w_0, w_1, \ldots, w_{n-1})$ are elements of $\mathscr{T}_n$, then

$$
T(z_0, z_1, \ldots, z_{n-1}) + T(w_0, w_1, \ldots, w_{n-1}) = T(z_0 + w_0, \zeta_1, \ldots, \zeta_{n-1}),
$$

for some $\zeta_j \in \mathbb{C}$, and

$$
T(z_0, z_1, \ldots, z_{n-1}) T(w_0, w_1, \ldots, w_{n-1}) = T(w_0, w_1, \ldots, w_{n-1}) T(z_0, z_1, \ldots, z_{n-1})
$$

$$
= T(z_0 w_0, \omega_1, \ldots, \omega_{n-1}),
$$

for some $\omega_j \in \mathbb{C}$. Thus, the map $\rho : \mathscr{T}_n \to \mathbb{C}$, in which $\rho\left(T(z_0, z_1, \ldots, z_{n-1})\right) = z_0$, is linear, multiplicative, unital, and bounded. That is, ρ is a character on $\mathscr{T}_n$.

Gelfand's Theorem asserts that the spectrum of each matrix in $\mathscr{T}_n$ is obtained by evaluating all characters on $\mathscr{T}_n$ at that matrix. Therefore, since the spectrum of $T(z_0, z_1, \ldots, z_{n-1})$ is the singleton set $\{z_0\}$, any other character on $\mathscr{T}_n$ necessarily has the same value as ρ at each matrix $T(z_0, z_1, \ldots, z_{n-1})$. Hence, $\mathscr{R}_{\mathscr{T}_n} = \{\rho\}$.

The radical Rad $\mathscr{T}_n$ of $\mathscr{T}_n$ is precisely the kernel of the Gelfand transform Γ. Gelfand's Theorem shows that the norm of Γ evaluated at $T(z_0, z_1, \ldots, z_{n-1})$ is the spectral radius of $T(z_0, z_1, \ldots, z_{n-1})$, namely $|z_0|$. Thus, $T(z_0, z_1, \ldots, z_{n-1}) \in \ker \Gamma$ if and only if $z_0 = 0$. Hence, the equivalence class in $\mathscr{T}_n / \text{Rad } \mathscr{T}_n$ of each matrix $T(z_0, z_1, \ldots, z_{n-1})$ is determined by z_0, and so $\mathscr{T}_n / \text{Rad } \mathscr{T}_n$ is a division ring, implying that $\mathscr{T}_n / \text{Rad } \mathscr{T}_n$ is isometrically isomorphic to $\mathbb{C}$, by the Gelfand-Mazur Theorem. $\square$

9.4 Absolutely Convergent Fourier Series

Any norm-closed subalgebra of $C(X)$ that contains the constant functions will be an example of a unital abelian Banach algebra. A rather different type of example is considered below, leading to a nontrivial application of the Gelfand theory of unital abelian Banach algebras to absolutely convergent Fourier series.

If $f : \mathbb{R} \to \mathbb{C}$ is a 2π-periodic continuous function, then the Fourier coefficients $\hat{f}(n)$ of f are computed according to the usual fashion in the Hilbert space $L^2(\mathbb{T})$:

$$\hat{f}(n) = \frac{1}{2\pi} \int_{-\pi}^{\pi} f(t) e^{-int} \, dt,$$

for $n \in \mathbb{Z}$.

Definition 9.35. The Fourier coefficients $\hat{f}(n)$ of a 2π-periodic continuous function $f : \mathbb{R} \to \mathbb{C}$ are *summable* if $\sum_{-\infty}^{\infty} |\hat{f}(n)| < \infty$.

The summability of the Fourier coefficients of continuous functions in a property that is not enjoyed by all continuous 2π-periodic functions.

Proposition 9.36. *If a 2π-periodic continuous function $f : \mathbb{R} \to \mathbb{C}$ has summable Fourier coefficients, then the series $\sum_{-\infty}^{\infty} \hat{f}(n) e^{int}$ converges uniformly on $\mathbb{R}$ to f.*

Proof. Exercise 9.60 □

Let $AC(\mathbb{T})$ denote the set of all 2π-periodic continuous functions $f : \mathbb{R} \to \mathbb{C}$ for which the Fourier coefficients of f are summable. By Proposition 9.36, such functions have *absolutely convergent Fourier series*. By the linearity of the Fourier-coefficient map $f \mapsto \hat{f}(n)$, the set $AC(\mathbb{T})$ is a vector space.

Proposition 9.37. *$AC(\mathbb{T})$ is a unital abelian Banach algebra with respect to the norm*

$$\|f\| = \sum_{-\infty}^{\infty} |\hat{f}(n)|,$$

for $f \in AC(\mathbb{T})$.

Proof. By the triangle inequality in $\mathbb{C}$, the formula for $\|f\|$ is indeed a norm on $AC(\mathbb{T})$. Thus, $AC(\mathbb{T})$ is a normed vector space and the function $W : AC(\mathbb{T}) \to \ell^1(\mathbb{Z})$ defined by $Wf = \left(\hat{f}(n) \right)_{n \in \mathbb{Z}}$ is a linear isometry. In fact this isometry is onto, for if $\alpha = (\alpha_n)_{n \in \mathbb{Z}} \in \ell^1(\mathbb{Z})$, then the series $\sum_{-\infty}^{\infty} \alpha_n e^{int}$ converges uniformly on $\mathbb{R}$ to a continuous 2π-periodic function f such that $\hat{f}(n) = \alpha_n$ for every $n \in \mathbb{Z}$. Hence, W

is an isometric isomorphism of $AC(\mathbb{T})$ and $\ell^1(\mathbb{Z})$, which proves that $AC(\mathbb{T})$ is a Banach space.

To show that the norm is submultiplicative, select $f, g \in AC(\mathbb{T})$. Since the series

$$\sum_{k\in\mathbb{Z}}\sum_{m\in\mathbb{Z}}\hat{f}(k)\hat{g}(m)e^{i(k+m)t}$$

converges uniformly to fg, we have that $f(t)g(t) = \sum_{n\in\mathbb{Z}}\sum_{k\in\mathbb{Z}}\hat{f}(k)\hat{g}(n-k)e^{int}$ and

$$\sum_{n\in\mathbb{Z}}|(\widehat{fg})(n)| \le \sum_{n\in\mathbb{Z}}\sum_{k\in\mathbb{Z}}|\hat{f}(k)\hat{g}(n-k)| = \left(\sum_{k\in\mathbb{Z}}|\hat{f}(k)|\right)\left(\sum_{n\in\mathbb{Z}}|\hat{g}(n)|\right).$$

Hence, $\|fg\| \le \|f\|\,\|g\|$, implying that $AC(\mathbb{T})$ is a Banach algebra. Lastly, the constant function $1 \in AC(\mathbb{T})$ has Fourier coefficient $\hat{1}(n) = 0$ for every nonzero $n \in \mathbb{Z}$ and $\hat{1}(0) = 1$. Thus, $\|1\| = 1$, which implies that $AC(\mathbb{T})$ is a unital Banach algebra. The fact that $AC(\mathbb{T})$ is abelian is obvious.   □

With the Gelfand theory one can compute the spectra of absolutely convergent Fourier series as follows.

Proposition 9.38. *If $f \in AC(\mathbb{T})$, then $\sigma(f) = \{f(t)\,|\,t \in \mathbb{R}\}$.*

Proof. The main issue is to identify the character space $\mathscr{R}_{AC(\mathbb{T})}$ of $AC(\mathbb{T})$. To begin, note that if $t_0 \in \mathbb{R}$, then the function $\rho_{t_0} : AC(\mathbb{T}) \to \mathbb{C}$ defined by $\rho_{t_0}(f) = f(t_0)$ is a character on $AC(\mathbb{T})$. Conversely, select any $\rho \in \mathscr{R}_{AC(\mathbb{T})}$ and let $\lambda = \rho(e^{it})$. Thus, $|\lambda| \le \|\rho\|\,\|e^{it}\| = 1$. Because ρ is a character, $\lambda^{-1} = \rho(e^{-it})$ and so $|\lambda^{-1}| \le 1$ also. Hence, $|\lambda| = 1$ and so $\lambda = e^{it_0}$ for some $t_0 \in \mathbb{R}$. If $f \in AC(\mathbb{T})$, then define for each $m \in \mathbb{N}$ the trigonometric polynomial

$$f_m(t) = \sum_{n=-m}^{m} \alpha_n e^{int}.$$

By linearity of ρ, $\rho(f_m) = f_m(t_0)$ for every m. Because $\lim_{m\to\infty}\|f - f_m\| = 0$ implies (i) that $\rho(f_m) \to \rho(f)$ and (ii) that $f_m \to f$ uniformly on $\mathbb{R}$, we deduce that $\rho(f) = f(t_0)$. Hence, $\sigma(f) = \{f(t)\,|\,t \in \mathbb{R}\}$.   □

Absolutely convergent Fourier series are of classical interest and have been studied by the methods of hard analysis. It is interesting, therefore, to note that by viewing these series as elements of the Banach algebra $AC(\mathbb{T})$ one can obtain some classically difficult results. A striking example is the following theorem of N. Wiener.

Corollary 9.39 (Wiener). *If $f : \mathbb{R} \to \mathbb{C}$ is a continuous 2π-periodic function such that the Fourier coefficients of f are summable, and if $f(t) \ne 0$ for every $t \in \mathbb{R}$, then the Fourier coefficients of $1/f$ are summable.*

Proof. If $f(t) \neq 0$ for every $t \in \mathbb{R}$, then $0 \notin \sigma(f)$ and so f is invertible in $AC(\mathbb{T})$. If $g = f^{-1}$, then $g(t)f(t) = 1$ for every $t \in \mathbb{R}$. Therefore, $g = 1/f$, implying that the continuous 2π-periodic function $1/f$ has summable Fourier coefficients. $\square$

9.5 The Exponential Function

The simplest of all Banach algebras is the unital abelian Banach algebra $\mathbb{C}$. Among the many holomorphic functions defined on all of $\mathbb{C}$, the exponential function is one of the most important. To begin this section, we shall see below how the exponential function and its usual property of sending sums to products extends to the level of Banach and abelian Banach algebras.

Proposition 9.40. *If A is a unital Banach algebra, and if for each $n \in \mathbb{N}$ the function $s_n : A \to A$ is defined by $s_n(x) = \sum_{k=0}^{n} \dfrac{1}{k!} x^k$, where $x^0 = 1$, then there exists a continuous function* $\exp : A \to A$ *such that*

1. *$\{s_n\}_{n \in \mathbb{N}}$ converges uniformly on $\{x \in A \,|\, \|x\| \leq r\}$ to $\exp$, for every $r > 0$, and*
2. *$\|\exp(x)\| \leq e^{\|x\|}$, for all $x \in A$.*

Proof. Fix $r > 0$. The sequence $\{t_n(r)\}_{n \in \mathbb{N}}$ of partial sums $t_n(r) = \sum_{k=0}^{n} \dfrac{r^k}{k!}$ of the series $e^r = \sum_{k=0}^{\infty} \dfrac{r^k}{k!}$ is a Cauchy sequence. If $x \in A$ satisfies $\|x\| \leq r$, then by the triangle inequality and the submultiplicativity of the norm,

$$\|s_\ell(x) - s_k(x)\| \leq \sum_{m=k+1}^{\ell} \frac{\|x\|^m}{m!} \leq \sum_{m=k+1}^{\ell} \frac{r^m}{m!} = |t_\ell(r) - t_k(r)|.$$

Hence, the sequence $\{s_n(x)\}_{n \in \mathbb{N}}$ is Cauchy and, therefore, converges in A to the element that we denote by $\exp(x)$. Observe that the inequalities above imply that the converge of the functions $s_n : A \to A$ to $\exp$ is uniform on the closed ball of radius r that is centred at $0 \in A$, and that $\|\exp(x)\| \leq e^{\|x\|}$. $\square$

Henceforth, we adopt the commonly used notation e^x to denote $\exp(x)$.

Proposition 9.41. *If A is a unital Banach algebra, then*

1. *$e^0 = 1$,*
2. *e^x is invertible and $(e^x)^{-1} = e^{-x}$, for all $x \in A$,*
3. *$e^{gxg^{-1}} = ge^x g^{-1}$, for all $x \in A$ and invertible $g \in A$, and*
4. *$e^{x+y} = e^x e^y$, if $xy = yx$.*

Proof. Suppose that $xy = yx$. Apply the binomial formula to obtain

$$\sum_{m=0}^{k} \frac{1}{m!}(x+y)^m = \sum_{m=0}^{k} \frac{1}{m!} \sum_{\ell=0}^{k} \frac{m!}{\ell!(m-\ell)!} x^\ell y^{m-\ell} = \sum_{\ell=0}^{k} \frac{1}{\ell!} x^\ell \sum_{j=0}^{k} \frac{1}{j!} y^j.$$

Therefore, upon passing to limits and using the continuity of multiplication in the algebra A, $e^{x+y} = e^x e^y$.

It is clear that $e^0 = 1$. Suppose that $x \in A$ is arbitrary. The elements x and $-x$ commute to produce $1 = e^0 = e^x e^{-x}$, and so e^x is invertible with inverse $(e^x)^{-1} = e^{-x}$.

Lastly, note that $s_n(gxg^{-1}) = gs_n(x)g^{-1}$ for all invertible g and all $n \in \mathbb{N}$, and so $e^{gxg^{-1}} = ge^x g^{-1}$, for all $x \in A$ and invertible $g \in A$.   □

The formula $e^{x+y} = e^x e^y$ may fail if $xy \neq yx$.

The following proposition is of interest for one-parameter continuous groups in A.

Proposition 9.42. *If A is a unital Banach algebra and if $x \in A$ is fixed, then the function $\Psi : \mathbb{R} \to A$ defined by $\Psi(t) = e^{tx}$ is a homomorphism of the additive group $(\mathbb{R}, +)$ into the multiplicative group $\mathrm{GL}(A)$ (that is, $\Psi(s+t) = \Psi(s)\Psi(t)$, for all $s, t \in \mathbb{R}$). Furthermore,*

$$\lim_{t \to 0} \frac{1}{t}(\Psi(t) - \Psi(0)) = x.$$

Proof. Proposition 9.41 shows that $\Psi(s+t) = \Psi(s)\Psi(t)$, as sx and tx commute. By the power series expansion of e^{tx} we have

$$\frac{1}{t}(\Psi(t) - \Psi(0)) = \frac{1}{t}\left(\sum_{k=0}^{\infty} \frac{1}{k!} t^k x^k - 1\right) = x + t\left(\sum_{k=2}^{\infty} \frac{t^{k-2}}{k!} x^k\right),$$

and so $\lim_{t \to 0} \frac{1}{t}(\Psi(t) - \Psi(0)) = x$.   □

Problems

9.43. Suppose that (X, Σ, μ) is a measure space.

1. Prove that $\mathscr{L}^\infty(X, \Sigma, \mu)$ is a complex, associative abelian algebra with respect to the product $\psi_1 \psi_2(t) = \psi_1(t)\psi_2(t)$, for all $\psi_1, \psi_2 \in L^\infty(X, \Sigma, \mu)$ and $t \in X$.
2. Prove that $L^\infty(X, \Sigma, \mu)$ is a Banach algebra, where the product on $L^\infty(X, \Sigma, \mu)$ is induced by the product on $\mathscr{L}^\infty(X, \Sigma, \mu)$.
3. Prove that the Banach algebra $L^\infty(X, \Sigma, \mu)$ is unital and abelian.

9.44. Determine the multiplicative identities of the unital abelian Banach algebras $\ell^\infty(\mathbb{N})$ and $\ell^\infty(n)$.

9.45. Prove that if $\{x_n\}_{n\in\mathbb{N}}$ and $\{y_n\}_{n\in\mathbb{N}}$ are convergent sequences in a Banach algebra A, and if x and y are the respective limits of these sequences, then the sequence $\{x_n y_n\}_{n\in\mathbb{N}}$ is convergent to xy.

9.46. Assume that A is a unital Banach algebra, and that $a_1, \ldots, a_n \in A$. Prove the following statements. (Suggestion: use induction.)

1. If $a_i a_j = 0$, for all i, j with $i < j$, then

$$\sigma\left(\sum_{j=1}^{n} a_j\right) \subseteq \bigcup_{j=1}^{n} \sigma(a_j).$$

2. If $a_i a_j = 0$, for all i, j with $i \neq j$, then

$$\sigma\left(\sum_{j=1}^{n} a_j\right) \setminus \{0\} = \left(\bigcup_{j=1}^{n} \sigma(a_j)\right) \setminus \{0\}.$$

9.47. Suppose that A is a unital Banach algebra, and that $a \in A$. Define functions $\ell_a : A \to A$ and $r_a : A \to A$ by $\ell_a(x) = ax$ and $r_a(x) = xa$, for $x \in A$.

1. Prove that ℓ_a and r_a are bounded linear operators on A.
2. In considering ℓ_a and r_a as elements in the unital Banach algebra $B(A)$, prove that $\sigma(\ell_a) \subseteq \sigma(a)$ and that $\sigma(r_a) \subseteq \sigma(a)$.
3. Prove that ℓ_a and r_b are commuting operators, for every $b \in A$.

9.48 (Sylvester). Suppose that A is a unital Banach algebra, and that $a, b \in A$. Prove that if $\sigma(a) \cap \sigma(b) = \emptyset$, then for every $y \in A$ there exists a unique $x \in A$ such that $ax - xb = y$. (Suggestion: Consider the operator D on A defined by $D = \ell_a - r_b$.)

9.49. Suppose that A is a unital Banach algebra, and that $a, b \in A$ are such that $\sigma(b) \subseteq \{\zeta \in \mathbb{C} \,|\, |\zeta| < \delta\}$ and $\sigma(a) \subseteq \{\zeta \in \mathbb{C} \,|\, |\zeta| > \delta\}$, for some $\delta > 0$.

1. Prove that, for each $y \in A$, the series $\sum_{n=0}^{\infty} a^{-n-1} y b^n$ converges in A.

2. Prove that if $y \in A$ and if $x = \sum_{n=0}^{\infty} a^{-n-1} y b^n$, then x is the unique solution to the Sylvestre equation $ax - xb = y$.

9.50. Suppose that A is a uniform algebra on a compact Hausdorff space X, and suppose that $f, g \in A$ are such that $\|1 + f + \overline{g}\| < 1$. Prove the following assertions for $h = f + g$:

1. $\|1 + \Re h\| < 1$;
2. there exists $\varepsilon > 0$ such that $1 + \varepsilon h(t) \in \mathbb{D}$, for every $t \in X$;
3. h is invertible in A.

9.51. Prove that if A is a unital abelian Banach algebra, and if $a \in A$ is a nonzero noninvertible element, then there exists a maximal ideal N of A such that $a \in N$.

9.52. Prove that if N is a maximal ideal of $C(X)$, where X is a compact Hausdorff space, then there exists a point $x_0 \in X$ such that $N = \{f \in C(X) \,|\, f(x_0) = 0\}$.

9.53. Suppose that A and B are unital abelian Banach algebras.

1. Prove that if $\phi : A \to B$ is a unital isometric isomorphism, then the map $\zeta_\phi : \mathscr{R}_B \to \mathscr{R}_A$, defined by $\zeta_\phi(\rho)[a] = \rho(\phi(a))$, for $a \in A$, is a homeomorphism of maximal ideal spaces.
2. Prove that if a continuous bijection $\zeta : \mathscr{R}_B \to \mathscr{R}_A$ is a homeomorphism, then the map $\phi_\zeta : C(\mathscr{R}_A) \to C(\mathscr{R}_B)$, defined by $\phi_\zeta(f) = f \circ \zeta$, for $f \in C(\mathscr{R}_A)$, is a unital isometric isomorphism.
3. Prove that A and B are unitally isometrically isomorphic if and only if the character spaces of A and B are homeomorphic.

9.54. An ideal K of a Banach algebra A is said to be an essential ideal if $K \neq A$ and $K \cap J \neq \{0\}$ for every nonzero ideal J of A. Prove that if X is a compact Hausdorff space, the K is an essential ideal of $C(X)$ if and only if there is a dense open subset U of X such that $K = \{f \in C(X) \,|\, f(x) = 0 \text{ for every } x \in U^c\}$.

9.55. If A is a unital abelian Banach algebra such that $\|x^2\| = \|x\|^2$ for each $x \in A$, then prove that the Gelfand transform Γ is an isometry.

9.56. Prove that if X is a locally compact Hausdorff space, then $C_b(X)$ is a semisimple Banach algebra.

9.57. If A is a unital abelian Banach algebra, then prove that

$$\operatorname{Rad} A = \{x \in A \,|\, \sigma(xy) = \{0\}, \text{ for every } y \in A\}.$$

9.58. Prove that $\operatorname{Rad}(A/\operatorname{Rad} A) = \{\dot{0}\}$, for every unital abelian Banach algebra A.

9.59. Determine the character space of the disc algebra $A(\mathbb{D})$.

9.60. Prove that if a 2π-periodic continuous function $f : \mathbb{R} \to \mathbb{C}$ has summable Fourier coefficients, then the series $\displaystyle\sum_{-\infty}^{\infty} \hat{f}(n) e^{int}$ converges uniformly on $\mathbb{R}$ to f.

9.61. Determine the matrix e^S, where S is the matrix

$$S = \begin{bmatrix} 0 & 1 & 0 & \cdots & 0 \\ 0 & 0 & 1 & \ddots & \vdots \\ 0 & 0 & \ddots & \ddots & 0 \\ \vdots & & \ddots & \ddots & 1 \\ 0 & \cdots & \cdots & 0 & 0 \end{bmatrix}.$$

Chapter 10
Hilbert Space Operators

The theory of bounded linear operators acting on Hilbert spaces has a special place in functional analysis. In many regards, it is a very specialised part of the subject; yet, it is impressively rich in both theory and application. While the results already established for operators acting on Banach spaces apply automatically to Hilbert space operators, there is at least one aspect in which there is a slight but important departure from Banach space operator theory, and it is the first issue addressed in the present chapter.

10.1 Hilbert Space Duality and Adjoint Operators

Proposition 6.16 concerning the dual spaces of ℓ^p describes, in the case $p = 2$, an isomorphism of Hilbert spaces. In fact, in the setting of abstract Hilbert space, all linear functionals are determined by vectors, which is the content of the final *Riesz Representation Theorem* in this book.

Theorem 10.1 (Riesz). *Suppose that H is a Hilbert space. For every $\varphi \in H^*$, there corresponds a unique $\eta \in H$ such that $\|\varphi\| = \|\eta\|$ and $\varphi(\xi) = \langle \xi, \eta \rangle$, for all $\xi \in H$. Conversely, for each $\eta \in H$, the formula $\varphi(\xi) = \langle \xi, \eta \rangle$, for $\xi \in H$, determines a unique $\varphi \in H^*$ of norm $\|\varphi\| = \|\eta\|$.*

Proof. First of all, suppose that $\varphi \in H^*$. If $\varphi = 0$, then take $\eta = 0$ and we obtain, trivially, that $\varphi(\xi) = \langle \xi, \eta \rangle$ for all $\xi \in H$. Therefore, assume that $\varphi \neq 0$. Because $H = \ker \varphi \oplus (\ker \varphi)^\perp$, the subspaces $\ker \varphi$ and $(\ker \varphi)^\perp$ form a complementary pair, and so the quotient space $H/ \ker \varphi$ and the Hilbert space $(\ker \varphi)^\perp$ are isomorphic (Proposition 8.17). Because the linear map $\dot{\varphi} : H/ \ker \varphi \to \mathbb{C}$ defined by $\dot{\varphi}(\dot{\xi}) = \varphi(\xi)$ is a well-defined linear isomorphism, the Banach space $H/ \ker \varphi$ is 1-dimensional; hence, so is $(\ker \varphi)^\perp$.

© Springer International Publishing Switzerland 2016
D. Farenick, *Fundamentals of Functional Analysis*, Universitext,
DOI 10.1007/978-3-319-45633-1_10

Thus, $\ker \varphi$ is spanned by some nonzero vector $w \in H$. Let $\eta = \overline{\varphi(w)} \|w\|^{-2} w$. Because each $\xi \in H$ has the form $\xi = v + \lambda w$ for some $v \in \ker \varphi$ and $\lambda \in \mathbb{C}$, we obtain $\varphi(\xi) = \lambda \varphi(w)$ and

$$\langle \xi, \eta \rangle = \left\langle v + \lambda w, \frac{\overline{\varphi(w)}}{\|w\|^2} w \right\rangle = \lambda \varphi(w) \frac{\langle w, w \rangle}{\|w\|^2} = \varphi(\xi).$$

Furthermore, $|\varphi(\xi)| = |\langle \xi, \eta \rangle| \le \|\xi\| \|\eta\|$ implies that $\|\varphi\| \le \|\eta\|$. On the other hand, if $\xi = \|\eta\|^{-1} \eta$, then $\|\xi\| = 1$ and $\varphi(\xi) = \|\eta\|$. Hence, $\|\varphi\| = \|\eta\|$.

To show the uniqueness of η, assume $\eta' \in H$ is another vector for which $\varphi(\xi) = \langle \xi, \eta' \rangle$ for every $\xi \in H$. Then, $\langle \xi, \eta' \rangle = \langle \xi, \eta \rangle$ implies $\langle \xi, \eta - \eta' \rangle = 0$ for every $\xi \in H$. Such is the case for $\xi = \eta - \eta'$ in particular; hence, $\|\eta - \eta'\|^2 = 0$.

The converse is clear.   □

As a consequence of the Riesz Representation Theorem, Hilbert spaces a *self-dual*. However, the natural notion of adjoint for Hilbert space operators is slightly different from the adjoint of Banach space operators because, in the Hilbert space setting, one needs to account for the fact that the inner product is not bilinear—rather, it is conjugate linear in the second variable. In this regard, the notion of adjoint in Theorem 8.1 is not the same as the adjoint that is shown to exist in the proposition below.

Proposition 10.2. *If T is an operator on a Hilbert space H, then there is a unique operator T^* on H such that*

$$\langle T\xi, \eta \rangle = \langle \xi, T^* \eta \rangle, \quad \forall \, \xi, \eta \in H. \tag{10.1}$$

Proof. Fix $\eta \in H$ and define $\varphi_\eta : H \to \mathbb{C}$ by $\varphi_\eta(\xi) = \langle T\xi, \eta \rangle$ for all $\xi \in H$. Because φ_η is plainly linear, the Riesz Representation Theorem (Theorem 10.1) states that there is a unique vector, which we will denote by η^*, such that

$$\langle T\xi, \eta \rangle = \varphi_\eta(\xi) = \langle \xi, \eta^* \rangle, \quad \forall \, \xi \in H. \tag{10.2}$$

Thus, η^* represents φ_η.

Now consider the function $T^* : H \to H$ that sends each $\eta \in H$ to $\eta^* \in H$. It is straightforward to verify that T^* is a linear transformation. Therefore equation (10.2) becomes

$$\langle T\xi, \eta \rangle = \langle \xi, T^* \eta \rangle, \quad \forall \, \xi, \eta \in H. \tag{10.3}$$

The Riesz Representation Theorem states that $\|\eta^*\| = \|\varphi_\eta\|$. Hence, for $\eta \in H$,

$$\|T^*\eta\| = \|\eta^*\| = \|\varphi_\eta\| = \sup_{\|\xi\|=1} |\langle T\xi, \eta\rangle|$$

$$\leq \left(\sup_{\|\xi\|=1} \|T\xi\| \right) \|\eta\| = \|T\| \, \|\eta\|.$$

That is, T^* is bounded and $\|T^*\| \leq \|T\|$.

All that remains now is to show that the transformation T^* that satisfies equation (10.3) is unique. Suppose that S is an operator on H such that $\langle T\xi, \eta\rangle = \langle \xi, S\eta\rangle$ for all $\xi, \eta \in H$. Then for any $\eta \in H$, $\langle T^*\eta - S\eta, \xi\rangle = 0$ for all $\xi \in H$. In particular, $T^*\eta - S\eta$ is orthogonal to itself and so $T^*\eta - S\eta = 0$. Therefore, $S = T^*$. $\qquad\square$

Definition 10.3. The operator T^* defined by equation (10.1) is called the *adjoint* of the operator T and the map $T \mapsto T^*$ on $B(H)$ is called the *involution* on $B(H)$.

A result related to Proposition 10.2 concerns bounded sesquilinear forms.

Definition 10.4. A function $\psi : H \times H \to \mathbb{C}$ is a *bounded sesquilinear form* on H if

1. $\psi(\alpha_1\xi_1 + \alpha_2\xi_2, \eta) = \alpha_1\psi(\xi_1, \eta) + \alpha_2\psi(\xi_2, \eta)$,
2. $\psi(\xi, \beta_1\eta_1 + \beta_2\eta_2) = \overline{\beta}_1\psi(\xi, \eta_1) + \overline{\beta}_2\psi(\xi, \eta_2)$, and
3. there exists $C > 0$ such that $|\psi(\xi, \eta)| \leq C\|\xi\| \, \|\eta\|$ for all $\xi, \eta \in H$.

for all $\xi, \xi_1, \xi_2, \eta, \eta, \eta_2 \in H$ and $\alpha_1, \alpha_2, \beta_1, \beta_2 \in \mathbb{C}$.

Proposition 10.5. *If ψ is a bounded sesquilinear form on H, then there exists a unique operator T such that $\psi(\xi, \eta) = \langle T\xi, \eta\rangle$, for all $\xi, \eta \in H$.*

Proof. As in the proof of Proposition 10.2, fixing $\eta \in H$ and defining $\varphi_\eta : H \to \mathbb{C}$ by $\varphi_\eta(\xi) = \psi(\xi, \eta)$, for all $\xi \in H$, results in an element $\varphi_\eta \in H^*$ which, by Theorem 10.1, has the form $\varphi_\eta(\xi) = \langle \xi, \eta^*\rangle$ for some unique $\eta^* \in H$. The function $T^* : H \to H$ that sends each $\eta \in H$ to $\eta^* \in H$ is a bounded linear operator, and yields $\psi(\xi, \eta) = \langle \xi, T^*\eta\rangle$, for all $\xi, \eta \in H$. Hence, by Proposition 10.2, $\psi(\xi, \eta) = \langle T\xi, \eta\rangle$, for all $\xi, \eta \in H$. $\qquad\square$

Proposition 10.6. *The involution on $B(H)$ has the following properties for all $S, T \in B(H)$ and $\alpha \in \mathbb{C}$:*

1. $T^{**} = T$;
2. $(\alpha T)^* = \overline{\alpha} T^*$;
3. $(S + T)^* = S^* + T^*$;
4. $(ST)^* = T^*S^*$.

Proof. To prove (1), the adjoint T^{**} of T^* is, by (10.3), the unique operator on H for which $\langle T^*\vartheta, \nu\rangle = \langle \vartheta, T^{**}\nu\rangle$—equivalently, $\overline{\langle T^*\vartheta, \nu\rangle} = \overline{\langle \vartheta, T^{**}\nu\rangle}$—for all $\vartheta, \nu \in H$. In setting $\nu = \xi$ and $\vartheta = \eta$, it follows that $\langle \xi, T^*\eta\rangle = \langle T^{**}\xi, \eta\rangle$, for all $\xi, \eta \in H$. Because $\langle \xi, T^*\eta\rangle = \langle T\xi, \eta\rangle$, if ξ is fixed, then for every $\eta \in H$ we have that

$\langle T\xi - T^{**}\xi, \eta \rangle = 0$. Thus, the vector $T\xi - T^{**}\xi$ is orthogonal to itself, which means that $T\xi - T^{**}\xi = 0$. As this is true for every ξ, we deduce that $T^{**} = T$. The proofs of the remaining algebraic statements are straightforward. □

Through the use of the Hilbert space inner product, the norm of an operator can be achieved as in Proposition 10.7 below. Another very important feature of the norm of $T \in B(H)$, which is also shown below, is that $\|T\|^2 = \|T^*T\|$.

Proposition 10.7. *If $T \in B(H)$, then*

$$\|T\| = \sup_{\|\xi\|=\|\eta\|=1} |\langle T\xi, \eta \rangle|.$$

Furthermore, $\|T^*\| = \|T\|$ and $\|T^*T\| = \|T\|^2$.

Proof. If $\omega \in H$, then $\|\omega\| = \sup\{|\varphi(\omega)| \mid \varphi \in H^*, \|\varphi\| = 1\}$ (Corollary 6.23). Thus, $\|\omega\| = \sup\{|\langle \omega, \eta \rangle| \mid \eta \in H, \|\eta\| = 1\}$, by the Riesz Representation Theorem. Hence,

$$\|T\| = \sup_{\|\xi\|=1} \|T\xi\| = \sup_{\|\xi\|=\|\eta\|=1} |\langle T\xi, \eta \rangle|.$$

Since $|\langle T\xi, \eta \rangle| = |\langle \xi, T^*\eta \rangle|$, for all $\xi, \eta \in H$, we obtain $\|T^*\| = \|T\|$ immediately. The norm on $B(H)$ satisfies $\|ST\| \le \|S\| \|T\|$ for every $S, T \in B(H)$. Therefore,

$$\|T^*T\| \le \|T^*\| \|T\| = \|T\| \|T\| = \|T\|^2.$$

Conversely, if $\xi, \eta \in H$ are unit vectors, then the Cauchy-Schwarz inequality yields

$$|\langle T\xi, \eta \rangle|^2 \le \|T\xi\|^2 \|\eta\|^2 = \langle T\xi, T\xi \rangle = \langle T^*T\xi, \xi \rangle \le \|T^*T\|.$$

Thus, $\|T\|^2 \le \|T^*T\|$. □

Proposition 10.8. *If $T \in B(H)$, then*

1. $\ker T = (\operatorname{ran} T^*)^{\perp}$ *and*
2. $\overline{\operatorname{ran} T} = (\ker T^*)^{\perp}$.

Proof. For the proof of the first assertion, assume that $\xi \in \ker T$. Any vector in $\operatorname{ran} T^*$ has the form $T^*\eta$, for some $\eta \in H$. Since $\langle \xi, T^*\eta \rangle = \langle T\xi, \eta \rangle = 0$, we conclude that $\xi \in (\operatorname{ran} T^*)^{\perp}$.

Conversely, suppose that $\xi \in (\operatorname{ran} T^*)^{\perp}$. Thus, for every $\eta \in H$, $0 = \langle \xi, T^*\eta \rangle$. In particular, if $\eta = T\xi$, then $0 = \langle \xi, T^*\eta \rangle = \langle \xi, T^*T\xi \rangle = \langle T\xi, T\xi \rangle = \|T\xi\|^2$, and so $\xi \in \ker T$.

The proof of the second assertion is left as an exercise (Exercise 10.105). □

Another aspect of the Hilbert space adjoint to be aware of—especially in light of what has come before in the study of operators on Banach spaces—is that the defect spectrum is characterised as follows:

Proposition 10.9. *If $T \in B(H)$, then $\lambda \in \sigma_d(T)$ if and only if $\overline{\lambda} \in \sigma_p(T^*)$.*

Proof. Exercise 10.107. □

The following little fact is very useful in the study of Hilbert space operators, as seen in Corollary 10.11 below.

Proposition 10.10 (Polarisation Identity). *For every pair of vectors ξ and η in an inner product space,*

$$\langle \xi, \eta \rangle = \frac{1}{4} \left(\langle \xi + \eta, \xi + \eta \rangle - \langle \xi - \eta, \xi - \eta \rangle + \langle \xi + i\eta, \xi + i\eta \rangle - \langle \xi - i\eta, \xi - i\eta \rangle \right).$$

Proof. A straightforward computation confirms the result. □

Corollary 10.11. *If $S, T \in B(H)$ satisfy $\langle S\xi, \xi \rangle = \langle T\xi, \xi \rangle$ for every $\xi \in H$, then $S = T$.*

Proof. By the Polarisation Identity, the hypothesis yields $\langle S\xi, \eta \rangle = \langle T\xi, \eta \rangle$ for every $\xi, \eta \in H$, and so $\langle (S - T)\xi, \eta \rangle = 0$ for every $\xi, \eta \in H$ yields $S - T = 0$ by Proposition 10.7. □

Isometries of Hilbert spaces are characterised by a succinct algebraic condition:

Proposition 10.12. *An operator $V \in B(H)$ is an isometry if and only if $V^*V = 1$.*

Proof. If V is an isometry and $\xi \in H$, then $\|\xi\|^2 = \|V\xi\|^2 = \langle V^*V\xi, \xi \rangle$ implies that $\langle (1 - V^*V)\xi, \xi \rangle = 0$ for every $\xi \in H$. By the Polarisation Identity, this implies that $V^*V = 1$.

The converse is clear. □

Definition 10.13. A surjective isometry $U \in B(H)$ is called a *unitary operator*.

The following facts about unitary operators are readily confirmed.

Proposition 10.14. *The following statements are equivalent for $U \in B(H)$:*

1. *U is a unitary operator;*
2. *$U^*U = UU^* = 1$;*
3. *for some orthonormal basis $\{\phi_i\}_i$ of H, $\{U\phi_i\}_i$ is also an orthonormal basis;*
4. *for every orthonormal basis $\{\phi_i\}_i$ of H, $\{U\phi_i\}_i$ is also an orthonormal basis.*

Proof. Exercise 10.108. □

Isometries and unitaries are among the most fundamental of all Hilbert space operators. Of equal importance are the projections.

Definition 10.15. An operator $P \in B(H)$ is a *projection* if $P^* = P = P^2$.

In contrast to Banach spaces, where there can be subspaces without complements, every subspace M of a Hilbert space H has a complement, the most important of which is $M^\perp$. Hence, Proposition 8.16 has the following form in Hilbert space:

Proposition 10.16. *For every subspace M of H there exists a projection $P \in B(H)$ such that $M = \operatorname{ran} P$ and $M^{\perp} = \ker P$. Conversely, if $P \in B(H)$ is a projection and if $M = \operatorname{ran} P$, then $\operatorname{ran}(1 - P) = \ker P = M^{\perp}$.*

Proof. Because the subspaces M and $M^{\perp}$ are a complementary pair, Proposition 8.16 shows that there exists an idempotent $E \in B(H)$ such that $\operatorname{ran} E = M$ and $\ker E = M^{\perp}$. Such an idempotent E necessarily satisfies $E^* = E$ because if $\xi \oplus \eta, \gamma \oplus \delta \in M \oplus M^{\perp}$, then

$$\langle E(\xi \oplus \eta), \gamma \oplus \delta \rangle = \langle \xi, \gamma \rangle = \langle \xi \oplus \eta, E(\gamma \oplus \delta) \rangle.$$

That is, $E^* = E$ and so the idempotent E is a projection.

Conversely, if $P \in B(H)$ is a projection, then it is also an idempotent operator, and so its range M is a subspace of H. If $\xi \in M$ and $\eta \in H$, then $P\xi = \xi$ and $P^* = P$ yield

$$\langle \xi, (1 - P)\eta \rangle = \langle \xi, \eta \rangle - \langle \xi, P\eta \rangle = \langle \xi, \eta \rangle - \langle P\xi, \eta \rangle = \langle \xi, \eta \rangle - \langle \xi, \eta \rangle = 0.$$

That is, $(1 - P)\eta \in M^{\perp}$ and so $\operatorname{ran}(1 - P) \subseteq M^{\perp}$. And, if $\gamma \in M^{\perp}$, then

$$\|\gamma - (1 - P)\gamma\|^2 = \langle P\gamma, P\gamma \rangle = \langle P^*P\gamma, \gamma \rangle = \langle P\gamma, \gamma \rangle = 0,$$

which shows that $\gamma = (1 - P)\gamma$ and, hence, that $M^{\perp} \subseteq \operatorname{ran}(1 - P)$.   □

Proposition 8.14 on the algebraic features of idempotents translates into the following proposition about projections.

Proposition 10.17. *The following properties hold for projections $P, Q \in B(H)$:*

1. *$P + Q$ is a projection if and only if $PQ = QP = 0$;*
2. *$P - Q$ is a projection if and only if $PQ = QP = Q$;*
3. *if $PQ = QP$, then QP is a projection with range $\operatorname{ran} P \cap \operatorname{ran} Q$ and kernel $\ker P + \ker Q$.*

Another convenient relationship between T and T^* occurs with invariant subspaces.

Proposition 10.18. *A subspace $M \subseteq H$ is invariant under an operator $T \in B(H)$ if and only if $M^{\perp}$ is invariant under T^*.*

Proof. If M is invariant under T and if $\eta \in M^{\perp}$, then for every $\xi \in M$ we have that $0 = \langle T\xi, \eta \rangle = \langle \xi, T^*\eta \rangle$, which yields $T^*\eta \in M^{\perp}$. That is, $M^{\perp}$ is T^*-invariant. The converse is proved in the obvious similar manner.   □

With regard to invariant subspaces, we have the following useful algebraic characterisations.

Proposition 10.19. *Suppose that $T, P \in B(H)$, where P is a projection.*

1. *The range of P is invariant under T if and only if $PTP = TP$.*
2. *The range of P is invariant under T and T^* if and only if $TP = PT$.*

Proof. For (1), if ran P is invariant under T, then for every $\xi \in H$, $TP\xi \in \operatorname{ran} P$ and so $P(\operatorname{ran} P\xi) = TP\xi$. Conversely, if $PTP = TP$, for each $\eta \in \operatorname{ran} P$ we have $T\eta = TP\eta = PTP\eta \in \operatorname{ran} P$, and so ran P is T-invariant.

To prove (2), if ran P is invariant under T and T^*, then by (1) we have that $PTP = TP$ and $PT^*P = T^*P$. Taking the adjoint of the second equation give $PTP = PT$; hence, $TP = PTP = PT$. Conversely, $TP = PT$ implies $PTP = P^2T = PT$ and, by taking adjoints, that $PT^* = T^*P$, whence $PT^*P = T^*P^2 = T^*P$. □

10.2 Examples

10.2.1 Matrix Representations

Suppose that $T \in B(H)$ and that $\mathscr{B} = \{\phi_k\}_{k \in \mathbb{N}}$ is an orthonormal basis of a separable Hilbert space H. If $\mathscr{T} = [\tau_{ij}]$ is the (infinite) matrix representation of T with respect to the orthonormal basis $\mathscr{B}$, then the (i,j)-entry of $\mathscr{T}$ is determined via

$$\tau_{ij} = \langle T\phi_j, \phi_i \rangle.$$

In using this for T^* in place of T we conclude that the (p,q)-entry of the matrix representation of T^* with respect to $\mathscr{B}$ is $\langle T^*\phi_q, \phi_p \rangle$. Furthermore, $\langle T^*\phi_q, \phi_p \rangle$ is given by

$$\langle T^*\phi_q, \phi_p \rangle = \overline{\langle \phi_p, T^*\phi_q \rangle} = \overline{\langle T\phi_p, \phi_q \rangle} = \overline{\tau_{qp}}.$$

Thus, the matrix representation $\mathscr{T}^*$ of T^* is determined by transposing $\mathscr{T}$, the matrix representation of T, and then conjugating each entry. In other words, $\mathscr{T}^*$ is the conjugate transpose of $\mathscr{T}$.

10.2.2 Multiplication Operators

Recall from Section 8.1 that if (X, Σ, μ) is a σ-finite measure space and if $\psi \in \mathscr{L}^\infty(X, \Sigma, \mu)$, then ψ induces an operator $M_\psi : L^2(X, \Sigma, \mu) \to L^2(X, \Sigma, \mu)$ via $M_\psi(\dot{f}) = (\dot{\psi f})$ for all $\dot{f} \in L^2(X, \Sigma, \mu)$. Because (X, Σ, μ) is σ-finite, Example 8.1.7 shows that $\|M_\psi\| = \operatorname{ess-sup} \psi$

To calculate the adjoint of M_ψ, note that if $f, g \in \mathscr{L}^2(X, \Sigma, \mu)$, then

$$\langle \dot{f}, M_{\overline{\psi}}\dot{g} \rangle = \int_X f\overline{\overline{\psi}g}\, d\mu = \int_X \psi f\overline{g}\, d\mu = \langle M_\psi \dot{f}, \dot{g} \rangle.$$

Hence, by the uniqueness of the Hilbert space adjoint, $M_\psi^* = M_{\overline{\psi}}$.

10.2.3 The Bilateral Shift Operator

Let $L^2(\mathbb{T})$ denote the Hilbert space $L^2([-\pi,\pi],\mathfrak{M},m)$. Recall that an orthonormal basis $\{\dot{\phi}_k\}_{k\in\mathbb{Z}}$ of $L^2(\mathbb{T})$ arises from the continuous functions $\phi_k : [-\pi,\pi] \to \mathbb{C}, k \in \mathbb{Z}$, whereby

$$\phi_k(t) = \frac{1}{\sqrt{2\pi}} e^{ikt}, \quad t \in [-\pi,\pi].$$

If $f \in \mathscr{L}^2(\mathbb{T})$, then $\dot{f} \in L^2(\mathbb{T})$ has a Fourier series decomposition

$$\dot{f} = \sum_{k\in\mathbb{Z}} \hat{f}(k)\, \dot{\phi}_k,$$

which is convergent in $L^2(\mathbb{T})$ and where

$$\hat{f}(k) = \langle \dot{f}, \dot{\phi}_k \rangle = \int_{-\pi}^{\pi} f(t) e^{-ikt}\, dm(t).$$

Let $\psi(t) = e^{it}$. The adjoint of the multiplication operator M_ψ is $M_{\overline{\psi}}$. Therefore, because $\overline{\psi}(t) = e^{-it}$, if $\dot{g} = M_\psi \dot{f}$ and $\dot{h}h = (M_\psi)^* \dot{f}$, then

$$\hat{g}(k) = \hat{f}(k-1) \quad \text{and} \quad \hat{h}(k) = \hat{f}(k+1), \quad \forall k \in \mathbb{Z}.$$

Thus, if $B = M_\psi$, then B shifts the Fourier coefficients of $\dot{f} \in L^2(\mathbb{T})$ forward by one position, and its adjoint shifts the Fourier coefficients backwards one position. For this reason, B is called the *bilateral shift operator* .

Put in terms of the action of B on the orthonormal basis of $L^2(\mathbb{T})$, we have

$$B\dot{\phi}_k = \dot{\phi}_{k+1} \quad \text{and} \quad B^*\dot{\phi}_k = \dot{\phi}_{k-1}, \quad \forall k \in \mathbb{Z}.$$

One last observation: because $BB^* = B^*B = 1$, the bilateral shift operator B is a unitary operator.

10.2.4 Toeplitz Operators

Continuing with the notation $L^2(\mathbb{T})$ for the Hilbert space $L^2([-\pi,\pi],\mathfrak{M},m)$, the linear submanifold

$$H^2(\mathbb{T}) = \left\{ \dot{f} \in L^2(\mathbb{T}) \,|\, \hat{f}(k) = 0, \quad \forall k < 0 \right\}$$

is a subspace of $L^2(\mathbb{T})$ called the *Hardy space*. Let $P_{H^2} \in B(L^2(\mathbb{T}))$ denote the projection whose range is the Hardy space $H^2(\mathbb{T})$. If $\psi \in \mathscr{L}^\infty(\mathbb{T})$, then the *Toeplitz operator* T_ψ on $H^2(\mathbb{T})$ is the operator $T_\psi = P_{H^2} M_\psi$. The function ψ is called the *symbol* of the operator T_ψ. Because $(M_\psi)^* = M_{\overline{\psi}}$, if $\dot{f}, \dot{g} \in H^2(\mathbb{T})$, then

$$\langle T_\psi \dot{f}, \dot{g} \rangle = \langle P_{H^2} M_\psi \dot{f}, \dot{g} \rangle = \langle M_\psi \dot{f}, P_{H^2} \dot{g} \rangle = \langle M_\psi \dot{f}, \dot{g} \rangle$$

$$= \langle \dot{f}, M_{\overline{\psi}} \dot{g} \rangle = \langle P_{H^2} \dot{f}, M_{\overline{\psi}} \dot{g} \rangle = \langle \dot{f}, P_{H^2} M_{\overline{\psi}} \dot{g} \rangle$$

$$= \langle \dot{f}, T_{\overline{\psi}} \dot{g} \rangle.$$

Thus, $(T_\psi)^* = T_{\overline{\psi}}$.

In particular, if $\psi(t) = e^{it}$ and $S = T_\psi$, then S shifts each element of the orthonormal basis $\{\dot{\phi}_k\}_{k \geq 0}$ forward one position, from $\dot{\phi}_k$ to $\dot{\phi}_{k+1}$. In this case the Toeplitz operator S is called the *unilateral shift operator*.

With respect to the orthonormal basis $\{\dot{\phi}_k\}_{k \geq 0}$ of $H^2(\mathbb{T})$, the matrix representation $\mathscr{T}_\psi$ of a Toeplitz operator T_ψ has a rather special form. Express ψ as a Fourier series in $L^2(\mathbb{T})$: $\psi = \sum_{n \in \mathbb{Z}} \alpha_n \dot{\phi}_n$. Observe that if $k, j \geq 0$, then k-th Fourier coefficient of the product $\psi(t) \phi_j(t)$ is the same as the k-th Fourier coefficient of $e^{ijt} \psi(t)$, which in $L^2(\mathbb{T})$ is the Fourier coefficient arising from

$$\langle (\psi \phi_j), \dot{\phi}_k \rangle = \langle B^j \psi, \dot{\phi}_k \rangle,$$

where B is the bilateral shift. Thus, the (k, j)-entry of the matrix for T_ψ is given by

$$\langle T_\psi \dot{\phi}_j, \dot{\phi}_k \rangle = \langle P_{H^2} M_\psi \dot{\phi}_j, \dot{\phi}_k \rangle = \langle M_\psi \dot{\phi}_j, \dot{\phi}_k \rangle = \langle (\psi \phi_j), \dot{\phi}_k \rangle$$

$$= \langle B^j \psi, \dot{\phi}_k \rangle \langle \psi, (B^j)^* \dot{\phi}_k \rangle = \langle \psi, \dot{\phi}_{k-j} \rangle = \alpha_{k-j}.$$

Note that the $(k+1, j+1)$-entry of T_ψ is also α_{k-j}. Hence, the matrix representation for T_ψ is given by

$$\mathscr{T}_\psi = \begin{bmatrix} \alpha_0 & \alpha_{-1} & \alpha_{-2} & \alpha_{-3} & \cdots \\ \alpha_1 & \alpha_0 & \alpha_{-1} & \alpha_{-2} & \ddots \\ \alpha_2 & \alpha_1 & \alpha_0 & \alpha_{-1} & \ddots \\ \alpha_3 & \alpha_2 & \alpha_1 & \alpha_0 & \ddots \\ \cdots & \ddots & \ddots & \ddots & \ddots \end{bmatrix}.$$

In particular, the unilateral shift S has matrix representation

$$\mathscr{S} = \begin{bmatrix} 0 & 0 & 0 & 0 & \cdots \\ 1 & 0 & 0 & 0 & \ddots \\ 0 & 1 & 0 & 0 & \ddots \\ 0 & 0 & 1 & 0 & \ddots \\ \cdots & \ddots & \ddots & \ddots & \ddots \end{bmatrix}.$$

10.2.5 Weighted Unilateral Shift Operators

The following class of operators were mentioned in Section 8.1.
If $\alpha = (\alpha)_{k \in \mathbb{N}} \in \ell^\infty(\mathbb{N})$ and $S_\alpha : \ell^2(\mathbb{N}) \to \ell^2(\mathbb{N})$ is defined by

$$S_\alpha v = S_\alpha \left(\begin{bmatrix} v_1 \\ v_2 \\ v_3 \\ \vdots \end{bmatrix} \right) = \begin{bmatrix} 0 \\ \alpha_1 v_1 \\ \alpha_2 v_2 \\ \vdots \end{bmatrix}, \quad v \in \ell^p(\mathbb{N}),$$

then S_α is is called a *weighted unilateral shift operator*. The (Hilbert space) adjoint $(S_\alpha)^*$ of S_α is given by

$$S_\alpha^* v = S_\alpha^* \left(\begin{bmatrix} v_1 \\ v_2 \\ v_3 \\ \vdots \end{bmatrix} \right) = \begin{bmatrix} \overline{\alpha}_1 v_2 \\ \overline{\alpha}_2 v_3 \\ \overline{\alpha}_3 v_4 \\ \vdots \end{bmatrix}, \quad v \in \ell^p(\mathbb{N}),$$

Note that the Hilbert space adjoint of S_α is slightly different from the Banach space adjoint of S_α that is given in equation (8.2) of Section 8.1.

10.2.6 Rank-1 Operators

An operator $T \in B(H)$ is of rank 1 if T has 1-dimensional range. Let $\phi \in H$ be a vector that spans the range of T. Then, for each $\xi \in H$, there is a unique $\alpha_\xi \in \mathbb{C}$ for which $T\xi = \alpha_\xi \phi$. The map $H \to \mathbb{C}$ given by $\xi \mapsto \alpha_\xi$ is easily seen to be linear and bounded, and so by the Riesz Representation Theorem there is a vector $\psi \in H$ for which $T\xi = \langle \xi, \psi \rangle \phi$, for every $\xi \in H$.

This form of a rank-1 operator T is expressed by $T = \phi \otimes \psi$. That is,

$$T\xi = (\phi \otimes \psi)[\xi] = \langle \xi, \psi \rangle \phi,$$

for every $\xi \in H$.

10.2.7 Direct Sums of Operators

If $T_j \in B(H_j)$, for $j = 1, \ldots, n$, and if H is the Hilbert space direct sum $H = \bigoplus_{j=1}^{n} H_j$ (see Proposition 5.81), then $T = T_1 \oplus \cdots \oplus T_n$ denotes the unique operator on H defined by $T\left(\bigoplus_{j=1}^{n} \xi_j\right) = \bigoplus_{j=1}^{n} T\xi_j$. Observe that T has norm $\|T\| = \max_j \|T_j\|$ and that the adjoint of T is given by $T^* = T_1^* \oplus \cdots \oplus T_n^*$. More generally, if $\bigoplus_{\alpha \in \Lambda} H_\alpha$ is the direct sum of a family $\{H_\alpha\}_{\alpha \in \Lambda}$ of Hilbert spaces (as in Proposition 5.98), and if $\{T_\alpha\}_{\alpha \in \Lambda}$ is a family of operators $T_\alpha \in B(H_\alpha)$ such that $\sup_\alpha \|T_\alpha\| < \infty$, then the linear transformation $\bigoplus_{\alpha \in \Lambda} T_\alpha$ on $\bigoplus_{\alpha \in \Lambda} H_\alpha$ defined by

$$\left(\bigoplus_{\alpha \in \Lambda} T_\alpha\right)[(\xi_\alpha)_\alpha] = (T\xi_\alpha)_\alpha,$$

for $(\xi_\alpha)_\alpha \in \bigoplus_{\alpha \in \Lambda} H_\alpha$, is an operator of norm $\sup_\alpha \|T_\alpha\|$ and adjoint $\bigoplus_{\alpha \in \Lambda} T_\alpha^*$.

10.3 Hermitian Operators

Much of the theory of Hilbert space operators is devoted to the ways in which T and T^* interact. The first case of interest occurs when T and T^* are in fact the same. Such operators are probably the most important in all of operator theory and its applications.

Definition 10.20. An operator $T \in B(H)$ is *hermitian* if $T^* = T$.

Example 10.21. A multiplication operator M_ψ on $L^2(X, \Sigma, \mu)$ is hermitian if and only if ess-ran $\psi \subset \mathbb{R}$.

Proof. Recall that if (X, Σ, μ) is a measure space, then $M_\psi^* = M_{\overline{\psi}}$. However, $M_\psi = M_{\overline{\psi}}$ if and only if $\psi(t) = \overline{\psi(t)}$ for almost all $t \in X$. $\square$

For any $T \in B(H)$, $T = \frac{1}{2}(T + T^*) + i\frac{1}{2i}(T - T^*)$. Since $T + T^*$ and $i(T - T^*)$ are hermitian, the hermitian operators span $B(H)$. This is one reason why hermitian operators are of importance.

If $T \in B(H)$ is hermitian, then $\langle T\xi, \eta \rangle = \langle \xi, T^*\eta \rangle = \langle \xi, T\eta \rangle$, for every $\xi, \eta \in H$. In particular, if $\eta = \xi$, this implies that $\overline{\langle T\xi, \xi \rangle} = \langle T\xi, \xi \rangle$, for all $\xi \in H$; that is, the form $\xi \mapsto \langle T\xi, \xi \rangle$ is necessarily real valued if T is hermitian. This necessary condition is also sufficient.

Proposition 10.22. *An operator $T \in B(H)$ is hermitian if and only if $\langle T\xi, \xi \rangle \in \mathbb{R}$ for all $\xi \in H$.*

Proof. If T is hermitian, then for every vector ξ,

$$\langle T\xi, \xi \rangle = \langle \xi, T^*\xi \rangle = \langle \xi, T\xi \rangle = \overline{\langle T\xi, \xi \rangle},$$

and therefore $\langle T\xi, \xi \rangle \in \mathbb{R}$.

Conversely, suppose that $\langle T\vartheta, \vartheta \rangle \in \mathbb{R}$ for all ϑ; then $\langle T\vartheta, \vartheta \rangle = \langle \vartheta, T\vartheta \rangle$. However, T^* also satisfies $\langle T\vartheta, \vartheta \rangle = \langle \vartheta, T^*\vartheta \rangle$, for every $\vartheta \in H$. Therefore, $\langle \vartheta, T\vartheta \rangle = \langle \vartheta, T^*\vartheta \rangle$, for every $\vartheta \in H$, and so $T^* = T$, by the Polarisation Identity (Corollary 10.11). $\square$

Corollary 10.23. *If $T \in B(H)$ is hermitian and if $M \subseteq H$ is a T-invariant subspace, then the restriction $T_{|M}$ of T to M is hermitian.*

Proof. If $\xi \in M$, then $\langle T_{|M}\xi, \xi \rangle = \langle T\xi, \xi \rangle \in \mathbb{R}$. Hence, by Proposition 10.22, $T_{|M}$ is hermitian. $\square$

The next set of propositions reveals some striking features of the spectra of hermitian operators.

Proposition 10.24. *If $T \in B(H)$ is a hermitian operator, then $\sigma(T) = \sigma_{\mathrm{ap}}(T) \subset \mathbb{R}$.*

Proof. Recall that $\sigma(T) = \sigma_{\mathrm{ap}}(T) \cup \sigma_{\mathrm{d}}(T)$. By Proposition 10.9, $\lambda \in \sigma_{\mathrm{d}}(T)$ if and only if $\overline{\lambda} \in \sigma_{\mathrm{p}}(T^*)$. As $T^* = T$, if we show that every eigenvalue of T is real, then we will obtain $\sigma_{\mathrm{d}}(T) \subset \mathbb{R}$. To this end, let $T\xi = \lambda\xi$ for some $\lambda \in \mathbb{C}$ and unit vector $\xi \in H$. Because $\langle T\xi, \xi \rangle \in \mathbb{R}$ (Proposition 10.22), we obtain

$$\lambda = \lambda\langle \xi, \xi \rangle = \langle \lambda\xi, \xi \rangle = \langle T\xi, \xi \rangle \in \mathbb{R}.$$

Hence, $\sigma_{\mathrm{d}}(T) = \sigma_{\mathrm{p}}(T) \subset \mathbb{R}$ and

$$\sigma(T) = \sigma_{\mathrm{ap}}(T) \cup \sigma_{\mathrm{d}}(T) = \sigma_{\mathrm{ap}}(T) \cup \sigma_{\mathrm{p}}(T) \subseteq \sigma_{\mathrm{ap}}(T) \subseteq \sigma(T).$$

We now show that $\sigma_{\mathrm{ap}}(T) \subset \mathbb{R}$. Suppose that $\lambda \in \mathbb{C} \setminus \mathbb{R}$. For any nonzero $\xi \in H$, we have

$$0 < |\lambda - \overline{\lambda}| \, \|\xi\|^2 = |\langle (T - \lambda 1)\xi, \xi \rangle - \langle (T - \overline{\lambda} 1)\xi, \xi \rangle|$$

$$= |\langle (T - \lambda 1)\xi, \xi \rangle - \langle \xi, (T - \lambda 1)\xi \rangle|$$

$$\le 2\|(T - \lambda 1)\xi\| \, \|\xi\|.$$

Hence, $\|(T - \lambda 1)\xi\| \ge \frac{1}{2}|\lambda - \overline{\lambda}| \, \|\xi\|$ for all $\xi \in H$. This proves that $(T - \lambda 1)$ is bounded below and, hence, $\lambda \notin \sigma_{\mathrm{ap}}(T)$. This concludes the proof of $\sigma(T) = \sigma_{\mathrm{ap}}(T) \subset \mathbb{R}$. $\qquad\square$

Proposition 10.25. *If $T \in B(H)$ is hermitian, then*

$$\|T\| = \max\{|\lambda| \, | \, \lambda \in \sigma(T)\}.$$

That is, $\|T\| = \mathrm{spr}\, T$.

Proof. Let $\alpha = \|T\|$. For any unit vector $\xi \in H$,

$$\|(T^2 - \alpha^2 1)\xi\|^2 = \langle (T^2 - \alpha^2 1)\xi, (T^2 - \alpha^2 1)\xi \rangle$$

$$= \|T^2\xi\|^2 - 2\alpha^2\|T\xi\|^2 + \alpha^4\|\xi\|^2$$

$$\le \alpha^2\|T\xi\|^2 - 2\alpha^2\|T\xi\|^2 + \alpha^4$$

$$= \alpha^4 - \alpha^2\|T\xi\|^2. \tag{10.4}$$

By definition of the norm of an operator, there are unit vectors $\xi_n \in H$ such that $\|T\xi_n\| \to \|T\|$. Hence, by inequality (10.4), $\lim_n \|(T^2 - \alpha^2 1)\xi_n\|^2$ exists and is equal to 0. Therefore, $\alpha^2 \in \sigma(T^2)$.

Because $T^2 - \alpha^2 1 = (T + \alpha 1)(T - \alpha 1)$, at least one of the two operators on the right-hand side of this expression must fail to be invertible. Thus, $\alpha \in \sigma(T)$ or $-\alpha \in \sigma(T)$. In either case, there is a $\lambda \in \sigma(T)$ such that $|\lambda| = \|T\|$. On the other hand, $|\lambda| \le \|T\|$, for all $\lambda \in \sigma(T)$ (Theorem 8.42), which completes the proof. $\qquad\square$

Proposition 10.26. *Assume that $T \in B(H)$ is hermitian and let*

$$m_\ell = \inf_{\|\xi\|=1} \langle T\xi, \xi \rangle \quad \text{and} \quad m_u = \sup_{\|\xi\|=1} \langle T\xi, \xi \rangle.$$

Then $m_\ell, m_u \in \sigma(T)$ and $\sigma(T) \subseteq [m_\ell, m_u]$.

Proof. If $T' = T - m_\ell 1$, then T' is hermitian and $m_\ell \in \sigma(T)$ if and only if $0 \in \sigma(T')$. Therefore, we assume without loss of generality that $m_\ell = 0$. Under this assumption, the sesquilinear form $[\cdot, \cdot] : H \times H \to \mathbb{C}$ defined by $[\xi, \eta] = \langle T\xi, \eta \rangle$ satisfies the

Cauchy-Schwarz inequality

$$|[\xi,\eta]| \le [\xi,\xi]^{1/2}[\eta,\eta]^{1/2}, \quad \forall \xi,\eta \in H.$$

Therefore, with $\eta = T\xi$,

$$\|T\xi\|^4 = |\langle T\xi, T\xi\rangle|^2 \le \langle T\xi,\xi\rangle \langle T^2\xi,T\xi\rangle \le \langle T\xi,\xi\rangle \|T\|^3 \|\xi\|.$$

Hence, $\inf\limits_{\|\xi\|=1} \|T\xi\| = 0$, which proves that T is not bounded below. That is, $0 \in \sigma(T)$.
On the other hand, if $\lambda < 0$, then

$$\|(T-\lambda 1)\xi\|^2 = \|T\xi\|^2 - 2\lambda\langle T\xi,\xi\rangle + \lambda^2\|\xi\|^2 \ge \lambda^2\|\xi\|^2$$

implies that $T - \lambda 1$ is bounded below. Hence, $\lambda \notin \sigma_{\mathrm{ap}}(T) = \sigma(T)$. This proves that $\sigma(T) \subseteq [0,\infty)$.

The proof that $m_u \in \sigma(T)$ and that $\sigma(T) \subseteq (-\infty, m_u]$ is left to the reader (Exercise 10.116). $\qquad\square$

Corollary 10.27. *If $T \in B(H)$ is hermitian, then* $\mathrm{Conv}\,\sigma(T) = \overline{\{\langle T\xi,\xi\rangle \mid \|\xi\| = 1\}}$.

Proof. The unit sphere $\mathscr{S}$ is a Hilbert space H is a path-connected set, and so the continuous map $\mathscr{S} \to \mathbb{R}$ given by $\xi \mapsto \langle T\xi,\xi\rangle$ has a path-connected range. Because the infimum and supremum of the range of this map are the minimum and maximum elements of the spectrum of T, we deduce the equality of the sets $\mathrm{Conv}\,\sigma(T)$ and $\overline{\{\langle T\xi,\xi\rangle \mid \|\xi\| = 1\}}$. $\qquad\square$

By way of the classical Weierstrass Approximation Theorem, the usual polynomial functional calculus $T \mapsto f(T) = \sum\limits_{j=0}^{n} \alpha_j T^j$, where T is a Banach space operator and $f(t) = \sum\limits_{j=0}^{n} \alpha_j t^j$ is a polynomial with complex coefficients, extends to continuous functions in cases where T is a hermitian Hilbert space operator. The main result in this direction is stated below, and it is a major tool in the analysis of Hilbert space operators.

Theorem 10.28 (Continuous Functional Calculus for Hermitian Operators). *If $T \in B(H)$ is a hermitian operator, then for every continuous function $f : \sigma(T) \to \mathbb{R}$ there is a unique hermitian operator $f(T)$ with the property that*

$$\|f(T) - q_n(T)\| \to 0 \tag{10.5}$$

for every sequence $\{q_n\}_{n\in\mathbb{N}}$ of polynomials $q_n \in \mathbb{R}[t]$ for which

$$\lim_{n\to\infty} \left(\max_{t\in\sigma(T)} |f(t) - q_n(t)| \right) = 0.$$

Furthermore, for all continuous functions $f, g : \sigma(T) \to \mathbb{R}$ and $\alpha \in \mathbb{R}$,

1. $\|f(T)\| = \max_{\lambda \in \sigma(T)} |f(\lambda)|$,
2. $\alpha f(T) = \alpha(f(T))$,
3. $(f + g)(T) = f(T) + g(T)$, *and*
4. $fg(T) = f(T)g(T)$.

Proof. Select a real-valued $f \in C(\sigma(T))$. By the Weierstrass Approximation Theorem (Exercise 5.109), there exists a sequence $\{q_n\}_{n \in \mathbb{N}}$ of polynomials with complex coefficients such that $\lim_n \left(\max_{t \in \sigma(T)} |f(t) - q_n(t)| \right) = 0$. By considering $\frac{1}{2}(q_n + \overline{q}_n)$, we may assume without loss of generality that each q_n has real coefficients. The convergent sequence $\{q_n\}_{n \in \mathbb{N}}$ is necessarily Cauchy in $C(\sigma(T))$. Furthermore, $q_m(T) - q_n(T)$ is hermitian, for all $m, n \in \mathbb{N}$, and so the Spectral Mapping Theorem shows that

$$\sigma(q_m(T) - q_n(T)) = \{q_m(\lambda) - q_n(\lambda) \,|\, \lambda \in \sigma(T)\}.$$

Thus, $\|q_m(T) - q_n(T)\| = \max_{\lambda \in \sigma(T)} |q_m(\lambda) - q_n(\lambda)|$ (Proposition 10.36) and therefore $\{q_n(T)\}_{n \in \mathbb{N}}$ is a Cauchy sequence of hermitian operators in $B(H)$. Denote the limit of this sequence by $f(T)$, and observe that $f(T)$ is independent of the choice of approximating sequence $\{q_n\}_{n \in \mathbb{N}}$.

Now if $\alpha \in \mathbb{R}$ and f and g are polynomials with real coefficients, then $\|f(T)\| = \max_{\lambda \in \sigma(T)} |f(\lambda)|$, $\alpha f(T) = \alpha(f(T))$, $(f + g)(T) = f(T) + g(T)$, and $fg(T) = f(T)g(T)$. Hence, by the approximation in equation (10.5), this properties also hold for continuous functions $f, g : \sigma(T) \to \mathbb{R}$. □

The map $f \mapsto f(T)$ in Theorem 10.28 is called *continuous functional calculus for* T. Not all Hilbert space operators admit continuous functional calculus (Exercise 10.131), and so Theorem 10.28 is quite particular to hermitian (and hermitian-like) operators.

A useful application of the continuous functional calculus concerns isolated points in the spectrum of a hermitian operator. Recall from Definition 1.68 that a limit point of a subset Y in a topological space X is a point $x \in X$ such that for every open set U containing x there is an element $y \in Y$ such that $y \in U$ and $y \neq x$. By Proposition 1.69, the closure of Y is given by $\overline{Y} = Y \cup L(Y)$, where $L(Y)$ denotes the set of limit points of Y.

Definition 10.29. If Y is a subset of a topological space X, then an element $y \in Y$ is an *isolated point* of Y if $y \in Y \setminus L(Y)$.

Proposition 10.30. *An isolated point of the spectrum of a hermitian operator T is necessarily an eigenvalue of T.*

Proof. The set $\{\lambda\}$ is a closed subset of $\sigma(T)$. Because λ is an isolated point of $\sigma(T)$, the characteristic function $f = \chi_{\{\lambda\}}$, as a map $\sigma(T) \to \mathbb{C}$, is a nonzero real-valued continuous function such that $f^2 = f$. Therefore, by the continuous functional calculus, the operator $P = f(T)$ is hermitian, nonzero, and satisfies $P^2 = P$; in other words, P is a nonzero projection. Now consider the function $h : \sigma(T) \to \mathbb{C}$ given by

$h(t) = tf(t)$, for $t \in \sigma(T)$. Because $h(t) = \lambda$ if $t = \lambda$ and $h(t) = 0$ otherwise, we see that $h(t) = \lambda f(t)$. Hence, the continuous functional calculus yields $TP = Tf(T) = \lambda f(T) = \lambda P$, and so every $\xi \in \operatorname{ran} P$ satisfies $T\xi = \lambda\xi$.   □

The spectrum of a compact operator K is finite or countable, and the nonzero points of $\sigma(K)$ correspond to eigenvalues of finite geometric multiplicity (Theorem 8.55). If a compact operator K acts on a Hilbert space H and if $K^* = K$, then the eigenvectors ξ and η corresponding to distinct eigenvalues λ and μ of K are orthogonal by the following computation:

$$(\lambda - \mu)\langle \xi, \eta \rangle = \langle \lambda\xi, \eta \rangle - \langle \xi, \mu\eta \rangle = \langle K\xi, \eta \rangle - \langle \xi, K\eta \rangle = \langle K\xi, \eta \rangle - \langle K\xi, \eta \rangle = 0,$$

which implies that $\langle \xi, \eta \rangle = 0$ because $\lambda - \mu \neq 0$. Hence, $\ker(K - \lambda 1) \perp \ker(K - \mu 1)$ for every pair of distinct eigenvalues λ and μ of K.

Theorem 10.31 (Spectral Theorem #1). *If $K \in B(H)$ is hermitian and compact, then H has an orthonormal basis consisting of eigenvectors of K.*

Proof. Express H as $H = \ker K \oplus (\ker K)^\perp = \ker K \oplus \overline{(\operatorname{ran} K^*)} = \ker K \oplus \overline{(\operatorname{ran} K)}$. If one finds orthonormal bases for each of $\ker K$ and $\overline{(\operatorname{ran} K)}$, then the union of these bases will be an orthonormal basis for H. Because $\ker K$ is the eigenspace of K corresponding to the eigenvalue $\lambda = 0$, it is enough to prove that $\overline{(\operatorname{ran} K)}$ has an orthonormal basis of eigenvectors of K. Furthermore, because $\overline{(\operatorname{ran} K)}$ is a Hilbert space invariant under K, we may assume without loss of generality that $\ker K = \{0\}$ and that the range of K is dense in H.

Because $\ker(K - \lambda 1) \perp \ker(K - \mu 1)$ for every pair of distinct eigenvalues λ and μ of K, we may form a direct sum (as in Proposition 5.98) of all the eigenspaces of K. To this end, let

$$M = \bigoplus_{\lambda \in \sigma(K), \lambda \neq 0} \ker(K - \lambda 1).$$

An orthonormal basis $\mathscr{B}_M$ for M is obtained by taking an orthonormal basis $\mathscr{B}_\lambda$ (necessarily finite) for each $\ker(K - \lambda 1)$ and then setting

$$\mathscr{B}_M = \bigcup_{\lambda \in \sigma(K), \lambda \neq 0} \mathscr{B}_\lambda.$$

The elements of $\mathscr{B}_M$ are obviously eigenvectors of K. Therefore, it remains to prove that $M^\perp = \{0\}$.

The subspace M is K-invariant, and so $M^\perp$ is K^*-invariant (Proposition 10.18); that is, $M^\perp$ is K-invariant. Because the restriction $K_{|M^\perp}$ of K to $M^\perp$ is a compact operator, the spectrum of $K_{|M^\perp}$ consists of 0 and nonzero eigenvalues. However, any eigenvalue of $K_{|M^\perp}$ would also be an eigenvalue of K, and since the eigenvectors of K corresponding to its nonzero eigenvalues all lie in M, it cannot be that $K_{|M^\perp}$ has nonzero eigenvalues. Therefore, because $K_{|M^\perp}$ is hermitian (Corollary 10.23) and because the norm and spectral radius of a hermitian operator coincide (Corollary 10.27), we deduce that $K_{|M^\perp} = 0$. But if $M^\perp$ were nonzero, then nonzero vectors in $M^\perp$ would be elements of $\ker K$, which would contradict $\ker K = \{0\}$. Hence, $M^\perp = \{0\}$ and, therefore, $M = H$. □

A second formulation of the spectral theorem may be viewed as a "spectral decomposition" of compact hermitian operators.

Theorem 10.32 (Spectral Theorem #2). *The following statements are equivalent for a compact operator K acting on a separable Hilbert space H:*

1. *K is hermitian;*
2. *there are a bounded sequence $\{\lambda_j\}_{j=1}^r$ of nonzero real numbers, where r is finite or infinite, and a sequence $\{\phi_j\}_{j=1}^r$ of pairwise-orthogonal unit vectors in H such that*

   a. $K\phi_j = \lambda_j\phi_j$, for each j,
   b. $\lim_j \lambda_j = 0$, if r is infinite, and
   c. $K\xi = \sum_{j=1}^r \lambda_j \langle \xi, \phi_j \rangle \phi_j$, for every $\xi \in H$.

Proof. Assume (1). Theorem 10.31 asserts that H has an orthonormal basis consisting of eigenvectors of K. The proof of Theorem 10.31 indicate that this basis consists of two parts: (i) an orthonormal basis $\{\psi_k\}_{k=1}^s$ of $\ker K$ and an orthonormal basis $\{\phi_j\}_{j=1}^r$ of

$$M = \bigoplus_{\lambda \in \sigma(K), \lambda \neq 0} \ker(K - \lambda 1),$$

where $M = \overline{\operatorname{ran} K}$. Let $\lambda_j \in \mathbb{R}$ denote the eigenvalue corresponding to the eigenvector ϕ_j, which thereby yields property (2a). If r is infinite, then there must be infinitely distinct points in the sequence $\{\lambda_j\}_{j=1}^r$ because, by Theorem 8.55, (i) the eigenspace associated with any individual nonzero eigenvector is finite dimensional (meaning an individual point λ_j in the sequence is repeated in the sequence at most finitely many times) and (ii) the point $0 \in \mathbb{R}$ is the only cluster point of $\sigma(K)$. Hence, property (2b) holds.

To prove property (2c), express H as $H = \ker K \oplus M$ so that $\xi \in H$ has the form $\xi = \gamma + \eta$, where $\gamma \in \ker K$ and $\eta \in M$. Write η in its Fourier decomposition:

$$\eta = \sum_{j=1}^r \langle \eta, \phi_j \rangle \phi_j.$$

Note that $K\xi = K\gamma + K\eta = K\eta$ and that $\langle \xi, \phi_j \rangle = \langle \eta, \phi_j \rangle$ for every j. Thus,

$$K\xi = K\eta = \sum_{j=1}^r \langle \eta, \phi_j \rangle \, K\phi_j = \sum_{j=1}^r \lambda_j \langle \eta, \phi_j \rangle \phi_j = \sum_{j=1}^r \lambda_j \langle \xi, \phi_j \rangle \phi_j,$$

which proves property (2c).

Conversely, if an operator $K \in B(H)$ satisfies (2), then, for every $\xi \in H$,

$$\langle K\xi, \xi \rangle = \left\langle \sum_{j=1}^{r} \lambda_j \langle \xi, \phi_j \rangle \phi_j, \sum_{i=1}^{r} \langle \xi, \phi_i \rangle \phi_i \right\rangle = \sum_{j=1}^{r} \lambda_j |\langle \xi, \phi_j \rangle|^2 \in \mathbb{R}$$

because each λ_j is real. Hence, $K^* = K$.   □

10.4 Normal Operators

The relationship between an operator T and its adjoint T^* is an important one, and indeed T^* shares many of the properties that T possesses (such as the norm). However, there may be very little relation between the two operators with regard to how they act on the Hilbert space. This section considers one case in which there is a rather strong algebraic link between T and T^*, and this specific relationship is called normality. The class of normal operators includes every hermitian operator and every unitary operator.

Definition 10.33. An operator $N \in B(H)$ is a *normal operator* if $N^*N = NN^*$.

Example 10.34. *The bilateral shift operator B on $\ell^2(\mathbb{Z})$ is normal, but the unilateral shift operator S on $\ell^2(\mathbb{N})$ is nonnormal.*

Proof. Because the bilateral shift operator B on $\ell^2(\mathbb{Z})$ is unitary, B is plainly normal. However, the unilateral shift operator S on $\ell^2(\mathbb{N})$ is nonnormal, as $S^*Se_1 = e_1$ but $SS^*e_1 = 0$ (where e_1 is the first canonical basis vector of $\ell^2(\mathbb{N})$).   □

The unilateral shift operator S is a good example of the challenges in quantifying nonnormality. On the one hand, S is rather close to being a normal operator because $S^*S - SS^* = e_1 \otimes e_1$ is a rank-1 operator. On the other hand, S is rather far from normal in the sense of norm, given that $\|S^*S - SS^*\| = 1$.

Proposition 10.35. *The following statements are equivalent for $T \in B(H)$:*

1. *T is normal;*
2. *$\|T^*\xi\| = \|T\xi\|$, for all $\xi \in H$;*
3. *the real and imaginary parts of T commute.*

Proof. Exercise 10.119.   □

The third equivalent statement of Proposition 10.35 above points to an intimate relationship between normality and the property of being hermitian. The next set of results illustrate how some of the spectral features of hermitian operators are shared by normal operators.

Proposition 10.36. *If $N \in B(H)$ is normal, then* $\operatorname{spr} N = \|N\|$.

Proof. By the spectral radius formula (Theorem 8.45),

$$\operatorname{spr} N = \lim_n \|N^n\|^{1/n}.$$

In particular,

$$\operatorname{spr} N = \lim_k \|N^{2k}\|^{\frac{1}{2k}}. \tag{10.6}$$

Because N is normal, $(N^2)^*(N^2) = (N^*N)^2$. Thus,

$$\|N^2\|^2 = \|(N^2)^*(N^2)\| = \|(N^*N)(N^*N)\| = \|N^*N\|^2 = \|N\|^4,,$$

which implies that $\|N^2\| = \|N\|^2$. By induction, $\|N^{2k}\| = \|N\|^{2k}$, for all $k \in \mathbb{N}$. Hence, by (10.6), $\operatorname{spr} N = \|N\|$. $\qquad\square$

Proposition 10.37. *If N is a normal Hilbert space operator, then* $\sigma(N) = \sigma_{\mathrm{ap}}(N)$.

Proof. Because $\sigma(N) = \sigma_{\mathrm{ap}}(N) \cup \sigma_{\mathrm{d}}(N) = \sigma_{\mathrm{ap}}(N) \cup \sigma_{\mathrm{p}}(N^*)^*$, it is sufficient to show that if $\overline{\lambda} \in \sigma_{\mathrm{p}}(N^*)$, then $\lambda \in \sigma_{\mathrm{p}}(N)$. Therefore, assume that $\overline{\lambda} \in \sigma_{\mathrm{p}}(N^*)$ and let $\xi \in H$ be nonzero with $N\xi = \overline{\lambda}\xi$. Because N is normal, it is also true that $N - \lambda 1$ is a normal operator. Consequently, Proposition 10.35 implies that

$$0 = \|(N^* - \overline{\lambda}1)\xi\| = \|(N - \lambda 1)^*\xi\| = \|(N - \lambda 1)\xi\|.$$

Hence, $\lambda \in \sigma_{\mathrm{p}}(N) \subseteq \sigma_{\mathrm{ap}}(N)$. $\qquad\square$

Using the properties of normal operators above, we can deduce precise information concerning multiplication operators.

Example 10.38. *If (X, Σ, μ) is a σ-finite measure space, then every multiplication operator M_ψ on $L^2(X, \Sigma, \mu)$, where $\psi \in \mathscr{L}^\infty(X, \Sigma, \mu)$, is a normal operator with spectrum*

$$\sigma(M_\psi) = \text{ess-ran}\,\psi$$

and norm

$$\|M_\psi\| = \operatorname{spr} M_\psi = \text{ess-sup}\,\psi.$$

Proof. The multiplication operator M_ψ on $L^2(X, \Sigma, \mu)$ has adjoint $M_\psi^* = M_{\overline{\psi}}$, and it is clear that M_ψ commutes with every multiplication operator M_ϱ, not just with $M_{\overline{\psi}}$. In any case, M_ψ is a normal operator.

Example 8.1.7 shows that $\|M_\psi\| = $ ess-sup ψ, whereas Example 8.53 proves that $\sigma_{ap}(M_\psi) = $ ess-ran ψ. The normality of M_ψ yields $\sigma_{ap}(M_\psi) = \sigma(M_\psi)$ (Proposition 10.37), and so $\|M_\psi\| = \mathrm{spr}\, M_\psi$. □

Example 10.38 is indicative of the general case, for if N is a normal operator acting on a separable Hilbert space H, then there is a σ-finite measure space (X, Σ, μ) and a surjective isometry $U : H \to L^2(X, \Sigma, \mu)$ such that UNU^{-1} is the multiplication operator M_ψ on $L^2(X, \Sigma, \mu)$, for some $\psi \in \mathscr{L}^\infty(X, \Sigma, \mu)$ (Corollary 11.34).

The proof of Proposition 10.37 reveals that $N\xi = \lambda\xi$ if and only if $N^*\xi = \bar{\lambda}\xi$. Therefore, if $N\xi = \lambda\xi$ and $N\eta = \mu\eta$ for distinct eigenvalues λ and μ of N and nonzero vectors ξ and η, then

$$(\lambda - \mu)\langle \xi, \eta \rangle = \langle \lambda\xi, \eta \rangle - \langle \xi, \bar{\mu}\eta \rangle = \langle N\xi, \eta \rangle - \langle \xi, N^*\eta \rangle$$

$$= \langle N\xi, \eta \rangle - \langle N\xi, \eta \rangle = 0,$$

which implies that $\langle \xi, \eta \rangle = 0$. Hence, $\ker(N - \lambda 1) \perp \ker(N - \mu 1)$ for distinct eigenvalues λ and μ of N. With this observation, the proofs of the spectral theorems for compact operators carry over verbatim, with only change being the requirement (in the hermitian case of the theorem) that the eigenvalues be real.

Theorem 10.39 (Spectral Theorem for Compact Normal Operators). *The following statements are equivalent for a compact operator K acting on a separable Hilbert space H:*

1. K is normal;
2. there are a bounded sequence $\{\lambda_j\}_{j=1}^r$ of nonzero complex numbers, where r is finite or infinite, and a sequence $\{\phi_j\}_{j=1}^r$ of pairwise-orthogonal unit vectors in H such that

   a. $K\phi_j = \lambda_j\phi_j$, for each j,
   b. $\lim_j \lambda_j = 0$, if r is infinite, and

   c. $K\xi = \displaystyle\sum_{j=1}^r \lambda_j \langle \xi, \phi_j \rangle \phi_j$, for every $\xi \in H$.

As with hermitian operators, normal operators admit a continuous functional calculus. Below, $\mathbb{C}[s,t]$ denotes the commutative ring of complex polynomials in two variables.

Theorem 10.40 (Continuous Functional Calculus for Normal Operators). *If N is a normal operator, then for every continuous function $f : \sigma(N) \to \mathbb{C}$ there is a unique hermitian operator $f(N)$ with the property that*

$$\|f(N) - q_n(N, N^*)\| \to 0 \tag{10.7}$$

for every sequence $\{q_n\}_{n \in \mathbb{N}}$ of polynomials $q_n \in \mathbb{C}[s,t]$ for which

$$\lim_{n \to \infty} \left(\max_{z \in \sigma(N)} |f(z) - q_n(z, \bar{z})| \right) = 0.$$

Furthermore, for all continuous functions $f, g : \sigma(N) \to \mathbb{C}$ and $\alpha \in \mathbb{C}$,

1. $\|f(N)\| = \max_{\lambda \in \sigma(N)} |f(\lambda)|$,
2. $\alpha f(N) = \alpha(f(N))$,
3. $(f + g)(N) = f(N) + g(N)$, and
4. $fg(N) = f(N)g(N)$.

Proof. The proof is identical to the proof of Theorem 10.28 except for the approximation indicated in (10.7). The use of polynomials q in two variables s and t is necessary to invoke the Stone-Weirerstrass Theorem. That is, if $\mathscr{S}$ is the set of all continuous functions on $\sigma(N)$ of the form $z \mapsto q(z, \bar{z})$, where $q \in \mathbb{C}[s,t]$, then $\mathscr{S}$ contains the constants, is self-adjoint, and separates the points of $\sigma(N)$. Hence, $\mathscr{S}$ is dense in $C(\sigma(N))$. □

Theorem 10.40 has an even stronger form, in which the use of continuous functions is extended to bounded Borel functions on the spectrum of N. This extension of continuous functional calculus to Borel functional calculus will not be needed for the topics studied in this text, and so we shall not develop it here.

With Theorem 10.40 at hand, one can prove assertions such as the following result.

Proposition 10.41. *If λ is an isolated point in the spectrum of a normal operator N, then λ is an eigenvalue of N.*

Proof. Exercise 10.120. □

If a normal operator N leaves the subspace $\mathrm{Span}\{\xi\}$ invariant, for some nonzero vector $\xi \in H$, then $\mathrm{Span}\{\xi\}$ is invariant under the action of N^* also. But this feature does not apply to all invariant subspaces of a normal operator.

Example 10.42. *The subspace $\ell^2(\mathbb{N})$ of $\ell^2(\mathbb{Z})$ is invariant for the bilateral shift operator B, but not for B^*.*

Proof. Because B is a forward shift, which is to say that $Be_k = e_{k+1}$ for every $k \in \mathbb{Z}$, it is clear that $B\xi \in \ell^2(\mathbb{N})$ for every $\xi \in \ell^2(\mathbb{N})$. However, with the vector $e_0 \in \ell^2(\mathbb{N})$, we have that $B^*e_0 = e_{-1} \notin \ell^2(\mathbb{N})$, and so $\ell^2(\mathbb{N})$ is not B^*-invariant. □

Motivated by the dual-invariance feature exhibited by normal operators with respect to the eigenvectors, one is led to the following class of operators.

Definition 10.43. An operator $T \in B(H)$ is a *reductive operator* if $M^\perp$ is T-invariant for every T-invariant subspace $M \subseteq H$.

Evidently every hermitian operator is reductive. However, as Example 10.42 demonstrates, not every normal operator is reductive. If we consider only compact normal operators, then the following positive result holds.

Proposition 10.44. *Every compact normal operator is reductive.*

Proof. Exercise 10.124.   □

Do there exist reductive operators that are nonnormal? This is an open question, equivalent to the long-standing open question of whether every operator on a separable Hilbert space has a nontrivial invariant subspace. Therefore, we might ask whether there exist compact reductive operators that are nonnormal.

Theorem 10.45 (Rosenthal). *If a compact operator K is reductive, then K is normal.*

Proof. The first step of the proof is to show that there is a unit vector $\gamma \in H$ such that γ is an eigenvector of both K and K^*. This is achieved by a Zorn's Lemma argument.

Because K is compact, K has a nontrivial invariant subspace (Theorem 8.59). Consider the family $\mathscr{F}$ of all nonzero K-invariant subspaces $M \subseteq H$ such that $\|K_{|M}\| = \|K\|$. The family $\mathscr{F}$ is nonempty because $H \in \mathscr{F}$. Furthermore, the relation $\precsim$ on $\mathscr{F}$ defined by

$$L \precsim M \quad \text{if and only if} \quad L \supseteq M$$

is a partial order on $\mathscr{F}$.

Let $\{M\}_{\alpha \in \Lambda}$ be linearly ordered chain in $\mathscr{F}$ and consider the subspace

$$N = \bigcap_{\alpha \in \Lambda} M_\alpha. \tag{10.8}$$

Note that the subspace N is reducing for K. If it can be shown that $N \in \mathscr{F}$, then N will be an upper bound in $\mathscr{F}$ for the linearly ordered chain $\{M\}_{\alpha \in \Lambda}$.

Because H is a separable metric space, every open covering of an open set admits a countable subcovering (by Exercise 2.103). Taking set-theoretic complements in equation (10.8) leads to

$$H \setminus N = \bigcup_{\alpha \in \Lambda} H \setminus M_\alpha,$$

which is a covering of the open set $H \setminus N$ by the family of open sets $H \setminus M_\alpha$. Hence, by Exercise 2.103, $\{H \setminus M_\alpha\}_\alpha$ admits a countable covering $\{H \setminus M_{\alpha_n}\}_{n \in \mathbb{N}}$ of $H \setminus N$. Therefore,

$$N = \bigcap_{n \in \mathbb{N}} M_{\alpha_n}. \tag{10.9}$$

Because the original descending chain was linearly ordered, the sequence of subspaces M_{α_n} can be assumed to be ordered so as to satisfy

$$M_{\alpha_n} \supseteq M_{\alpha_{n+1}}, \quad \text{for every } n \in \mathbb{N}. \tag{10.10}$$

Every compact operator achieves its norm on some unit vector. Therefore, for each $n \in \mathbb{N}$ there is a unit vector $\xi_n \in M_{\alpha_n}$ such that

$$\|K\xi_n\| = \|K_{|M_{\alpha_n}}\| = \|K\|.$$

Because K is compact, there is a subsequence $\{\xi_{n_k}\}_k$ of $\{\xi_n\}_n$ such that $\{K\xi_{n_k}\}_k$ converges to a vector $\eta \in H$. The norm of η is necessarily $\|K\|$, since $\|K\xi_{n_k}\| = \|K\|$ for every $k \in \mathbb{N}$. Equations (10.9) and (10.10) still hold if one replaces the sequence of subspaces M_{α_n} by the subsequence $\{M_{\alpha_{n_k}}\}_{k \in \mathbb{N}}$. Therefore, without loss of generality, it may be assumed that the original sequence $\{K\xi_n\}_{n \in \mathbb{N}}$ converges to $\eta \in H$.

The closed unit ball of H is weakly compact by the Banach-Alaoglu Theorem. Thus, the sequence $\{\xi_n\}_{n \in \mathbb{N}}$ admits a subsequence $\{\xi_{n_k}\}_{k \in \mathbb{N}}$ that is weakly convergent to some vector $\xi \in H$ of norm $\|\xi\| \leq 1$. Fix k. For every $j > k$, the vector ξ_{n_j} belongs to $M_{\alpha_{n_k}}$. Thus, the weak limit ξ is also the weak limit of the sequence $\{\xi_{n_j}\}_{j \geq k}$ in $M_{\alpha_{n_k}}$. Therefore the vector ξ belongs to $M_{\alpha_{n_k}}$. As this is true for any k, equation (10.10) yields $\xi \in N$.

For each $k \in \mathbb{N}$,

$$\|\eta - K\xi_{n_k}\|^2 = \|\eta\|^2 + \|K\xi_{n_k}\|^2 - 2\Re\langle K\xi_{n_k}, \eta \rangle = 2\|K\|^2 - 2\Re\langle K\xi_{n_k}, \eta \rangle.$$

As $k \to \infty$, the equation above yields $\Re\langle K\xi, \eta \rangle = \|K\|^2$. Thus,

$$\|K\|^2 \leq |\langle K\xi, \eta \rangle| \leq \|K\xi\|\,\|\eta\| = \|K\xi\|\,\|K\| \leq \|K\|^2,$$

and so $\|K\xi\| = \|K\|$. This shows that $\xi \in N$ is a nonzero vector (in fact it is a unit vector). Hence, the orthogonally reducing subspace N is at least one-dimensional and K achieves its norm on N. That is, $N \in \mathscr{F}$, and so N is an upper bound in $\mathscr{F}$ for the linearly ordered chain $\{M\}_{\alpha \in \Lambda}$. By Zorn's Lemma, $\mathscr{F}$ has a maximal element, say M.

Since $M \in \mathscr{F}$, M is nonzero. We shall show that M is one-dimensional. Suppose that $\dim M > 1$. Because $K_{|M}$ is compact, there is a nontrivial subspace $L \subseteq M$ that is invariant under $K_{|M}$. Since the subspace L is K-invariant and because K is a reductive operator, $L^\perp$ is K-invariant. Hence, L and $L^\perp \cap M$ are invariant under $K_{|M}$. This implies that $K_{|N}$ achieves its norm on L or on $L^\perp \cap M$. Either case contradicts the maximality of M in $\mathscr{F}$. Therefore, M must be one-dimensional. Thus, if $\gamma \in M$ is a unit vector, then γ is an eigenvector of K and K^*.

The second step in the proof shows that K is normal. Let $\mathscr{E}$ be a maximal family of orthonormal vectors $\gamma \in H$ for which γ is an eigenvector of K and K^*. If $\mathscr{E}$ is an orthonormal basis of H, then K has an orthonormal basis consisting of eigenvectors and is, therefore, normal. Suppose that $\mathscr{E}$ is not an orthonormal basis of H. Since the closure of the span of E is K-invariant, the subspace $\mathscr{E}^{\perp}$ is also K-invariant, since K is a reductive operator. By the arguments in the first step, there is a unit vector $\psi \in \mathscr{E}^{\perp}$ which is an eigenvector of $K_{|\mathscr{E}^{\perp}}$ and $(K_{|\mathscr{E}^{\perp}})^{*}$. Since K is reductive, this vector ψ is an eigenvector of K and K^{*}. But $\psi \in \mathscr{E}^{\perp}$, which is in contradiction to the maximality of the family $\mathscr{E}$. Hence, it must be that $\mathscr{E}$ is an orthonormal basis of H. $\square$

10.5   Positive Operators

To this point we have seen that hermitian operators have properties connected to the real numbers, while normal operators have the flavour of arbitrary complex numbers. In this section, the operator-theoretic analogue of a nonnegative real number is introduced, which is one of the most important features of Hilbert space operator theory.

Definition 10.46. An operator $A \in B(H)$ is *positive* if A is hermitian and $\sigma(A) \subset [0, \infty)$.

Corollary 10.27 provides an alternate criterion for the positivity of an operator:

Proposition 10.47. $A \in B(H)$ *is positive if and only if* $\langle A\xi, \xi \rangle \geq 0$ *for every* $\xi \in H$.

Corollary 10.48. $T^{*}T$ *is a postive operator, for every operator* T.

Positivity also captures information about the norm of an operator.

Proposition 10.49. *If* α *is a positive real number and* $T \in B(H)$, *then* $\alpha 1 - T^{*}T$ *is positive if and only if* $\sqrt{\alpha} \geq \|T\|$.

Proof. For each $\xi \in H$,

$$\langle (\alpha 1 - T^{*}T) \xi, \xi \rangle = \alpha \|\xi\|^{2} - \|T\xi\|^{2}.$$

Thus, $\langle (\alpha 1 - T^{*}T) \xi, \xi \rangle \geq 0$ for every $\xi \in H$ if and only if $\|T\xi\| \leq \sqrt{\alpha}\|\xi\|$ for all $\xi \in H$. That is, $\alpha 1 - T^{*}T$ is positive if and only if $\sqrt{\alpha} \geq \|T\|$. $\square$

One of the most useful features of positive operators is that they possess (unique) positive square roots.

Theorem 10.50. *If* $A \in B(H)$ *is positive, then*

1. *there is a positive operator* $R \in B(H)$ *such that* $R^{2} = A$, *and*
2. *if* $R_{1} \in B(H)$ *is a positive operator such that* $R_{1}^{2} = A$, *then* $R_{1} = R$.

Proof. Since $\sigma(A) \subset [0, \infty)$ and the function $f(t) = \sqrt{t}$ is continuous on the spectrum of A. Consider the hermitian operator $R = f(A)$, which satisfies, by Theorem 10.28, $R^{2} = A$. We now show that R is positive.

By scaling A we may assume without loss of generality that $\|A\| = 1$. Thus, $\sigma(A) \subseteq [0, 1]$. For each $n \in \mathbb{N}$, let q_n be the n-th Bernstein polynomial approximating $f(t) = \sqrt{t}$: namely,

$$q_n(t) = \sum_{k=1}^{n} \sqrt{\frac{k}{n}} \binom{n}{k} t^k (1-t)^{n-k}.$$

Therefore, $q_n(t) \geq 0$ for all $t \in [0, 1]$, which implies that $\sigma(q_n(A)) \subset [0, \infty)$, by the Spectral Mapping Theorem; furthermore, $\lim_n \left(\max_{t \in [0,1]} |q_n(t) - f(t)| \right) = 0$ ([14, §10.3]). Thus, the sequence $\{q_n(A)\}_n$ of positive operators converges to R. Proposition 10.26 implies that the smallest element m_ℓ in $\sigma(R)$ has the form

$$m_\ell = \inf_{\|\xi\|=1} \langle R\xi, \xi \rangle.$$

If $m_\ell < 0$, then there must be a unit vector ξ and an $n \in \mathbb{N}$ such that $\langle q_n(A)\xi, \xi \rangle < 0$. On the other hand, as $q_n(A)$ is positive, $\langle q_n(A)\xi, \xi \rangle \geq 0$ by Proposition 10.26. This contradiction implies that $m_\ell \geq 0$ and so R is positive. Thus, $\sigma(R) \subset [0, \infty)$. Furthermore, since $0 \leq q_n(t) \leq 1$ for all $t \in [0, 1]$, each $\|q_n(A)\| \leq 1$ and so $\|R\| \leq 1$.

Assume that R_1 is positive and $R_1^2 = A$. Since $[0, 1] \supseteq \sigma(A) = \{\lambda^2 \mid \lambda \in \sigma(R_1)\}$, we see that $\sigma(R_1) \subseteq [0, 1]$. Note that $q_n(t) \to \sqrt{t}$ uniformly on $[0, 1]$ implies that $q_n(t^2) \to \sqrt{t^2} = t$ uniformly. Thus, $q_n(R_1^2) \to R_1$. That is,

$$R_1 = \lim_n q_n(R_1^2) = \lim_n q_n(A) = R,$$

which proves that A has a unique positive square root. $\qquad\square$

Notational Convention If $A \in B(H)$ is positive, then $A^{1/2}$ will denote the unique positive square root of A.

The proof of Theorem 10.50 establishes the following result, which we record here formally for future use.

Proposition 10.51. *If A is a positive operator and if $f : \sigma(A) \to \mathbb{R}$ is a nonnegative continuous function, then $f(A)$ is a positive operator.*

Definition 10.52. The *Loewner ordering* on the set $B(H)_{\mathrm{sa}}$ of hermitian operators is the partial ordering $\leq$ in which $S \leq T$, for hermitian S and T, if and only if $T - S$ is positive.

Note that $S \leq T$, for hermitian operators S and T, if and only if $\langle S\xi, \xi \rangle \leq \langle T\xi, \xi \rangle$, for every $\xi \in H$. An elementary but useful fact about the Loewner ordering is:

Proposition 10.53. *If S and T are hermitian operators for which $S \leq T$, then $X^*SX \leq X^*TX$ for every operator $X \in B(H)$.*

Proof. Exercise 10.127. $\qquad\square$

The continuous functional calculus of Theorem 10.28 demonstrates that algebraic features of continuous maps f and g carry over to operators $f(T)$ and $g(T)$ (that is, the continuous functional calculus preserves sums, products, and scalar multiplication). The situation is rather different when considering the preservation of the Loewner order (see Exercise 10.132). There are, however, some positive results, and the first of these (below) is amongst the most important.

Proposition 10.54. *If A and B are positive operators and if $A \leq B$, then $A^{1/2} \leq B^{1/2}$.*

Proof. By the uniqueness of the positive square root of a positive operator, it is enough to prove that if $A, B \in B(H)$ are positive operators such that $A^2 \leq B^2$, then $A \leq B$. Under these assumptions, note that if $\xi \in H$ is a unit vector and if $\lambda \in \mathbb{R}$, then $\langle B^2 \xi, \xi \rangle - \lambda \langle B\xi, \xi \rangle$ is a real number and $\langle B\xi, A\xi \rangle$ is a real or complex number such that

$$\Re \langle B\xi, A\xi \rangle \leq |\langle B\xi, A\xi \rangle| \leq \|B\xi\| \, \|A\xi\| = \langle B^2 \xi, \xi \rangle^{1/2} \langle A^2 \xi, \xi \rangle^{1/2} \leq \langle B^2 \xi, \xi \rangle.$$

Now, to show that $B - A$ is positive, it is sufficient, by Proposition 10.26, to prove that $\lambda \geq 0$ for each $\lambda \in \sigma(B - A)$. To this end, select $\lambda \in \sigma(B - A)$. Because $B - A$ is hermitian, λ is necessarily real and an approximate eigenvalue (Proposition 10.24). Thus, there is a sequence of unit vectors ξ_n such that $\lim_n \|(B - A)\xi_n - \lambda \xi_n\| = 0$. For every n we have that

$$|\langle B\xi_n, (B - \lambda 1)\xi_n - A\xi_n \rangle| \leq \|B\| \, \|(B - A)\xi_n - \lambda \xi_n\|,$$

and therefore $\lim_n \langle B\xi_n, (B - \lambda 1)\xi_n - A\xi_n \rangle = 0$. Because every ξ_n is a unit vector, each of the sequences $\{\langle B^2 \xi_n, \xi_n \rangle\}_{n \in \mathbb{N}}$, $\{\langle B\xi_n, \xi_n \rangle\}_{n \in \mathbb{N}}$, and $\{\langle B\xi_n, A\xi_n \rangle\}_{n \in \mathbb{N}}$ lies in a compact subset of $\mathbb{C}$. Hence, there is a subsequence $\{\xi_{n_j}\}_{j \in \mathbb{N}}$ of $\{\xi_n\}_{n \in \mathbb{N}}$ such that $\{\langle B^2 \xi_{n_j}, \xi_{n_j} \rangle\}_{j \in \mathbb{N}}$, $\{\langle B\xi_{n_j}, \xi_{n_j} \rangle\}_{j \in \mathbb{N}}$, and $\{\langle B\xi_{n_j}, A\xi_{n_j} \rangle\}_{j \in \mathbb{N}}$ are convergent. Thus,

$$\lim_{j \to \infty} \left(\langle B^2 \xi_{n_j}, \xi_{n_j} \rangle - \lambda \langle B\xi_{n_j}, \xi_{n_j} \rangle \right) = \lim_{j \to \infty} \langle B\xi_{n_j}, A\xi_{n_j} \rangle$$

$$= \lim_{j \to \infty} \Re \langle B\xi_{n_j}, A\xi_{n_j} \rangle$$

$$\leq \lim_{j \to \infty} \langle B^2 \xi_{n_j}, \xi_{n_j} \rangle,$$

which implies that

$$\lim_{j \to \infty} \left(-\lambda \langle B\xi_{n_j}, \xi_{n_j} \rangle \right) \leq 0.$$

If $\lim_j \langle B\xi_{n_j}, \xi_{n_j} \rangle \neq 0$, then necessarily $\lambda \geq 0$. If, however, $\lim_j \langle B\xi_{n_j}, \xi_{n_j} \rangle = 0$, then

$$0 = \lim_{j \to \infty} \langle B\xi_{n_j}, \xi_{n_j} \rangle = \lim_{j \to \infty} \langle B^{1/2} \xi_{n_j}, B^{1/2} \xi_{n_j} \rangle = \lim_{j \to \infty} \|B^{1/2} \xi_{n_j}\|^2$$

yields

$$0 = \lim_{j \to \infty} \|B^{1/2}(B^{1/2}\xi_{n_j})\|^2 = \lim_{j \to \infty} \langle B^2 \xi_{n_j}, \xi_{n_j} \rangle.$$

Because, for every $j \in \mathbb{N}$,

$$\langle B^2 \xi_{n_j}, \xi_{n_j} \rangle \geq \langle A^2 \xi_{n_j}, \xi_{n_j} \rangle = \|A\xi_{n_j}\|^2 \geq 0,$$

we deduce from the inequality and limit equation above that $\lim_j \|A\xi_{n_j}\|^2$ exists and is equal to 0. Therefore, $\lim_j \|(B-A)\xi_{n_j} - \lambda \xi_{n_j}\| = 0$ holds only if $\lambda = 0$. $\square$

If A is a positive operator, then the compactness of the spectrum implies that two scenarios are possible: either (i) $0 \in \sigma(A)$ or (ii) there exists $\delta > 0$ such that $\lambda \geq \delta$ for all $\lambda \in \sigma(A)$. In the latter case, Proposition 10.26 yields $0 < \delta \leq \langle A\xi, \xi \rangle$ for every unit vector $\xi \in H$—in other words, $\delta 1 \leq A$. This leads to a simple criterion for the invertibility of positive operators:

Proposition 10.55. *A positive operator $A \in B(H)$ is invertible if and only if there exists a real number $\delta > 0$ such that $\delta 1 \leq A$.*

The next proposition asserts that the function $t \mapsto t^{-1}$ on $(0, \infty)$ is operator monotone.

Proposition 10.56. *If A and B are invertible positive operators such that $A \leq B$, then A^{-1} and B^{-1} are positive operators and $B^{-1} \leq A^{-1}$.*

Proof. If T is an invertible positive operator, then for every $\eta \in H$ we have, using $\xi = T^{-1}\eta$,

$$\langle T^{-1}\eta, \eta \rangle = \langle T^{-1}(T\xi), T\xi \rangle = \langle \xi, T\xi \rangle \geq 0.$$

Hence, T^{-1} is a positive operator, by Proposition 10.47. (Alternatively, one could argue via Proposition 10.51.)

By hypothesis, $\delta 1 \leq A \leq B$ for some real number $\delta > 0$. Hence, $\sqrt{\delta} 1 \leq A^{1/2} \leq B^{1/2}$, by Proposition 10.54, which implies that $B^{1/2}$ is invertible. Let $T = A^{1/2}B^{-1/2}$ and choose any $\eta \in H$. Thus, there is a unique $\xi \in H$ for which $\eta = B^{1/2}\xi$, and so

$$\|T\eta\|^2 = \|A^{1/2}\xi\|^2 = \langle A\xi, \xi \rangle \leq \langle B\xi, \xi \rangle = \|B^{1/2}\xi\|^2 = \|\eta\|^2.$$

Therefore, $\|T\| \leq 1$ and, hence, $\|T^*\| \leq 1$. Thus, $\|B^{-1/2}A^{1/2}\vartheta\|^2 \leq \|\vartheta\|^2$ for every $\vartheta \in H$.

Select any $\gamma \in H$ and let $\vartheta \in H$ denote the unique vector for which $\gamma = A^{1/2}\vartheta$. Thus,

$$\langle B^{-1}\gamma, \gamma \rangle = \|B^{-1/2}\gamma\|^2 = \|B^{-1/2}A^{1/2}\vartheta\|^2 \leq \|\vartheta\|^2 = \|A^{-1/2}\gamma\|^2 = \langle A^{-1}\gamma, \gamma \rangle,$$

which proves that $B^{-1} \leq A^{-1}$. $\square$

Propositions 10.54 and 10.56 belong to a wider set of results on operator monotone functions. These propositions say that the functions $t \mapsto \sqrt{t}$ and $t \mapsto t^{-1}$ are operator monotone on $[0, \infty]$ and $(0, \infty)$, respectively. In contrast $t \mapsto t^2$ is not operator monotone in the sense that there exist positive operators A and B such that $A \leq B$ but $A^2 \not\leq B^2$ (Exercise 10.132).

A common technique in measure theory is to write an arbitrary real-valued function as a difference of two nonnegative functions whose product is zero. That idea carries over, via functional calculus, to hermitian operators.

Proposition 10.57. *If A is a hermitian operator, then there are positive operators A_+ and $A_- \in A_+$ such that $A = A_+ - A_-$ and $A_+A_- = A_-A_+ = 0$.*

Proof. Let $X = [-\|A\|, \|A\|]$, which is a compact set that contains $\sigma(A)$ and 0. Consider the functions $f, g \in C(X)$ defined by $f(t) = (t + |t|)/2$ and $g(t) = f(-t)$. The functions f and g are nonnegative and vanish at 0; thus, by Proposition 10.51, the operators $f(A)$ and $g(A)$ are positive. Let $A_+ = f(A)$ and $A_- = g(A)$. Because $t = f(t) - g(t)$ and $f(t)g(t) = 0$ for all $t \in X$, the continuous functional calculus yields $A = A_+ - A_-$ and $A_+A_- = A_-A_+ = 0$. $\qquad\Box$

For compact positive operators, the min-max variational principle exhibited by equation (10.11) below is very useful in the analysis and estimation of eigenvalues.

Theorem 10.58 (Courant-Fischer Theorem). *If $A \in B(H)$ is a positive compact operator, then there are a bounded sequence $\{\lambda_j\}_{j=1}^r$ of real numbers, where r is finite or infinite, and a sequence $\{\phi_j\}_{j=1}^r$ of pairwise-orthogonal unit vectors in H such that*

1. $A\phi_j = \lambda_j\phi_j$, for each j,
2. $\lambda_j \geq \lambda_{j+1} > 0$, for all j,
3. $\lim_j \lambda_j = 0$, if r is infinite, and
4. $A\xi = \displaystyle\sum_{j=1}^{r} \lambda_j \langle \xi, \phi_j \rangle \phi_j$, for every $\xi \in H$.

Furthermore, for each j such that $1 \leq j \leq r$,

$$\lambda_j = \min_{L \subseteq H, \, \dim L = j-1} \left(\max_{\phi \in L^\perp, \, \|\phi\|=1} \langle A\phi, \phi \rangle \right). \tag{10.11}$$

Proof. Theorem 10.32 provides the spectral decomposition of A. By relabelling the indices, we may assume that the elements of the sequence $\{\lambda_j\}_{j=1}^r$ are ordered so that $\lambda_j \geq \lambda_{j+1}$, for every j. Therefore, all that remains is to prove equation (10.11). By Proposition 10.26 the spectral radius λ_1 of A is given by

$$\lambda_1 = \sup_{\|\xi\|=1} \langle A\xi, \xi \rangle.$$

But since A is compact, the unit eigenvector ϕ_1 corresponding to the eigenvalue λ_1 also satisfies $\langle A\phi_1, \phi_1 \rangle = \lambda_1$ and the supremum above is in fact a maximum:

$$\lambda_1 = \max_{\|\xi\|=1} \langle A\xi, \xi \rangle.$$

Let $M_1 = \mathrm{Span}\{\phi_1\}$, which is A-invariant. Thus, $M_1^\perp$ is A-invariant and the restriction $A_{|M_1^\perp}$ of A to $M_1^\perp$ is positive and compact. Therefore, the spectral radius of $A_{|M_1^\perp}$ is given by

$$\mathrm{spr} A_{|M_1^\perp} = \max_{\xi \in M_1^\perp, \|\xi\|=1} \langle A_{|M_1^\perp}\xi, \xi \rangle = \max_{\xi \in M_1^\perp, \|\xi\|=1} \langle A\xi, \xi \rangle.$$

Note that $\phi_j \in M_1^\perp$ for all $j \geq 2$, and so $\mathrm{spr} A_{|M_1^\perp} \geq \lambda_j$ for all $j \geq 2$. But since $\mathrm{spr} A_{|M_1^\perp}$ is also an eigenvalue of A, it must be that $\mathrm{spr} A_{|M_1^\perp} = \lambda_2$. By induction, if $M_{j-1} = \mathrm{Span}\{\phi_1, \ldots, \phi_{j-1}\}$, then

$$\lambda_j = \max_{\xi \in M_{j-1}^\perp, \|\xi\|=1} \langle A\xi, \xi \rangle.$$

Suppose now that $L \subset H$ is a subspace of dimension $j-1$ and that $\{\psi_1, \ldots, \psi_{j-1}\}$ is an orthonormal basis of L. The matrix $\mathscr{Z} = [\langle \phi_r, \psi_s \rangle]_{1 \leq r \leq j, 1 \leq s \leq j-1}$ is a linear map of $\mathbb{C}^j$ into $\mathbb{C}^{j-1}$, and so $\ker \mathscr{Z} \neq \{0\}$. Select a unit vector $\alpha \in \ker \mathscr{Z}$ and let $\xi = \sum_{\ell=1}^{j} \alpha_\ell \phi_\ell$, which is a unit vector in M_j. The condition $\mathscr{Z}\alpha = 0$ implies that $\xi \in L^\perp$. Furthermore,

$$\langle A\xi, \xi \rangle = \sum_{k=1}^{j}\sum_{\ell=1}^{j} \alpha_k \overline{\alpha}_\ell \langle A\phi_k, \phi_\ell \rangle = \sum_{\ell=1}^{j} |\alpha_\ell|^2 \lambda_\ell \geq \lambda_j \sum_{\ell=1}^{j} |\alpha_\ell|^2 = \lambda_j.$$

Hence, $\lambda_j \leq \max\{\langle A\xi, \xi \rangle \mid \xi \in L^\perp, \|\xi\| = 1\}$ for every subspace L of dimension $j-1$. This completes the proof of equation (10.11). $\qquad\square$

Definition 10.59. The *unit operator interval* is the subset $I(H)$ of $B(H)$ defined by

$$I(H) = \{A \in B(H) \mid 0 \leq A \leq 1\}.$$

Thus, $I(H)$ is the set of positive operators A of norm $\|A\| \leq 1$—or, equivalently, the set of positive operators A for which $1 - A$ is positive (Proposition 10.49).

The unit operator interval is plainly a convex set that contains every projection.

Proposition 10.60. *An operator is an extreme point of the unit operator interval $I(H)$ if and only if it is a projection.*

Proof. Suppose that P is a projection and express P as a proper convex combination $P = \tau A_1 + (1 - \tau)A_2$ of $A_1, A_2 \in I(H)$, for some $\tau \in (0, 1)$. If $\xi \in \ker P$, then $0 = \langle P\xi, \xi \rangle = \tau \langle A_1 P\xi, \xi \rangle + (1 - \tau)\langle A_2\xi, \xi \rangle \geq 0$ gives $\|A_j^{1/2}\xi\|^2 = \langle A_j\xi, \xi \rangle = 0$ for $j = 1, 2$, and so $A_1\xi = A_2\xi = 0$. Because it is also true that the projection $1 - P$ is given by $1 - P = \tau(1 - A_1) + (1 - \tau)(1 - A_2)$, the same argument shows that $\ker(1 - P) \subseteq \ker(1 - A_1) \cap \ker(1 - A_2)$. Thus, the action of A_j on each of $\ker P$ and $(\ker P)^\perp = \ker(1 - P)$ coincides with the action of P on these subspaces. Because $H = \ker P \oplus \ker(1 - P)$, we deduce that $A_1 = A_2 = P$. Hence, every projection P is an extreme point of $I(H)$.

Conversely, suppose that $A \in I(H)$ is not a projection. This means, by Proposition 10.30 that A has a point of spectrum in the open interval $(0, 1)$, say λ. Select $f \in C(\sigma(A))$ such that $0 \leq t \pm f(t) \leq 1$ for every $t \in \sigma(A)$ and $f(\lambda) \neq 0$. Thus, $f(A) \neq 0$ and $A_1 = A + f(A)$ and $A_2 = A - f(A)$ are elements of $I(H)$ such that $\frac{1}{2}A_1 + \frac{1}{2}A_2 = A$ but neither A_1 nor A_2 equal A. Thus, A is not an extreme point of $I(H)$. Hence, every extreme point of $I(H)$ must be a projection. $\square$

If H is finite-dimensional, then $I(H)$ is a compact convex set and so, by the Kreĭn-Milman Theorem, $I(H)$ is the closed convex hull of the set of projections. However, a much sharper statement can be made.

Proposition 10.61. *If H has finite dimension, then the unit operator interval $I(H)$ is the convex hull of the set of projections on H.*

Proof. Select $A \in I(H)$ and write A in its spectral decomposition: $A = \sum_{j=1}^{m} \lambda_j P_j$, where $\lambda_1, \ldots, \lambda_m$ are the distinct eigenvalues of A and each P_j is a projection with range $\ker(A - \lambda_j 1)$. If ξ_j is a unit eigenvector of A corresponding to the eigenvalue λ_j, then $\lambda_j = \langle \lambda_j \xi_j, \xi_j \rangle = \langle A\xi_j, \xi_j \rangle \in [0, 1]$. Thus, we may assume the eigenvalues of A are ordered so that $1 \geq \lambda_1 > \lambda_2 > \cdots > \lambda_m \geq 0$. Set $\tau_i = \lambda_i - \lambda_{i+1}$ for $1 \leq i < m$, $\tau_m = \lambda_m$, and $\tau_{m+1} = 1 - \lambda_1$; thus, $\tau_1, \ldots, \tau_{m+1}$ are convex coefficients such that $\sum_{i=j}^{m} \tau_i = \lambda_j$. For $i = 1, \ldots, m$, let $Q_i = \sum_{j=1}^{i} P_j$ and let $Q_{m+1} = 0$. Thus, $Q_1, \ldots, Q_{m+1}$ are projections and $\sum_{i=1}^{m+1} \tau_i Q_i = A$. $\square$

10.6   Polar Decomposition

In working with complex numbers z, it is sometimes advantageous to express z in its polar form $z = e^{i\theta}|z|$, where θ is the argument of z. One can do the same with operators on Hilbert space, and the result is a major structure theorem for arbitrary operators called Polar Decomposition.

Definition 10.62. For any $T \in B(H)$, the *modulus* $|T|$ of an operator $T \in B(H)$ is the positive operator $|T| = (T^*T)^{1/2}$.

One could of course elect to have defined $|T|$ by $(TT^*)^{1/2}$, which results in a different operator than $(T^*T)^{1/2}$ (if, for example, T is the unilateral shift operator). The adoption of $(T^*T)^{1/2}$ for $|T|$ is made so that the polar form $T = U|T|$ appears, at least on the surface, in exactly the same form as the traditional way of expressing the polar form of a complex number z—namely, as $z = e^{i\theta}|z|$.

Definition 10.63. An operator $V \in B(H)$ is a *partial isometry* if there exists a subspace M of H such that $V_{|M}$ is an isometry and $V_{|M^\perp} = 0$.

The subspace M is called the *initial space* and of V, and the range of V is called the *final space* of V.

Proposition 10.64. *If $V \in B(H)$ is a partial isometry with initial space M and final space $V(M)$, then*

1. *V^*V is a projection with range M, and*
2. *VV^* is a projection with range $V(M)$.*

Proof. Let $P \in B(H)$ be the projection with $\operatorname{ran} P = M$. To show that $V^*V = P$, first note that V is a contraction because, for every $\xi \in H$, $(1 - P)\xi \in M^\perp = \ker V$ and $P\xi \in M$ and so for every $\xi \in M$,

$$\|V\xi\| = \|V(P\xi + (1-P)\xi)\| = \|VP\xi\| = \|P\xi\| \le \|\xi\|.$$

Therefore, if $\xi \in M$ is a unit vector, then

$$1 = \|V\xi\|^2 = \langle V^*V\xi, \xi \rangle \le \|V^*V\xi\| \, \|\xi\| \le \|V^*V\| \le 1$$

gives a case of equality in the Cauchy-Schwarz inequality; hence, $V^*V\xi = \xi$, which shows that V^*V and P agree on $\operatorname{ran} P$. If $\eta \in M^\perp = (\operatorname{ran} P)^\perp = \ker P = \operatorname{ran}(1 - P)$, then $V^*V\eta = 0$ since $M^\perp = \ker V$. Thus, V^*V and P agree on $(\operatorname{ran} P)^\perp$. Hence, V^*V and P agree on H.

Let $Q = VV^*$. Because $V = V(V^*V + (1 - V^*V)) = VV^*V = QV$, the range of V is contained in the range of Q. But $Q = VV^*$ implies that the range of Q is contained in the range of V. Thus, $\operatorname{ran} Q = \operatorname{ran} V = V(M)$. Lastly, Q is plainly hermitian and $Q^2 = (VV^*)(VV^*) = (VV^*V)V^* = VV^* = Q$.   □

It is also true that if V is an operator such that V^*V is a projection, then V is a partial isometry (Exercise 10.138).

Theorem 10.65 (Polar Decomposition). *For every $T \in B(H)$ there exists a partial isometry $V \in B(H)$ such that*

1. *the initial space of V is $\overline{\operatorname{ran}|T|}$,*
2. *the final space of V is $\overline{\operatorname{ran}T}$, and*
3. *$T = V|T|$.*

Furthermore, if $T = V_1 R_1$ for some positive operator R_1 and partial isometry V_1 with initial space $\overline{\operatorname{ran}R_1}$, then $R_1 = |T|$ and $V_1 = V$.

Proof. For every $\xi \in H$,

$$\| \, |T|\xi \, \|^2 = \langle |T|\xi, |T|\xi \rangle = \langle |T|^2\xi, \xi \rangle = \langle T^*T\xi, \xi \rangle = \|T\xi\|^2. \tag{10.12}$$

Therefore, $\ker|T| = \ker T$.

Let $V_0 : \mathrm{ran}|T| \to H$ be the function that maps each $|T|\xi \in \mathrm{ran}|T|$ to $T\xi \in \mathrm{ran}\,T$. Since $|T|\xi_1 = |T|\xi_2$ only if $\xi_1 - \xi_2 \in \ker|T| = \ker T$, V_0 is a well-defined linear surjection $\mathrm{ran}|T| \to \mathrm{ran}\,T$. Because $\|V_0\psi\| = \|\psi\|$ for all $\psi \in \mathrm{ran}|T|$, V_0 extends (by continuity) to an isometry $\overline{\mathrm{ran}|T|} \to H$, denoted again by V_0. Therefore, the range of the isometry V_0 is closed and coincides with $\overline{\mathrm{ran}\,T}$. Now extend V_0 to a partial isometry $V \in B(H)$ by defining $V\eta = 0$ for all $\eta \in \overline{\mathrm{ran}|T|}^{\perp} = \ker|T| = \ker T$. Hence, V is a partial isometry with initial space $\overline{\mathrm{ran}|T|}$, final space $\overline{\mathrm{ran}\,T}$, and satisfies $V|T| = T$.

Suppose next that $T = V_1 R_1$ for some positive operator R_1 and partial isometry V_1 with initial space $\overline{\mathrm{ran}\,R_1}$. Because, for every $\xi \in H$,

$$\langle T^*T\xi, \xi \rangle = \|T\xi\|^2 = \|V_1 R_1 \xi\|^2 = \|R_1 \xi\|^2 = \langle R_1^2 \xi, \xi \rangle,$$

$T^*T = R_1^2$ by the Polarisation Identity. Thus, $|T| = (T^*T)^{1/2} = (R_1^2)^{1/2} = R_1$ by the uniqueness of the positive square root. Hence, $V|T| = V_1|T|$. That is, V and V_1 agree on $\overline{\mathrm{ran}|T|}$. But since the initial space of R_1 is $\overline{\mathrm{ran}|T|}$, R_1 is zero on $\overline{\mathrm{ran}|T|}^{\perp}$. Hence, V and V_1 agree on all of H. □

Definition 10.66. The *polar decomposition* of an operator $T \in B(H)$ is the unique decomposition of T as $T = V|T|$, where V is a partial isometry with initial space $\overline{\mathrm{ran}|T|}$ and final space $\overline{\mathrm{ran}\,T}$.

Two properties of the polar decomposition are noted below as corollaries for future reference.

Corollary 10.67. *If $T = V|T|$ is the polar decomposition of T, then $V^*T = |T|$.*

Proof. The operator V^*V is a projection with range $\overline{\mathrm{ran}\,T}$, and so $V^*V\xi = \xi$ for every $\xi \in \mathrm{ran}\,T$. □

Corollary 10.68. *If $T \in B(H)$ is invertible, then the partial isometry V in the polar decomposition $T = V|T|$ of T is a unitary operator.*

Proof. Equation (10.12) shows that there is a sequence of unit vectors $\{\xi_n\}_{n\in\mathbb{N}}$ with $\lim_n \|T\xi_n\| = 0$ if and only if $\lim_n \| |T|\xi_n\| = 0$—that is, $0 \in \sigma_{\mathrm{ap}}(T)$ if and only if $0 \in \sigma_{\mathrm{ap}}(|T|)$. Thus, if $0 \notin \sigma_{\mathrm{ap}}(T)$, then $0 \notin \sigma_{\mathrm{ap}}(|T|) = \sigma(|T|)$, which implies that $|T|$ is invertible. Therefore, $V = T|T|^{-1}$ is invertible and the initial space of V is H, which means that $V^*V = 1$. Hence, $V^* = V^{-1}$, which implies that V is unitary. □

The polar decomposition informs the theory of Hilbert space operators in a variety of manners. For example, the polar decomposition yields the following information about the geometry of the closed unit ball of $B(H)$.

Proposition 10.69. *If an invertible contraction T is not unitary, then T is the average of two unitaries.*

Proof. Corollary 10.68 shows that the polar decomposition of T is of the form $T = U|T|$, for a unitary operator U. Because $T^*T \leq 1$, $1 - |T|^2 = 1 - T^*T$ is a positive operator. Note that $T^*T \neq 1$, because T is invertible but nonunitary. Hence, $1 - |T|^2$ is a nonzero positive operator. Define an operator W by $W = |T| + i(1 - |T|^2)^{1/2}$ and observe that

$$W^*W = |T|^2 + \left(1 - |T|^2\right) + i\left(|T|(1 - |T|^2)^{1/2} - (1 - |T|^2)^{1/2}|T|\right), \text{ and}$$

$$WW^* = |T|^2 + \left(1 - |T|^2\right) + i\left(-|T|(1 - |T|^2)^{1/2} + (1 - |T|^2)^{1/2}|T|\right).$$

The operators $|T|$ and $1 - |T|^2$ obviously commute, and so $|T|$ and $f\left(1 - |T|^2\right)$ commute for every polynomial f. Hence, by the continuous functional calculus, $|T|$ and $(1 - |T|)^{1/2}$ commute, which implies that $W^*W = WW^* = 1$. Thus, the operators $U_1 = UW$ and $U_2 = UW^*$ are unitary and distinct, and

$$\frac{1}{2}(U_1 + U_2) = U\left(\frac{1}{2}(W + W^*)\right) = U\Re(W) = U|T| = T$$

expresses T as an average of unitaries U_1 and U_2. $\qquad\square$

Returning to properties of the polar decomposition, a useful fact about the modulus of a complex number z is that the real part $\Re z$ of z satisfies $\Re z \leq |z|$. This is not true verbatim in the case of operators, but a very closely related property holds.

Proposition 10.70. *If $Z \in B(H)$, then $\Re(Z) \leq V|Z|V^*$ for some isometry V.*

Proof. Let $Z = U|Z|$ be the polar decomposition of Z, where U is a partial isometry. Decompose the hermitian operator $\Re(Z)$ as a difference $\Re(Z) = Y_+ - Y_-$ of positive operators Y_+ and Y_- such that $Y_+Y_- = Y_-Y_+ = 0$. Let $Q \in B(H)$ denote the projection with range $\overline{\operatorname{ran} Y_+}$. Note that $QY_- = Y_-Q = 0$. Now let $R = Q(Z + |Z|)$ and let $R = W|R|$ be the polar decomposition of R in which W is a partial isometry with final space $\overline{\operatorname{ran} R}$. Note that $\operatorname{ran} R \subseteq \operatorname{ran} Q$, and so $\overline{\operatorname{ran} R} \subseteq \operatorname{ran} Q$. Therefore, because the projection WW^* has range $\overline{\operatorname{ran} R}$, we deduce that $\operatorname{ran}(WW^*) \subseteq \operatorname{ran} Q$. The orthogonal complement of $\operatorname{ran}(WW^*)$ in H is $\left(\overline{\operatorname{ran} R}\right)^\perp = \ker R^*$. If $\xi \in \ker R^* \cap \operatorname{ran} Q$, then

$$0 = \langle \xi, R^*\xi \rangle = \langle R\xi, \xi \rangle = \langle Q(Z + |Z|)\xi, \xi \rangle = \langle (Z + |Z|)\xi, Q\xi \rangle = \langle (Z + |Z|)\xi, \xi \rangle$$

$$= \Re\left(\langle (Z + |Z|)\xi, \xi \rangle\right) = \langle (\Re Z + |Z|)\xi, \xi \rangle \geq \langle \Re(Z)\xi, \xi \rangle = \langle Y_+\xi, \xi \rangle,$$

where the final inequality is on account of $\xi \in \operatorname{ran} Q = \overline{\operatorname{ran} Y_+}$ and $QY_- = 0$. Thus, $\langle Y_+\xi, \xi \rangle = 0$ and Y_+ positive yield $Y_+\xi = 0$. However, $\ker Y_+ \perp \overline{\operatorname{ran} Y_+}$ implies that $Y_+\xi = 0$ only if $\xi = 0$. Hence,

$$\left(\operatorname{ran} WW^*\right)^\perp \cap \operatorname{ran} Q = \ker R^* \cap \operatorname{ran} Q = \{0\}.$$

Therefore, $WW^* \leq Q$ and $\operatorname{ran} WW^* = \operatorname{ran} Q$ together yield $WW^* = Q$.

Let $P = W^*W$, the projection with range $(\ker W)^{\perp} = \overline{\operatorname{ran}|R|}$. Thus,

$$\ker W^*W \cap \operatorname{ran} Q = (\overline{\operatorname{ran}|R|})^{\perp} \cap \operatorname{ran} Q = \ker R^* \cap \operatorname{ran} Q = \{0\}.$$

Thus, $W^*W_{|\operatorname{ran} Q}$ is injective. Therefore, if $\xi \in \ker(1 - WW^*) \cap \operatorname{ran}(1 - W^*W)$, then $\xi = WW^*\xi \in \operatorname{ran} Q$ and $W^*W\xi = 0$ imply that $\xi = 0$. Hence, the projection $1 - Q$ is injective on the range of $1 - P$.

Consider $(1 - Q)(1 - P)$. Because $\ker(1 - Q)(1 - P) = \ker(1 - P)$, in the polar decomposition $(1 - Q)(1 - P) = W_0|(1 - Q)(1 - P)|$ the range of the projection $W_0^*W_0$ is $\operatorname{ran}(1 - P)$ and the range of $W_0W_0^*$ is contained in $\operatorname{ran}(1 - Q)$. Because $\operatorname{ran} W = \overline{\operatorname{ran} R} \subseteq \operatorname{ran} Q$, we have $(\operatorname{ran} Q)^{\perp} \subseteq (\operatorname{ran} W)^{\perp}$. Thus,

$$\operatorname{ran} W_0 \subseteq \operatorname{ran}(1 - Q) = \ker Q = (\operatorname{ran} Q)^{\perp} \subseteq (\operatorname{ran} W)^{\perp},$$

which implies that $\langle W_0^*W\xi, \eta \rangle = 0$ for all $\xi, \eta \in H$. Hence, if $V = W + W_0$, then

$$V^*V = W^*W + W^*W_0 + W_0W^* + W_0^*W_0 = P + 0 + 0 + (1 - P) = 1.$$

That is, V is an isometry.

Recall that $Z = U|Z|$, $\Re(Z) = Y_+ - Y_-$, $Q\Re(Z)Q = Y_+$, $R = Q(Z + |Z|)$, and $R = W|R|$. Thus,

$$4Y_+ = 2Q(Z + Z^*)Q = 2Q(U|Z| + |Z|U^*)$$

$$= Q[(1 + U)|Z|(1 + U)^* - (1 - U)|Z|(1 - U)^*]$$

$$\leq Q(1 + U)|Z|(1 + U)^*Q.$$

Note that $[Q(1 + U)|Z|(1 + U)^*Q]^2 = Q(1 + U)|Z|(1 + U)^*Q(1 + U)|Z|(1 + U)^*Q$. Therefore, by the uniqueness of the positive square root, we obtain

$$4Y_+ \leq \big(Q(1 + U)|Z|(1 + U)^*Q(1 + U)|Z|(1 + U)^*Q\big)^{1/2}.$$

Let $X = (1 + U^*)Q$ and note that $\|X\| \leq \|Q\| + \|U^*\|\|Q\| \leq 2$. Thus, $X^*X \leq \|X^*X\|1 = \|X\|^21 \leq 4 \cdot 1$, and so

$$Q(1 + U)|Z|(1 + U)^*Q(1 + U)|Z|(1 + U)^*Q = Q(1 + U)|Z|X^*X|Z|(1 + U)^*Q$$

$$\leq 4\big(Q(1 + U)|Z|^2(1 + U)^*Q\big).$$

As the square root function is operator monotone, we now have that

$$4Y_+ \leq (Q(1+U)|Z|(1+U)^*Q(1+U)|Z|(1+U)^*Q)^{1/2}$$

$$\leq 2\left(Q(1+U)|Z|^2(1+U)^*Q\right)^{1/2}$$

$$= 2\,(RR^*)^{1/2}.$$

Because $\operatorname{ran} W_0 \subseteq (\operatorname{ran} W)^\perp = \overline{(\operatorname{ran} R)}^\perp = \ker R^*$, the operator $R^*W_0 = 0$; thus, $W_0^*R = 0$ and $RV^* = R(W^* + W_0^*) = RW^*$. By passing to adjoints, $VR^* = WR^*$. Therefore,

$$RR^* = WV|R|^2W^* = W(R^*R)W^* = V(R^*R)V^* = V|Z|(1+U)^*Q(1+U)|Z|V^*,$$

and so $RR^* \leq 4V|Z|^2V^*$. Hence, using that the square root is operator monotone,

$$4Y_+ \leq 2\,(RR^*)^{1/2} \leq 4(V|Z|^2V^*)^{1/2}$$

$$= 4(V|Z|V^*V|Z|V^*)^{1/2}$$

$$= 4V|Z|V^*.$$

Hence, $\Re Z \leq Y_+ \leq V|Z|V^*$. □

An important consequence of the proposition above is the following triangle inequality for Hilbert space operators.

Theorem 10.71 (Triangle Inequality). *If $S, T \in B(H)$, then there are isometries $V, W \in B(H)$ such that*

$$|S+T| \leq V|S|V^* + W|T|W^*.$$

Proof. Let $S + T = U|S + T|$ be the polar decomposition of $S + T$, where $U \in B(H)$ is a partial isometry. Therefore, $U^*(S+T) = |S+T|$ and so

$$|S+T| = \Re\,(|S+T|) = \Re(U^*S) + \Re(U^*T).$$

Because $\|U^*\| = 1$, we have that $UU^* \leq 1$ and therefore $X^*UU^*X \leq X^*X$ for every $X \in B(H)$. Hence, by Proposition 10.54, $|U^*X| = (X^*UU^*X)^{1/2} \leq (X^*X)^{1/2} = |X|$ for each $X \in B(H)$. Further, Proposition 10.70 asserts that there exist isometries V and W such that $\Re(U^*S) \leq V|U^*S|V^*$ and $\Re(U^*T) \leq W|U^*T|W^*$. Hence,

$$|S+T| = \Re(U^*S) + \Re(U^*T) \leq V|U^*S|V^* + W|U^*T|W^* \leq V|S|V^* + W|T|W^*,$$

which completes the proof. □

10.7 Strong and Weak Operator Topologies

Recall from Proposition 1.88 that if X is a set, $\{(Y_\xi, \mathscr{T}_\xi)\}_{\xi \in \Lambda}$ is a family of topological spaces, and if $g_\xi : X \to Y_\xi$ is a function, for each $\xi \in \Lambda$, then there is a coarsest topology on X in which each function $g_\xi : X \to Y_\xi$ is continuous. If we let the set X be $B(H)$, $\Lambda = H$, $(Y_\xi, \mathscr{T}_\xi) = H$ (in the usual topology of H), and if each $g_\xi : B(H) \to H$ is the function $g_\xi(T) = T\xi$, then we obtain a topology on $B(H)$ called the strong operator topology.

Definition 10.72. The *strong operator topology* on $B(H)$ is the coarsest topology on $B(H)$ in which the functions $g_\xi : B(H) \to H$ defined by $g_\xi(T) = T\xi$, for $T \in B(H)$, are continuous for every $\xi \in H$.

Observe that in the strong operator topology (SOT) a basic open set containing a given operator $T_0 \in B(H)$ is a set of the form

$$U_{\xi_1,\dots,\xi_m;\varepsilon_1,\dots\varepsilon_m} = \{T \in B(H) \mid \|T\xi_k - T_0\xi_k\| < \varepsilon_k \text{ for all } k = 1,\dots,m\},$$

for some $m \in \mathbb{N}$, $\xi_1,\dots,\xi_m \in H$, and $\varepsilon_1,\dots\varepsilon_m \in (0,\infty)$. In particular, if $\{T_k\}_{k\in\mathbb{N}}$ is a sequence of operators such that, for some operator $T \in B(H)$, $\lim_k \|T_k\xi - T\xi\| = 0$ for every $\xi \in H$, then T is the limit of the sequence $\{T_k\}_k$ in the strong operator topology. (Here, "T is the limit of $\{T_k\}_k$" means that for every SOT-open set U there is an $n_0 \in \mathbb{N}$ such that $T_k \in U$ for every $k \geq n_0$.)

Another application of Proposition 1.88 leads to the weak operator topology.

Definition 10.73. The *weak operator topology* on $B(H)$ is the coarsest topology on $B(H)$ in which the functions $f_{\xi,\eta} : B(H) \to \mathbb{C}$ defined by $f_{\xi,\eta}(T) = \langle T\xi, \eta \rangle$, for $T \in B(H)$, are continuous for every $(\xi,\eta) \in H \times H$.

In the weak operator topology, a basic open set containing $T_0 \in B(H)$ is a set of the form

$$W_{\xi_1,\dots,\xi_m;\eta_1,\dots,\eta_m;\varepsilon_1,\dots\varepsilon_m} = \{T \in B(H) \mid |\langle T\xi_k - T_0\xi_k, \eta_k \rangle| < \varepsilon_k \text{ for all } k = 1,\dots,m\},$$

for some $m \in \mathbb{N}$, $\xi_1,\dots,\xi_m, \eta_1,\dots,\eta_m \in H$, and $\varepsilon_1,\dots\varepsilon_m \in (0,\infty)$. Thus, if $T \in B(H)$ and if $\{T_k\}_{k\in\mathbb{N}}$ is a sequence of operators such that $\lim_k |\langle T_k\xi - T\xi, \eta \rangle| = 0$ for every $\xi, \eta \in H$, then T is the limit of the sequence $\{T_k\}_k$ in the weak operator topology.

One of the most useful features of the weak operator topology is the compactness of the closed unit ball, which is an Alaoglu-type theorem in both its statement and its method of proof.

Theorem 10.74. *The set $\{T \in B(H) \mid \|T\| \leq 1\}$ is compact and Hausdorff in the weak operator topology.*

Proof. For each ordered pair $(\xi,\eta) \in H \times H$, let $K_{(\xi,\eta)} = \{\lambda \in \mathbb{C} \mid |\lambda| \leq \|\xi\|\,\|\eta\|\}$. Consider the space $K = \prod_{(\xi,\eta)\in H\times H} K_{(\xi,\eta)}$, endowed with the product topology. By

Tychonoff's Theorem (Theorem 2.14), K is a compact set. Furthermore, K is Hausdorff because each $K_{(\xi,\eta)}$ is Hausdorff.

Let $X = \{T \in B(H) \mid \|T\| \le 1\}$ and consider X as a topological space in which the topology on X is induced by the weak operator topology of $B(H)$. Define $f : X \to K$ by $f(T) = (\langle T\xi, \eta \rangle)_{(\xi,\eta)}$, and note that f is an injective function. Select $T \in X$ and consider an open set $W \subset K$ that contains $f(T)$. Thus, there are open subsets $W_{(\xi,\eta)} \subseteq K_{(\xi,\eta)}$ such that $\langle T\xi, \eta \rangle \in W_{(\xi,\eta)}$, for every $(\xi,\eta) \in H \times H$, and $W_{(\xi,\eta)} = K_{(\xi,\eta)}$ for all but at most a finite number of elements in $H \times H$—say $(\xi_1, \eta_1), \ldots, (\xi_n, \eta_n)$—and $W = \prod_{(\xi,\eta) \in H \times H} W_{(\xi,\eta)}$. Hence there are positive real numbers $\varepsilon_1, \ldots, \varepsilon_n$ such that

$$W_{(\xi_j, \eta_j)} = \{z \in \mathbb{C} \mid |z - \langle T\xi, \eta_j \rangle| < \varepsilon_j\}$$

for every $j = 1, \ldots, n$. Therefore,

$$f^{-1}(W) = \bigcap_{j=1}^{n} \{S \in X \mid |\langle (S - T)\xi_j, \eta_j \rangle| < \varepsilon_j\},$$

which is a basic WOT-open neighbourhood of $T \in X$. Hence, f is continuous at every $T \in X$, which implies that f is a continuous function on X.

On the other hand, if $U \subseteq X$ is an arbitrary open set and if $T \in U$, then there is a basic WOT-open set B such that $T \in B \subseteq U$. By definition, there are $(\xi_1, \eta_1), \ldots, (\xi_n, \eta_n) \in H \times H$ and positive real numbers $\varepsilon_1, \ldots, \varepsilon_n$ such that, for $S \in X$, we have $S \in B$ if and only if $|\langle (S - T)\xi_j, \eta_j \rangle| < \varepsilon_j$ for each $j = 1, \ldots, n$. Hence, if $W_{(\xi_j, \eta_j)} = \{\lambda \in K_{(\xi_j, \eta_j)} \mid |\lambda - \langle T\xi_j, \eta_j \rangle| < \varepsilon_j\}$ and if $W_{(\xi,\eta)} = K_{(\xi,\eta)}$ for every $(\xi, \eta) \in H \times H \setminus \{(\xi_1, \eta_1), \ldots, (\xi_n, \eta_n)\}$, then $W_T = \prod_{(\xi,\eta) \in H \times H} W_{(\xi,\eta)}$ is open in K and $f(T) \in W_T \subseteq f(U)$. Thus, $f(U) = \bigcup_{T \in U} W_T$, which shows that $f(U)$ is open. Hence, $f^{-1} : f(X) \to X$ is continuous, and therefore f is a homeomorphism between X and $f(X)$.

We now show that $f(X)$ is a closed subset of K. Let $\lambda = (\lambda_{(\xi,\eta)})_{(\xi,\eta)} \in K$ be in the closure of $f(X)$, and define a function $\psi : H \times H \to \mathbb{C}$ by $\psi(\xi, \eta) = \lambda_{(\xi,\eta)}$. Claim: ψ is a bounded sesquilinear form. To prove this claim, select vectors $\xi_0, \xi_1, \xi_2, \eta_0, \eta_2, \eta_2 \in H$ and scalars $\alpha_1, \alpha_2, \beta_1, \beta_2 \in \mathbb{C}$. Let $\varepsilon > 0$ be arbitrary. Define subsets $W_{(\xi,\eta)} \subseteq K_{(\xi,\eta)}$ as follows:

$$W_{(\alpha_1\xi_1+\alpha_2\xi_2,\eta_0)} = \left\{ z \in K_{(\alpha_1\xi_1+\alpha_2\xi_2,\eta_0)} \mid |z - \lambda_{(\alpha_1\xi_1+\alpha_2\xi_2,\eta_0)}| < \varepsilon \right\};$$

$$W_{(\xi_0,\beta_1\eta_1+\beta_2\eta_2)} = \left\{ z \in K_{(\xi_0,\beta_1\eta_1+\beta_2\eta_2)} \mid |z - \lambda_{(\xi_0,\beta_1\eta_1+\beta_2\eta_2)}| < \varepsilon \right\};$$

$$W_{(\xi_0,\eta_j)} = \left\{ z \in K_{(\xi_0,\eta_j)} \mid |z - \lambda_{(\xi_0,\eta_j)}| < \tfrac{\varepsilon}{|\beta_j|} \right\}, \text{ for } j = 1, 2;$$

$$W_{(\xi_j,\eta_0)} = \left\{ z \in K_{(\xi_j,\eta_0)} \mid |z - \lambda_{(\xi_j,\eta_0)}| < \tfrac{\varepsilon}{|\alpha_j|} \right\}, \text{ for } j = 1, 2;$$

$$W_{(\xi,\eta)} = K_{(\xi,\eta)} \text{ in all other cases.}$$

Thus, $W = \displaystyle\prod_{(\xi,\eta)\in H\times H} W_{(\xi,\eta)}$ is open in K and contains λ. Because λ is in the closure of $f(X)$, there is an operator $S \in X$ with $f(S) \in W$. Therefore,

$$|\alpha_1\psi(\xi_1,\eta_0) - \alpha_1\langle S\xi_1,\eta_0\rangle| < \varepsilon, \quad |\alpha_2\psi(\xi_2,\eta_0) - \alpha_2\langle S\xi_2,\eta_0\rangle| < \varepsilon,$$

and

$$|\psi(\alpha_1\xi_1 + \alpha_2\xi_2,\eta_0) - \langle S(\alpha_1\xi_1 + \alpha_2\xi_2),\eta_0\rangle| < \varepsilon.$$

Hence,

$$|\psi(\alpha_1\xi_1 + \alpha_2\xi_2,\eta_0) - \alpha_1\psi(\xi_1,\eta_0) - \alpha_2\psi(\xi_2,\eta_0)| < 3\varepsilon.$$

A similar argument shows that

$$|\psi(\xi_0,\beta_1\eta_1 + \beta_2\eta_2) - \overline{\beta}_1\psi(\xi_0,\eta_1) - \overline{\beta}_2\psi(\xi_0,\eta_2)| < 3\varepsilon.$$

The choice of $\varepsilon > 0$ being arbitrary yields

$$\psi(\alpha_1\xi_1 + \alpha_2\xi_2,\eta_0) = \alpha_1\psi(\xi_1,\eta_0) + \alpha_2\psi(\xi_2,\eta_0), \text{ and}$$

$$\psi(\xi_0,\beta_1\eta_1 + \beta_2\eta_2) = \overline{\beta}_1\psi(\xi_0,\eta_1) + \overline{\beta}_2\psi(\xi_0,\eta_2).$$

Hence, ψ is a sesquilinear form.

The boundedness of ψ is immediate from $|\psi(\xi,\eta)| \leq \|\xi\| \|\eta\|$. By Proposition 10.5, there is a unique $T \in B(H)$ such that $\psi(\xi,\eta) = \langle T\xi,\eta\rangle$ for all $\xi, \eta \in H$. Because $\|T\|$ is the supremum of all $|\langle T\xi,\eta\rangle|$ as ξ and η range through unit vectors, $\|T\| \leq 1$. This proves that $T \in X$ and, hence, that $\lambda = f(T) \in f(X)$.

Because $f(X)$ is closed in K and since K is compact and Hausdorff, we deduce that $f(X)$ is compact and Hausdorff; hence, X is compact and Hausdorff. $\qquad\square$

The following example helps distinguish the two topologies on $B(H)$.

Example 10.75. *The involution $T \mapsto T^*$ is continuous with respect to the weak operator topology, but not with respect to the strong operator topology.*

Proof. The proof of the first assertion is left as an exercise (Exercise 10.142).

Let S denote the unilateral shift operator on the Hilbert space $\ell^2(\mathbb{N})$, and let $T_n = (S^*)^n$, for every $n \in \mathbb{N}$. Note that if $\xi = (\xi_k)_{k \in \mathbb{N}} \in H$ and $n \in \mathbb{N}$, then

$$\|T_n \xi\| = \left(\sum_{k=n+1}^{\infty} |\xi_k|^2 \right)^{1/2}.$$

Thus, $\lim_{n \to \infty} \|T_n \xi\| = 0$, for every $\xi \in \ell^2(\mathbb{N})$; that is, the sequence $\{T_n\}_n$ converges to 0 with respect to the strong operator topology. However, $\|T_n^* \xi\| = \|S^n \xi\| = \|\xi\|$, because S is an isometry, and so 0^* is not the SOT-limit of the sequence $\{T_n^*\}_n$, implying that the involution fails to be continuous with respect to the strong operator topology. $\square$

In contrast to Example 10.75, $B(H)$ admits the same set of continuous linear maps into $\mathbb{C}$ regardless of whether $B(H)$ has the strong operator topology or the weak operator topology.

Proposition 10.76. *The following statements are equivalent for a linear transformation $\varphi : B(H) \to \mathbb{C}$:*

1. *φ is continuous with respect to the weak operator topology on $B(H)$;*
2. *φ is continuous with respect to the strong operator topology on $B(H)$;*
3. *there exist $n \in \mathbb{N}$ and nonzero vectors $\xi_1, \ldots, \xi_n, \eta_1, \ldots, \eta_n \in H$ such that*

$$\varphi(T) = \sum_{j=1}^{n} \langle T\xi_j, \eta_j \rangle,$$

for every $T \in B(H)$.

Proof. (1) $\Rightarrow$ (2). Assume that φ is continuous with respect to the weak operator topology on $B(H)$. Suppose that $V \subseteq \mathbb{C}$ is a nonempty open set, and select $T_0 \in B(H)$ and $\varepsilon > 0$ so that $B_\varepsilon(z_0) \subseteq V$, where $z_0 = \varphi(T_0)$. Because φ is weakly continuous, there is a basic WOT-open set W about T_0, say

$$W = \{T \in B(H) \,|\, |\langle T\xi_j - T_0\xi_j, \eta_j \rangle| < \varepsilon_j \text{ for all } j = 1, \ldots, n\},$$

for some nonzero vectors $\xi_1, \ldots, \xi_n, \eta_1, \ldots, \eta_n$ and positive real numbers $\varepsilon_1, \ldots \varepsilon_n$, such that $\varphi(W) \subseteq B_\varepsilon(z_0)$. For each j let $\tilde{\varepsilon}_j = \|\eta_j\|^{-1} \varepsilon$, and consider the SOT-open set

$$U_{T_0} = \{T \in B(H) \,|\, \|T\xi_j - T_0\xi_j\| < \tilde{\varepsilon}_j \text{ for all } j = 1, \ldots, n\},$$

By the Cauchy-Schwarz inequality, if $T \in W$ for every $T \in U_{T_0}$. Hence, $\varphi^{-1}(V)$ is a union of SOT-open sets U_{T_0}, and therefore φ is continuous with respect to the strong operator topology.

(2) $\Rightarrow$ (3). Assume that φ is continuous with respect to the strong operator topology on $B(H)$. Therefore, using the open unit disc $\mathbb{D}$ in $\mathbb{C}$, the set $\varphi^{-1}(\mathbb{D})$ is SOT-open, and hence there exists a basic SOT-open set U about $0 \in B(H)$ of the form $U = U_{\xi_1,\ldots,\xi_n;\varepsilon_1,\ldots,\varepsilon_n}$ for some nonzero $\xi_j \in H$ and $\varepsilon_j > 0$. Let $\varepsilon = \min_j \varepsilon_j$; thus, if $T \in B(H)$ satisfies $\|T\xi_j\| < \varepsilon$ for each j, then $|\varphi(T)| < 1$.

Let $C = \frac{2}{\varepsilon}$ and suppose that $R \in B(H)$ satisfies $R\xi_j \neq 0$ for at least one j. Let

$$\alpha = C \left(\sum_{k=1}^{n} \|R\xi_k\|^2 \right)^{1/2}.$$

Thus, for any j,

$$\left\| \frac{1}{\alpha} R\xi_j \right\| \leq \frac{\varepsilon \|R\xi_j\|}{2 \left(\sum_{k=1}^{n} \|R\xi_k\|^2 \right)^{1/2}} \leq \frac{\varepsilon}{2} < \varepsilon,$$

and therefore $|\varphi(R)| < \alpha$. Hence, for every $R \in B(H)$,

$$|\varphi(R)| \leq C \left(\sum_{k=1}^{n} \|R\xi_k\|^2 \right)^{1/2}.$$

By replacing each ξ_k with $C\xi_k$ in the inequality above, we may assume without further change of notation that

$$|\varphi(R)| \leq \left(\sum_{k=1}^{n} \|R\xi_k\|^2 \right)^{1/2},$$

for every $R \in B(H)$.

In the Hilbert space $H^{(n)} = \bigoplus_1^n H$ (the direct sum of n copies of H), consider the linear submanifold $L_0 = \{ \bigoplus_j T\xi_j \mid T \in B(H) \}$. If $S, T \in B(H)$ are such that $S\xi_j = T\xi_j$ for $j = 1, \ldots, n$, then

$$|\varphi(S) - \varphi(T)| = |\varphi(S-T)| \leq \left(\sum_{k=1}^{n} \|(S-T)\xi_k\|^2 \right)^{1/2} = 0.$$

Therefore, the map $\left(\bigoplus_j T\xi_j \right) \mapsto \varphi(T)$ is well defined, linear, and contractive. Thus, by the Hahn-Banach Theorem, there is a contractive linear functional ψ on H such

$$\psi \left(\bigoplus_j T\xi_j \right) = \varphi(T),$$

for all $T \in B(H)$. The Riesz Representation Theorem yields a vector $\eta = \bigoplus_j \eta_j$ in $H^{(n)}$ that implements ψ. In particular, for each $T \in B(H)$,

$$\varphi(T) = \left\langle \bigoplus_{j=1}^{n} T\xi_j, \bigoplus_{j=1}^{n} \eta_j \right\rangle = \sum_{j=1}^{n} \langle T\xi_j, \eta_j \rangle.$$

The proof of (3) $\Rightarrow$ (1) is obvious.   □

Equipped with Proposition 10.76 and the Hahn-Banach Separation Theorem, the following fundamental fact about $B(H)$ is deduced.

Proposition 10.77. *If $K \subset B(H)$ is a convex set, then $\overline{C}^{WOT} = \overline{C}^{SOT}$.*

Proof. It is clear that $\overline{C}^{SOT} \subseteq \overline{C}^{WOT}$. To prove the inclusion $\overline{C}^{WOT} \subseteq \overline{C}^{SOT}$, select $T \in \overline{C}^{WOT}$. If, contrary to what we aim to prove, $T \notin \overline{C}^{SOT}$, then the Hahn-Banach Separation Theorem implies that there are a SOT-continuous $\varphi : B(H) \to \mathbb{C}$ and a $\gamma \in \mathbb{R}$ such that

$$\Re(\varphi(R)) \leq \gamma < \varphi(T), \ \forall R \in C.$$

But Proposition 10.76 implies that φ is also WOT-continuous, and so the inequality above implies that $T \notin \overline{C}^{WOT}$, which is a contradiction.   □

10.8 Matrices of Operators

Through the use of matrices of operators, a number of properties concerning individual operators are revealed.

Proposition 10.78. *If $T \in B(H)$, then the operator*

$$A = \begin{bmatrix} 1 & T \\ T^* & 1 \end{bmatrix}$$

is a positive operator on $H \oplus H$ if and only if $\|T\| \leq 1$.

Proof. By Exercise 10.128, if Q is hermitian and X is invertible, then Q is positive if and only if X^*QX is positive. Factor the hermitian operator $A = \begin{bmatrix} 1 & T \\ T^* & 1 \end{bmatrix}$ as

$$A = \begin{bmatrix} 1 & 0 \\ T^* & 1 \end{bmatrix} \begin{bmatrix} 1 & 0 \\ 0 & 1 - T^*T \end{bmatrix} \begin{bmatrix} 1 & T \\ 0 & 1 \end{bmatrix} = X^*QX.$$

Because $X = \begin{bmatrix} 1 & T \\ 0 & 1 \end{bmatrix}$ is invertible, we have that A is positive if and only if the

matrix $Q = \begin{bmatrix} 1 & 0 \\ 0 & 1 - T^*T \end{bmatrix}$ is positive. But Q is positive if and only if $1 - T^*T$ is

positive, which is equivalent to saying that Q is positive if and only if $\|T\| \leq 1$, by
Proposition 10.49.   □

A 3×3 version of Proposition 10.78 is:

Proposition 10.79. *If $T_1, T_2 \in B(H)$, then the operator*

$$A = \begin{bmatrix} 1 & T_1 & 0 \\ T_1^* & 1 & T_2 \\ 0 & T_2^* & 1 \end{bmatrix}$$

is a positive operator on $H \oplus H \oplus H$ if and only if $1 - T_1^*T_1 - T_2T_2^*$ is positive.

Proof. Factor A as

$$A = \begin{bmatrix} 1 & 0 & 0 \\ T_1^* & 1 & 0 \\ 0 & 0 & 1 \end{bmatrix} \begin{bmatrix} 1 & 0 & 0 \\ 0 & (1 - T_1^*T_1) & T_2 \\ 0 & T_2^* & 1 \end{bmatrix} \begin{bmatrix} 1 & T_1 & 0 \\ 0 & 1 & 0 \\ 0 & 0 & 1 \end{bmatrix}.$$

Thus, A is positive if and only if the middle factor is positive, which in turn is

positive if and only if $\begin{bmatrix} (1 - T_1^*T_1) & T_2 \\ T_2^* & 1 \end{bmatrix}$ is positive. This matrix is equal to

$$\begin{bmatrix} 0 & 1 \\ 1 & 0 \end{bmatrix} \left(\begin{bmatrix} 1 & 0 \\ T_2 & 1 \end{bmatrix} \begin{bmatrix} 1 & 0 \\ 0 & (1 - T_1^*T_1 - T_2T_2^*) \end{bmatrix} \begin{bmatrix} 1 & T_2^* \\ 0 & 1 \end{bmatrix} \right) \begin{bmatrix} 0 & 1 \\ 1 & 0 \end{bmatrix},$$

which is positive if and only if $1 - T_1^*T_1 - T_2T_2^*$ is positive.   □

The next theorem is one of the first ever *dilation*, or matrix completion,
theorems established for Hilbert space operators. Below, given a contraction T,

the unspecified entries in the 2×2 operator matrix $\begin{bmatrix} T & * \\ * & * \end{bmatrix}$ are determined so that

the completed matrix is a unitary operator.

Proposition 10.80 (Halmos). *If $T \in B(H)$ satisfies $\|T\| \leq 1$, then the matrix*

$$U = \begin{bmatrix} T & (1 - TT^*)^{1/2} \\ (1 - T^*T)^{1/2} & -T^* \end{bmatrix}$$

is a unitary operator on $H \oplus H$.

Proof. By Proposition 10.49, the condition $\|T\| \leq 1$ is equivalent to the positivity of $1 - T^*T$. Because $\|T^*\| = \|T\| \leq 1$, we also have that $1 - TT^*$ is positive. Therefore, in the definition of the matrix U, the (1,2) and (2,1) entries are well defined.

Computation of U^*U and UU^* leads to

$$U^*U = \begin{bmatrix} T^*T + (1 - T^*T) & T^*(1 - TT^*)^{1/2} - (1 - T^*T)^{1/2}T^* \\ (1 - TT^*)^{1/2}T^* - T(1 - TT^*)^{1/2} & (1 - TT^*) + TT^* \end{bmatrix}$$

and

$$UU^* = \begin{bmatrix} TT^* + (1 - TT^*) & T(1 - T^*T)^{1/2} - (1 - TT^*)^{1/2}T \\ (1 - T^*T)^{1/2}T^* - T^*(1 - TT^*)^{1/2} & (1 - T^*T) + T^*T \end{bmatrix}.$$

Therefore, it is enough to prove that $T(1 - T^*T)^{1/2} = (1 - TT^*)^{1/2}T$. Let A and B denote the positive contractions $A = (1 - T^*T)^{1/2}$ and $B = (1 - TT^*)^{1/2}$. Observe that

$$TA^2 = T - TT^*T = B^2T.$$

Thus, $Tf(A^2) = f(B^2)T$ for every polynomial $f \in \mathbb{C}[t]$. Hence, if $\{f_n\}_{n \in \mathbb{N}}$ is a sequence of polynomials converging uniformly on the interval $[0,1]$ to the square-root function $h(t) = \sqrt{t}$, then $Th(A^2) = h(B^2)T$; that is, $T(1 - T^*T)^{1/2} = (1 - TT^*)^{1/2}T$, which proves that $U^*U = UU^* = 1 \in B(H \oplus H)$. $\square$

Proposition 10.80 has numerous interesting applications, one of which concerns the weak operator topology.

Proposition 10.81. *If H is an infinite-dimensional Hilbert space, then the closure of the set $\{U \in B(H) \mid U \text{ is unitary}\}$ in the weak operator topology of $B(H)$ is the set of all $T \in B(H)$ for which $\|T\| \leq 1$.*

Proof. Select $T \in B(H)$ such that $\|T\| \leq 1$. Consider a basic WOT-open set W containing T, which by the definition of the weak operator topology is a set of the form

$$W = \bigcap_{j=1}^{m} \{S \in B(H) \mid |\langle (S - T)\xi_j, \eta_j \rangle| < \varepsilon_j\},$$

for some $m \in \mathbb{N}$, $\xi_1, \dots, \xi_m, \eta_1, \dots, \eta_m \in H$, and $\varepsilon_1, \dots \varepsilon_m \in (0, \infty)$. We aim to prove that W contains some unitary operator U.

Let $H_0 = \text{Span}\{\xi_1, \dots, \xi_m, \eta_1, \dots, \eta_m\}$. Because H has infinite dimension, we may consider the finite-dimensional Hilbert space $H_0 \oplus H_0$ as a subspace of H; hence, H decomposes as $H = (H_0 \oplus H_0) \oplus H_1$, where $H_1 = (H_0 \oplus H_0)^{\perp}$. Let $P \in B(H)$ denote the projection with range H_0 and consider the contraction PTP acting on H_0. By Proposition 10.80, there is a unitary operator $U_0 \in B(H_0 \oplus H_0)$ such that

$U_0 = \begin{bmatrix} (PTP) & X \\ Y & Z \end{bmatrix}$. Extend U_0 to a unitary operator U acting on $H = (H_0 \oplus H_0) \oplus H_1$, where

$$U = \begin{bmatrix} U_0 & 0 \\ 0 & 1_{|H_1} \end{bmatrix} = \begin{bmatrix} (PTP) & X & 0 \\ Y & Z & 0 \\ 0 & 0 & 1_{|H_1} \end{bmatrix}.$$

Note that $PUP = PTP$ and so, for each $j = 1, \ldots, m$,

$$\langle U\xi_j, \eta_j \rangle = \langle UP\xi_j, P\eta_j \rangle = \langle PUP\xi_j, \eta_j \rangle = \langle PTP\xi_j, \eta_j \rangle = \langle TP\xi_j, P\eta_j \rangle = \langle T\xi_j, \eta_j \rangle.$$

Thus, $U \in W$. Hence, $\{T \in B(H) \mid \|T\| \leq 1\} \subseteq \overline{\{U \in B(H) \mid U \text{ is unitary}\}}^{\text{WOT}}$.

Conversely, if $T \in B(H)$ satisfies $\|T\| > 1$, then there are unit vectors $\xi, \eta \in H$ such that $|\langle T\xi, \eta \rangle| > 1$. On the other hand, $|\langle U\xi, \eta \rangle| \leq \|U\xi\| \|\eta\| = \|\xi\| \|\eta\| = 1$ for every unitary operator U. Hence, $T \notin \overline{\{U \in B(H) \mid U \text{ is unitary}\}}^{\text{WOT}}$.   □

The following two results of this section are in the spirit of Propositions 10.80 and 10.81.

Proposition 10.82. *If $A \in B(H)$ is positive and $\|A\| \leq 1$, then*

$$P = \begin{bmatrix} A & (A(1-A))^{1/2} \\ (A(1-A))^{1/2} & 1-A \end{bmatrix}$$

is a projection operator on $H \oplus H$.

Proof. Exercise 10.133.   □

Proposition 10.83. *If H is an infinite-dimensional Hilbert space, then the closure of the set $\{P \in B(H) \mid P \text{ is a projection}\}$ in the weak operator topology of $B(H)$ is the unit operator interval $I(H)$.*

Proof. Exercise 10.135.   □

One of the most striking applications of matrices of operators involves an infinite matrix. Let H be a Hilbert space and suppose that $T \in B(H)$ is a contraction. Consider the Hilbert space $\ell_H^2(\mathbb{Z})$ of sequences $\xi = (\xi_n)_{n \in \mathbb{Z}}$ of vectors $\xi_n \in H$ for which

$$\sum_{n \in \mathbb{Z}} \|\xi_n\|^2 = \lim_{k \to \infty} \sum_{n=-k}^{k} \|\xi_n\|^2 < \infty.$$

(The inner product is $\langle (x_n)_n, (\eta_n)_n \rangle = \sum_{n \in \mathbb{Z}} \langle \xi_n, \eta_n \rangle$.) With respect to this sequence space, consider the operator $U : \ell_H^2(\mathbb{Z}) \to \ell_H^2(\mathbb{Z})$ defined by the following lower-triangular infinite matrix, with entries indexed by $\mathbb{Z} \times \mathbb{Z}$, of operators acting on H:

$$U = \begin{bmatrix} \ddots & & & & & \\ & \ddots & 0 & & & \\ & & 1 & 0 & & \\ & & X & T & & \\ & & -T^* & Y & 0 & \\ & & & 1 & 0 & \\ & & & & \ddots & \ddots \end{bmatrix},$$

where $X = (1 - TT^*)^{1/2}$ and $Y = (1 - T^*T)^{1/2}$, and where the operator T is the $(0,0)$-entry of the matrix U, all remaining diagonal entries are 0, all subdiagonal entries are 1 except for X and Y, and all other matrix entries are 0. By using arguments like those employed in the proof of Proposition 10.80, one sees that the operator U satisfies $U^*U = UU^* = 1$.

Let P_0 be the projection on $\ell^2_H(\mathbb{Z})$ with range given by the 0-th copy of H in $\ell^2_H(\mathbb{Z})$. Thus, $P_0 U|_{\operatorname{ran} P_0} = T$. Moreover, because U is in lower-triangular form, $P_0(U^k)|_{\operatorname{ran} P_0} = T^k$ for every positive integer k. This leads to the following important theorem.

Theorem 10.84 (Sz.-Nagy Dilation Theorem). *If $T \in B(H)$ satisfies $\|T\| \leq 1$, then there is a Hilbert space $\tilde{H}$ that contains H as a subspace and a unitary operator U on $\tilde{H}$ such that*

$$P(U^k)|_H = T^k$$

for every positive integer k, where $P \in B(\tilde{H})$ is the projection with range H.

Corollary 10.85 (von Neumann's Inequality). *If $T \in B(H)$ satisfies $\|T\| \leq 1$, then $\|f(T)\| \leq 1$ for every polynomial $f \in \mathbb{C}[t]$ for which $\max_{|z| \leq 1} |f(z)| \leq 1$.*

Proof. Exercise 10.136. □

10.9 Singular Values and Trace-Class Operators

The spectral theory of compact hermitian operators leads to a general structure theorem for arbitrary compact Hilbert space operators known as the singular value decomposition.

Theorem 10.86 (Singular Value Decomposition). *If $K \in B(H)$ is a compact operator of rank $r \in \mathbb{N} \cup \{\infty\}$, then there exist a sequence $\{s_j\}_{j=1}^r$ of real numbers and orthonormal sets $\{\phi_j\}_{j=1}^r$ and $\{\psi_j\}_{j=1}^r$ of vectors such that*

1. $s_j \geq s_{j+1} > 0$, for all j,

2. $0 = \lim_j s_j$, *if* $r = \infty$, *and*

3. $K\xi = \sum_{j=1}^{r} s_j \langle \xi, \phi_j \rangle \psi_j$, *for every* $\xi \in H$.

Proof. Apply Theorem 10.32 to the compact positive operator K^*K to obtain a bounded sequence $\{\lambda_j\}_{j=1}^{r'}$ of (nonzero) positive real numbers and a sequence $\{\phi_j\}_{j=1}^{r'}$ of pairwise-orthogonal unit vectors in H such that $K^*K\phi_j = \lambda_j\phi_j$, for each

j, $\lim_j \lambda_j = 0$, if r' is infinite, and $K^*K\xi = \sum_{j=1}^{r'} \lambda_j \langle \xi, \phi_j \rangle \phi_j$, for every $\xi \in H$. Note

that $r' = \operatorname{rank}(K^*K)$. Furthermore, if $R \in B(H)$ is given by $R\xi = \sum_{j=1}^{r'} \sqrt{\lambda_j} \langle \xi, \phi_j \rangle \phi_j$,

for every $\xi \in H$, then R is compact, positive, and $R^2 = K^*K$. Therefore, by the uniqueness of the positive square root, $R = (K^*K)^{1/2}$; thus, if $s_j = \sqrt{\lambda_j}$ for each j, then

$$|K|\xi = \sum_{j=1}^{r'} \sqrt{\lambda_j} \langle \xi, \phi_j \rangle \phi_j,$$

for every $\xi \in H$. Hence, if $K = V|K|$ is the polar decomposition of K, and if $\psi_j = V\phi_j$ for each j, then

$$K\xi = \sum_{j=1}^{r'} s_j \langle \xi, \phi_j \rangle \psi_j,$$

for all $\xi \in H$. Because the range of V is isometric on $\operatorname{ran}|K|$ and has range $\overline{\operatorname{ran}K}$, $\{\psi_k\}_{k=1}^{r'}$ is a set of orthonormal vectors and $r' = \operatorname{rank}K$.   □

The *singular decomposition* of a compact operator K refers to the representation in (3) of Theorem 10.86 of the action of K on the Hilbert space H.

Definition 10.87. The *singular values* of a compact operator K of rank $r \in \mathbb{N} \cup \{\infty\}$ acting on a separable Hilbert space of dimension $d \in \mathbb{N} \cup \{\infty\}$ are the nonnegative real numbers $s_j(K)$ defined by

$$s_j(K) = s_j,$$

if $1 \leq j \leq r$ and where $\{s_j\}_{j=1}^{r}$ are the positive numbers arising in the singular value decomposition (3) of K, and by

$$s_j(K) = 0,$$

if $j \in \{1, \ldots, d\}$ is such that $j > r$.

Using the notation $\phi \otimes \psi$ to denote rank-1 operators (of the form $\xi \mapsto \langle \xi, \psi \rangle \phi$), the singular value decomposition of a compact operator K can be expressed as

$$K = \sum_{j=1}^{d} s_j(K) \phi_j \otimes \psi_j. \qquad (10.13)$$

In the case where H has infinite dimension, the series $\sum_{j=1}^{d} s_j(K) \phi_j \otimes \psi_j$ converges in the norm of $B(H)$ to K. Indeed, if $\xi \in H$ and $N \in \mathbb{N}$, then

$$\left\| K\xi - \sum_{j=1}^{N} s_j(K)(\phi_j \otimes \psi_j)[\xi] \right\|^2 = \sum_{j>N} s_j(K)^2 |\langle \xi, \psi_j \rangle|^2 \le s_N(K)^2 \|\xi\|^2.$$

Thus,

$$\left\| K - \sum_{j=1}^{N} s_j(K)(\phi_j \otimes \psi_j) \right\| \le s_N(K),$$

which converges to zero because $\lim_{N \to \infty} s_N(K) = \lim_{N \to \infty} \sqrt{\lambda_N(K^*K)} = 0$. Hence, the following proposition has been proved.

Proposition 10.88. *If H is a separable Hilbert space, then the algebraic ideal $F(H)$ of finite-rank operators on H is dense in the ideal $K(H)$ of compact operators on H.*

The following elegant application of Proposition 10.88 returns us to the notion of complementation (see Definition 8.15) in Banach space theory.

Proposition 10.89 (Conway). *If H is an infinite-dimensional separable Hilbert space, then the subspace $K(H)$ is not complemented in $B(H)$.*

Proof. Fix an orthonormal basis $\{\phi_n\}_{n \in \mathbb{N}}$ of H. Let ℓ^∞ and c_0 denote $\ell^\infty(\mathbb{N})$ and $c_0(\mathbb{N})$. For each $\psi \in \ell^\infty$, define M_ψ on H by

$$M_\psi \xi = \sum_{n=1}^{\infty} \psi(n) \langle \xi, \phi_n \rangle,$$

for $\xi \in H$. The linear map $\pi : \ell^\infty \to B(H)$ in which $\pi(\psi) = M_\psi$ is plainly linear and isometric. Now if $f \in \ell^\infty$ has the property that $f(n) \ne 0$ for at most a finite number of $n \in \mathbb{N}$, then M_f has finite rank. Such functions are dense in c_0; therefore, if $f \in c_0$ and if $\{f_k\}_{k \in \mathbb{N}}$ is sequence in ℓ^∞ converging to f and such that, for each k, $f_k(n) \ne 0$ for at most a finite number of $n \in \mathbb{N}$, then $\lim_k \|M_f - M_{f_k}\| = \lim_k \|f - f_k\| = 0$ shows that M_f is a compact operator.

376 10 Hilbert Space Operators

Let $\beta : B(H) \to \ell^\infty$ be the contractive linear map defined by $\beta(T) = (\langle T\phi_n, \phi_n \rangle)_n$. Suppose that K is a rank-1 operator, say $K = \gamma \otimes \psi$, for some nonzero vectors γ and ψ. Then, for a fixed n, $\langle K\phi_n, \phi_n \rangle = \langle \phi_n, \psi \rangle \langle \gamma, \phi_n \rangle$. Because $\langle \psi, \phi_n \rangle$ and $\langle \gamma, \phi_n \rangle$ of the n-th Fourier coefficients of ψ and γ, respectively, these complex numbers converge to 0 as $n \to \infty$. Hence, $\beta(K) \in c_0$ for every rank-1 operator K, and so by linearity $\beta(K) \in c_0$ for all finite-rank operators K. Because $F(H)$ is dense in $K(H)$, and because β is continuous and c_0 is closed in ℓ^∞, we deduce that $\beta(K) \in c_0$ for every $K \in K(H)$.

Assume, contrary to what we aim to prove, that $K(H)$ is complemented in $B(H)$. Hence, there exists an idempotent operator $\mathscr{E} : B(H) \to B(H)$ with range $K(H)$ (Proposition 8.16). Define now an operator $E : \ell^\infty \to \ell^\infty$ by $E = \beta \circ \mathscr{E} \circ \pi$ and observe that E is an idempotent with range c_0. Therefore, by Proposition 8.16, c_0 is a complemented subspace of ℓ^∞, which is in contradiction to Proposition 8.20. Therefore, it cannot be that $K(H)$ is complemented in $B(H)$.   □

Returning to the study of singular values, we begin with two basic properties.

Proposition 10.90. *If $K, K_1, K_2, S, T \in B(H)$ and if K, K_1, K_2 are compact, then*

1. $s_j(|K|) = s_j(K) = s_j(K^*)$ *and*
2. $s_j(SKT) \le \|S\| \, \|T\| s_j(K)$

for every j.

Proof. Note that the singular values of K are simply the square roots of the eigenvalues of K^*K, labelled in non-ascending order. Thus,

$$s_j(|K|) = \lambda_j(|K|) = s_j(K)$$

for every j.

Suppose that $K = V|K|$ is the polar decomposition of K. Thus, $KK^* = V(K^*K)V^*$. Because

$$\operatorname{ran} K^*K = \operatorname{ran} |K|^2 \subseteq \operatorname{ran} |K|,$$

$V^*V(K^*K) = K^*K$. Hence, $KK^* = V(K^*K)V^*$ yields $f(K^*K) = Vf(K^*K)V^*$ for all polynomials $f \in \mathbb{R}[t]$, and so $|K^*| = V|K|V^*$ by continuous functional calculus.

Suppose now that $|K|\xi = \lambda\xi$ for some $\lambda \in (0, \infty)$ and nonzero $\xi \in H$. Thus, $\xi \in \operatorname{ran} |K|$. Hence, if $\psi = V\xi$, then $\|\psi\| = \|\xi\|$ and $V^*V\xi = \xi$. Thus,

$$|K^*|\psi = V|K|V^*\psi = V|K|V^*V\xi = V|K|\xi = \lambda V\xi = \lambda\psi.$$

That is, λ is a nonzero eigenvalue of $|K|$ if and only if λ is a nonzero eigenvalue of $|K^*|$.

The argument above shows that if $\{\xi_1, \ldots, \xi_m\}$ is an orthonormal basis for $\ker(|K| - \lambda 1)$, then $\{V\xi_1, \ldots, V\xi_m\}$ is an orthonormal subset of $\ker(|K^*| - \lambda 1)$, and so the multiplicity of λ as an eigenvalue of $|K|$ is bounded above by the multiplicity of

λ as an eigenvalue of $|K^*|$. By letting $L = K^*$ and invoking the argument again shows that $\ker(|K| - \lambda 1)$ and $\ker(|K^*| - \lambda 1)$ have the same dimension. Hence, $s_j(K) = s_j(K^*)$ for every j.

To prove the second assertion, because

$$|SK|^2 = (SK)^*(SK) = K^*(S^*S)K \leq \|S\|^2 K^*K = \|S\|^2 |K|^2,$$

Theorem 10.54 implies that $|SK| \leq \|S\| |K|$. Therefore, the min-max variation principle (equation (10.11)) in the Courant-Fischer Theorem (Theorem 10.58) yields

$$s_j(SK) = \lambda_j(|SK|) \leq \|S\|\lambda_j(|K|) = s_j(K)$$

for every j. Thus,

$$s_j(SKT) \leq \|S\|s_j(KT) = \|S\|s_j(T^*K^*) \leq \|S\| \|T^*\|s_j(K^*) = \|S\| \|T\|s_j(K)$$

for every j. $\square$

Definition 10.91. A compact operator K acting on a separable Hilbert space of dimension $d \in \mathbb{N} \cup \{\infty\}$ is a *trace-class operator* if

$$\sum_{j=1}^{d} s_j(K) < \infty.$$

Let $T(H)$ denote the set of trace-class operators acting on a separable Hilbert space H. Thus, we have

$$T(H) \subseteq K(H) \subseteq B(H).$$

If H has infinite dimension, then the inclusions above are sharp. The proper inclusion of $K(H)$ into $B(H)$ has already been noted (as the identity operator is not compact), and so consider the inclusion $T(H) \subseteq K(H)$. Select an orthonormal basis $\{\phi_j\}_{j \in \mathbb{N}}$ and consider the compact positive operator K for which $K\phi_j = j^{-1}\phi_j$ for all $j \in \mathbb{N}$. Because $s_j(K) = \frac{1}{j}$ for each j, the sum $\sum_{j=1}^{\infty} s_j(K)$ diverges, and so $K \notin T(H)$.

On the other hand, observe that

$$K \in T(H) \Longleftrightarrow |K| \in T(H) \Longleftrightarrow K^* \in T(H)$$

and that $RKS \in T(H)$ if $K \in T(H)$ and $R, S \in B(H)$.

The use of the adjective "trace" in Definition 10.91 above will be explained shortly, but note that every operator of finite rank is a trace-class operator; in particular, if H has finite dimension, then every operator on H is a trace-class operator.

In linear algebra, the trace of an $n \times n$ matrix $T = [t_{ij}]_{i,j=1}^n$ is defined by $\operatorname{Tr} T = \sum_{i=1}^n t_{ii}$, the sum of the diagonal elements of T. If T is the matrix representation of an operator on an n-dimensional Hilbert space with respect to some orthonormal basis $\{\phi_i\}_{i=1}^n$ of H, then

$$\operatorname{Tr} T = \sum_{i=1}^n \langle T\phi_i, \phi_i \rangle.$$

This motivates the definition of trace for operators on infinite-dimensional separable Hilbert spaces, starting first with the cone $B(H)_+$ of positive operators.

Definition 10.92. Let $\mathscr{B} = \{\phi_i\}_{i=1}^\infty$ denote an orthonormal basis of an infinite-dimensional separable Hilbert space H. The function $\tau_{\mathscr{B}} : B(H)_+ \to [0, \infty]$ defined by

$$\tau_{\mathscr{B}}(A) = \sum_{i=1}^\infty \langle A\phi_i, \phi_i \rangle.$$

is called a *canonical tracial weight* on $B(H)_+$.

Besides its linearity, a distinguishing property of the trace of matrices is that $\operatorname{Tr}(ST) = \operatorname{Tr}(TS)$ for all matrices S and T.

Proposition 10.93. *If $\mathscr{B}$ is a given orthonormal basis of an infinite-dimensional separable Hilbert space H, then*

1. $\tau_{\mathscr{B}}(\alpha_1 A_1 + \alpha_2 A_2) = \alpha_1 \tau_{\mathscr{B}}(A_1) + \alpha_2 \tau_{\mathscr{B}}(A_2)$, *for all $A_j \in B(H)_+$ and $\alpha_j \in \mathbb{R}_+$,*
2. $\tau_{\mathscr{B}}(TT^*) = \tau_{\mathscr{B}}(T^*T)$, *for every $T \in B(H)$,*
3. $\tau_{\mathscr{B}}(U^*AU) = \tau_{\mathscr{B}}(A)$, *for every positive operator A and unitary operator U, and*
4. $\tau_{\mathscr{B}} = \tau_{\mathscr{B}'}$ *for every orthonormal basis $\mathscr{B}'$ of H.*

Proof. It is clear that the property $\tau_{\mathscr{B}}(\alpha_1 A_1 + \alpha_2 A_2) = \alpha_1 \tau_{\mathscr{B}}(A_1) + \alpha_2 \tau_{\mathscr{B}}(A_2)$ holds, for all $A_1, A_2 \in B(H)_+$ and $\alpha_1, \alpha_2 \in \mathbb{R}_+$.

Let $\mathscr{B} = \{\phi_i\}_{i=1}^\infty$. By way of the Fourier series decompositions of $T^*\phi_i$ and $T\phi_j$,

$$\tau_{\mathscr{B}}(TT^*) = \sum_i \langle TT^*\phi_i, \phi_i \rangle = \sum_i \left\langle \sum_j \langle T^*\phi_i, \phi_j \rangle T\phi_j, \phi_i \right\rangle = \sum_i \sum_j |\langle T\phi_j, \phi_i \rangle|^2$$

$$= \sum_j \sum_i |\langle T\phi_j, \phi_i \rangle|^2 = \sum_j \left\langle T\phi_j, \sum_i \langle T\phi_j, \phi_i \rangle \phi_i \right\rangle = \sum_i \langle T^*T\phi_i, \phi_i \rangle$$

$$= \tau_{\mathscr{B}}(T^*T).$$

Hence, $\tau_{\mathscr{B}}(TT^*) = \tau_{\mathscr{B}}(T^*T)$.

Suppose that A is positive and U is unitary; let $T = A^{1/2}U$. Thus,

$$\tau_{\mathscr{B}}(U^*AU) = \tau_{\mathscr{B}}(U^*A^{1/2}A^{1/2}U) = \tau_{\mathscr{B}}(T^*T)$$

$$= \tau_{\mathscr{B}}(TT^*) = \tau_{\mathscr{B}}(A^{1/2}UU^*A^{1/2})$$

$$= \tau_{\mathscr{B}}(A).$$

Lastly, if $\mathscr{B}' = \{\phi_i'\}_{i=1}^d$ is an orthonormal basis of H, then let $U \in B(H)$ be the unitary operator for which $U\phi_i = \phi_i'$ for every $i \in \{1,\ldots,d\}$. Hence, for every $A \in B(H)_+$,

$$\tau_{\mathscr{B}}(A) = \tau_{\mathscr{B}}(U^*AU) = \tau_{\mathscr{B}'}(A),$$

which completes the proof. $\square$

In light of Proposition 10.93, we may drop the reference to the orthonormal basis $\mathscr{B}$ when discussing a canonical tracial weight $\tau_{\mathscr{B}}$—since there is exactly one such function—and denote the canonical tracial weight on $B(H)$ by τ.

The domain of definition of τ can be extended to the $\mathbb{R}$-vector space $B(H)_{\mathrm{sa}}$ of hermitian operators as follows. For any difference $C = A_1 - A_2$ of positive operators A_1 and A_2, define $\tau(C)$ by

$$\tau(C) = \tau(A_1) - \tau(A_2).$$

This is a well-defined function on $B(H)_{\mathrm{sa}}$ because, if $A_1 - A_2 = A_1' - A_2'$, then $A_1' + A_2 = A_1 + A_2' \in B(H)_+$ and therefore

$$\tau(A_1') + \tau(A_2) = \tau(A_1' + A_2) = \tau(A_1 + A_2') = \tau(A_1) + \tau(A_2');$$

hence, $\tau(A_1) - \tau(A_2) = \tau(A_1') - \tau(A_2')$ in the extended real number system.

Definition 10.94. If H is a separable Hilbert space, then the *canonical trace* on $B(H)$ is the function Tr on $B(H)$ defined by

$$\mathrm{Tr}\,((A_1 - A_2) + i(B_1 - B_2)) = (\tau(A_1) - \tau(A_2)) + i(\tau(B_1) - \tau(B_2)),$$

for all positive operators A_1, A_2, B_1, B_2.

Alternatively, the canonical trace on $B(H)$ is the map $T \mapsto \sum_{i=1}^{\infty} \langle T\phi_i, \phi_i \rangle$ for some (and every) orthonormal basis $\{\phi_i\}_{i=1}^{\infty}$ of H.

Proposition 10.95. *If H is an infinite-dimensional separable Hilbert space and if*
K ∈ B(H) is a trace-class operator, then the sum

$$\sum_{\omega \in \mathscr{B}} \langle K\omega, \omega \rangle \tag{10.14}$$

is absolutely convergent in $\mathbb{C}$ *for every orthonormal basis* $\mathscr{B}$ *of H.*

Proof. By the singular value decomposition of K, there exist orthonormal sets
$\{\phi_j\}_{j=1}^{\infty}$ and $\{\psi_j\}_{j=1}^{\infty}$ in H such that

$$K = \sum_{j=1}^{\infty} s_j(K)\phi_j \otimes \psi_j.$$

If the set $\mathscr{B}_0 = \{\phi_j\}_{j=1}^{\infty}$, which is an orthonormal basis for $\overline{\mathrm{ran}\,|K|}$, is not already an
orthonormal basis of H, then it may be extended to one, say $\mathscr{B} = \mathscr{B}_0 \cup \mathscr{B}_1$, where
$\mathscr{B}_1$ is an orthonormal basis of $(\overline{\mathrm{ran}\,|K|})^{\perp} = \ker|K| = \ker K$. Thus, $K\omega = 0$ for every
$\omega \in \mathscr{B}_1$, and so

$$\left| \sum_{\omega \in \mathscr{B}} \langle K\omega, \omega \rangle \right| = \left| \sum_{\omega \in \mathscr{B}_0} \langle K\omega, \omega \rangle \right| = \left| \sum_{i=1}^{r} \langle K\phi_i, \phi_i \rangle \right|$$

$$= \left| \sum_{i=1}^{r} \left\langle \sum_{j=1}^{r} s_j(K)\langle \phi_i, \phi_j \rangle \psi_j, \phi_i \right\rangle \right|$$

$$\leq \sum_{i=1}^{\infty} s_i(K) |\langle \psi_i, \phi_i \rangle|$$

$$\leq \sum_{i=1}^{d} s_i(K) < \infty.$$

Hence, the sum (10.14) is absolutely convergent.   □

Corollary 10.96. *If K is a trace-class operator acting on a separable Hilbert space*
H of dimension $d \in \mathbb{N} \cup \{\infty\}$, *then*

$$|\mathrm{Tr}\,K| \leq \mathrm{Tr}\,|K| = \sum_{j=1}^{d} s_j(K).$$

Proof. The proof of Proposition 10.95 shows that

$$|\mathrm{Tr}\,K| \leq \sum_{i=1}^{r} s_i(K).$$

However, because $\lambda_i(|K|) = s_i(|K|) = s_i(K)$ for all i such that $1 \leq i \leq r$, this latter sum is precisely $\text{Tr}\,|K|$ by the spectral theorem (Theorem 10.32). Since $s_j(|K|) = 0$ for $r < j \leq d$, $\displaystyle\sum_{j=1}^{d} s_j(|K|) = \text{Tr}\,|K|$. □

Every rank-1 operator $\phi \otimes \psi$ is a trace-class operator, and therefore $T(\phi \otimes \psi)$ is also a trace-class operator, for every $T \in B(H)$. The traces of such operators (of rank 0 or 1) are easily determined as follows.

Example 10.97. *If $\phi, \psi \in H$ are nonzero and if $T \in B(H)$ is arbitrary, then*

$$\text{Tr}\,(T(\phi \otimes \psi)) = \langle T\phi, \psi \rangle.$$

Proof. We may assume that $\|\psi\| = 1$, for if not we could replace ϕ with $\tilde{\phi} = \|\psi\|\phi$ and ψ with $\tilde{\psi} = \|\psi\|^{-1}\psi$ to obtain $\phi \otimes \psi = \tilde{\phi} \otimes \tilde{\psi}$. Select an orthonormal basis $\{\phi_j\}_j$ of H in which $\phi_1 = \psi$. Thus,

$$\text{Tr}\,(T(\phi \otimes \psi)) = \sum_j \langle T\left(\langle \phi_j, \psi \rangle\right)\phi, \phi_j \rangle = \sum_j \langle \phi_j, \psi \rangle \langle T\phi, \phi_j \rangle = \langle T\phi, \psi \rangle,$$

which completes the calculation. □

In particular, the computation above yields $\|\phi \otimes \psi\|_1 = |\langle \phi, \psi \rangle|$. The first major result about trace-class operators is the following theorem.

Theorem 10.98. *The set $T(H)$ of all trace-class operators acting on a separable Hilbert space H of dimension $d \in \mathbb{N} \cup \{\infty\}$ is an algebraic ideal of $B(H)$ and the function $\|\cdot\|_1 : T(H) \to \mathbb{R}$ defined by*

$$\|K\|_1 = \sum_{j=1}^{d} s_j(K) \tag{10.15}$$

is a norm on $T(H)$. Furthermore, with respect to the norm $\|\cdot\|_1$, $T(H)$ is a separable Banach space and the algebraic ideal $F(H)$ of finite-rank operators is dense in $T(H)$.

Proof. We shall assume for the proof that H has infinite dimension.

Proposition 10.90 shows that $T(H)$ is closed under scalar multiplication, the involution $*$, and under products of the form RKS, where $K \in T(H)$ and $R, S \in B(H)$. All that remains, therefore, to show that $T(H)$ is an algebraic ideal is to show that $K_1 + K_2 \in T(H)$ for every $K_1, K_2 \in T(H)$. To this end, let $K_1, K_2 \in T(H)$ and consider $|K_1 + K_2|$. By the Triangle Inequality (Theorem 10.71), there are isometries $V, W \in B(H)$ such that

$$|K_1 + K_2| \leq V|K_1|V^* + W|K_2|W^* = X^*X + Y^*Y,$$

where $X = |K_1|^{1/2} V^*$ and $Y = |K_2|^{1/2} W^*$. Proposition 10.93 asserts that $\tau(G^*G) = \tau(GG^*)$ for every $G \in B(H)$, and therefore

$$\tau(|K_1 + K_2|) \leq \tau(X^*X + Y^*Y) = \tau(X^*X) + \tau(Y^*Y) = \tau(XX^*) + \tau(YY^*).$$

Because $XX^* = |K_1|$ and $YY^* = |K_2|$, the inequality above yields

$$\text{Tr}\,|K_1 + K_2| \leq \text{Tr}\,|K_1| + \text{Tr}\,|K_2|.$$

By Corollary 10.96,

$$\sum_{j=1}^{d} s_j(K_1 + K_2) = \text{Tr}\,|K_1 + K_2| \leq \text{Tr}\,|K_1| + \text{Tr}\,|K_2| = \sum_{j=1}^{d} s_j(K_1) + \sum_{j=1}^{d} s_j(K_2).$$

Thus, $K_1 + K_2 \in T(H)$ and $\|K_1 + K_2\|_1 \leq \|K_1\|_1 + \|K_2\|_1$. Hence, $T(H)$ is an algebraic ideal of $B(H)$ and $\|\cdot\|_1$ satisfies the triangle inequality on $T(H)$.

If $K \in T(H)$ satisfies $\|K\|_1 = 0$, then the only eigenvalue of $|K|$ is 0, and so the spectral radius of $|K|$ is 0. Hence, $|K| = 0$ and therefore $K = 0$. The property $\|\alpha K\|_1 = |\alpha|\,\|K\|_1$ is trivial. Since the triangle inequality was established in the previous paragraph, $\|\cdot\|_1$ is a norm on $T(H)$.

Suppose that $\{K_n\}_{n \in \mathbb{N}}$ is a Cauchy sequence in $T(H)$. Because

$$\|K_n - K_m\|_1 \geq s_1(K_n - K_m) = \|\,|K_n - K_m|\,\| = \|K_n - K_m\|,$$

$\{K_n\}_{n \in \mathbb{N}}$ is a Cauchy sequence in $B(H)$ and is, hence, convergent to some $K \in B(H)$. Because each K_n is compact and $K(H)$ is norm-closed, the limit operator K must also be compact. Furthermore,

$$\|K_n^* K_n - K^* K\| = \|K_n^*(K_n - K) + (K_n - K)^* K\| \leq \|K_n\|\|K_n - K\| + \|K_n^* - K^*\|\|K\|$$

implies that $K_n^* K_n$ converges to $K^* K$. Thus, by the continuous functional calculus, $\|\,|K_n| - |K|\,\| \to 0$.

Let $\{\phi_j\}_j$ be an orthonormal set of eigenvectors of $|K|$ corresponding to the nonzero eigenvalues $\lambda_j(|K|)$ of K. If $N \in \mathbb{N}$, then

$$\sum_{j=1}^{N} s_j(K) = \sum_{j=1}^{N} \lambda_j(|K|) = \sum_{j=1}^{N} \langle |K|\phi_j, \phi_j \rangle$$

$$= \lim_{n \to \infty} \sum_{j=1}^{N} \langle |K_n|\phi_j, \phi_j \rangle \leq \liminf_{n} \text{Tr}\,|K_n|$$

$$= \liminf_{n} \|K_n\|_1.$$

Now since $\lim_n \|K_n\|_1$ exists, $\liminf_n \|K_n\|_1 = \lim_n \|K_n\|_1$, and so the sum $\sum_{j=1}^{N} s_j(K)$ is bounded above for all N. Hence, K is a trace-class operator.

If $\varepsilon > 0$, then there exists $N \in \mathbb{N}$ such that $\|K_n - K_\ell\|_1 < \varepsilon$ for all $\ell, n \geq N$. Thus, if $\ell \geq N$, then

$$\|K - K_\ell\|_1 = \lim_{n \to \infty} \|K_n - K_\ell\|_1 < \varepsilon.$$

Hence, $\{K_n\}_{n \in \mathbb{N}}$ converges in $T(H)$ to K, thus proving that $T(H)$ is a Banach space.

To show that $F(H)$ is dense in $T(H)$, let $K \in T(H)$ have infinite rank; thus, the positive trace-class operator $|K|$ also has positive rank. Express $|K|$ in its spectral decomposition:

$$|K| = \sum_{j=1}^{\infty} \lambda_j(|K|)\phi_j \otimes \phi_j,$$

for some orthonormal set $\{\phi_j\}_{j=1}^{\infty}$. Consider $|K|_N = \sum_{j=1}^{N} s_j(K)\lambda_j(|K|)\phi_j \otimes \phi_j$, which is a positive operator of finite rank and satisfies $|K|_N \leq |K|$ in the Loewner ordering. Thus,

$$\||K| - |K|_N\|_1 = \mathrm{Tr}\,(\,|\,|K| - |K|_N\,|) = \mathrm{Tr}\,(|K| - |K|_N) = \sum_{j>N} \lambda_j(|K|).$$

Because $\left(\lambda_j(|K|)\right)_{j=1}^{\infty} \in \ell^1(\mathbb{N})$, the partial sums of the eigenvalues converge to the sum of the eigenvalues; hence, $\lim_{N \to \infty} \sum_{j>N} \lambda_j(|K|) = 0$ and $\lim_{N \to \infty} \||K| - |K|_N\|_1 = 0$.

Express K in its polar decomposition: $K = V|K|$ and let $G_N = V|K|_N$, which is an operator of finite rank. Thus,

$$\|K - G_N\|_1 = \|V(|K| - |K|_N)\|_1 \leq \|V\|\,\||K| - |K|_N\|_1 = \||K| - |K|_N\|_1,$$

and so $\lim_{N \to \infty} \|K - G_N\|_1 = 0$.

The proof of the separability of $T(H)$ is left as an exercise (Exercise 10.145). □

The norm $\|\cdot\|_1$ on $T(H)$ is called the *trace norm*. Note that the proof of Theorem 10.98 shows that if

$$K = \sum_{j=1}^{\infty} s_j(K)\phi_j \otimes \psi_j$$

is the singular value decomposition of $K \in T(H)$, then the sum converges in the trace norm.

Because $T(H)$ is a Banach space, it is of interest to understand its dual. The following theorem shows that the dual space of $T(H)$ is (isometrically isomorphic to) $B(H)$.

Theorem 10.99. *If H is a separable Hilbert space, then for each $T \in B(H)$ the function $\varphi_T : T(H) \to \mathbb{C}$ defined by $\varphi_T(K) = \mathrm{Tr}\,(TK)$ is a bounded linear functional. Furthermore, the map $\Omega : B(H) \to T(H)^*$ defined by $\Omega(T) = \varphi_T$ is a linear isometric isomorphism.*

Proof. Only the case in which H has infinite dimension will be treated.

If $T \in B(H)$ and $K \in T(H)$, then

$$|\mathrm{Tr}\,(TK)| \leq \mathrm{Tr}\,|TK| = \sum_{j=1}^{\infty} s_j(TK) \leq \sum_{j=1}^{\infty} \|T\| s_j(K) = \|T\|\,\|K\|_1.$$

Furthermore, $K \mapsto \mathrm{Tr}\,TK$ is plainly linear in K. Hence, the function φ_T is a bounded linear functional on $T(H)$ of norm $\|\varphi_T\| \leq \|T\|$. If $\varepsilon > 0$, then by Proposition 10.7 there exist unit vectors $\phi, \psi \in H$ such that $\|T\| - \varepsilon \leq |\langle T\phi, \psi \rangle|$. Thus,

$$\|T\| - \varepsilon \leq |\langle T\phi, \psi \rangle| = |\varphi_T(\phi \otimes \psi)| \leq \|\varphi_T\|\,\|\phi \otimes \psi\| = \|\varphi_T\|\,\|\phi\|\,\|\psi\| = \|\varphi_T\|,$$

which proves that $\|\varphi_T\| = \|T\|$. Because the map $\Omega : B(H) \to T(H)^*$ defined by $\Omega(T) = \varphi_T$ is linear in T, we deduce that Ω is a linear isometry.

To show that Ω is surjective, let $\varphi \in T(H)^*$ and define $\psi_\varphi : H \times H \to \mathbb{C}$ by $\psi_\varphi(\xi, \eta) = \varphi(\xi \otimes \eta)$. Because $|\psi_\varphi(\xi, \eta)| \leq \|\varphi\|\,\|\xi \otimes \eta\| = \|\varphi\|\,\|\xi\|\,\|\eta\|$ and ψ_φ is plainly a sesquilinear form, Proposition 10.5 implies that there exists an operator $T_\varphi \in B(H)$ such that $\varphi(\xi \otimes \eta) = \langle T_\varphi \xi, \eta \rangle$, for all $\xi, \eta \in H$.

If $K = \sum_{j=1}^{\infty} s_j(K)\phi_j \otimes \psi_j$ is the singular value decomposition of a trace-class operator K, then the sum converges in $T(H)$ and, by the continuity of φ and the trace,

$$\varphi(K) = \sum_{j=1}^{\infty} s_j(K)\varphi(\phi_j \otimes \psi_j) = \sum_{j=1}^{\infty} s_j(K)\langle T_\varphi \phi_j, \psi_j \rangle$$

$$= \sum_{j=1}^{\infty} s_j(K)\mathrm{Tr}\,\left(T_\varphi(\phi_j \otimes \psi_j)\right)$$

$$= \mathrm{Tr}\,(T_\varphi K).$$

In the notation established above, this means that $\Omega(T_\varphi) = \varphi$. Hence, the linear isometry $\Omega : B(H) \to T(H)^*$ is surjective. □

As a dual space, $B(H)$ has a weak*-topology, which is also commonly referred to as the ultraweak topology or the σ-weak topology.

Definition 10.100. The *ultraweak topology* on $B(H)$, where H is a separable Hilbert space, is the weak*-topology on $B(H)$ induced by the isometric isomorphism $B(H) \cong T(H)^*$.

We noted in Proposition 10.76 that a linear transformation $\varphi : B(H) \to \mathbb{C}$ is weakly continuous if and only if it is strongly continuous, and that such linear maps have the form

$$\varphi(T) = \sum_{j=1}^{n} \langle T\xi_j, \eta_j \rangle,$$

for some finite sets $\{\xi_1, \ldots, \xi_n\}$ and $\{\eta_1, \ldots, \eta_n\}$ of nonzero vectors. The situation is slightly different with the ultraweak topology.

Proposition 10.101. *If H is an infinite-dimensional separable Hilbert space, then a linear transformation $\varphi : B(H) \to \mathbb{C}$ is continuous with respect to the ultraweak topology of $B(H)$ if and only if there are sequences $\{\xi_j\}_{j\in\mathbb{N}}$ and $\{\eta_j\}_{j\in\mathbb{N}}$ of vectors such that*

$$\sum_{j=1}^{\infty} \|\xi_j\|^2 \ and \ \sum_{j=1}^{\infty} \|\eta_j\|^2$$

converge and

$$\varphi(T) = \sum_{j=1}^{\infty} \langle T\xi_j, \eta_j \rangle,$$

for every $T \in B(H)$.

Proof. Exercise 10.147. □

It so happens that $T(H)$ is itself a dual space.

Theorem 10.102. *If H is a separable Hilbert space, then for each $S \in T(H)$ the function $\varphi_S : K(H) \to \mathbb{C}$ defined by $\varphi_S(K) = \mathrm{Tr}(SK)$ is a bounded linear functional. Furthermore, the map $\omega : T(H) \to K(H)^*$ defined by $\omega(S) = \varphi_S$ is a linear isometric isomorphism.*

Proof. If $S \in T(H)$ and $K \in K(H)$, then

$$|\mathrm{Tr}(SK)| \leq \mathrm{Tr}(|SK|) = \sum_{j=1}^{\infty} s_j(SK) \leq \|K\| \sum_{j=1}^{\infty} s_j(S) = \|K\| \|S\|_1.$$

Therefore, the linear transformation φ_S on $K(H)$ is bounded of norm $\|\varphi_S\| \leq \|S\|_1$. Hence, ω is a contractive linear map of $T(H)$ into the dual space $K(H)^*$.

If $\varphi \in K(H)^*$, then define a function $\Psi : H \times H \to \mathbb{C}$ by $\Psi(\xi, \eta) = \varphi(\xi \otimes \eta)$. Because $\xi \otimes (\alpha \eta) = \overline{\alpha}(\xi \otimes \eta)$ for all $\alpha \in \mathbb{C}$, the function ψ is a sesquilinear form and satisfies $|\Psi(\xi, \eta)| \leq \|\varphi\| \|\xi\| \|\eta\|$. Hence, by Proposition 10.5, there exists an operator $S \in B(H)$ such that $\Psi(\xi, \eta) = \langle S\xi, \eta \rangle$, for all $\xi, \eta \in H$.

Recall from Example 10.97 that if $\phi, \psi \in H$, then $\mathrm{Tr}\,(S(\phi \otimes \psi)) = \langle S\phi, \psi \rangle$. Thus, if F is a finite-rank operator expressed as $F = \sum_{j=1}^{n} \phi_j \otimes \psi_j$, then

$$\mathrm{Tr}\,(SF) = \sum_{j=1}^{n} \mathrm{Tr}\,(S(\phi_j \otimes \psi_j)) = \sum_{j=1}^{n} \langle S\phi_j, \psi_j \rangle = \sum_{j=1}^{n} \varphi\left(\phi_j \otimes \psi_j\right) = \varphi(F).$$

Let $S = V|S|$ denote the polar decomposition of S, and write $|S|$ as $|S| = V^*S$. If $\{\phi_j\}_{j \in \mathbb{N}}$ is an orthonormal basis of H, and if $P_k = \sum_{j=1}^{k} \phi_j \otimes \phi_j$, which is the projection with range $\mathrm{Span}\{\phi_1, \dots, \phi_k\}$, then the operator $P_k W^*$ is a finite-rank contraction. Because $(\phi_j \otimes \phi_j)V^* = \phi_j \otimes (V\phi_j)$, we have that

$$\|\varphi\| \geq |\varphi(P_k W^*)| = \left| \sum_{j=1}^{k} \langle S\phi_j, V\phi_j \rangle \right| = \left| \sum_{j=1}^{k} \langle V^* S\phi_j, \phi_j \rangle \right| = \mathrm{Tr}\,(|S|P_k).$$

Therefore, $\lim_{k \to \infty} \mathrm{Tr}\,(|S|P_k) = \mathrm{Tr}\,(|S|)$ exists, which implies that $|S|$ and S are trace-class operators of norm $\|S\|_1 \leq \|\varphi\|$. Hence, ω is surjective and isometric. $\qquad \square$

To conclude, the final result describes the extreme points of the closed unit ball of trace-class operators.

Theorem 10.103. *The following statements are equivalent for $S \in \mathscr{T}(H)$:*

1. S is an extreme point of the closed unit ball of $\mathscr{T}(H)$;
2. $\mathrm{rank}\, S = 1$ and $\mathrm{Tr}(S^*S) = 1$.

Proof. Denote the closed unit balls of $\mathscr{T}(H)$ and $\ell^1(\mathbb{N})$ by $\mathscr{T}(H)_1$ and $\left(\ell^1(\mathbb{N})\right)_1$, respectively.

Assume that (1) holds. Express S in its singular value decomposition:

$$T\xi = \sum_{j=1}^{r} s_j \langle \xi, \phi_j \rangle \psi_j, \quad \text{for every } \xi \in H, \tag{10.16}$$

where $r = \mathrm{rank}\, S$, $s_1 \geq \cdots \geq s_r > 0$, and $\{\phi_1, \dots, \phi_r\}$ and $\{\psi_1, \dots, \psi_r\}$ are orthonormal systems in H. Let $\{e_n\}_{n \in \mathbb{N}}$ denote the canonical coordinate vectors of $\ell^1(\mathbb{N})$. The extreme points of the closed unit ball of $\ell^1(\mathbb{N})$ are precisely the vectors of the form $e^{i\theta} e_n$, for some $\theta \in \mathbb{R}$ and $n \in \mathbb{N}$ (Exercise 7.35).

Now let $\mathbf{s} \in \ell^1(\mathbb{N})$ be the vector of the singular values of S. We aim to show that $\mathbf{s}$ is an extreme point of the closed unit ball $\left(\ell^1(\mathbb{N})\right)_1$ of $\ell^1(\mathbb{N})$. To this end, assume that $\mathbf{s} = \tau\alpha + (1-\tau)\beta$, for some $\alpha, \beta \in \left(\ell^1(\mathbb{N})\right)_1$ and $\tau \in (0, 1)$. Let A and B be operators defined by the equations

$$A\xi = \sum_{j=1}^{d} \alpha_j \langle \xi, \phi_j \rangle \, \psi_j \quad \text{and} \quad B\xi = \sum_{j=1}^{d} \beta_j \langle \xi, \phi_j \rangle \, \psi_j ,$$

Thus, $S = \tau A + (1-\tau)B$. Furthermore, $A, B \in \mathscr{T}(H)_1$ because $\|A\|_1 = \|\alpha\|_1$ and $\|B\|_1 = \|\beta\|_1$. Hence, $A = B = S$, as S is an extreme point of $\mathscr{T}(H)_1$. But this occurs only if $\alpha = \beta = \mathbf{s}$, which implies that $\mathbf{s}$ is an extreme point of the unit ball of $\ell^1(\mathbb{N})$. Thus, $r = 1$ and $s_1 = 1$, which yields $\operatorname{rank} S = 1$ and $\operatorname{Tr}(S^*S) = 1$.

Conversely, assume that (2) holds: namely, S has rank 1 and $\operatorname{Tr}(S^*S) = 1$. The polar decomposition $S = U|S|$, for some partial isometry W, shows that S and $|S|$ have the same rank. By the Spectral Theorem, $|S|$ and $S^*S = |S|^2$ have the same rank. Now since S^*S has exactly one nonzero eigenvalue and $\operatorname{Tr}(S^*S) = 1$, this sole nonzero eigenvalue is 1, which implies that S^*S is a rank-1 projection.

Let $P = S^*S$ and let $\phi \in H$ be a unit vector that spans the range of P. Define a linear functional $\psi : \mathscr{T}(H) \to \mathbb{C}$ by

$$\psi(T) = \operatorname{Tr}\left(PW^*T\right), \ T \in \mathscr{T}(H).$$

Thus, for any $T \in \mathscr{T}(H)_1$,

$$|\psi(T)| = |\operatorname{Tr}(PW^*T)| = |\langle W^*T\phi, \phi \rangle| \le \|W^*\| \|T\| \le \|T\|_1 \le 1.$$

In particular, $\Re(T) \le 1$ for every $T \in \mathscr{T}(H)$ with $\|T\|_1 \le 1$. The value of ψ at S is $\psi(S) = \operatorname{Tr}(PW^*S) = \operatorname{Tr}(S^*S) = 1$. Suppose now that $S = \tau A + (1-\tau)B$ for some $A, B \in \mathscr{T}(H)_1$ and $\tau \in (0, 1)$. Thus,

$$1 = \Re\psi(S) = \tau\Re\psi(A) + (1-\tau)\Re\psi(B) \le \tau + (1-\tau) = 1.$$

Hence, $\Re\psi(A) = \Re\psi(B) = 1$; however, because $|\psi(A)|$ and $|\psi(B)|$ are at most 1, we deduce that in fact $\psi(A) = \psi(B) = 1$. In particular for A, this means that $\langle W^*A\phi, \phi \rangle = 1$, which is a case of equality in the Cauchy-Schwarz inequality and so $W^*A\phi = \phi$, whence $A\phi = W\phi = WP\phi = W|S|\phi = S\phi$. Furthermore, the equation $A\phi = W\phi$ implies that $\|A\phi\| = \|\phi\| = 1$ and so the operator norm of A is $\|A\| = 1$. Thus, $1 \le \|A\| \le \|A\|_1 \le 1$ yields $1 = \|A\| = \|A\|_1 = 1$. Because $\|A\|$ is the spectral radius of $|A|$ and $\| \, |A| \, \|_1$ is the trace of $|A|$, the equation $\operatorname{spr}|A| = \operatorname{Tr}(|A|) = 1$ implies that $|A|$ has exactly one nonzero eigenvalue (namely, 1). Thus, $|A|$ is a rank-1 projection and $A = V|A|$ (polar decomposition) has rank-1. Hence, if $\omega = S\phi$, then $A\xi = \langle \xi, \phi \rangle \omega = S\xi$ for every $\xi \in H$. By a similar argument, $B = S$ as well. Hence, S is an extreme point of $\mathscr{T}(H)_1$. $\qquad \square$

Problems

10.104. Prove that $T, S \in B(H)$ are equal if and only if $\langle T\xi, \xi \rangle = \langle S\xi, \xi \rangle$, for all $\xi \in H$.

10.105. Prove that $\overline{\operatorname{ran} T} = (\ker T^*)^{\perp}$, for every $T \in B(H)$.

10.106. If S is the unilateral shift operator on $\ell^2(\mathbb{N})$, and if R is an operator on $\ell^2(\mathbb{N})$ for which $RS = SR$ and $RS^* = S^*R$, then prove that R has the form $R = \lambda 1$ for some $\lambda \in \mathbb{C}$.

10.107. For $T \in B(H)$, prove that $\lambda \in \sigma_d(T)$ if and only if $\overline{\lambda} \in \sigma_p(T^*)$.

10.108. Prove that the following statements are equivalent for an operator $U \in B(H)$:

1. U is unitary;
2. $U^*U = UU^* = 1$;
3. for some orthonormal basis $\{\phi_i\}_i$ of H, $\{U\phi_i\}_i$ is also an orthonormal basis;
4. $\{U\phi_i\}_i$ is an orthonormal basis for H for every orthonormal basis $\{\phi_i\}_i$ of H.

10.109. Suppose that $V \in B(H)$ is an isometry.

1. Prove that $\sigma_{\mathrm{ap}}(T) \subseteq \mathbb{T}$.
2. If V is unitary, prove that $\sigma(T) \subseteq \mathbb{T}$.

10.110. Prove that if $P \in B(H)$ is a projection different from 0 and 1, then $\sigma(P) = \{0, 1\}$.

10.111. Suppose that $P \in B(H)$ is a nonzero projection acting on a separable Hilbert space H and that $\{\phi_k\}_{k=1}^d$ is an orthonormal basis for the range of P, where d is either finite or infinite. Prove that

$$P\xi = \sum_{k=1}^{d} \langle \xi, \phi_k \rangle \phi_k,$$

for every $\xi \in H$.

10.112. Assume that $T \in B(H)$ is an operator of rank $m \in \mathbb{N}$.

1. Prove that if $m = 1$, then there are unit vectors $\gamma, \eta \in H$ such that $T\xi = \langle \xi, \gamma \rangle \eta$, for all $\xi \in H$.
2. Prove that the rank of T^* is m.

10.113. Prove that if $\{\phi_k\}_{k=0}^{\infty}$ is the canonical orthonormal basis of the Hardy space $H^2(\mathbb{T})$ and if $S \in B(H^2(\mathbb{T}))$ is the unilateral shift operator, then S^* satisfies $S\phi_k = \phi_{k-1}$, for all $k \in \mathbb{N}$ and $S^*\phi_0 = 0$.

10.114. With respect to the canonical orthonormal basis $\{\phi_k\}_{k=0}^{\infty}$ of the Hardy space $H^2(\mathbb{T})$, find the matrix representation $\mathscr{S}$ of the unilateral shift operator

$S \in B(H^2(\mathbb{T}))$. Viewing $\mathscr{S}$ as acting on $\ell^2(\mathbb{N} \cup \{0\})$, determine the action of the matrix $\mathscr{S}$ on a vector $\xi \in \ell^2(\mathbb{N} \cup \{0\})$.

10.115. Prove that if $T \in B(H)$ is hermitian, then

$$\|T\| = \sup_{\|\xi\|=1} |\langle T\xi, \xi \rangle|.$$

10.116. Assume that $T \in B(H)$ is hermitian and let

$$m_\ell = \inf_{\|\xi\|=1} \langle T\xi, \xi \rangle \quad \text{and} \quad m_u = \sup_{\|\xi\|=1} \langle T\xi, \xi \rangle.$$

Complete the proof of Proposition 10.26 by proving the following statements.

1. $m_u \in \sigma(T)$.
2. $\sigma(T) \subseteq (-\infty, m_u]$.
3. $\sigma(T) \subseteq [m_\ell, m_u]$.

10.117. Suppose that T is a hermitian operator and that $\sigma(T) = \sigma_1 \cup \sigma_2$, where σ_1 and σ_2 are compact subsets of $\mathbb{R}$ with $\sigma_1 \cap \sigma_2 = \emptyset$. Prove that there are subspaces M_1 and M_2 of H such that

1. $H = M_1 \oplus M_2$,
2. M_j is invariant under T, for $j = 1, 2$,
3. the operator $T_{|M_j}$ is hermitian and has spectrum σ_j, for $j = 1, 2$.

10.118. Assume that H is a separable Hilbert space and that for an operator $K \in B(H)$ there are a bounded sequence $\{\lambda_j\}_{j=1}^r$ of nonzero real numbers, where r is finite or infinite, and a sequence $\{\phi_j\}_{j=1}^r$ of pairwise-orthogonal unit vectors in H such that

1. $K\phi_j = \lambda_j \phi_j$, for each j,
2. $\lim_j \lambda_j = 0$, if r is infinite, and
3. $K\xi = \sum_{j=1}^r \lambda_j \langle \xi, \phi_j \rangle \phi_j$, for every $\xi \in H$.

Prove that K is a compact operator.

10.119. Prove that the following statements are equivalent for $T \in B(H)$:

1. T is normal;
2. $\|T^*\xi\| = \|T\xi\|$, for all $\xi \in H$;
3. $T^*T = TT^*$.

10.120. Prove that if λ is an isolated point in the spectrum of a normal operator N, then λ is an eigenvalue of N.

10.121. Let B denote the bilateral shift operator on $L^2(\mathbb{T})$. Prove that the Hardy space $H^2(\mathbb{T})$ is invariant under B and that, with respect to the decomposition

$L^2(\mathbb{T}) = H^2(\mathbb{T}) \oplus H^2(\mathbb{T})^\perp$, B is represented by an operator matrix of the form

$$\begin{bmatrix} S & B_{12} \\ 0 & B_{22} \end{bmatrix},$$

where S is the unilateral shift operator on $H^2(\mathbb{T})$ and $B_{12} \neq 0$.

10.122. Prove that the following statements are equivalent for an operator T and projection P on a Hilbert space H:

1. $\operatorname{ran} P$ is reducing for T;
2. $(1-P)T^*P = (1-P)TP = 0$;
3. $TP = PT$.

10.123. A subspace $L \subset H$ is said to be *nontrivial* if L is neither $\{0\}$ nor H. Suppose that $T \in B(H)$.

1. Prove that if H has finite dimension, then T has a nontrivial invariant subspace. (Hint: think about eigenvectors.)
2. Prove that if H is nonseparable, then T has a nontrivial invariant subspace. (Hint: if $\xi \in H$ is nonzero, consider the subspace generated by $T^k \xi$ for $k \in \mathbb{N}$.)

10.124. Prove that every compact normal operator is reductive.

10.125. Let S be the unilateral shift operator on $\ell^2(\mathbb{N})$.

1. Prove that if $T \in B(\ell^2(\mathbb{N}))$ satisfies $TS = ST$ and $TS^* = S^*T$, then $T = \lambda 1$ for some $\lambda \in \mathbb{C}$.
2. Prove that the only subspaces $L \subseteq \ell^2(\mathbb{N})$ that are invariant under both S and S^* are $L = \{0\}$ and $L = \ell^2(\mathbb{N})$.

10.126. Assume that $N \in B(H)$ is normal. If $\lambda_1, \lambda_2 \in \sigma_{\mathrm{ap}}(N)$ are distinct, and if $\xi_n, \eta_n \in H$ are unit vectors for which

$$\lim_n \|(N - \lambda_1 1)\xi_n\| = \lim_n \|(N - \lambda_2 1)\eta_n\| = 0,$$

then prove that

$$\lim_n \langle \xi_n, \eta_n \rangle = 0.$$

10.127. Prove that if S and T are hermitian operators for which $S \leq T$, then $X^*SX \leq X^*TX$ for every operator $X \in B(H)$.

10.128. Prove that if A is hermitian and X is invertible, then A is positive if and only if X^*AX is positive.

10.129. Prove that, for every operator $T \in B(H)$, the operator $1 + T^*T$ is invertible and positive, and

$$(1 + T^*T)^{-1/2} T (1 + T^*T)^{1/2} = T.$$

10.130. Suppose that A and B are positive operators such that $AB = BA = 0$. Prove that if A' and B' are positive operators such that $A' \leq A$ and $B' \leq B$, then $A'B' = B'A' = 0$.

10.131. Let $T = \begin{bmatrix} 0 & 1 \\ 0 & 0 \end{bmatrix} \in B(\mathbb{C}^2)$.

1. Prove that $\sigma(T) = \{0\}$.
2. Prove that there are no operators $R \in B(\mathbb{C}^2)$ such that $R^2 = T$.

10.132. Let $H = \mathbb{C}^2$ and consider hermitian operators $A, B \in B(H)$.

1. If $A = \begin{bmatrix} \alpha & \gamma \\ \overline{\gamma} & \beta \end{bmatrix} \in B(H)$, where $\alpha, \beta, \gamma \in \mathbb{C}$, then prove that A is positive if and only if α and β are nonnegative real numbers such that $\alpha\beta \geq \|\gamma\|^2$.
2. Given an example of positive $A, B \in B(H)$ for which $A \leq B$ but $A^2 \not\leq B^2$.

10.133. Prove that if $A \in B(H)$ is positive and $\|A\| \leq 1$, then

$$P = \begin{bmatrix} A & (A(1-A))^{1/2} \\ (A(1-A))^{1/2} & 1-A \end{bmatrix}$$

is a projection operator on $H \oplus H$.

10.134. The *commutant* of a nonempty subset $\mathscr{S} \subseteq B(H)$ is the set

$$\mathscr{S}' = \{T \in B(H) \mid ST = TS \ \forall \, s \in \mathscr{S}\}.$$

1. Prove that $\mathscr{S}'$ is closed in the weak operator topology.
2. Prove that $\mathscr{S}'$ is an associative subalgebra of $B(H)$.
3. Prove that if $S^* \in \mathscr{S}$ for every $S \in \mathscr{S}$, then $T^* \in \mathscr{S}'$ for every $T \in \mathscr{S}'$.

10.135. Prove that if H is an infinite-dimensional Hilbert space, then the closure of the set $\{P \in B(H) \mid P \text{ is a projection}\}$ in the weak operator topology of $B(H)$ is the set of all positive operators $A \in B(H)$ for which $\|A\| \leq 1$.

10.136 (von Neumman's Inequality). Use the Sz.-Nagy Dilation Theorem to prove that if $T \in B(H)$ satisfies $\|T\| \leq 1$, then $\|f(T)\| \leq 1$ for every polynomial $f \in \mathbb{C}[t]$ for which $\max_{|z| \leq 1} |f(z)| \leq 1$.

10.137. Prove that if $T \in B(H)$ is invertible, then $T = U|T|$ for some unitary operator $U \in B(H)$.

10.138. Prove that if $V \in B(H)$ is a partial isometry, then V^*V is a projection.

10.139. Suppose that $T_1, T_2 \in B(H)$ are hermitian operators such that $T_1 T_2 = T_2 T_1 = 0$. Prove that $\langle \xi, \eta \rangle = 0$ for all $\xi \in \operatorname{ran} T_1$ and $\eta \in \operatorname{ran} T_2$.

10.140. Assume that $T \in B(H)$ and $\lambda \in \mathbb{C}$. Prove that

$$\ker(T - \lambda 1) \in \operatorname{Lat} T \quad \text{and} \quad \overline{\operatorname{ran}(T - \lambda 1)} \in \operatorname{Lat} T.$$

10.141. Prove that, for every $T \in B(H)$,

$$M \in \operatorname{Lat} T \quad \text{if and only if} \quad M^\perp \in \operatorname{Lat} T^* .$$

10.142. Prove that the involution $T \mapsto T^*$ is a continuous function on $B(H)$ with respect to the weak operator topology.

10.143. Let $\psi : [0,1] \to [0,1]$ be given by $\psi(t) = t$ and consider the (hermitian) multiplication operator M_ψ on $L^2([0,1], \mathfrak{M}, m)$.

1. Prove that M_ψ has no eigenvalues.
2. Prove that M_ψ has no finite-dimensional invariant subspaces.
3. Find one nontrivial subspace of $L^2([0,1], \mathfrak{M}, m)$ that is invariant under M_ψ.

10.144. Prove that the space $K(H)$ compact operators on a separable Hilbert space H is separable.

10.145. Prove that the space $T(H)$ trace-class operators on a separable Hilbert space H is separable with respect to the trace norm $\| \cdot \|_1$.

10.146. Assume that T_{ij}, for $i,j = 1,2$, are trace-class operators acting on a separable Hilbert space H.

1. Prove that $T = \begin{bmatrix} T_{11} & T_{12} \\ T_{21} & T_{22} \end{bmatrix}$ is a trace-class operator on $H \oplus H$.
2. Prove that $\|T_{11}\|_1 + \|T_{22}\|_1 \leq \|T\|_1$.

10.147. Assume that H is an infinite-dimensional separable Hilbert space and that $\varphi : B(H) \to \mathbb{C}$ is a linear transformation.

1. Prove that if $B(H)$ has the weak operator topology, then φ is continuous if and only if there are finite sets $\{\xi_1, \dots, \xi_n\}$ and $\{\eta_1, \dots, \eta_n\}$ of vectors such that

$$\varphi(T) = \sum_{j=1}^{n} \langle T\xi_j, \eta_j \rangle,$$

for every $T \in B(H)$.
2. Prove that if $B(H)$ has the ultraweak operator topology, then φ is continuous if and only if there are sequences $\{\xi_j\}_{j \in \mathbb{N}}$ and $\{\eta_j\}_{j \in \mathbb{N}}$ of vectors such that

$$\sum_{j=1}^{\infty} \|\xi_j\|^2 \quad \text{and} \quad \sum_{j=1}^{\infty} \|\eta_j\|^2$$

converge and

$$\varphi(T) = \sum_{j=1}^{\infty} \langle T\xi_j, \eta_j \rangle,$$

for every $T \in B(H)$.

Chapter 11
Algebras of Hilbert Space Operators

Collections of Hilbert space operators lead to a rich palette of algebraic structures. Of principal interest in this chapter are certain associative algebras of operators, called $*$-algebras, that are closed under the involution $T \mapsto T^*$. However, one could also quite readily consider other algebraic structures, such as semigroups of operators, Lie algebras of operators, or vector spaces of operators. Our focus on $*$-algebras of Hilbert space operators (and their abstractions known as C^*-algebras) stems from the fact that such algebras are widely employed and studied, and exhibit special features that are not present in more generic algebraic structures. The monograph of Paulsen [42] has a good treatment of the theory of general operator algebras and discusses a wide variety of applications to operator theory.

Operator algebras, in the sense formulated in this chapter, are sometimes considered as the basis for noncommutative topology and noncommutative measure theory, thereby completing the arc of this text by bringing us back to the book's topological and measure-theoretic beginnings.

11.1 Examples

Definition 11.1. A $*$-algebra of operators is a subset $A \subseteq B(H)$ that is closed under addition, product, scalar multiplication, and the adjoint operation. Furthermore, a $*$-algebra A is called:

1. a C^*-algebra of operators, if A is closed with respect to the norm topology of $B(H)$, and
2. a von Neumann algebra if A is closed with respect to the strong operator topology of $B(H)$.

By the term *operator algebra* we shall henceforth mean a $*$-algebra of Hilbert space operators.

© Springer International Publishing Switzerland 2016
D. Farenick, *Fundamentals of Functional Analysis*, Universitext,
DOI 10.1007/978-3-319-45633-1_11

One might wonder why there is not a third category of operator algebra, defined by the requirement that it be closed with respect to the weak operator topology. The reason for there being no third category of $*$-algebra is that nothing new is gained: if A is a $*$-algebra of operators, then the closure of A in the weak operator topology coincides with the closure of A in the strong operator topology, by the convexity of A and Proposition 10.77. Thus, every von Neumann algebra is closed with respect to the weak operator topology.

Notational Convention In keeping with notation that is standard in the operator algebra literature, lowercase letters are used in this chapter to denote individual operators, whereas uppercase letters denote algebras of operators.

Definition 11.2. A $*$-algebra A of operators is *abelian* if $xy = yx$ for all $x, y \in A$.

Before considering operator algebra theory, a few basic examples are considered.

Example 11.3. *Group Algebras.*

Proof. Suppose that G is a countable discrete group, with identity element e. Define a Hilbert space $\ell^2(G)$ by

$$\ell^2(G) = \{f : G \to \mathbb{C} \mid \sum_{h \in G} |f(h)|^2 < \infty\}.$$

An orthonormal basis for this Hilbert space is given by the set $\{\delta_g\}_{g \in G}$, where $\delta_g(h) = 1$ if $h = g$ and 0 if $h \neq g$. Thus,

$$\ell^2(G) = \left\{ \sum_{g \in G} \alpha_g \delta_g \mid \alpha_g \in \mathbb{C}, \sum_{g \in G} |\alpha_g|^2 < \infty \right\},$$

and the inner product on $\ell^2(G)$ is given by

$$\left\langle \sum_g \alpha_g \delta_g, \sum_{g \in G} \beta_g \delta_g \right\rangle = \sum_{g \in G} \alpha_g \overline{\beta}_g.$$

For each $h \in G$, let $\lambda_h : \ell^2(G) \to \ell^2(G)$ be the operator that sends $f \in \ell^2(G)$ to the function whose value at $k \in G$ is $f(h^{-1}k)$. Note that λ_h is an isometry, and that $\lambda_{h^{-1}} = \lambda_h^{-1} = \lambda_h^*$; thus, λ_h is a unitary operator, for each $h \in G$.

The action of λ_h on the basis elements of $\ell^2(G)$ is given by $\lambda_h[\delta_g] = \delta_{hg}$, and so λ, considered as a map $G \to B\left(\ell^2(G)\right)$ in which $h \mapsto \lambda_h$, is called the *left regular representation* of G.

The group algebra $\mathbb{C}[G]$, which is the set all products of finite linear combinations of elements from the set $\{\lambda_h \mid h \in G\}$, is a $*$-algebra of operators. The norm-closure in $B\left(\ell^2(G)\right)$ of $\mathbb{C}[G]$ is denoted by $C_\lambda^*(G)$, and is called the *reduced group C^*-algebra of G*. The SOT-closure of $\mathbb{C}[G]$ is denoted by $V_\lambda(G)$, and is called the *group von Neumann algebra of G.* Evidently, $C^*_\lambda(G)$ is a C*-algebra of operators, and $V_\lambda(G)$ is a von Neumann algebra of operators.

There is also the possibility of allowing G to act on the right, leading to operators ρ_h on $\ell^2(G)$ for which $\rho_h(\delta_g) = \delta_{gh}$, for every $g \in G$. Hence, once again each ρ_h is a unitary operator, and the map $\rho : G \to B(H)$ is the *right regular representation* of G. As with the left regular representation, one obtains a C*-algebra $C^*_\rho(G)$ and a von Neumann algebra $V_\rho(G) = \overline{C^*_\rho(G)}^{SOT}$. □

Observe that $C^*_\lambda(G)$ (equivalently, $V_\lambda(G)$) is an abelian operator algebra if and only if G is an abelian group. We shall mainly concern ourselves with $C^*_\lambda(G)$ and $V_\lambda(G)$, using $V_\rho(G)$ only for the purpose of studying $V_\lambda(G)$.

Example 11.4. *The C*-algebra $K(H)$ of compact operators.*

Proof. Theorem 8.35 shows that $K(H)$ is a Banach algebra of operators, while Proposition 8.31 implies that $K(H)$ is closed under the adjoint. Thus, $K(H)$ is a C*-algebra of operators.

Note, however, that $K(H)$ is not a von Neumann algebra if H has infinite dimension, as $1 \notin K(H)$ but 1 is the SOT-limit of a net of compact operators (Exercise 11.107). □

Example 11.5. *Matrix operator algebras.*

Proof. Suppose that A is a *-algebra of operators acting on H, and consider the set $M_n(A)$ of $n \times n$ matrices with entries from A. Given a matrix $X = [a_{ij}]^n_{i,j=1} \in M_n(A)$, define X^* by $X = [a^*_{ji}]^n_{ij,=1}$. The map $X \mapsto X^*$ is an involution on $M_n(A)$ and, thus under the usual algebra operations on matrices, $M_n(A)$ is a *-algebra of operator acting on the direct sum $H^{(n)}$ of n copies of H.

If A is norm closed, then so is $M_n(A)$. To prove this, let $X = [a_{ij}]_{i,j}$ be in the norm-closure of $M_n(A)$. Thus, there is a sequence of elements $X_k = [a^{(k)}_{ij}]_{i,j}$ such that $\|X - X_k\| \to 0$ as $k \to \infty$. Hence, for a given pair of i and j, $\|a_{ij} - a^{(k)}_{ij}\| \to 0$ as $k \to \infty$. Thus, $a_{ij} \in A$, and therefore $X \in M_n(A)$.

By analogy, the same type of argument in which a SOT-convergent net of matrices is reduced to n^2 SOT-convergent nets of operators shows that $M_n(A)$ is SOT-closed if A is SOT-closed. □

Definition 11.6. A *-algebra A of operators acting on a Hilbert space H is *unital* if the identity operator $1 \in B(H)$ is an element of A.

In the examples above, the group algebra $\mathbb{C}[G]$ is unital, where the identity on the Hilbert space $\ell^2(G)$ is given by the identity element of the group G; consequently, $C^*_\lambda(G)$ and $V_\lambda(G)$ are unital operator algebras. If H has infinite dimension, then $K(H)$ is not unital, whereas if A is a unital *-algebra, then so is the matrix algebra $M_n(A)$, where the identity of $M_n(A)$ is the diagonal matrix in which each entry is the identity of A.

Definition 11.7. If $\mathscr{S} \subseteq B(H)$, then the *-algebra generated by $\mathscr{S}$ is the smallest *-subalgebra of $B(H)$ that contains $\mathscr{S}$, and is denoted by *-Alg $\mathscr{S}$.

The definitions of C^*-algebra and von Neumann algebra generated by a set of operators are straightforward; however, note that the definition of von Neumann algebra requires that it be unital.

Definition 11.8. Assume that $\mathscr{S}$ is a set of operators acting on a Hilbert space H.

1. The C^*-*algebra generated by* $\mathscr{S}$ is denoted by $C^*(\mathscr{S})$ and is defined to be the norm-closure of $*$-$\mathrm{Alg}\,\mathscr{S}$.
2. The *unital* C^*-*algebra generated by* $\mathscr{S}$ is denoted by $C^*(\mathscr{S}, 1)$ and is defined to be the norm-closure of $*$-$\mathrm{Alg}\,(\mathscr{S} \cup \{1\})$.
3. The *von Neumann algebra generated by* $\mathscr{S}$ is denoted by $W^*(\mathscr{S})$ and is defined to be the SOT-closure of $*$-$\mathrm{Alg}\,(\mathscr{S} \cup \{1\})$.

If $\mathscr{S}$ consists of a single operator, x, then

$$*\text{-}\mathrm{Alg}(x) = \left\{ \sum_{i=0}^{n}\sum_{j=1}^{m}\alpha_{ij}x^i(x^*)^j + \sum_{k=0}^{s}\sum_{\ell=1}^{t}\beta_{k\ell}(x^*)^k x^\ell \,\middle|\, n,m,s,t \in \mathbb{N},\, \alpha_{ij},\beta_{k\ell} \in \mathbb{C} \right\}$$

and

$$*\text{-}\mathrm{Alg}(x,1) = \left\{ \sum_{i=0}^{n}\sum_{j=0}^{m}\alpha_{ij}x^i(x^*)^j + \sum_{k=0}^{s}\sum_{\ell=0}^{t}\beta_{k\ell}(x^*)^k x^\ell \,\middle|\, n,m,s,t \in \mathbb{N},\, \alpha_{ij},\beta_{k\ell} \in \mathbb{C} \right\}.$$

In particular, any one of the algebras $*$-$\mathrm{Alg}(x)$, $C^*(x)$, or $W^*(x)$ is abelian if and only if $x^*x = xx^*$.

Example 11.9. *Algebras of multiplication operators.*

Proof. Assume that (X, Σ, μ) is a σ-finite measure space and consider the Hilbert space $L^2(X, \Sigma, \mu)$. For every essentially bounded Borel function $\psi : X \to \mathbb{C}$, the multiplication operator M_ψ on H has norm $\|M_\psi\| = \|\dot{\psi}\|$ (see Section 8.1) and adjoint $M_\psi^* = M_{\overline{\psi}}$ (see Section 10.2). It is clear that $M_{\psi_1\psi_2} = M_{\psi_1}M_{\psi_2} = M_{\psi_2}M_{\psi_1}$ and that $M_{\alpha_1\psi_1+\alpha_2\psi_2} = \alpha_1 M_{\psi_1} + \alpha_2 M_{\psi_2}$, and so the set

$$\{M_\psi \,|\, \dot{\psi} \in L^\infty(X, \Sigma, \mu)\}$$

is an abelian $*$-algebra of operators. Because the map $L^\infty(X, \Sigma, \mu) \to B(L^2(X, \Sigma, \mu))$ given by $\dot{\psi} \mapsto M_\psi$ is multiplicative and a linear isometry, this set is a norm closed algebra, isometrically isomorphic to $L^\infty(X, \Sigma, \mu)$. (This isomorphism is also compatible with the conjugation on $L^\infty(X, \Sigma, \mu)$ and involution on $B(L^2(X, \Sigma, \mu))$ in the sense that $M_{\overline{\psi}} = M_\psi^*$.) Hence, the set of all such multiplication operators on $L^2(X, \Sigma, \mu)$ form an abelian C^*-algebra. □

11.2 von Neumann Algebras

One of the most immediate ways by which examples of von Neumann algebras are obtained is through the use of commutants. Recall from Exercise 10.134 that the *commutant* of a nonempty subset $\mathscr{S} \subseteq B(H)$ is the set

$$\mathscr{S}' = \{y \in B(H) \,|\, sy = ys, \ \forall\, s \in \mathscr{S}\}.$$

Furthermore, Exercise 10.134 shows that if $s^* \in \mathscr{S}$ for every $s \in \mathscr{S}$, then $\mathscr{S}'$ is a unital C*-subalgebra of $B(H)$, closed in the weak operator topology.

Example 11.10. *If G is a countable discrete group, then $V_\rho(G) \subseteq V_\lambda(G)'$.*

Proof. Select any basis element δ_g of $\ell^2(G)$. If $h, k \in G$, then

$$\lambda_h \rho_k[\delta_g] = \lambda_h[\delta_{gk}] = \delta_{hgk} = \rho_k[\delta_{hg}] = \rho_k \lambda_h[\delta_g].$$

Hence, $\lambda_h \rho_k = \rho_k \lambda_h$. As the choice of $h, k \in G$ is arbitrary, and because commutants are closed in the weak operator topology, $xy = yx$ for $x \in V_\lambda(G)$ and $y \in V_\rho(G)$. $\square$

Definition 11.11. The *double commutant* of a nonempty subset $\mathscr{S} \subseteq B(H)$ is the set denoted by $\mathscr{S}''$ and defined by $\mathscr{S}'' = (\mathscr{S}')'$.

Observe that it always the case that $\mathscr{S} \subseteq \mathscr{S}''$.

The following example shows how the use of commutants can play a role in showing that certain C*-algebras are von Neumann algebras.

Example 11.12. *If (X, Σ, μ) is a finite measure space, then $L^\infty(X, \Sigma, \mu)$ is a von Neumann algebra, when considered as a C*-algebra of multiplication operators acting on the Hilbert space $L^2(X, \Sigma, \mu)$.*

Proof. Let $H = L^2(X, \Sigma, \mu)$ and let $M = \{M_\psi \,|\, \psi \in \mathscr{L}^\infty(X, \Sigma, \mu)\}$. Example 11.9 shows that M is a unital C*-algebra of operators acting on H. As M is abelian, $M \subseteq M'$.

Conversely, suppose that $z \in M'$. Let $\dot{f} \in H$ by given by $\dot{f} = z(\dot{\chi}_X)$, where $\chi_X \in \mathscr{L}^\infty(X, \Sigma, \mu)$ is the characteristic function of X. Select any $\psi \in \mathscr{L}^\infty(X, \Sigma, \mu)$; because $M_\psi z = z M_\psi$, we have that

$$(\dot{\psi f}) = M_\psi \dot{f} = M_\psi z(\dot{\chi}_X) = z M_\psi(\dot{\chi}_X) = z(\dot{\psi}).$$

Thus, $\|(\dot{\psi f})\|^2 = \|z(\dot{\psi})\|^2 \leq \|z\|^2 \|\dot{\psi}\|^2$. Now if $E \in \Sigma$, then

$$\int_E |f|^2 \, d\mu = \|z(\dot{\chi}_E)\|^2 \leq \|z\|^2 \mu(E).$$

For each $\alpha > 0$, let $E_\alpha = |f|^{-1}((\alpha, \infty))$. Thus, the inequality above yields

$$\alpha^2 \mu(E_\alpha) \leq \int_{E_\alpha} |f|^2 \, d\mu \leq \|z\|^2 \mu(E_\alpha).$$

Therefore, if $\mu(E_\alpha) > 0$, then $\alpha \leq \|z\|$. Hence, $f \in \mathscr{L}^\infty(X, \Sigma, \mu)$ and the multiplication operator M_f and the operator z agree on $L^\infty(X, \Sigma, \mu)$. Because $L^\infty(X, \Sigma, \mu)$ is dense in $L^2(X, \Sigma, \mu)$, the operators M_f and z necessarily agree on all of H, which shows that $z \in M$. Hence, $M' \subseteq M$, which proves that $M' = M$.

Because $M = M'$, M is necessarily closed in the weak operator topology, by Exercise 10.134. Hence, M is also closed in the strong operator topology, and so M is a von Neumann algebra.   □

It is also true that $L^\infty(X, \Sigma, \mu)$ is a von Neumann algebra if (X, Σ, μ) is a σ-finite measure space (Exercise 11.108). However, there exist examples of measure spaces (X, Σ, μ) that are not σ-finite and for which the C^*-algebra $L^\infty(X, \Sigma, \mu)$ of multiplication operators is not a von Neumann algebra.

The role of the commutant in Example 11.12 is not accident, as the following fundamental theorem of von Neumann demonstrates.

Theorem 11.13 (Double Commutant Theorem). *The following statements are equivalent for a unital C^*-subalgebra $M \subseteq B(H)$:*

1. $M = M''$;
2. M is a von Neumann algebra.

Proof. By Exercise 10.134, if $M = M''$, then M is necessarily closed in the weak operator topology; hence, M is also closed in the strong operator topology, thereby implying that M is a von Neumann algebra.

Conversely, suppose that M is closed with respect to the strong operator topology. Because $M \subseteq M''$, we need only prove that $M'' \subseteq M = \overline{M}^{SOT}$. Let $y \in M''$ and assume that $U \subset B(H)$ is a SOT-open set containing y. Thus, there exist $\varepsilon > 0$ and $\xi_1, \ldots, \xi_n \in H$ such that

$$B_{\varepsilon, \xi_1, \ldots, \xi_n}(y) = \{x \in B(H) \mid \|y\xi_k - x\xi_k\| < \varepsilon, \ k = 1, \ldots, n\} \subseteq U.$$

Consider the Hilbert space $H^{(n)}$ obtained as the n-fold direct sum of H, and let $\xi = \xi_1 \oplus \cdots \oplus \xi_n \in H^{(n)}$. Let $\tilde{M} \subseteq B(H^{(n)})$ be the set of all $n \times n$ diagonal matrices D of operators in which each diagonal entry d_{jj} is a fixed element of M: that is, $\tilde{M}$ consists of all operators of the form $D_x = x \oplus \cdots \oplus x$, where $x \in M$. The commutant of $\tilde{M}$ consists of all $n \times n$ matrices of operators such that each entry of the matrix is an element of M', whereas $\tilde{M}''$ consists of all $n \times n$ diagonal operators D whose entries come from M'' and which satisfy $d_{ii} = d_{jj}$, for all $1 \leq i, j \leq n$. Hence, $\tilde{M} = \tilde{M}''$.

Let $p \in B(H^{(n)})$ be the projection onto the closure L_1 of the linear submanifold $L_0 = \{D_x \xi \mid x \in M\}$. If $z \in M$, then $D_z(D_x \xi) \in L_0$ for all $x \in M$, and so L_0 and its closure, L_1, are invariant under the algebra $\tilde{M}$. Because $\tilde{M}$ is self-adjoint, the invariance of L_1 under $\tilde{M}$ implies that $p \in \tilde{M}'$; hence, $pD_y = D_yp$. In addition, $\xi \in L_0$ because $1 \in M$, and so $D_y\xi = D_y(p\xi) = p(D_y\xi) \in L_1$, which implies that $D_y\xi$ is within ε of some vector in L_0. That is, $\|D_y\xi - D_x\xi\|^2 < \varepsilon^2$ for some $x \in M$, which implies that the basic SOT-open neighbourhood $B_{\varepsilon, \xi}(y)$ of y intersects M. Hence, $S \in \overline{M}^{SOT} = M$, thereby proving that $M'' \subseteq M$.   □

The Double Commutant Theorem has an interesting consequence for the polar decomposition of operators.

Proposition 11.14. *If M is a von Neumann algebra and if the polar decomposition of $x \in M$ is denoted by $x = v|x|$, then both $|x|$ and v are elements of M.*

Proof. Because $|x| = (x^*x)^{1/2} = \lim_n f_n(x^*x)$ for some sequence of polynomials f_n for which $f_n(0) = 0$ for every $n \in \mathbb{N}$, we see that $|x| \in C^*(x) \subseteq M$. Therefore, it remains to show that $v \in M$.

Suppose that $y \in C^*(x, 1)'$. If $\xi \in \ker x$, then $xy\xi = yx\xi = 0$ implies that $yx\xi \in \ker x$, and so $\ker x$ is invariant under y. Equation (10.12) in the proof of the polar decomposition (Theorem 10.65) shows that $\| \, |x|\xi \, \| = \|x\xi\|$ for all $\xi \in H$, and so $\ker|x| = \ker x$. Therefore, the restriction of y to the y-invariant subspace $\ker|x|$ commutes with the restriction of x to x-invariant subspace $\ker|x|$.

Suppose now that $\eta \in \operatorname{ran}|x|$, say $\eta = |x|\gamma$ for some $\gamma \in H$. Thus,

$$vy\eta = vy|x|\gamma = v|x|(y\gamma) = xy\gamma = yx\gamma = yv(|x|\gamma) = yv\eta.$$

Therefore, v and y commute on the linear submanifold $\operatorname{ran}|x|$, and so they commute on the subspace $\overline{\operatorname{ran}|x|}$. Because $H = \overline{\operatorname{ran}|x|} \oplus \ker|x|$, we deduce that y and v commute on H.

Hence, v lies in the commutant of $C^*(x, 1)'$, and so by the Double Commutant Theorem,

$$v \in C^*(x, 1)'' = \overline{C^*(x, 1)}^{\,SOT} \subseteq M,$$

which proves the result. $\square$

Note that the proof of Proposition 11.14 in fact shows that v and $|x|$ belong to the von Neumann algebra $W^*(x)$ generated by x.

The final fundamental result in von Neumann algebra theory considered in this section concerns approximation. If A is a untial C^*-algebra of operators, and if one considers the von Neumann algebra $M = A'' = \overline{A}^{\,SOT}$ generated by A, then it far from apparent, for example, that an operator in the closed unit ball of M is the SOT-limit of a net of operators from the closed unit ball of A. That such a fact is true is part of what the density theorem of Kaplansky (Theorem 11.16) asserts below.

The proof of Theorem 11.16 requires the following technical fact.

Lemma 11.15. *Let $B(H)_{\mathrm{sa}} = \{x \in B(H) \,|\, x^* = x\}$ and define $f : B(H)_{\mathrm{sa}} \to B(H)_{\mathrm{sa}}$ by $f(x) = (1 + x^2)^{-1/2}(2x)(1 + x^2)^{-1/2}$, for $x \in B(H)_{\mathrm{sa}}$. Then:*

1. $f(x) = 2x(1 + x^2)^{-1}$ and $\|f(x)\| \leq 1$, for every $x \in B(H)_{\mathrm{sa}}$, and
2. f is continuous with respect to the strong operator topology.

Proof. If $x^* = x$, then $\sigma(1 + x^2) = \{1 + \lambda^2 \,|\, \lambda \in \sigma(x)\}$, by the Spectral Mapping Theorem. Thus, as $\sigma(x) \subset \mathbb{R}$, $0 \notin \sigma(1 + x^2)$. Because x commutes with $1 + x^2$, x also commutes with any continuous function in x, including $(1 + x^2)^{-1/2}$. Hence, $f(x) = (1 + x^2)^{-1/2}(2x)(1 + x^2)^{-1/2} = 2x(1 + x^2)^{-1}$.

The spectral radius of $1 + x^2$ is at least 1, and so the spectral radius of $(1 + x^2)^{-1}$ is at most 1; therefore, $\|(1 + x^2)^{-1}\| \leq 1$. Furthermore, elementary calculus shows that the real-valued function $t \mapsto \frac{2t}{1+t^2}$ is strictly increasing on the interval $[-1, 1]$ and has an absolute maximum value of 1 over all $t \in \mathbb{R}$. Therefore, the equality of the norm and spectral radius for hermitian operators yields

$$\sup_{x \in B(H)_{\mathrm{sa}}} \|f(x)\| \leq 1.$$

To prove (2), note that if $y, z \in B(H)_{\mathrm{sa}}$, then

$$f(y) - f(z) = 2(1 + y^2)^{-1} \left(y(1 + z^2) - (1 + y^2)z\right)(1 + z^2)^{-1}$$

$$= 2(1 + y^2)^{-1} (y - z + y(z - y)z)(1 + z^2)^{-1}.$$

Hence, for every $\xi \in H$,

$$\|f(y)\xi - f(z)\xi\| \leq \left\|2(1 + y^2)^{-1}(y - z)(1 + z^2)^{-1}\xi\right\|$$

$$+ \left\|2(1 + y^2)^{-1}(y(z - y)z)(1 + z^2)^{-1}\xi\right\|.$$

Now fix $z \in B(H)_{\mathrm{sa}}$ and suppose that $(y_\alpha)_{\alpha \in \Lambda}$ is a net of hermitian operators that converges in the strong operator topology to z. Because $\|(1 + y_\alpha^2)^{-1}\| \leq 1$, for each $\xi \in H$ we obtain

$$\left\|2(1 + y_\alpha^2)^{-1}(y_\alpha - z)(1 + z^2)^{-1}\xi\right\| \leq 2\|(y_\alpha - z)((1 + z^2)^{-1}\xi)\|,$$

and so $\lim_\alpha \left\|2(1 + y_\alpha^2)^{-1}(y_\alpha - z)(1 + z^2)^{-1}\xi\right\| = 0$. Similarly,

$$\left\|2(1 + y_\alpha^2)^{-1}(y_\alpha(z - y_\alpha)z)(1 + z^2)^{-1}\xi\right\| \leq \|f(y_\alpha)\| \left\|(z - y_\alpha)[z(1 + z^2)^{-1}\xi]\right\|$$

$$\leq \left\|(z - y_\alpha)[z(1 + z^2)^{-1}\xi]\right\|.$$

Thus, $\lim_\alpha \left\|2(1 + y_\alpha^2)^{-1}(y_\alpha - z)(1 + z^2)^{-1}\xi\right\| = 0$.

Combining the two limiting argument above yields $\lim_\alpha \|f(y_\alpha)\xi - f(z)\xi\| = 0$, which shows that f is continuous with respect to the strong operator topology. $\square$

Theorem 11.16 (Density Theorem). *Assume that A is a unital C^*-subalgebra of $B(H)$ and that $M = \overline{A}^{SOT}$. If*

$$A_1 = \{a \in A \mid \|a\| \leq 1\}, \quad A_{1,\mathrm{sa}} = \{a \in A \mid a^* = a, \|a\| \leq 1\},$$

and if

$$M_1 = \{x \in M \mid \|x\| \leq 1\}, \quad M_{1,\mathrm{sa}} = \{x \in M \mid x^* = x, \|x\| \leq 1\},$$

then $M_{1,\mathrm{sa}} = \overline{A_{1,\mathrm{sa}}}^{\,SOT}$ *and* $M_1 = \overline{A_1}^{\,SOT}$.

Proof. Let $M_{\mathrm{sa}} = \{x \in M \mid x^* = x\}$ and $A_{\mathrm{sa}} = \{a \in A \mid a^* = a\}$. Example 10.75 asserts that the involution is continuous with respect to the weak operator topology, and so $M_{\mathrm{sa}} = \overline{M_{\mathrm{sa}}}^{\,WOT}$. The set A_{sa} is convex, and so $\overline{A_{\mathrm{sa}}}^{\,SOT} = \overline{A_{\mathrm{sa}}}^{\,WOT}$, by Proposition 10.77. Therefore, the inclusions $M = \overline{A}^{\,SOT} = \overline{A}^{\,WOT} \subseteq M$ imply that

$$M_{\mathrm{sa}} = \overline{A_{\mathrm{sa}}}^{\,WOT} = \overline{A_{\mathrm{sa}}}^{\,SOT}.$$

Likewise, the set $A_{1,\mathrm{sa}}$ is convex, and so $\overline{A_{1,\mathrm{sa}}}^{\,SOT} = \overline{A_{1,\mathrm{sa}}}^{\,WOT}$, by Proposition 10.77. Continuity of the involution with respect to the weak operator topology yields $M_{1,\mathrm{sa}} = \overline{M_{1,\mathrm{sa}}}^{\,WOT}$ and so

$$\overline{A_{1,\mathrm{sa}}}^{\,SOT} \subseteq \overline{M_{1,\mathrm{sa}}}^{\,SOT} = \overline{M_{1,\mathrm{sa}}}^{\,WOT} = M_{1,\mathrm{sa}}.$$

Conversely, select $x \in M_{1,\mathrm{sa}}$. Let $g : [-1, 1] \to [-1, 1]$ denote the inverse of the strictly increasing function $f(t) = \frac{2t}{1+t^2}$ on the interval $[-1, 1]$. Thus, $g(x) \in M_{1,\mathrm{sa}}$. Because $M_{\mathrm{sa}} = \overline{A_{\mathrm{sa}}}^{\,SOT}$, there exists a net $(a_\alpha)_{\alpha \in \Lambda}$ of hermitian operators $a_\alpha \in A_{\mathrm{sa}}$ that converges in the strong operator topology to $g(x)$. Hence, by Lemma 11.15, the net $(f(a_\alpha))_{\alpha \in \Lambda}$ of hermitian contractions $f(a_\alpha) \in A$ converges in the strong operator topology to $f(g(x)) = x$, which proves that $M_{1,\mathrm{sa}} \subseteq \overline{A_{1,\mathrm{sa}}}^{\,SOT}$.

The proof that $M_1 = \overline{A_1}^{\,SOT}$ is outlined in Exercise 11.113. $\square$

11.3   Irreducible Operator Algebras

If A is a $*$-algebra of operators acting on a Hilbert space H, and if $L \subseteq H$ is a subspace invariant under A, which is to say that $x\xi \in L$ for every $\xi \in L$ and $x \in A$, then for every $x \in A$ the orthogonal complement $L^\perp$ is invariant under x^*. Because A is closed under the involution, we see that if L is invariant under A, then so is $L^\perp$. Writing $H = L \oplus L^\perp$ and each operator $x \in B(H)$ as a matrix $\begin{bmatrix} x_{11} & x_{12} \\ x_{21} & x_{22} \end{bmatrix}$ of operators, each element $a \in A$ therefore is a diagonal operator matrix:

$$a = \begin{bmatrix} a_1 & 0 \\ 0 & a_2 \end{bmatrix},$$

where $a_1 = a_{|L}$ and $a_2 = a_{|L^\perp}$. Because $A_1 = \{a_{|L} \,|\, a \in A\}$ and $A_2 = \{a_{|L^\perp} \,|\, a \in A\}$ are $*$-subalgebras of $B(L)$ and $B(L^\perp)$, respectively, the existence of a nontrivial invariant subspace for A yields two $*$-algebras A_1 and A_2 that constitute A and for which $A \subseteq A_1 \oplus A_2 \subseteq B(L) \oplus B(L^\perp)$.

If A cannot be reduced in this way, then A is said to be irreducible.

Definition 11.17. A $*$-subalgebra $A \subseteq B(H)$ is an *irreducible operator algebra* if H has dimension at least 2 and the only subspaces L of H that are invariant under A are $L = \{0\}$ and $L = H$.

The following condition characterise irreducible operator algebras.

Proposition 11.18. *If H has dimension at least 2, then the following statements are equivalent for a $*$-subalgebra $A \subseteq B(H)$:*

1. *A is an irreducible operator algebra;*
2. *if $p \in A'$ is a projection, then $p = 0$ or $p = 1$; and*
3. *$A' = \{\lambda 1 \,|\, \lambda \in \mathbb{C}\}$.*

Proof. The equivalence of (1) and (2) is a simple exercise (Exercise 11.110). Evidently, if (3) holds, then the projections in $A' = \{\lambda 1 \,|\, \lambda \in \mathbb{C}\}$ occur when $\lambda = 0$ or $\lambda = 1$, and so (2) holds. Therefore, assume that (2) holds, and we shall prove that (3) holds.

If $A' \neq \{\lambda 1 \,|\, \lambda \in \mathbb{C}\}$, then there exists a hermitian operator $y \in A'$ with at least two points of spectrum; hence, by translation by a scalar if necessary, we may assume that both $\sigma(y) \cap (-\infty, 0)$ and $\sigma(y) \cap (0, \infty)$ are nonempty. Therefore, in writing $y = y_+ - y_-$, for positive $y_+, y_- \in C^*(y, 1) \subseteq A'$ with $y_+ y_- = y_- y_+ = 0$, it must be that neither y_+ nor y_- is the zero operator. Thus, $y_+ y_- = 0$ implies that the range of y_- is contained in the kernel of y_+, and so the range of y_+ is not dense. Therefore, the partial isometry v in A' with final space $\overline{\mathrm{ran}\, y_+}$ yields a projection $vv^* \in A'$ such that vv^* is neither 0 nor 1, in contradiction to assumption (2). $\square$

Example 11.19. *The C^*-algebra generated by the unitaleral shift operator is an irreducible operator algebra.*

Proof. If s denotes the unilateral shift operator on $\ell^2(\mathbb{N})$, then Exercise 10.106 shows that the only operators y that commute with both s and s^* are operators of the form $y = \lambda 1$, for $\lambda \in \mathbb{C}$. Therefore, $C^*(s)' = \{\lambda 1 \,|\, \lambda \in \mathbb{C}\}$, and so $C^*(s)$ is an irreducible operator algebra. $\square$

The following concept, that of transitivity, is an import from the subject of pure algebra, where analysis usually has no role.

Definition 11.20. A $*$-subalgebra $A \subseteq B(H)$ is *transitive* if, for each pair of vectors $\xi, \eta \in H$ in which $\xi \neq 0$, there is an operator $x \in A$ such that $x\xi = \eta$.

Quite remarkably, irreducible C^*-algebras are transitive.

Theorem 11.21. *If $A \subseteq B(H)$ is an irreducible C^*-algebra, then A is a transitive C^*-algebra.*

Proof. By Proposition 11.18, the commutant of A is $\{\lambda 1 \mid \lambda \in \mathbb{C}\}$, and so the double commutant of A is $B(H)$. Therefore, $\overline{A}^{SOT} = B(H)$ by the Double Commutant Theorem. Hence, if $\varepsilon > 0$ is given and if $y \in B(H)$ and $\xi \in H$, then by the Density Theorem (Theorem 11.16) there is an element $a \in A$ of norm $\|a\| \le \|y\|$ such that $\|a\xi - y\xi\| < \varepsilon$. We shall make use of this last fact repeatedly in the argument below.

Select any $\xi, \eta \in H$ with $\xi \ne 0$; because A is closed under scalar multiplication, we may assume without loss of generality that $\|\xi\| = 1$. If $\eta = 0$, then $a = 0 \in A$ has the property that $a\xi = \eta$. Therefore, assume that $\eta \ne 0$. Consider the operator $y_0 \in B(H)$ given by $y_0 = \eta \otimes \xi$; thus, $y_0 \gamma = \langle \gamma, \xi \rangle \eta$ for all $\gamma \in H$, and so $\|y_0\| = \|\eta\|$ and $y_0\xi = \eta$. As explained in the previous paragraph, there is an operator $a_0 \in A$ of norm $\|a_0\| \le \|\eta\|$ such that $\|a_0 - \eta\| < \frac{1}{2}$.

Now consider the operator $y_1 \in B(H)$ given by $y_1 = (\eta - a_0\xi) \otimes \xi$, which has the properties $\|y_1\| \le \frac{1}{2}$ and $y_1\xi = \eta - a_0\xi$. Therefore, there is an operator $a_1 \in A$ of norm $\|a_1\| \le \|y_1\| < \frac{1}{2}$ such that $\|a_1\xi - (\eta - a_0\xi)\| < \frac{1}{4}$. Continue this construction inductively so that at the completion of the n-th step one has $a_0, \dots, a_n \in A$, each of norm $\|a_j\| \le 2^{-j}$, and such that

$$\left\| \left(\sum_{j=0}^{n} a_j \right) \xi - \eta \right\| < \frac{1}{2^{n+1}}.$$

Now let $a = \sum_{j=0}^{\infty} a_j$, which converges in A because $\sum_{j=0}^{\infty} \|a_j\|$ is bounded above by a convergent geometric series. Because $\left\| \left(\sum_{j=0}^{n} a_j \right) \xi - \eta \right\| \to 0$ as $n \to \infty$, the operator $a \in A$ satisfies $a\xi = \eta$, which proves that A is a transitive C*-algebra. $\square$

Kadison's theorem on transitivity leads to the following useful result.

Proposition 11.22. *If an irreducible C*-subalgebra $A \subseteq B(H)$ contains a nonzero compact operator, then A contains every compact operator.*

Proof. Let $x \in A$ be a nonzero compact operator. Thus, at least one of $\Re(x)$ and $\Im(x)$ is nonzero, which implies that A contains a nonzero hermitian operator h. Therefore, h has a nonzero eigenvalue ω. Express h in its spectral decomposition as $h = \omega q + g$, where q is the projection onto $\ker(h - \omega 1)$ and where $g = h - \omega q$ has eigenvalues different from ω. The characteristic function χ_ω of the set $\{\omega\}$ is continuous on $\sigma(h)$, because $\{\omega\}$ is an open subset of $\sigma(h)$. Hence, the continuous functional calculus yields the operator $q = \chi_\omega(h) \in A$. Therefore, A contains an operator of finite rank.

Let $y \in A$ be an operator of least positive finite rank. If y is not a rank-1 operator, then there are $\xi, \eta \in H$ such that $y\xi$ and $y\eta$ are linearly independent. Because A is an irreducible C*-algebra, Theorem 11.21 shows that A is transitive. Hence, there exists an operator $a \in A$ such that $ay\xi = \eta$. The vectors $yay\xi$ and $y\xi$ are, therefore, linearly independent. The operator ya plainly leaves the finite-dimensional subspace $\operatorname{ran} y$

invariant. Therefore, the linear mapping $ya : \operatorname{ran} y \to \operatorname{ran} y$ has an eigenvalue λ, and so the rank of the restriction of $(ya - \lambda 1)$ to $\operatorname{ran} y$ is less than the dimension of $\operatorname{ran} y$, and so the rank of the operator $yay - \lambda y \in A$ is less than the rank of y. But y has the smallest positive rank in A, and so it must be that $yay - \lambda y = 0$; hence, $yay\xi = \lambda y\xi$, which is in contradiction of the fact that $yay\xi$ and $y\xi$ are linearly independent. It must be, therefore, that y is indeed a rank-1 operator.

As a rank-1 operator, y has the form $y = \xi \otimes \eta$, for some $\xi, \eta \in H$. Without loss of generality, we may assume that $\langle \eta, \xi \rangle = 1$. Let $L = \{ \vartheta \in H \,|\, \xi \otimes \vartheta \in A \}$ and suppose that there exists a nonzero $\delta \in L^{\perp}$. Because A is transitive, there is an operator $b \in A$ such that $b\delta = \eta$. Because the rank-1 operator $yb \in A$ has the form $yb = \xi \otimes (b\delta)$, the vector $b\delta$ is orthogonal to L; in particular, $0 = \langle b\delta, \eta \rangle = \langle \eta, \eta \rangle \neq 0$, which is a contradiction. Therefore, it must be that $L = H$. Hence, A contains all rank-1 operators of the form $\xi \otimes \vartheta$, for all $\vartheta \in H$.

Select an arbitrary rank-1 operator, say $\xi_0 \otimes \vartheta_0$, on H. By the transitivity of A, there exists an operator $c \in A$ with $\xi_0 = c\xi$. Because both c and $\xi \otimes \vartheta_0$ belong to A, so does $c(\xi \otimes \vartheta_0) = \xi_0 \otimes \vartheta_0$. Hence, A contains every rank-1 operator, and so A also contains the norm-closure of the linear span of the set of rank-1 operators; that is, A contains $K(H)$. $\Box$

Corollary 11.23. *If A is an irreducible C^*-algebra acting on a finite-dimensional Hilbert space H, then $A = B(H)$.*

Example 11.24. *The C^*-algebra generated by the unilateral shift operator is unital and contains every compact operator.*

Proof. If s denotes the unilateral shift operator on $\ell^2(\mathbb{N})$, then s is an isometry, which implies that $s^*s = 1$. Given that $s^*s \in *\text{-Alg}(s)$, we deduce that $1 \in C^*(s)$. Example 11.19 shows that $C^*(s)$ is an irreducible C^*-algebra; hence, $C^*(s)$ is also a transtive C^*-algebra.

The operator ss^* fixes every component of a vector in $\ell^2(\mathbb{N})$, except for the first component, which is sent to zero. Thus, the element $p = 1 - ss^* \in C^*(s)$ is the projection of rank 1 onto the subspace spanned by the canonical unit basis vector e_1. Therefore, $C^*(s)$ is a transitive C^*-algebra that contains a nonzero compact operator, and so Proposition 11.22 shows that $C^*(s)$ contains every compact operator on $\ell^2(\mathbb{N})$. $\Box$

If A_1 and A_2 are operator algebras acting on Hilbert spaces H_1 and H_2, respectively, then $A = A_1 \oplus A_2 = \{ a_1 \oplus a_2 \,|\, a_j \in A_j \}$ acts on $H = H_1 \oplus H_2$. However, if it were to happen that $A_2 = \{0\}$, then the presence of H_2 is somewhat artificial, making H overly large relative to the action of A upon it. For this reason, one is most commonly interested in nondegenerate algebras.

Definition 11.25. A $*$-subalgebra $A \subseteq B(H)$ is *nondegenerate* if the only $\xi \in H$ for which $x\xi = 0$ for every $x \in A$ is $\xi = 0$.

In contrast to degeneracy is the following concept.

Definition 11.26. A vector $\xi \in H$ is a *cyclic vector* for an operator algebra $A \subseteq B(H)$ if the linear submanifold $\{x\xi \mid x \in A\}$ is dense in H.

Example 11.27. *If G is a countable discrete group, then the unit vector δ_e of $\ell^2(G)$ is a cyclic vector for the group C^*-algebra $C^*_\lambda(G)$.*

Proof. For each $h \in G$, the action of the unitary operator λ_h on δ_e produces the unit vector δ_g. Therefore, the linear submanifold $\{x\delta_e \mid x \in C^*_\lambda(G)\}$ contains the orthonormal basis vectors for $\ell^2(G)$. Hence, δ_e is a cyclic vector for $C^*_\lambda(G)$.   □

The relevance of cyclic vectors to operator algebra theory is indicated by the next result.

Proposition 11.28. *If H is a separable Hilbert space and if $A \subseteq B(H)$ is a nondegenerate C^*-subalgebra, then there exist a finite or countable family $\{H_n\}_n$ of pairwise-orthogonal subspaces $H_n \subseteq H$ and unit vectors $\xi_n \in H_n$ such that*

1. each H_n is invariant under A,
2. $\{x\xi_n \mid x \in A\}$ is dense in H_n, and
3. $H = \bigoplus_n H_n$.

Proof. Select a unit vector $\xi \in H$ and denote the A-invariant subspace $\overline{\{x\xi_n \mid x \in A\}}$ by H_ξ. Suppose that $p \in B(H)$ is the projection with range H_ξ. The invariance of H_ξ under $x, x^* \in A$ implies that $xp = px$ (Proposition 10.19). Thus, for every $x \in A$, $(1-p)x\xi = x(1-p)\xi$, and so

$$(1-p)\xi \in \bigcap_{x \in A} \ker x = \{0\},$$

where the equality of the intersection of kernels $\ker x$ with $\{0\}$ is because of the nondegeneracy of A. Hence, $\xi = p\xi \in H_\xi$, and this is the case for every unit vector $\xi \in H$.

Let $\mathfrak{S}$ consist of all sets $\mathcal{O}$ of unit vectors from H such that $H_\xi \perp H_\eta$ for any pair of distinct $\xi, \eta \in \mathcal{O}$. Order $\mathfrak{S}$ by set inclusion, and apply a Zorn's Lemma argument to produce a maximal element $\mathcal{O}$ in $\mathfrak{S}$. Because H is separable, the set $\mathcal{O}$ is finite or countably infinite, and so we denote $\mathcal{O}$ by $\{\xi_n\}_n$. Let $H_0 = \bigoplus_n H_{\xi_n}$. If $H_0 \neq H$, then there is a subspace $H_1 \subset H$ such that $H = H_0 \oplus H_1$ and H_1 is A-invariant. Therefore, applying our arguments to $A_{|H_1}$ would yield a unit vector $\eta \in H_1$ which is cyclic on some A-invariant subspace H_{11} of H_1, which would therefore imply that $\mathcal{O} \cup \{\eta\} \in \mathfrak{S}$, in contradiction to the maximality of $\mathcal{O}$.   □

Corollary 11.29. *If A is an irreducible operator algebra, then every nonzero vector is a cyclic vector for A.*

Proof. Because A is irreducible, the direct sum decomposition in Proposition 11.28 is trivial in the sense that n must be 1. However, the algebra A_1 was constructed in the proof of Proposition 11.28 by selecting any unit vector $\xi \in H$ and considering the subspace $H_1 = \overline{\{x\xi \mid x \in A\}}$. As $H_1 = H$, the vector ξ is cyclic for A.   □

11.4   Abelian Operator Algebras

The Gelfand Representation Theorem (Theorem 9.27) for abelian Banach algebras certainly applies to any abelian C^*-algebra. However, the presence of the Hilbert space adjoint yields an even stronger form of Gelfand's theorem.

Definition 11.30. If A and B are C^*-algebras of operators, then a $*$-*homomorphism* from A to B is a map $\rho : A \to B$ such that, for all $x, y \in A$ and $\alpha, \beta \in \mathbb{C}$,

1. $\rho(\alpha x + \beta y) = \alpha \rho(x) + \beta \rho(y)$,
2. $\rho(xy) = \rho(x) \rho(y)$, and
3. $\rho(x^*) = \rho(x)^*$.

Moreover,

4. If A and B are unital C^*-algebras of operators, and if ρ maps the identity operator in A to the identity operator in B, then ρ is said to be a *unital* map.
5. If a $*$-homomorphism ρ is bijective, then ρ is called an $*$-*isomorphism*.

In what follows, we shall be considering the complex number system $\mathbb{C}$ as a 1-dimensional C^*-algebra of operators (with adjoint $\alpha^* = \overline{\alpha}$ and norm $\|\alpha\| = |\alpha|$). Moreover, if X is a compact Hausdorff space, then we shall regard $C(X)$ as a unital abelian C^*-algebra acting on the Hilbert space $L^2(X, \Sigma, \mu)$, where μ is a fixed regular Borel probability measure on the Borel sets Σ of X, and where $f \in C(X)$ is identified with the multiplication operator M_f on $L^2(X, \Sigma, \mu)$. Under this identification, $\overline{f}$ is identified with $M_{\overline{f}} = M_f^*$ and the norms of f and M_f coincide.

Theorem 11.31 (Gelfand). *If $A \subseteq B(H)$ is a unital abelian C^*-algebra and if $\mathscr{R}_A$ is the maximal ideal space of A, then the Gelfand transform $\Gamma : A \to C(\mathscr{R}_A)$ is an isometric $*$-isomorphism of the C^*-algebras A and $C(\mathscr{R}_A)$.*

Proof. If $x \in A$, then

$$\|x^2\|^2 = \|(x^2)^*(x^2)\| = \|(x^*x)^*(x^*x)\| = \|x^*x\|^2 = \left(\|x\|^2\right)^2.$$

Thus, $\|x^2\| = \|x\|^2$ for each $x \in A$ and, therefore, Γ is an isometry by Exercise 9.55.

The maximal ideal space $\mathscr{R}_A$ consists of all unital continuous linear maps $\rho : A \to \mathbb{C}$ for which $\rho(xy) = \rho(x)\rho(y)$ for all $x, y \in A$. Because A is abelian, for all $z_1, z_2 \in A$, the exponential map $z \mapsto e^z = \sum_{n=0}^{\infty} \frac{1}{n!} z^n$ satisfies $e^{z_1 + z_2} = e^{z_1} e^{z_2}$ (Proposition 9.41).

In particular, if $h \in A$ is a hermitian operator and $\theta \in \mathbb{R}$, then $e^{-i\theta h} e^{i\theta h} = 1$ implies

$$1 = \rho(1) = \rho\left(e^{-i\theta h} e^{i\theta h}\right) = e^{-i\theta \rho(h)} e^{i\theta \rho(h)} = |e^{i\theta \rho(h)}|^2.$$

As the equation above is true for every $\theta \in \mathbb{R}$, $\rho(h)$ must be a real number. Thus, if $x \in A$ is arbitrary, then expressing $x = h + ig$, where $h, g \in A$ are the real and imaginary parts of x, gives

$$\Gamma(x^*) = \Gamma(h) - i\Gamma(g) = (\Gamma(h) + i\Gamma(g))^* = (\Gamma(x))^*,$$

implying that the homomorphism Γ is a $*$-homomorphism.

Because Γ is a unital $*$-preserving isometric homomorphism, the range of Γ is a unital self-adjoint Banach subalgebra of $C(\mathcal{R}_A)$. Because $\Gamma(A)$ clearly separates the points of $\mathcal{R}_A$, the Stone-Weierstrass Theorem asserts that $\Gamma(A) = C(\mathcal{R}_A)$, thereby proving that Γ is surjective. $\square$

Theorem 11.31 demonstrates that the Gelfand transform of a unital abelian C*-algebra of operators is an isometric $*$-isomorphism with $C(X)$, for an appropriate compact Hausdorff space X. However, if the abelian C*-algebra of operators is a von Neumann algebra M, say $M = L^\infty(X, \Sigma, \mu)$, then expressing M as a C*-algebra of multiplication operators by continuous functions on some compact Hausdorff space Y is slightly unnatural since M is already an algebra of multiplication operators. Therefore, the Gelfand theory for an abelian von Neumann algebra should take a measure-theoretic form, which is accomplished here in Theorem 11.33 below.

For simplicities of cardinal arithmetic, we will assume that the Hilbert spaces upon which these abelian von Neumann algebras act are separable.

Theorem 11.32. *Assume that M is an abelian von Neumann algebra acting on a separable Hilbert space H. If M has a cyclic vector, then there exist a compact metrisable space X, a regular probability measure on the σ-algebra Σ of Borel sets of X, and a surjective isometry $u : H \to L^2(X, \Sigma, \mu)$ such that the linear map $\Phi : L^\infty(X, \Sigma, \mu) \to B(H)$ defined by*

$$\Phi(M_\psi) = u^{-1} M_\psi u,$$

is an isometric $*$-isomorphism of the von Neumann algebra $L^\infty(X, \Sigma, \mu)$ of multiplication operators on $L^2(X, \Sigma, \mu)$ and the von Neumann algebra M. Furthermore, Φ is continuous with respect to the strong operator topology on each of $L^\infty(X, \Sigma, \mu)$ and M.

Proof. Because H is separable, Exercise 11.114 shows that the strong operator topology on the closed unit ball of $B(H)$ is separable and metrisable; hence, the same is true of the closed unit ball M_1 of M with respect to the strong operator topology. Let $\{a_n\}_{n \in \mathbb{N}}$ be a countable SOT-dense subset of M_1, and let $A = C^*(\{a_n\}_n, 1)$, which is a separable unital abelian C*-algebra for which $\overline{A}^{SOT} = M$. By the Density Theorem, $\overline{A_1}^{SOT} = M_1$; and via the Gelfand transform Γ, A is isometrically $*$-isomorphic to $C(X)$, where because of the separability of A the space X is necessarily compact and metrisable (Theorem 5.57).

Let $\xi \in H$ be a unit cyclic vector for M and define a linear functional φ on $C(X)$ by $\varphi(f) = \langle \Gamma^{-1}(f)\xi, \xi \rangle$. By the Riesz Representation Theorem (Theorem 6.51), there exists a regular Borel measure μ on the σ-algebra Σ of Borel sets of X such that

$$\varphi(f) = \int_X f \, d\mu,$$

for every $f \in C(X)$. Define $\pi : C(X) \to B\big(L^2(X, \Sigma, \mu)\big)$ by $\pi(f) = M_f$, the operator of multiplication by f. The map π is plainly a unital $*$-homomorphism, but it is also isometric by Example 10.38. Thus, A is isometrically $*$-isomorphic with the operator algebra $\{M_f \,|\, f \in C(X)\}$ acting on $L^2(X, \Sigma, \mu)$.

If $\eta \in H$ and $\varepsilon > 0$ is given, then there exists $x \in M$ with $\|\eta - x\xi\| < \varepsilon/2$, and there exists $a \in A$ with $\|x\xi - a\xi\| < \varepsilon/2$. Therefore, the linear submanifold $H_0 = \{a\xi \,|\, a \in A\}$ is dense in H. If $a_1, a_2 \in A$ satisfy $a_1\xi = a_2\xi$, then for every $a \in A$ we have that $(a_1 - a_2)a\xi = a(a_1 - a_2)\xi = 0$, which implies that $a_1\eta = a_2\eta$ for every $\eta \in H$. Thus, the function $u_0 : H_0 \to C(X)$ defined by $u_0(a\xi) = \Gamma(a)$ is well defined, linear, and surjective. In viewing $C(X)$ as a linear submanifold of $L^2(X, \Sigma, \mu)$, we have that the norm of $u_0(a\xi)$ in $L^2(X, \Sigma, \mu)$ satisfies

$$\|u_0(a\xi)\|^2 = \int_X |\Gamma(a)|^2 \, d\mu = \int_X \Gamma(a^*a) \, d\mu = \varphi\big(\Gamma(a^*a)\big) = \langle a^*a\xi, \xi \rangle = \|a\xi\|^2.$$

Therefore, u_0 is a linear isometry of H_0 onto $C(X)$, and so passing to closures in each of H and $L^2(X, \Sigma, \mu)$ yields a surjective isometry

$$u : H \to \overline{C(X)}^{\,\|\cdot\|_2} = L^2(X, \Sigma, \mu).$$

The map u also has the following property: given $a \in A$, then for every $b \in A$,

$$ua(b\xi) = u(ab\xi) = \Gamma(ab) = \Gamma(a)\Gamma(b) = M_{\Gamma(a)}\big(\Gamma(b)\big) = M_{\Gamma(a)}u(b\xi).$$

Thus, $ua = M_{\Gamma(a)}u$, for every $a \in A$. If $x \in M$ and if $(a_\alpha)_{\alpha \in \Lambda}$ is a net of operators $a_\alpha \in A$ converging strongly to x, then $(ua_\alpha u^{-1})_\alpha$ converges strongly to uxu^{-1}. Therefore, the net $\big(M_{\Gamma(a_\alpha)}\big)_\alpha$ is strongly convergent in $B\big(L^2(X, \Sigma, \mu)\big)$ to an operator in $\overline{C(X)}^{SOT} = L^\infty(X, \Sigma, \mu)$ (Exercise 11.109). The same argument shows that if $\big(M_{\Gamma(a_\alpha)}\big)_\alpha$ is strongly convergent to M_ψ, then $(a_\alpha)_\alpha$ converges strongly to $u^{-1}M_\psi u$. Hence, the isometric $*$-isomorphism $\Phi_0 : C(X) \to A$ given by $\Phi_0(M_f) = \Gamma^{-1}f$ has the property that $u\Phi_0(M_f) = M_f u$ and extends to an isometric $*$-isomorphism $\Phi : \overline{C(X)}^{SOT} \to \overline{A}^{SOT}$ and satisfies $u\Phi(M_\psi) = M_\psi u$ for all essentially bounded measurable functions ψ.   □

The case of noncyclic abelian von Neumann algebras may now be examined.

Theorem 11.33. *If M is an abelian von Neumann algebra acting on a separable Hilbert space H, then there exists a σ-finite measure space (X, Σ, μ) and a surjective isometry $u : H \to L^2(X, \Sigma, \mu)$ such that the linear map $\Phi : L^\infty(X, \Sigma, \mu) \to B(H)$ defined by*

$$\Phi(M_\psi) = u^{-1}M_\psi u,$$

is an isometric $*$-isomorphism of the von Neumann algebra $L^\infty(X, \Sigma, \mu)$ of multiplication operators on $L^2(X, \Sigma, \mu)$ and the von Neumann algebra M.

Proof. By Proposition 11.28, there exist a finite or countable family $\{H_n\}_n$ of pairwise-orthogonal subspaces $H_n \subseteq H$ and unit vectors $\xi_n \in H_n$ such that each H_n is invariant under M, $\{x\xi_n \,|\, x \in M\}$ is dense in H_n, and $H = \bigoplus_n H_n$. Moreover, for each n the restriction of M to H_n is an abelian von Neumann algebra acting on H_n with cyclic vector ξ_n. Therefore, by Theorem 11.32, $M_{|H_n} \cong L^\infty(X_n, \Sigma_n, \mu_n)$, for some Borel probability measure μ_n on the Borel sets Σ_n of some compact metrisable space X_n. Let $X = \dot{\bigcup}_n X_n$, the disjoint union of the family $\{X_n\}_{n \in \mathbb{N}}$, and let $\Sigma = \bigcup_n \Sigma_n$. Define $\mu : \Sigma \to [0, \infty]$ by $\mu(E) = \sum_n \mu(E_n)$. Because $\mu(X_n) = \mu_n(X_n) = 1$ for every $n \in \mathbb{N}$, the measure space (X, Σ, μ) is σ-finite. Furthermore, $L^2(X, \Sigma, \mu)$ is given by $\bigoplus_n L^2(X_n, \Sigma_n, \mu_n) \cong \bigoplus_n H_n = H$, where the Hilbert space isomorphism between each H_n and $L^2(X_n, \Sigma_n, \mu_n)$ is implemented by a surjective linear isometry $u_n : H_n \to L^2(X_n, \Sigma_n, \mu_n)$ and where $u = \bigoplus_n u_n$ is a surjective isometry $H \to L^2(X, \Sigma, \mu)$. Hence, $\Phi(M_\psi) = u^{-1} M_\psi u = \bigoplus_n u_n^{-1} M_{\psi|X_n} u_n$ is an isometric $*$-homomorphism of $L^\infty(X, \Sigma, \mu)$ onto M.   □

Theorem 11.33 has the following important consequence: every normal operator is unitarily equivalent to a multiplication operator.

Corollary 11.34 (Spectral Theorem for Normal Operators). *If N is a normal operator acting on a separable Hilbert space H, then there is a σ-finite measure space (X, Σ, μ) and a surjective isometry $U : H \to L^2(X, \Sigma, \mu)$ such that UNU^{-1} is the multiplication operator M_ψ on $L^2(X, \Sigma, \mu)$, for some $\psi \in L^\infty(X, \Sigma, \mu)$.*

Proof. Let $M = W^*(N)$, the von Neumann algebra generated by N. Because N is normal, $W^*(N)$ is abelian. Thus, apply Theorem 11.33 to obtain the result.   □

A more specific form of Theorem 11.34 is possible when $W^*(N)$ admits a cyclic vector; see Exercise 11.115.

11.5 C*-Algebras

While the study of operator algebras has to this point been quite satisfactory, there is some limit to what one can achieve using a purely operator-theoretic approach. For example, if H is an infinite-dimensional Hilbert space, then the compact operators form a proper ideal $K(H)$ of $B(H)$, and so one can consider the quotient Banach algebra $B(H)/K(H)$. One would be correct in thinking that the involution on $B(H)$ (and on the ideal $K(H)$) would lead to an involution on the quotient space $B(H)/K(H)$; further, as we shall see, the quotient norm behaves just like the operator norm in the sense that $\|\dot{x}^*\dot{x}\| = \|\dot{x}\|^2$, for every $\dot{x} \in B(H)/K(H)$. Therefore, it is both natural and necessary to consider Banach algebras that share the involutive and norm properties of $B(H)$.

Definition 11.35. A complex associate algebra A is said to be *involutive* if there exists a function $x \mapsto x^*$ (called an *involution*) such that, for all $x, y \in A$ and $\alpha \in \mathbb{C}$,

1. $(x^*)^* = x$,
2. $(x+y)^* = x^* + y^*$,
3. $(\alpha x)^* = \bar{\alpha} x^*$, and
4. $(xy)^* = y^* x^*$.

Definition 11.36. An involutive Banach algebra A is called a C^*-*algebra* if, for every $x \in A$, $\|x^* x\| = \|x\|^2$. If, in addition, A admits a multiplicative identity 1, then A is said to be a *unital C^*-algebra*.

Evidently, every C^*-algebra of Hilbert space operators is a C^*-algebra as defined above. Algebras of continuous functions offer another example, and they can be considered in their original form rather than in the guise of a C^*-algebra of multiplication operators.

Example 11.37. *If X is a locally compact Hausdorff space, then the algebra $C_0(X)$ of continuous functions $f : X \to \mathbb{C}$ that vanish at infinity is a C^*-algebra under the norm $\|f\| = \max_{x \in X} |f(x)|$ and involution $f^*(t) = \overline{f(t)}$, for $t \in X$.*

Let us now explore a few more consequences of the axioms. If A is a C^*-algebra and if $x \in A$, then

$$\|x\|^2 = \|x^* x\| \le \|x^*\| \|x\| \quad \text{and} \quad \|x^*\|^2 = \|x^{**} x^*\| \le \|x^{**}\| \|x^*\| = \|x\| \|x^*\|,$$

implying that $\|x\| \le \|x^*\|$ and $\|x^*\| \le \|x\|$. That is, $\|x^*\| = \|x\|$ and so the involution on a C^*-algebra is isometric.

If A is a C^*-algebra with multiplicative identity $1 \in A$, then $\|1\| = 1$ by an argument that is similar to the one above. Thus, C^*-algebras with identity are unital Banach algebras in the sense of Definition 5.25. Furthermore, if $x \in A$, then

$$1^* x = (1^* x)^{**} = (x^* 1)^* = x^{**} = x.$$

Likewise, $x 1^* = x$. By the uniqueness of the multiplicative identity in a unital ring, $1^* = 1$.

Definition 11.38. If A is a C^*-algebra, then a subset $B \subseteq A$ is a C^*-*subalgebra* of A if B is a C^*-algebra with respect to the sum, product, involution, and norm of A. If A is unital and if the multiplicative identity of A belongs to B, then B is said to be a *unital C^*-subalgebra* of A.

If $\mathscr{F} \subset A$ is a subset of a C^*-algebra A, then the C^*-algebra *generated* by $\mathscr{F}$ is the smallest C^*-subalgebra of A that contains $\mathscr{F}$ and is denoted by $C^*(\mathscr{F})$. Of special interest is the case in which $\mathscr{F} = \{x\}$ for some $x \in A$. In this regard, each of the elements of the following type will generate abelian C^*-algebras.

Definition 11.39. Assume that A is a C*-algebra and $x \in A$.

1. If $x^* = x$, then x is said to be *hermitian*.
2. If $x^*x = xx^*$, then x is *normal*.
3. If A is unital and if $x^*x = xx^* = 1$, then x is *unitary*.

As noted earlier, if A is a unital C*-algebra, then the multiplicative identity $1 \in A$ is hermitian.

Definition 11.40. The set of hermitian elements in a C*-algebra A is denoted by A_{sa}.

As with Hilbert space operators, real and imaginary parts of every $x \in A$ are defined

$$\Re x = \frac{1}{2}(x + x^*) \text{ and } \Im x = \frac{1}{2i}(x - x^*).$$

Hence, A_{sa} is a real vector space and $\text{Span}_{\mathbb{C}} A_{sa} = A$.

In the category of C*-algebras, the natural maps between C*-algebras are called $*$-homomorphisms.

Definition 11.41. If A and B are C*-algebras, then a $*$-*homomorphism* from A to B is a map $\rho : A \to B$ such that, for all $x, y \in A$ and $\alpha, \beta \in \mathbb{C}$,

1. $\rho(\alpha x + \beta y) = \alpha \rho(x) + \beta \rho(y)$,
2. $\rho(xy) = \rho(x)\rho(y)$, and
3. $\rho(x^*) = \rho(x)^*$.

Moreover,

4. If A and B are unital C*-algebras, and if ρ maps the identity of A to the of B, then ρ is said to be a *unital* map.
5. If a $*$-homomorphism ρ is bijective, then ρ is called an $*$-*isomorphism*.

The Gelfand Theorem for unital abelian C*-algebras has exactly the same form as the version for abelian C*-algebras of operators (and has exactly the same proof).

Theorem 11.42 (Gelfand). *If A is a unital abelian C*-algebra, then the Gelfand transform $\Gamma : A \to C(\mathscr{R}_A)$ is an isometric $*$-isomorphism of A and $C(\mathscr{R}_A)$.*

Gelfand's Theorem has many consequences, including a determination of the C*-algebra generated by a normal operator (see Theorem 11.47).

Example 11.43. *If A is a unital C*-algebra, and if $x \in A$ is normal operator, then the character space of $C^*(x, 1)$ is the spectrum of x, and the unital abelian C*-algebra $C^*(x, 1)$ is isometrically $*$-isomorphic to $C(\sigma(x))$.*

Proof. As $x^*x = xx^*$, the algebra $C^*(x, 1)$ is abelian. Let $\Gamma : C^*(x, 1) \to C(\mathscr{R}_{C^*(x,1)})$ denote the Gelfand transform. Therefore,

$$\sigma(x) = \{\rho(x) \mid \rho \in \mathscr{R}_{C^*(x,1)}\},$$

and the function $\psi : \mathscr{R}_{C^*(x,1)} \to \sigma(x)$ defined by $\psi(\rho) = \rho(x)$ is surjective. To prove that ψ is injective, suppose that $\rho_1(x) = \rho_2(x)$. Then it is also true that $\rho_1(y) = \rho_2(y)$ for every $y \in C^*(x,1)$, because $\{x, x^*, 1\}$ generates $C^*(x,1)$ and ρ_1 and ρ_2 are unital $*$-homomorphisms. Thus, ψ is injective

If $\{\rho_\alpha\}_{\alpha \in \Lambda}$ is a net in $\mathscr{R}_{C^*(x,1)}$ converging to $\rho \in \mathscr{R}_{C^*(x,1)}$, then, by definition of weak*-topology, $\rho(y) = \lim_\alpha \rho_\alpha(y)$, for every $y \in C^*(x,1)$. Hence, ψ is a continuous function. By Proposition 2.9, any continuous bijection from a compact space onto a Hausdorff space is a homeomorphism.   □

The following example is of interest from the point of view of topology.

Example 11.44. *If (X, Σ, μ) is a σ-finite measure space, then there exists a compact Hausdorff space Y such that $L^\infty(X, \Sigma, \mu)$ and $C(Y)$ are isometrically $*$-isomorphic.*

Proof. Let Y be the character space of the abelian von Neumann algebra $L^\infty(X, \Sigma, \mu)$, and apply Gelfand's Theorem.   □

In reference to the example above, we know that $L^\infty(X, \Sigma, \mu)$ has a multitude of projections, whereas a projection in $C(Y)$ corresponds to the characteristic function of some subset E of Y. Therefore, because continuous functions preserve connectivity, the space Y must possess a high degree of disconnectivity. The next example shows this fact explicitly, since $\beta\mathbb{N}$ is a nonmetrisable, totally disconnected compact Hausdorff space (Proposition 2.79).

Example 11.45. *The C^*-algebras $\ell^\infty(\mathbb{N})$ and $C(\beta\mathbb{N})$ are isometrically $*$-isomorphic.*

Proof. By Example 9.31, the maximal ideal space of $\ell^\infty(\mathbb{N})$ is homeomorphic to $\beta\mathbb{N}$. Therefore, by Theorem 11.42, the C^*-algebras $\ell^\infty(\mathbb{N})$ and $C(\beta\mathbb{N})$ are isometrically $*$-isomorphic.   □

The next result shows that the spectrum of a C^*-algebra element x does not depend on the particular C^*-algebra that contains x.

Proposition 11.46 (Spectral Permanence). *If B is a unital C^*-subalgebra of a unital C^*-algebra A, then $\sigma_A(x) = \sigma_B(x)$ for every $x \in B$.*

Proof. Let $x \in B$. The inclusion $\sigma_A(x) \subseteq \sigma_B(x)$ has already been noted in Proposition 9.10. To prove the containment $\sigma_B(x) \subseteq \sigma_A(x)$ it is sufficient to show that $0 \in \sigma_B(x)$ implies $0 \in \sigma_A(x)$. This is most simply done by proving the contrapositive: if $x \in B$ has an inverse x^{-1} in A, then $x^{-1} \in B$.

Therefore, assume that $x \in B$ is invertible in A. Hence, x^* is invertible as well, since $1 = xz = zx$ implies that $1 = z^*x^* = x^*z^*$. Consequently, $x^*x \in B$ is invertible in A.

Let $C = C^*(x^*x, 1)$, the unital abelian C^*-subalgebra of B (and of A) generated by x^*x. Applying the Gelfand transform on C to the hermitian element x^*x yields $\sigma_C(x^*x) \subset \mathbb{R}$. Proposition 9.10 on spectral permanence in abelian Banach algebras now implies the following inclusions:

$$\sigma_C(x^*x) = \partial\sigma_C(x^*x) \subseteq \sigma_A(x^*x) \subseteq \sigma_C(x^*x).$$

Therefore, $\sigma_C(x^*x) = \sigma_A(x^*x)$. Consequently, the invertibility of x^*x in A implies the invertibility of x^*x in $C \subseteq B$. A left inverse for x in B is $[(x^*x)^{-1}x^*]x$; because x is in fact invertible in A, this left inverse is necessarily the inverse $x^{-1} \in A$ of x. Hence, $x^{-1} = (x^*x)^{-1}x^* \in B$. □

The Gelfand Theorem (Theorem 11.42) carries the continuous functional calculus for normal operators on Hilbert spaces to the more abstract setting of normal elements in C*-algebras. Note that Proposition 11.46 allows us to adopt the notation $\sigma(x)$, for the spectrum of x, unambiguously.

Theorem 11.47 (Continuous Functional Calculus). *If x is a normal element of a unital C*-algebra A, then the unital C*-subalgebra $C^*(x, 1)$ of A generated by x is abelian and*

1. *the character space $\mathscr{R}_{C^*(x,1)}$ of $C^*(x, 1)$ is homeomorphic to the spectrum of x,*
2. *there is an isometric isomorphism $\Phi : C(\sigma(x)) \to C^*(x, 1)$ such that $\Phi(\iota) = x$, where $\iota \in C(\sigma(x))$ is the function $\iota(t) = t$, and*
3. *(Spectral Mapping Theorem) for each $f \in C(\sigma(x))$, the spectrum of $\Phi(f) \in C^*(x, 1)$ is*

$$\sigma(\Phi(f)) = \{f(\lambda) \mid \lambda \in \sigma(x)\}.$$

Proof. Example 11.43 shows that the character space of $C^*(x, 1)$ is homeomorphic to $\sigma(x)$, via the homeomorphism $\psi : \mathscr{R}_{C^*(x,1)} \to \sigma(x)$ defined by $\psi(\rho) = \rho(x)$. Let $\Omega : C(\sigma(x)) \to C(\mathscr{R}_{C^*(x,1)})$ be defined, for $f \in C(\sigma(x))$, by

$$\Omega(f)[\rho] = f(\psi(\rho)) = f(\rho(x)),$$

for each $\rho \in \mathscr{R}_{C^*(x,1)}$. The map Ω is evidently a unital *-isomorphism and

$$\|\Omega(f)\| = \max_{\rho \in \mathscr{R}_{C^*(x,1)}} |\Omega(f)[\rho]| = \max_{\rho \in \mathscr{R}_{C^*(x,1)}} |f(\rho(x))| = \max_{\lambda \in \sigma(x)} |f(\lambda)| = \|f\|$$

for every $f \in C(\sigma(x))$.

The map $\Phi : C(\sigma(x)) \to C^*(x, 1)$ defined by $\Phi = \Gamma^{-1} \circ \Omega$ is a unital isometric *-isomorphism. If $\iota \in C(\sigma(x))$ denotes the function $\iota(t) = t$, then $\iota^* \in C(\sigma(x))$ is the function $\iota^*(t) = \bar{t}$. The Gelfand transform evaluated at $x \in C^*(x, 1)$ yields the function $\Gamma(x) \in C(\mathscr{R}_{C^*(x,1)})$ defined by $\Gamma(x)[\rho] = \rho(x)$ for every character ρ. Hence, Φ^{-1} maps x to ι, implying that $x = \Phi(\iota)$ and $x^* = \Phi(\iota^*)$. More generally,

$$\Phi\left(\sum_{k=0}^{m}\sum_{j=0}^{n} \alpha_{kj}\iota^k\bar{\iota}^j\right) = \sum_{k=0}^{m}\sum_{j=0}^{n} \alpha_{kj}x^k(x^*)^j. \tag{11.1}$$

The spectrum of the element given in (11.1) is the range of the continuous function

$$\Gamma\left(\sum_{k=0}^{m}\sum_{j=0}^{n}\alpha_{kj}x^{k}(x^{*})^{j}\right) \in C(\mathscr{R}_{C^{*}(x,1)}),$$

namely,

$$\left\{\sum_{k=0}^{m}\sum_{j=0}^{n}\alpha_{kj}\rho(x)^{k}\overline{\rho(x)}^{j}\,|\,\omega \in \mathscr{R}_{C^{*}(x,1)}A\right\} = \left\{\sum_{k=0}^{m}\sum_{j=0}^{n}\alpha_{kj}\lambda^{k}\overline{\lambda}^{j}\,|\,\lambda \in \sigma(x)\right\}.$$

Since the ring of all polynomials in commuting variables t and $\bar{t}$ is uniformly dense in $C(\sigma(x))$ (by the Stone-Weierstrass Theorem), equation (11.1) shows that

$$\sigma(\Phi(f)) = \{f(\lambda)\,|\,\lambda \in \sigma(x)\},$$

which completes the proof.   □

Notational Convention In applications of Theorem 11.47 it is customary to denote $\Phi(f)$ by $f(x)$, for each $f \in C(\sigma(x))$.

Proposition 11.48. *Suppose that A is a unital C^{*}-algebra and $h \in A$ is hermitian. Let $X \subset \mathbb{R}$ be a compact set such that $X \supseteq \sigma(h) \cup \{0\}$. If $f \in C(X)$ satisfies $f(0) = 0$, then $f(h) \in C^{*}(h)$.*

Proof. By Exercise 11.123, the condition $f(0) = 0$ implies that there is a sequence of polynomials f_{n} for which $f_{n}(0) = 0$ and $|f(t) - f_{n}(t)| \to 0$ uniformly on X (and, thus, on $\sigma(h)$ as well). Since $f_{n}(h) \in C^{*}(h)$,

$$\lim_{n\to\infty}\|f(h) - f_{n}(h)\| = \lim_{n\to\infty}\left(\max_{t\in\sigma(h)}|f(t) - f_{n}(t)|\right) = 0,$$

and so $f(h) \in C^{*}(h)$.   □

In Proposition 11.48 above, the algebra $C^{*}(h)$ does not necessarily contain the identity of A; thus, the conclusion $f(h) \in C^{*}(h)$ is sharper than the conclusion of Theorem 11.47, which is that $f(h) \in C^{*}(h, 1)$.

The use of nonunital C^{*}-algebras is necessarily in many settings; however, each such algebra may be realised as a C^{*}-subalgebra, of co-dimension 1, of a unital C^{*}-algebra.

Proposition 11.49. *If A is a nonunital C^{*}-algebra, then on the set*

$$A^{1} = A \times \mathbb{C} = \{(a,\alpha)\,|\,a \in A,\, \alpha \in \mathbb{C}\}$$

define an involution and vector space operations through the involution and the vector space operations in each coordinate, and define multiplication by

$$(a,\alpha)\cdot(b,\beta) = (ab+\alpha b+\beta a, \alpha\beta).$$

Furthermore, let $\|\cdot\|' : A^1 \to \mathbb{R}$ *be defined by*

$$\|z\|' = \sup\{\|zb\| \,|\, b \in A, \|b\| \leq 1\}, \tag{11.2}$$

for $z \in A^1$. *Then:*

1. $\|\cdot\|'$ *is a norm;*
2. A^1 *is a C*-algebra with respect to* $\|\cdot\|'$;
3. *the ordered pair* $(0, 1)$ *is a multiplicative identity for* A^1; *and*
4. $\|a\|' = \|a\|$, *for every* $a \in A$.

Proof. It is clear that A^1 is a Banach space and that $(0, 1)$ is a multiplicative identity for A^1. Identify A with the subalgebra $\{(a,0)\,|\,a \in A\}$ of A^1. As vector spaces, $A^1/A \cong \mathbb{C}$, and so A has codimension 1 in A^1. Moreover, if $z \in A^1$ and $a \in A$, then $za \in A$.

Suppose that $z \in A^1$ satisfies $\|z\|' = 0$. If $z \in A$, then $\|z\|' = 0$ implies that $\|zb\| = 0$ for every $b \in A$. In particular, $\|zz^*\| = 0$, whence $\|z^*\| = 0$. Since the involution on A is an isometry, $\|z\| = 0$. This proves that $z = 0$ if $z \in A$ and $\|z\|' = 0$.

Next, consider the possibility that $z \neq 0$ yet $\|z\|' = 0$. The paragraph above shows that $z \notin A$ (for otherwise z would be 0). Thus, $z = (a, \lambda)$ for some $a \in A$ and nonzero $\lambda \in \mathbb{C}$. The hypothesis $\|z\|' = 0$ again implies that $z \cdot b = 0$ for all $b \in A$—that is, $ab + \lambda b = 0$ for every $b \in A$. Hence, $-\lambda^{-1}a$ is a left multiplicative identity for A. By passing to adjoints, $(-\lambda^{-1}a)^*$ is a right multiplicative identity of A. Thus,

$$-\lambda^{-1}a = \left(-\lambda^{-1}a\right)\left(-\lambda^{-1}a\right)^* = \left(-\lambda^{-1}a\right)^*.$$

In other words, $-\lambda^{-1}a$ is a multiplicative identity for A, which is in contradiction to the hypothesis that A is a nonunital algebra. Therefore, it must be that $\|z\|' = 0$ only if $z = 0$. The remaining properties required for $\|\cdot\|'$ to be a submultiplicative norm are straightforward to verify and, thus, are omitted.

To verify the property $\|z^*z\|' = (\|z\|')^2$ for all $z \in A^1$, let $z \in A^1$ and $b \in A$. Because A is an algebraic ideal of A^1, $zb \in A$; thus,

$$\|zb\|^2 = \|(zb)^*(zb)\| = \|b^*(z^*z)b\| = \|b^*(z^*z)b\|' \leq \|b\|^2 \|z^*z\|'. \tag{11.3}$$

To show that $\|z^*z\|' \geq (\|z\|')^2$, note that for each $\varepsilon > 0$ there is a $b \in A$ with $\|b\| \leq 1$ such that $\|zb\| > (1 - \varepsilon)\|z\|'$. Thus, $\|z^*z\|' > (1 - \varepsilon)^2(\|z\|')^2$ by (11.3). As $\varepsilon > 0$ is arbitrary, the inequality $\|z^*z\|' \geq (\|z\|')^2$ must hold. Conversely, because $\|\cdot\|'$ is submultiplicative and $*$ is an isometry on A^1 with respect to $\|\cdot\|'$, the inequality $\|z^*z\|' \leq \|z^*\|'\|z\|' = (\|z\|')^2$ leads to the conclusion that $\|z^*z\|' = (\|z\|')^2$.

To show $\|\cdot\|'$ extends the norm $\|\cdot\|$ on A, let $a \in A$. For every $b \in A$ with $\|b\| \leq 1$, $\|ab\| \leq \|a\| \|b\| \leq \|a\|$; thus, $\|a\|' \leq \|a\|$. On the other hand, if a is normalised so as to have norm $\|a\| = 1$, then $\|a\|' \geq \|aa^*\| = \|a^*\|^2 = \|a\|^2 = 1 = \|a\|$. This proves that $\|a\|' = \|a\|$, for every $a \in A$.   □

Definition 11.50. If A is a nonunital C*-algebra, then the C*-algebra A^1 is called the *unitisation* of A and the norm $\|\cdot\|'$ on A^1 is simply denoted by $\|\cdot\|$.

Proposition 11.46 indicates that the notation $\sigma(x)$ for the spectrum of x is unambiguous. Therefore, one can define the spectrum for elements of nonunital C*-algebras.

Definition 11.51. If A is a nonunital C*-algebra, and if $x \in A$, then the *spectrum* of x is the set

$$\sigma(x) = \{\lambda \in \mathbb{C} \,|\, x - \lambda 1 \text{ is not invertible in } A^1\}.$$

Another consequence of spectral permanence is that the norm on a C*-algebra is necessarily unique.

Proposition 11.52. *If A is a C*-algebra with norm $\|\cdot\|$ and if $\|\cdot\|'$ on a norm such that A is a C*-algebra with respect to $\|\cdot\|'$, then $\|x\|' = \|x\|$ for every $x \in A$.*

Proof. By (4) of Proposition 11.49, we may suppose without loss of generality that A is a unital C*-algebra.

Let $x \in A$ and consider x^*x. The Gelfand transform Γ corresponding to the unital abelian C*-algebra C of A^1 generated by x^*x is an isometry. Hence, $\|x^*x\|$ is the norm of $\Gamma(x^*x)$ in $C(\mathscr{R}_C)$—that is, $\|x^*x\|$ is the maximum modulus of the elements in the range of $\Gamma(x^*x)$ and is, therefore, given by the spectral radius of x^*x (since $\sigma_C(x^*x) = \sigma(x^*x)$ by spectral permanence). The invertibility of elements in A^1 is based upon a purely algebraic criterion; therefore, the spectrum of x^*x is independent of the norm on A. Consequently, $\|x^*x\|' = \mathrm{spr}\,(x^*x)$ as well. Thus, for each $x \in A$,

$$\|x\|'^2 = \|x^*x\|' = \mathrm{spr}\,(x^*x) = \|x^*x\| = \|x\|^2,$$

which completes the argument.   □

11.6   Positive Elements and Functionals

The definition of a positive operator T acting on a Hilbert space H is not purely algebraic in that the definition takes into the action of T on H, as well as the way the inner product is defined. Therefore, in the abstract setting of C*-algebras a different approach is required.

Definition 11.53. An element $h \in A$ is *positive* if $h^* = h$ and $\sigma(h) \subset [0, \infty)$.

In addition to positive operators acting on a Hilbert space, one has the following example, which is a consequence of the definitions.

Example 11.54. *If X is a compact Hausdorff space, then $f \in C(X)$ is a positive element of the unital C^*-algebra $C(X)$ if and only if $f(t) \geq 0$ for every $t \in X$.*

Let A_+ be the set

$$A_+ = \{h \in A \mid h \text{ is a positive element of } A\}.$$

Note that if A is a C^*-subalgebra of a C^*-algebra B, then $A_+ \subseteq B_+$ (since the spectrum of an element x is independent of the C^*-algebra that contains x).

The first main objectives of this section are to show (i) that every positive element has a unique positive square root, (ii) that A_+ is a pointed convex cone, and (iii) that $x^*x \in A_+$ for every $x \in A$. This latter fact is not as obvious as one might expect. The final goal of this section is to examine those linear functionals on A that take on nonnegative real values on positive elements of A.

Proposition 11.55. *The following statements are equivalent for a hermitian element h in a unital C^*-algebra A:*

1. $h \in A_+$;
2. $h = b^2$ for some $b \in A_+$ such that $b \in C^*(h)$;
3. $\|\alpha 1 - h\| \leq \alpha$, for every $\alpha \geq \|h\|$;
4. $\|\alpha_0 1 - h\| \leq \alpha_0$, for some $\alpha_0 \geq \|h\|$.

Proof. (1) $\Rightarrow$ (2). Because h is positive, $\sigma(h) \subset \mathbb{R}_+$. Let $X = [0, \|h\|]$ and let $f \in C(X)$ be given by $f(t) = \sqrt{t}$. By Proposition 11.48, the hermitian element $b = f(h)$ is an element of $C^*(h)$. Furthermore, $\sigma(b) = \{f(\sqrt{\lambda}) \mid \lambda \in \sigma(h)\} \subset [0, \infty)$, which implies that $b \in C^*(h)_+$, and so $b \in A_+$.

(2) $\Rightarrow$ (3). Assume that $h = b^2$ for some positive $b \in C^*(h)$. Choose any $\alpha \geq \|h\|$. By spectral mapping, $\sigma(b^2) = \{\lambda^2 \mid \lambda \in \sigma(b)\}$; thus, $\sigma(b^2) \subset \mathbb{R}_+$. Since the norm of a positive element is its spectral radius, $0 \leq \lambda \leq \|h\| \leq \alpha$ implies that $\alpha - \lambda = |\alpha - \lambda| \leq \alpha$ for every $\lambda \in \sigma(h)$. Hence, $\alpha \geq \mathrm{spr}(\alpha 1 - h) = \|\alpha 1 - h\|$.

(3) $\Rightarrow$ (4). This is trivial.

(4) $\Rightarrow$ (1). Assume that $\alpha_0 \geq \|h\|$ satisfies $\alpha_0 \geq \|\alpha_0 1 - h\|$. Thus, if $\lambda \in \sigma(h)$, then $|\lambda| \leq \alpha_0$ and $|\alpha_0 - \lambda| \leq \alpha_0$; that is, $\lambda \geq 0$. $\square$

If $h \in A$ is positive, then the positive element $b \in C^*(h)$ that satisfies $b^2 = h$ in assertion (2) of Proposition 11.55 is unique, as shown by the following proposition.

Proposition 11.56. *If $b_1, b_2 \in A_+$ are such that $b_1^2 = b_2^2$, then $b_1 = b_2$.*

Proof. Let $\beta > 0$ be large enough so that $\sigma(h) \cup \sigma(b_1) \cup \sigma(b_2) \subseteq [0, \beta]$, where $h = b_1^2 = b_2^2$. Therefore, for any $g \in C([0, \beta])$,

$$\|g(b_j)\| = \max_{\lambda \in \sigma(b_j)} |g(\lambda)| \leq \max_{0 \leq t \leq \beta} |g(t)|.$$

By Exercise 11.123, there is a sequence of polynomials f_n such that $f_n(0) = 0$, for all $n \in \mathbb{N}$, and $|f_n(t) - \sqrt{t}| \to 0$ uniformly on $[0, \beta]$ as $n \to \infty$. Thus, each $f_n(h) \in A$ and $\|f_n(h) - f(h)\| \to 0$. Note that $f_n(h) = f_n(b_j^2) \in A$, and so

$$\|f(h) - b_j\| = \lim_n \|f_n(h) - b_j\| = \lim_n \|f_n(b_j^2) - b_j\| = \lim_n \|g_n(b_j)\|,$$

where $g_n(t) = f_n(t^2) - t$, for each n. Since $g_n \to 0$ uniformly on $[0, \beta]$ as $n \to \infty$, we have that $\|g_n(b_j)\| \to 0$, whence $f(h) = b_1 = b_2$.   □

Definition 11.57. If $h \in A_+$, then the unique $b \in A_+$ for which $b^2 = h$ is called the *positive square root* of h and is denoted by $h^{1/2}$.

The decomposition of a hermitian operator as a difference of positive operators with product 0 applies to the C*-algebra setting as well. The result is proved again below because of the need to account for nonunital algebras.

Proposition 11.58. *If h is a hermitian element of a C*-algebra A, then there are positive $h_+, h_- \in A_+$ such that $h = h_+ - h_-$ and $h_+h_- = h_-h_+ = 0$.*

Proof. First assume that A is unital. The C*-algebra $C^*(h, 1)$ is a unital, abelian C*-subalgebra of A; moreover, $C^*(h, 1)$ and $C(\sigma(h))$ are isometrically *-isomorphic. Let $X = [-\|h\|, \|h\|]$, a compact set that contains $\sigma(h)$ and 0. Consider the functions $f, g \in C(X)$ defined by $f(t) = (t + |t|)/2$ and $g(t) = f(-t)$. The functions f and g are nonnegative and vanish at 0; thus, by the Spectral Mapping Theorem and Proposition 11.48 the elements $f(h)$ and $g(h)$ are positive and belong to $C^*(h)$. Let $h_+ = f(h)$ and $f_- = g(h)$. Because $t = f(t) - g(t)$ and $f(t)g(t) = 0$ for all $t \in X$, the Continuous Functional Calculus yields $h = h_+ - h_-$ and $h_+h_- = h_-h_+ = 0$.

If A is nonunital, then consider A as a C*-subalgebra of its minimal unitisation A^1. The argument above yields $h_+, h_- \in C^*(h)_+ \subseteq A_+ \subset (A^1)_+$ such that $h = h_+ - h_-$ and $h_+h_- = h_-h_+ = 0$, thereby completing the proof.   □

Our second objective for this section is achieved by the next result.

Proposition 11.59. *If A is a C*-algebra, then A_+ is a pointed convex cone. That is, if $\gamma, \delta \in [0, \infty)$ and if $h, k \in A_+$, then*

1. $\gamma h + \delta k \in A_+$, and
2. $-h \in A_+$ only if $h = 0$.

Proof. Exercise 11.125.   □

We now arrive at our third objective for this section.

Theorem 11.60. $A_+ = \{x^*x \,|\, x \in A\}$, *for every C*-algebra A.*

Proof. If $h \in A_+$, then assertion (2) of Proposition 11.55 yields a positive element $b \in C^*(h)$ such that $b^2 = h$. Hence, $h = b^*b \in \{x^*x \,|\, x \in A\}$.

Conversely, let $x \in A$. By Proposition 11.58, the hermitian element $x^*x \in A$ may be expressed as $x^*x = b_+ - b_-$, where $b_+, b_- \in A_+$ and $b_+b_- = b_-b_+ = 0$. To show that $x^*x \in A_+$ it is sufficient to prove that $b_- = 0$.

Let $c = (b_-)^{1/2} \in A_+$ and $a = xc$. By Exercise 11.123, $f(t) = \sqrt{t}$ can be approximated uniformly on $\sigma(b_-)$ by polynomials p such that $p(0) = 0$. Since $p(b_-)b_+ = b_+p(b_-) = 0$ for any polynomial for which $p(0) = 0$, we conclude that $cb_+ = b_+c = 0$. Hence,

$$-a^*a = -cx^*xc = -c(b_+ - b_-)c = cb_-c = b_-^2,$$

which implies that $\sigma(-aa^*) \subset \mathbb{R}_+$. Thus,

$$\sigma(a^*a) \subset -(\mathbb{R}_+). \tag{11.4}$$

Let $u, v \in A$ be the real and imaginary parts of a; thus, $a = u + iv$. By the Spectral Mapping Theorem, u^2 and v^2 are positive and so, by Proposition 11.59, $u^2 + v^2 \in A_+$. Therefore, $a^*a + aa^* \in A_+$ as well, since $a^*a + aa^* = 2(u^2 + v^2)$. By Proposition 11.59 once again, we have that $a^*a + aa^* + b_-^2 \in A_+$. But

$$a^*a + aa^* + b_-^2 = a^*a + aa^* - a^*a = aa^*;$$

this shows that $aa^* \in A_+$ and so

$$\sigma(aa^*) \subset \mathbb{R}. \tag{11.5}$$

Theorem 9.3 asserts that $\sigma(aa^*) \cup \{0\} = \sigma(a^*a) \cup \{0\}$. Therefore, (11.4) and (11.5) combine to give

$$\sigma(a^*a) \subseteq (-\mathbb{R}_+) \cap \mathbb{R}_+ = \{0\}.$$

Therefore, the spectral radius of a^*a is 0. Since the spectral radius and norm coincide for hermitian elements, $a^*a = 0$. That is, $0 = \|a^*a\| = \|a\|^2$, which proves that $a = 0$. Since $b_-^2 = -a^*a = 0$ and b_- is positive, we obtain $b_- = [b_-^2]^{1/2} = 0^{1/2} = 0$. □

Definition 11.61. If $h, k \in A_{sa}$, then $h \leq k$ if $k - h \in A_+$.

The relation "$\leq$" on A_{sa} has the following properties (Exercise 11.126). If $a, b, c \in A_{sa}$, then:

1. $a \leq a$;
2. if $a \leq b$ and $b \leq a$, then $b = a$; and
3. if $a \leq b$ and $b \leq c$, then $a \leq c$.

That is, "$\leq$" is a partial order on the $\mathbb{R}$-vector space A_{sa}.

Proposition 11.62. *If $h, k \in A_{sa}$ satisfy $h \leq k$, then $x^*hx \leq x^*kx$ for every $x \in A$.*

Proof. If $x \in A$, then

$$x^*kx - x^*hx = x^*(k-h)x = x^*(k-h)^{1/2}(k-h)^{1/2}x = z^*z \in A_+,$$

where $z = (k-h)^{1/2}x$. □

As with Hilbert space operators, one has the following definition.

Definition 11.63. If A is a C^*-algebra and if $x \in A$, then the *modulus* of x is the element $|x| \in A$ defined by $(x^*x)^{1/2}$.

If X is a compact Hausdorff space and if μ is a probability measure on the Borel sets of X, then the linear functional $\varphi : C(X) \to \mathbb{C}$ defined by

$$\varphi(f) = \int_X f \, d\mu,$$

satisfies $\varphi(f) \geq 0$ whenever $f(t) \geq 0$ for all $t \in X$. This example motivates the following definition.

Definition 11.64. A linear functional $\varphi : A \to \mathbb{C}$ on a C^*-algebra A is a *positive linear functional* if $\varphi(h) \geq 0$ for every $h \in A_+$. If, in addition, $\|\varphi\| = 1$, then the positive linear functional φ is called a *state* on A. The *state space* of A is the set $S(A)$ of all states on A.

Example 11.65. *If $\{\xi_n\}_{n \in \mathbb{N}}$ is a sequence of vectors for which $\sum_{n=1}^{\infty} \|\xi_n\|^2$ converges, then the function $\varphi : B(H) \to \mathbb{C}$ defined by*

$$\varphi(x) = \sum_{n=1}^{\infty} \langle x\xi_n, \xi_n \rangle$$

is a positive linear functional on $B(H)$.

Positive linear functionals on A necessarily map A_{sa} onto $\mathbb{R}$, which can be seen via writing $h \in A_{sa}$ as $h = h_+ - h_-$, where $h_+, h_- \in A_+$. Therefore, by expressing any $x \in A$ in terms of its real and imaginary parts, we obtain

$$\varphi(x^*) = \overline{\varphi(x)}, \quad \forall x \in A, \ \forall \varphi \in S(A).$$

Proposition 11.66 (Schwarz Inequality). *If $\varphi \in S(A)$ and $x, y \in A$, then*

$$|\varphi(y^*x)|^2 \leq \varphi(x^*x) \varphi(y^*y). \tag{11.6}$$

Proof. The equation $[x, y] = \varphi(y^*x)$ defines a sesquilinear form on $A \times A$. Therefore, the proof of the inequality can be achieved by arguing as in the proof of the Cauchy-Schwarz inequality in Hilbert space.

Choose $x, y \in A$. If $[x, y] = 0$, then the inequality holds trivially. Thus, assume that $[x, y] \neq 0$. Note that $x^*x, y^*y \in A_+$ imply that $[x, x], [y, y] \in \mathbb{R}_+$. For any $\lambda \in \mathbb{C}$,

$$0 \leq [x - \lambda y, x - \lambda y] = [x, x] - 2\Re(\lambda[y, x]) + |\lambda|^2[y, y].$$

For

$$\lambda = \frac{[x, x]}{[y, x]},$$

the inequality above becomes

$$0 \leq -[x,x] + \frac{[x,x]^2[y,y]}{|[x,y]|^2},$$

which yields inequality (11.6).   □

If A is a unital C*-algebra, then there is a relatively simple criterion for a linear functional to be a state.

Proposition 11.67. *The following statements are equivalent for a linear functional φ of norm $\|\varphi\| = 1$ on a unital C*-algebra A:*

1. φ is a state on A;
2. $\varphi(1) = 1$.

Proof. Assume that φ is a state on A. Because $1 = 1^*1 \in A_+$ and $\|1\| = 1$, we have that $0 \leq \varphi(1^*1) = \varphi(1) \leq \|\varphi\| \|1\| = 1$. To show that $1 \leq \varphi(1)$, choose any $x \in A$ with $\|x\| \leq 1$. Thus, $\|x^*x\| \leq 1$. Since $\|x^*x\| = r(x^*x)$ and $\sigma(x^*x) \subset \mathbb{R}_+$, the hermitian element $1 - x^*x$ is positive in A. Thus, $0 \leq \varphi(1 - x^*x) = \varphi(1) - \varphi(x^*x)$, which implies that $\varphi(x^*x) \leq \varphi(1)$. Therefore, by an application of the Schwarz inequality,

$$|\varphi(x)| = |\varphi(1^*x)| \leq \varphi(x^*x)\varphi(1^*1) \leq \varphi(1)^2 \leq 1,$$

since $\varphi(1) \leq 1$. Hence $|\varphi(x)| \leq 1$, for all $x \in A$ with $\|x\| \leq 1$, implies that $\|\varphi\| \leq \varphi(1)$. By hypothesis, $\|\varphi\| = 1$; therefore, $\varphi(1) = 1$.

Conversely, suppose that $\varphi(1) = 1$; thus, $\|\varphi\| = \varphi(1) = 1$. It must happen that $\varphi(A_{\text{sa}}) = \mathbb{R}$, for if not then there is a hermitian element $h \in A_{\text{sa}}$ such that $\varphi(h) = \alpha + i\beta$, where $\alpha, \beta \in \mathbb{R}$ and $\beta \neq 0$. Therefore, with $k = \beta^{-1}(h - \alpha 1) \in A_{\text{sa}}$, we would have that $\varphi(k) = i$ and, for each $\gamma \in \mathbb{R}$,

$$(\gamma + 1)^2 = |i + \gamma i|^2 = |\varphi(k + \gamma i1)|^2 \leq \|\varphi\|^2 \|k + \gamma i1\|^2$$

$$= \|(k + \gamma i1)^*(k + \gamma i1)\| = \|k^2 + \gamma^2 1\| = \|k^2\| + \gamma^2.$$

Thus, $(2\gamma + 1) \leq \|k^2\|$ for all $\gamma \in \mathbb{R}$. But this is impossible; therefore, it must be that $\varphi(h)$ is real for every $h \in A_{\text{sa}}$. If $h \in A_+$, then $\|\varphi\| = 1$ and $\varphi(h) \in \mathbb{R}$ imply that $\varphi(h) \in [-\|h\|, \|h\|]$. Thus, $\|h\| \geq \|h\| - \varphi(h) \geq 0$, which implies that $\varphi(h) \geq 0$.   □

For every $\varphi \in S(A)$ and $x \in A$ we have the basic inequality $|\varphi(x)| \leq \|x\|$. If x is positive, then equality is achieved for some state φ, as shown by the following result.

Proposition 11.68. *For every $h \in A_+$ there is a state φ on A with $\varphi(h) = \|h\|$.*

Proof. If A is nonunital, then consider the unitisation A^1 of A; otherwise, let A^1 denote A in the case where A is unital.

If $h \in A_+$, then $h \in (A^1)_+$ as well. Consider the unital abelian C^*-algebra $C^*(h,1)$ generated by h. Via the Gelfand transform, there is a character $\rho : C^*(h,1) \to \mathbb{C}$ such that $\rho(h) = \|h\|$. Of course, $\rho(1) = \|\rho\| = 1$. By the Hahn-Banach Theorem, ρ extends to a linear function $\Phi : A^1 \to \mathbb{C}$. Since $\|\Phi\| = \Phi(1) = 1$, Φ is a state on A^1 by Proposition 11.67. Thus, if A is unital, we may take $\varphi = \Phi$. If A is nonunital, then let $\varphi = \Phi_{|A}$. Note that $\varphi(k) \geq 0$ for all $k \in A_+$ and that $\|\varphi\| \leq 1$. With $k = \|h\|^{-1}h \in A_+$, we have $\|k\| = 1$ and $\varphi(k) = 1$. Hence $\|\varphi\| = 1$, and so φ is a state on A. $\square$

11.7 Ideals and Quotients

Ideals of C^*-algebras inherit many properties of the ambient C^*-algebra. First and foremost of these is that every ideal of a C^*-algebra is itself a C^*-algebra, which is proved as Theorem 11.70 below.

Lemma 11.69. *If J is an ideal of a C^*-algebra A and if $x \in J$, then there is a sequence $\{e_n\}_{n \in \mathbb{N}} \subset J_+$ such that $\sigma(e_n) \subset [0,1]$, for all $n \in \mathbb{N}$, and $\|xe_n - x\| \to 0$.*

Proof. First suppose that $x \in A$. If A is unital and if $e \in A_+$ satisfies $\sigma(e) \subset [0,1]$, then $\|1 - e\| \leq 1$ (Exercise 11.132). Thus, $\|x - xe\|^2 = \|(1-e)x^*x(1-e)\| \leq \|x^*x(1-e)\| = \|x^*x - x^*xe\|$. If A is nonunital, then one can embed A into A^1 to produce the same inequality. Therefore, regardless of whether A is unital or not,

$$\|x - xe\|^2 \leq \|x^*x - x^*xe\|, \quad \forall e \in A_+ \text{ with } \sigma(e) \subseteq [0,1]. \tag{11.7}$$

Suppose now that $x \in J$. Because J is an ideal, $x^*x \in J$. Let $h = x^*x$. For each $n \in \mathbb{N}$, let $f_n(t) = nt/(1+nt)$; thus, $f_n \in C(\sigma(h))$, $0 \leq f_n(t) \leq 1$, for all t, and $f_n(0) = 0$. Let $e_n = f_n(h)$. Theorem 11.47 and Proposition 11.48 show that $e_n \in J_+$ and $\sigma(e_n) \subset [0,1]$. We aim to verify that $\|h - he_n\| \to 0$. To this end, note that if $t \in \sigma(h)$, then

$$t - tf_n(t) = \frac{t}{1+nt} = \left(\frac{nt}{1+nt}\right)\left(\frac{1}{n}\right) < \frac{1}{n}, \quad \forall t \in \sigma(h).$$

Therefore, by the fact that continuous functional calculus is an isometric $*$-homomorphism, $\|h - he_n\| < 1/n$. Hence, by inequality (11.7),

$$\|x - xe_n\|^2 \leq \|x^*x - x^*e_n\| < \frac{1}{n}.$$

That is, $\lim_{n \to \infty} \|xe_n - x\| = 0$. $\square$

Theorem 11.70. *If J is an ideal of a C^*-algebra A, then J is a C^*-subalgebra of A.*

Proof. All that needs to be verified is that $x^* \in J$ for every $x \in J$. By Lemma 11.69, there is a sequence $\{e_n\}_{n \in \mathbb{N}} \subset J_+$ such that $\sigma(e_n) \subset [0,1]$, for all $n \in \mathbb{N}$, and $\|xe_n - x\| \to 0$. Note that $e_n x^* \in J$ for every $n \in \mathbb{N}$. The C^*-norm is isometric, and so

$$\lim_{n\to\infty} \|e_n x^* - x^*\| = \lim_{n\to\infty} \|x e_n - x\| = 0.$$

Because J is closed and because each $e_n x^* \in J$, we conclude that $x^* \in J$. □

Define a function $*$ on A/J by

$$\left(\dot{x}\right)^* = \left(\dot{x^*}\right).$$

It is clear that this definition above yields an involution on the associative algebra A/J.

Recall that A/J is a Banach algebra under the quotient norm

$$\|\dot{x}\| = \inf\{\|x - b\| \,|\, b \in J\}.$$

The new fact that is proved below is that the quotient norm satisfies the C*-norm axiom $\|\dot{x}\|^2 = \|\dot{x}^* \dot{x}\|$.

Theorem 11.71. *If J is an ideal of a C*-algebra A, then the quotient Banach algebra A/J is a C*-algebra with respect to quotient norm and the involution $\dot{x}^* = (\dot{x^*})$.*

Proof. Because J is closed under the involution, the function $\dot{x} \mapsto (\dot{x^*})$ is a well-defined involution on the quotient A/J. Because A/J is a Banach algebra in the quotient norm, the only issue remaining to be verified is that the quotient norm satisfies $\|\dot{x}\|^2 = \|\dot{x}^* \dot{x}\|$. To this end, fix $x \in A$ and define

$$E = \{e \in J_+ \,|\, \sigma(e) \subseteq [0,1]\}.$$

If A is unital and if $e \in E$, then $\|1 - e\| \le 1$ and, for any $b \in J$, $\|x + b\| \ge \|(x + b)(1 - e)\| = \|(x - xe) + (b - be)\|$. If A is nonunital, then one can embed A into A^1 to produce the same inequality. Hence,

$$\|x + b\| \ge \|(x - xe) + (b - be)\|, \quad \forall e \in E, b \in J,$$

regardless of whether A is unital or not. By definition of the quotient norm,

$$\|\dot{x}\| \le \inf\{\|x - xe\| \,|\, e \in E\}. \tag{11.8}$$

To show that equality holds in (11.8), let $b \in J$. By Lemma 11.69, there is a sequence $\{e_n\}_{n\in\mathbb{N}} \subset E$ such that $\|be_n - b\| \to 0$. Thus, for every $n \in \mathbb{N}$,

$$\|x + b\| \ge \|(x - xe_n) + (b - be_n)\|,$$

and so

$$\|x + b\| \ge \liminf_n \|x - xe_n\| \ge \inf_{e \in E} \|x - xe\| \ge \|\dot{x}\|.$$

Therefore,

$$\|\dot{x}\| = \inf_{b \in J} \|x + b\| \geq \|\dot{x}\|$$

implies that $\|\dot{x}\| = \inf\{\|x - e\| \,|\, e \in E\}$, for every $x \in A$. Consequently, by invoking inequality (11.7) we obtain

$$\|\dot{x}\|^2 = \inf_{e \in E} \|x - xe\|^2 \leq \inf_{e \in E} \|x^*x - x^*xe\| = \|\dot{x}^*\dot{x}\|,$$

implying that

$$\|\dot{x}\|^2 \leq \|\dot{x}^*\dot{x}\| \leq \|\dot{x}^*\| \,\|\dot{x}\|. \qquad (11.9)$$

Conversely, $\|\dot{x}^*\| = \inf\{\|x^* - b^*\| \,|\, b \in J\} = \inf\{\|x - b\| \,|\, b \in J\} = \|\dot{x}\|$, since J is ∗-closed. Therefore, inequality (11.9) is an equality. □

Quotient algebras occur by way of the kernels of ∗-homomorphisms. The main features of ∗-homomorphisms are described by the following result.

Proposition 11.72. *If A and B are C^*-algebras, and if $\rho : A \to B$ is a ∗-homomorphism, then*

1. *ρ is continuous and $\|\rho\| \leq 1$,*
2. *ρ is an isometry if and only if $\ker \rho = \{0\}$,*
3. *the kernel of ρ is an ideal of A, and*
4. *the range of ρ is a C^*-subalgebra of B.*

Proof. By Exercise 11.136, $\mathrm{spr}\,\rho(x^*x) \leq \mathrm{spr}\,(x^*x)$, for all $x \in A$. Thus,

$$\|\rho(x)\|^2 = \|\rho(x)^*\rho(x)\| = \mathrm{spr}\,\rho(x^*x) \leq \mathrm{spr}\,(x^*x) = \|x^*x\| = \|x\|^2.$$

That is, ρ is bounded and $\|\rho\| \leq 1$, which proves (1).

For (2), it is trivial that isometries are injective, and so only the converse is proved here. Thus, assume that $\ker \rho = \{0\}$. Assume, contrary to what we aim to prove, that there is an element $x \in A$ with $\|\rho(x)\| < \|x\|$. Then, $\|\rho(h)\| < \|h\|$, where $h = x^*x \in A_+$. Let $f : [0, \|h\|] \to \mathbb{R}$ be any continuous function such that $f(t) = 0$ for $t \in [0, \|\rho(h)\|]$ and $f(\|h\|) = 1$. By the Spectral Mapping Theorem, $\|f(\rho(h))\| = 0$ and $\|f(h)\| \geq 1$. Because $f(\rho(h)) = \rho(f(h))$ (by the continuity of ρ and the Weierstrass Approximation Theorem), it must be that $\|\rho(f(h))\| = 0$. Since ρ is injective, this means that $f(h) = 0$—in contradiction of $\|f(h)\| \geq 1$. Therefore, it must be that ρ is isometric if $\ker \rho = \{0\}$, which proves (2).

Since ρ is continuous, $\ker \rho$ is closed. As the kernel of any ∗-homomorphism is an algebraic ideal, we conclude that $\ker \rho$ is an ideal, thereby proving (3).

For the proof of (4), consider the quotient C^*-algebra $A/\ker \rho$ and let $\phi : A/\ker \rho \to B$ be defined by $\phi(\dot{x}) = \rho(x)$, for every $x \in A$. Then ϕ is a well-defined ∗-homomorphism with trivial kernel and range equal to the range of ρ. Thus, by (2), ϕ is an isometry, and so the range of ϕ is closed. Hence, the range of ρ is closed. □

Corollary 11.73. *If two C*-algebras are ∗-isomorphic, then they are isometrically* ∗-*isomorphic.*

One of the most important quotient C*-algebras occurs in operator theory.

Definition 11.74. If H is an infinite-dimensional Hilbert space, then the *Calkin algebra* is the quotient C*-algbera $Q(H) = B(H)/K(H)$.

Example 11.65 shows that $B(H)$ admits positive linear functionals φ of the form

$$\varphi(x) = \sum_{n=1}^{\infty} \langle x\xi_n, \xi_n \rangle,$$

for sequences $\{\xi_n\}_n$ in H in which $\sum_{n=1}^{\infty} \|\xi_n\|^2$ converges. With nonzero positive linear functionals of this form, one can always find a compact operator $k \in B(H)$ such that $\varphi(k) \neq 0$ (Exercise 11.139). In light of the following example, not all states on $B(H)$ are given by such a formula.

Example 11.75. *If H is an infinite-dimensional Hilbert space, then there exists a state φ on $B(H)$ such that $\varphi(k) = 0$ for every compact operator $k \in B(H)$.*

Proof. Let $q : B(H) \to Q(H)$ be the quotient map $q(x) = \dot{x}$, mapping $B(H)$ onto the Calkin algebra $Q(H)$. Because H has infinite dimension, $Q(H) \neq \{0\}$. Select a nonzero positive $h \in Q(H)$. By Proposition 11.68, there is a state ψ on $Q(H)$ with $\psi(h) = \|h\|$. Let $\varphi = \psi \circ q$, which is a positive linear functional such that $\varphi(1) = 1$. Because $q(K(H)) = \{\dot{0}\}$ in $Q(H)$, the state φ annihilates every compact operator in $B(H)$. □

11.8 Representations and Primitive Ideals

Some C*-algebras, such as the Calkin algebra, occur abstractly rather than as a C*-algebra of Hilbert space operators. The goal of this section is to show, for any C*-algebra A, the existence of ∗-homomorphisms $\pi : A \to B(H)$ (for an appropriate choice of Hilbert space H) where by π is isometric; in so doing, A and the C*-algebra $\pi(A)$ of operators acting on H are isometrically ∗-isomorphic.

Definition 11.76. A *representation* of a C*-algebra A on a Hilbert space H is a ∗-homomorphism $\pi : A \to B(H)$. Further, π is:

1. *unital*, if A is a unital C*-algebra and $\pi(1) = 1$;
2. *nondegenerate*, if the only $\xi \in H$ that satisfies $\pi(a)\xi = 0$ for all $a \in A$ is $\xi = 0$;
3. *cyclic*, if there is a vector $\xi \in H$ such that $\{\pi(a)\xi \,|\, a \in A\}$ is dense in H;
4. *irreducible*, if the commutant of $\pi(A)$ in $B(H)$ is $\{\lambda 1 \,|\, \lambda \in \mathbb{C}\}$.

The following theorem reveals a close relationship between states and representations of C*-algebras.

Theorem 11.77 (Gelfand-Naimark-Segal). *Assume that φ is a state on a unital C^*-algebra A.*

1. *(Existence) There exists a unital representation π of A on a Hilbert space H_π and a unit vector $\xi \in H_\pi$ such that*

   a. *ξ is a cyclic vector for $\pi(A)$, and*
   b. *$\varphi(x) = \langle \pi(x)\xi, \xi \rangle$ for every $x \in A$.*

2. *(Uniqueness) Given a triple (H_π, π, ξ) as in (1), if $\rho : A \to B(H_\rho)$ is a unital representation of A, if $\eta \in H_\rho$ is a unit cyclic vector for $\rho(A)$, and if $\varphi(x) = \langle \rho(x)\eta, \eta \rangle$, for every $x \in A$, then there is a surjective isometry $u : H_\pi \to H_\rho$ such that $u\xi = \eta$ and $u\pi(x) = \rho(x)u$, for every $x \in A$.*

Proof. Let $L = \{x \in A \mid \varphi(y^*x) = 0 \text{ for every } y \in A\}$. By the Schwarz inequality for states, $|\varphi(b^*a)| \leq \varphi(a^*a)\varphi(b^*b)$ for every $a, b \in A$, This inequality implies that L is a closed set; hence, because L is also a vector space, L is a subspace of A.

For each $x \in L$ and $a, y \in A$,

$$\varphi\big(y^*(ax)\big) = \varphi\big((y^*a)x\big) = \varphi\big((a^*y)^*x\big) = 0,$$

implying that $ax \in L$.

Define a sesquilinear form $\langle \cdot, \cdot \rangle$ on the quotient vector space A/L by

$$\langle \dot{a}, \dot{b} \rangle = \varphi(b^*a)$$

A straightforward computation shows that this form is well defined. Moreover, if $\langle \dot{a}, \dot{a} \rangle = 0$, then $\varphi(a^*a) = 0$ and, by the Schwarz inequality, $\varphi(y^*a) = 0$ for every $y \in A$. Thus, $\langle \dot{a}, \dot{a} \rangle = 0$ only if $\dot{a} = \dot{0}$, which proves that $\langle \cdot, \cdot \rangle$ is an inner product on A/L. In the metric on A/L induced by the norm $\|\dot{x}\| = \langle \dot{x}, \dot{x} \rangle^{1/2}$, let H_π be the completion of A/L. Thus, H_π is a Hilbert space that contains A/L as a dense linear submanifold.

For each $x \in A$, let $\pi_0(x) : A/L \to A/L$ be the (well-defined) linear transformation $\pi_0(x)[\dot{a}] = (\dot{xa})$. Because

$$\|\pi_0(x)[\dot{a}]\|^2 = \varphi(a^*(x^*x)a) \leq \|x^*x\|\varphi(a^*a) = \|x\|^2\|\dot{a}\|^2,$$

the linear transformation $\pi_0(x)$ extends to an operator $\pi(x)$ on H_π of norm at most $\|x\|$. Furthermore,

$$\langle \pi_0(x)\dot{a}, \dot{b} \rangle = \varphi\big(b^*(xa)\big) = \varphi\big((x^*b)^*a\big) = \langle \dot{a}, \pi_0(x^*)\dot{b} \rangle$$

implies that $\pi(x)^* = \pi(x)$. Because π_0 is a contractive $*$-homomorphism of A into $B(H_\pi)$, π is also a contractive $*$-homomorphism.

Let $\xi = \dot{1} \in A/L \subset H_\pi$ and note that $\langle \pi(x)\xi, \xi \rangle = \varphi(1^*x) = \varphi(x)$ for every $x \in H_\pi$ and that

$$\{\pi(x)\xi_\varphi \mid a \in A\} = A/L.$$

Hence, ξ_φ is a cyclic vector for $\pi(A)$, which establishes the existence of the trip (H_π, π, ξ) with stated properties in (1).

The proof of the uniqueness assertion (2) is left as Exercise 11.138. □

Definition 11.78. If φ is a state on a unital C*-algebra, then a triple (H_π, π, ξ) consisting of a Hilbert space H_π, a representation π of A on H_π, and a unit vector $\xi \in H_\pi$ is called a *GNS-triple* for φ if (i) ξ is a cyclic vector for $\pi(A)$ and (ii) $\varphi(x) = \langle \pi(x)\xi, \xi \rangle$, for every $x \in A$.

The main fundamental fact about representations C*-algebras is the following theorem.

Theorem 11.79 (Gelfand-Naimark). *For every unital C*-algebra A there exists a Hilbert space H and a unital representation $\pi : A \to B(H)$ such that π is injective. Moreover, if A is separable, then H can be taken to be a separable Hilbert space.*

Proof. Fix $z \in A$. By Proposition 11.68, there is a state φ_0 on A such that $\varphi_0(z^*z) = \|z^*z\|$. Let $C_z = \{\varphi \in S(A) \,|\, \varphi(z^*z) = \|z^*z\|\}$, which is a convex and weak*-closed subset of the unit sphere in the dual space of A. Therefore, the Kreĭn-Milman Theorem (Theorem 7.18) asserts that C_z has an extreme point, say φ. This state φ is also an extreme point of $S(A)$, for if $\varphi = \frac{1}{2}(\varphi_1 + \varphi_2)$, then the fact that states are contractive implies that $\varphi_1(z^*z) = \varphi_2(z^*z) = \|z^*z\|$; hence, $\varphi_1, \varphi_2 \in C_z$, implying that $\varphi_1 = \varphi_2 = \varphi$.

Suppose now that (H, π, ξ) is a GNS-triple for φ. If π is not an irreducible representation of A, then there exists a projection $p \in B(H)$ such that $p \neq 0$, $p \neq 1$, and $p\pi(x) = \pi(x)p$ for all $x \in A$. If $p\xi$ were 0, then it would be true that $p(\pi(x)\xi) = \pi(x)p = \xi$ for every $x \in A$; however, such vectors form a dense subspace of H and this would imply that $p = 0$, contrary to the assumption on p. Likewise, $1 - p \neq 0$. Let $t = \|p\xi\|^2$ so that $1 - t = \|(1-p)\xi\|^2$ and $t \in (0,1)$. Define states φ_1 and φ_2 by

$$\varphi_1(x) = \frac{1}{t}\langle \pi(x)p\xi, p\xi \rangle \quad \text{and} \quad \varphi_2(x) = \frac{1}{1-t}\langle \pi(x)(1-p)\xi, (1-p)\xi \rangle.$$

Because $\varphi = t\varphi_1 + (1-t)\varphi_2$, we have that $\varphi_1 = \varphi_2 = \varphi$. In particular, the equation

$$\langle \pi(x)\xi, \xi \rangle = t^{-1}\langle \pi(x)p\xi, p\xi \rangle = t^{-1}\langle \pi(x)\xi, p\xi \rangle$$

for every $x \in A$ implies that

$$\langle \pi(x)\xi, p\xi - t^{-1}\xi \rangle = 0$$

for every $x \in A$. Because the $\{\pi(x)\xi \,|\, x \in A\}$ is dense, the equation above yields $p\xi = t^{-1}\xi$, which implies t^{-1} is an eigenvalue of p. However, $t^{-1} \notin \{0, 1\} = \sigma(p)$ and, therefore, it must be that the only projections that commute with $\pi(A)$ are 0 and 1. Hence, π is a unital irreducible representation and it has the property that

$$\|z\|^2 \geq \|\pi(z)\|^2 = \|\pi(z^*z)\| \geq \langle \pi(z^*z)\xi, \xi \rangle = \varphi(z^*z) = \|z^*z\| = \|z\|^2.$$

Hence, $\|\pi(z)\| = \|z\|$.

Denote the GNS-triple associated with the irreducible representation π in the previous paragraph by (H_z, π_z, ξ_z). Consider the Hilbert space $H = \bigoplus_{z \in A} H_z$ and the representation $\pi = \bigoplus_{z \in A} \pi_z$ and note that $\pi : A \to B(H)$ is a unital representation of A for which $\pi(x) = 0$ only if $x = 0$.

If A is separable and if $\{a_n\}_{n \in \mathbb{N}}$ is a countable dense subset of A, then with $H = \bigoplus_{n \in \mathbb{N}} H_{a_n}$ and $\pi = \bigoplus_{n \in \mathbb{N}} \pi_{a_n}$ we have a representation π of A on the separable Hilbert space H for which $\pi(x) = 0$ only if $x = 0$.   □

Corollary 11.80. *Every C^*-algebra is isometrically $*$-isomorphic to a C^*-algebra of Hilbert space operators.*

Proof. If A is a unital C^*-algebra, then Theorem 11.79 applies immediately to achieve the assertion. If A is nonunital, then apply Theorem 11.79 to the unitisation A^1 to achieve an isometric $*$-isomorphism of A with a C^*-algebra of Hilbert space operators.   □

A noteworthy fact that is a consequence of the proof of Theorem 11.79 is:

Proposition 11.81. *If (H_π, π, ξ) is a GNS-triple for a state φ on a unital C^*-algebra A, and if φ is an extreme point of the state space of A, then π is an irreducible representation.*

The converse of Proposition 11.81 is also true.

We begin our consideration of primitive ideals by noting a basic relationship between irreducible representations of A and its ideals J.

Proposition 11.82. *Assume that J is an ideal of a C^*-algebra A.*

1. *If $\pi : A \to B(H)$ is an irreducible representation of A and if $J \not\subseteq \ker \pi$, then $\pi_{|J}$ is an irreducible representation of J.*
2. *If $\rho : J \to B(H)$ is an irreducible representation of J, then ρ extends to an irreducible representation $\pi : A \to B(H)$ of A.*

Proof. Assume that $\pi : A \to B(H)$ is irreducible. Let $\xi \in H$ be any unit vector and consider $H_J = \overline{\{\pi(x)\xi \,|\, x \in J\}}$. Since J is an ideal of A, H_J is $\pi(A)$-invariant. But $\pi(A)$ is an irreducible operator algebra, and so $H_J = \{0\}$ or $H_J = H$. We show that only the latter condition holds. If it were true that $H_J = \{0\}$, then $\pi(x)\pi(a)\xi = 0$ for every $x \in J$ and $a \in A$; but vectors of the form $\pi(a)\xi$, $a \in A$, are dense in H, and so $\pi(x)$ would be zero for every $x \in J$, in contradiction to $J \not\subseteq \ker \pi$. Thus, $H_J = H$. The choice of $\xi \in H$ being arbitrary shows that $\pi_{|J}$ is an irreducible representation of J.

For the second statement, assume $\rho : J \to B(H)$ is an irreducible representation of J. Choose any unit vector $x \in H$. For $a \in A$ define $\pi(a)$ on the dense linear submanifold $\{\rho(x)\xi \,|\, x \in J\}$ by $\pi(a)[\rho(x)\xi] = \rho(ax)\xi$. Then $\pi(a)$ extends to an operator on H and the map $a \mapsto \pi(a)$ determines an irreducible representation of A on H.   □

Definition 11.83. Assume that J is an ideal of a C*-algebra A.

1. J is a *primitive* ideal if $J = \ker \pi$ for some irreducible representation π of A.
2. J is a *prime* ideal if, for any ideals I and K of A, the inclusion $I \cap K \subseteq J$ holds only if $I \subseteq J$ or $K \subseteq J$.

The set of all primitive ideals of A is denoted by $\mathrm{Prim}A$.

Proposition 11.84. *J is prime, for every $J \in \mathrm{Prim}A$.*

Proof. Suppose that $J \in \mathrm{Prim}A$ and assume that I and K are ideals of A such that $I \cap K \subseteq J$. Suppose that $K \nsubseteq J$. Then $K \nsubseteq \ker \pi$, where $\pi : A \to B(H)$ is an irreducible representation of A with $\ker \pi = J$. Hence, $\pi_{|K} : K \to B(H)$ is an irreducible representation of K. Therefore, if $\xi \in H$ is a fixed unit vector, then $H = [\pi(K)\xi]$. In particular if $x \in I$ and $y \in K$, then $\pi(x)(\pi(y)\xi) = \pi(xy)\xi = 0$, as $xy \in I \cap K \subseteq J = \ker \pi$. But vectors of the form $\pi(y)\xi$, $y \in K$, are dense in H; thus, $\pi(x) = 0$ for every $x \in I$, which proves that $I \subseteq J$. $\square$

The converse to Proposition 11.84 is false in general, although it is true if A is separable.

Proposition 11.85. *If I is an ideal of A, then*

$$I = \bigcap_{I \subseteq J, \, J \in \mathrm{Prim}A} J.$$

Proof. Clearly I is a subset of the ideal on the right-hand side of the equation above. To show the other inclusion, assume that $x \notin I$. Thus, $0 \neq \dot{x} \in A/I$ and so there is an irreducible representation ρ of A/I such that $\|\rho(\dot{x})\| = \|\dot{x}\|$. If $q : A \to A/I$ is the canonical quotient homomorphism, then $\pi = \rho \circ q$ is an irreducible representation of A with $I \subseteq \ker \pi$ and $x \notin \ker \pi$. Therefore, x is not an element of the right-hand side of the equation above. $\square$

Definition 11.86. Assume that $\mathscr{F}$ is a nonempty subset of $\mathrm{Prim}A$. The closure of $\mathscr{F}$, which is denoted by $\overline{\mathscr{F}}$, is the set

$$\overline{\mathscr{F}} = \{J \in \mathrm{Prim}A \mid \bigcap_{I \in \mathscr{F}} I \subseteq J\}.$$

A subset $\mathscr{F} \subseteq \mathrm{Prim}A$ is closed if $\overline{\mathscr{F}} = \mathscr{F}$.

The closure operation satisfies the following properties:

(i) $\overline{\emptyset} = \emptyset$;
(ii) $\mathscr{F} \subseteq \overline{\mathscr{F}}$;
(iii) $\overline{\overline{\mathscr{F}}} = \overline{\mathscr{F}}$;
(iv) $\overline{\mathscr{F}_1 \cup \mathscr{F}_2} = \overline{\mathscr{F}_1} \cup \overline{\mathscr{F}_2}$.

Proposition 11.87. *There exists a unique topology $\mathscr{T}$ on $\mathrm{Prim}A$ in which the closed sets $\mathscr{F}$ of $\mathrm{Prim}A$ are precisely those in which $\overline{\mathscr{F}} = \mathscr{F}$, where $\overline{\mathscr{F}}$ is given by Definition 11.86.*

Proof. Exercise 11.140.   □

The topology $\mathscr{T}$ in Proposition 11.87 is called the *Jacobson topology*, or the *hull-kernel topology*, on Prim$A$.

Definition 11.88. Two representations π_1 and π_2 of a C^*-algebra A on Hilbert spaces H_1 and H_2, respectively, are *equivalent*, which is denoted by $\pi_1 \sim \pi_2$, if there is a surjective isometry $u : H_1 \to H_2$ such that $\pi_2(a)u = u\pi_1(a)$, for all $a \in A$.

The next proposition is basically self evident.

Proposition 11.89. *If* Irr$A$ *denotes the set of irreducible representations of A, then $\sim$ is an equivalence relation on* Irr$A$. *Furthermore, if $\pi_1 \sim \pi_2$, then* ker $\pi_1 =$ ker π_2.

Because equivalent irreducible representations have the same kernels, it is convenient to identify such representations by passing to the space Irr$A/\sim$.

Definition 11.90. The *spectrum* of A, denoted by $\hat{A}$, is the set Irr$A/\sim$ of equivalence classes of irreducible representations of A.

Elements of $\hat{A}$ will be denoted by $\dot{\pi}$, where $\pi \in$ Irr$A$.

Definition 11.91. Assume that J is an ideal of a C^*-algebra A.

1. J is a primitive ideal J if $J =$ ker π for some irreducible representation π of A.
2. J is a prime ideal if, for all ideals $I_1, I_2 \subseteq A$, the inclusion $I_1 \cap I_2 \subseteq J$ holds only if $I_1 \subseteq J$ or $I_2 \subseteq J$.

The set of primitive ideals of A will be denoted by Prim$A$. Note that the map

$$\hat{A} \to \text{Prim}A, \quad \dot{\pi} \mapsto \text{ker}\,\pi,$$

is a surjection, and via this surjection one endows the spectrum of A with a topology as follows.

Definition 11.92. A subset $U \subseteq \hat{A}$ is an *open set* if $\{\text{ker}\,\pi \mid \dot{\pi} \in U\}$ is open in the Jacobson topology of Prim$A$.

Thus, the surjection $\hat{A} \to$ Prim$A$ is an open, continuous map.

For the remainder of this section, the spectra and primitive ideal spaces defined above are used to describe $*$-homomorphisms between abelian C^*-algebras and to analyse the structure of ideals in such algebras.

Proposition 11.93. *If $A = C_0(X)$, where X is locally compact and Hausdorff, then $X \simeq \hat{A} \simeq$ Prim$A$.*

Proof. For each $t_0 \in X$ let $\rho_{t_0} : A \to \mathbb{C}$ be given by $\rho_{t_0}(f) = f(t_0)$, for all $f \in A$. Although $\hat{A} = \{\rho_t \mid t \in X\}$ as sets, it is not yet obvious that $\hat{A}$ and $\mathscr{R}_A$ have the same topologies. We first show that X is homeomorphic to Prim$A$ by identifying the closed sets of Prim$A$ with closed sets of X.

Note that $J \in$ Prim$A$ if and only if there is a $t_0 \in X$ such that $J = \{f \in A \mid f(t_0) = 0\}$. Suppose that $\mathscr{F} \subset$ Prim$A$ is arbitrary. Thus, there is a subset $F \subseteq X$ such that

$\mathscr{F} = \{\ker \rho_t \mid t \in F\}$. Claim: $\overline{\mathscr{F}} = \overline{F}$. To prove the claim, let $t_0 \in \overline{F}$. By definition, $\ker \rho_t \in \overline{\mathscr{F}}$ if $\ker \rho_t \supseteq \bigcap_{I \in \mathscr{F}} I$. But this is true, since (by continuity) we have

$$\bigcap_{I \in \mathscr{F}} I = \{f \in C_0(T) \mid f(t) = 0, \ \forall t \in F\} \subseteq \{f \in C_0(T) \mid f(t_0) = 0\}.$$

Conversely, assume that $J = \ker \rho_{t_0} \in \mathrm{Prim}A$ is such that $J \notin \{\ker \rho_t \mid t \in \overline{F}\}$. Thus, $t_0 \notin \overline{F}$. By Theorem 2.43, there is an $f \in C_0(X)$ such that $f(t_0) = 1$ and $f(\overline{F}) = \{0\}$. That is, $f \in \bigcap_{I \in \mathscr{F}} I$ but $f \notin J$; hence, $J \not\supseteq \bigcap_{I \in \mathscr{F}} I$, which is to say that $J \notin \overline{\mathscr{F}}$. This proves that T and $\mathrm{Prim}A$ have the same closed sets, and so $\mathrm{Prim}A$ is locally compact and Hausdorff.

Because the open, continuous, surjective map $\hat{A} \to \mathrm{Prim}A$ is injective (for if $\rho_{t_1} \neq \rho_{t_2}$, then there is an $f \in C_0(X)$ such that $f(t_1) \neq f(t_2)$, and so $g = f - f(t_1)$ belongs to $\ker \rho_{t_1}$ but $g \notin \ker \rho_{t_2}$), it is also true that $\hat{A}$ and $\mathrm{Prim}A$ are homeomorphic. □

Proposition 11.94 (Poincaré Duality). *Assume that X and Y are locally compact Hausdorff spaces. Then a map $\pi : C_0(X) \to C_0(Y)$ is a $*$-homomorphism if and only if there is a continuous map $\psi : Y \to X$ such that*

1. $\psi^{-1}(K)$ is compact in Y for all every compact $K \subset X$, and
2. $\pi(f) = f \circ \psi$, for all $f \in C_0(X)$.

Moreover, ψ is injective if and only if π is surjective, and ψ is surjective if and only if π is injective.

Proof. The sufficiency of the two conditions is clear, as the second defines a $*$-homomorphism $C_0(X) \to C(Y)$, and the first condition shows that $f \circ \psi$ vanishes at infinity, so that indeed $f \circ \psi \in C_0(Y)$.

Conversely, given a $*$-homomorphism $\pi : C_0(X) \to C_0(Y)$, when π is composed with a point evaluation $\rho_y : C_0(Y) \to \mathbb{C}$ the result is a nonzero $*$-homomorphism $C_0(X) \to \mathbb{C}$. Hence, $\rho_y \circ \pi = \rho_x$ for some (uniquely determined) $x \in X$. Let $\psi : Y \to X$ be the function that sends y to x. To show that ψ is continuous, we take advantage of the homeomorphisms $X \simeq \mathrm{Prim}C_0(X)$ and $Y \simeq \mathrm{Prim}C_0(Y)$.

Let $C \subset X$ be a closed set; thus, $\{\ker \rho_x \mid x \in C\}$ is closed in $\mathrm{Prim}C_0(X)$. Choose $y_0 \in \psi^{-1}(C)$ and let $x_0 \in X$ be such that $\rho_{x_0} = \rho_{y_0} \circ \pi$. Because $Y \simeq \mathrm{Prim}C_0(Y)$, $\ker \rho_{y_0}$ is in the closure of $\{\ker \rho_y \mid y \in \psi^{-1}(C)\}$ in $\mathrm{Prim}C_0(Y)$. That is,

$$\bigcap_{y \in \psi^{-1}(C)} \ker \rho_y \subseteq \ker \rho_{y_0}.$$

Hence, it is also true that

$$\bigcap_{y \in \psi^{-1}(C)} \ker \rho_y \circ \pi \subseteq \ker \rho_{y_0} \circ \pi.$$

That is, $\ker \rho_{x_0}$ is in the closure of $\{\ker \rho_x \mid x \in C\}$ in $\mathrm{Prim} C_0(X)$. But this is equivalent to $x_0 \in \overline{C}$. As $\overline{C} = C$, we obtain $y_0 \in \psi^{-1}(C)$, which proves that $\psi^{-1}(C)$ is closed. Hence, $\psi : Y \to X$ is a continuous function.

Let $K \subset X$ be compact. As X is Hausdorff, K is closed; thus, $\psi^{-1}(K)$ is closed in Y, by the continuity of ψ. By Theorem 2.43, there is a function $f \in C_0(X)$ such that $f(K) = \{1\}$. Thus,

$$\{y \in Y \mid |\pi(f)(y)| \geq \frac{1}{2}\} \supseteq \psi^{-1}(K).$$

Because $\pi(f)$ vanishes at infinity, the set on the left-hand side above is compact, which implies that $\psi^{-1}(K)$ is compact.

Because $\pi(f) = f \circ \psi$, for all $f \in C_0(X)$, we obtain the desired formula for π. The remaining assertions of the proposition are straightforward to deduce. □

Proposition 11.95. *Assume that* $I \subset C_0(X)$ *is an ideal. Let* $Z = \bigcap_{f \in I} f^{-1}\{0\}$. *Then*

$$I = \{g \in C_0(X) \mid g_{|Z} = 0\}.$$

Proof. Clearly $I \subseteq \{g \in C_0(T) \mid g_{|Z} = 0\}$. To prove the converse, assume that $h \notin I$. Every ideal is the intersection of primitive ideals that contain it, and so

$$I = \bigcap_{\ker \rho_t \supseteq I} \ker \rho_t.$$

Therefore, $h \notin I$ implies that $h \notin \ker \rho_{t_0}$ for some $t_0 \in X$ with $I \subseteq \ker \rho_{t_0}$. Thus, $t_0 \in Z$ and $h(t_0) \neq 0$ imply that $h \notin \{g \in C_0(X) \mid g_{|Z} = 0\}$. □

Proposition 11.96. *If* $I \subset C_0(X)$ *is an ideal, then* $I \cong C_0(U)$, *where* $U = X \setminus Z$ *and* $Z = \bigcap_{f \in I} f^{-1}\{0\}$.

Proof. The C*-algebra $C_0(X)/I$ is abelian. Let Y denote its maximal ideal space. We obtain a surjective homomorphism $\pi : C_0(X) \to C_0(Y)$ via

$$C_0(X) \to C_0(X)/I \cong C_0(Y).$$

By Poincaré Duality, there is a continuous injective function $\psi : Y \to X$ such that $\pi(f) = f \circ \psi$ for every $f \in C_0(X)$. Note that $I = \ker \pi$, which means (by Theorem 2.43) that $Z = \psi(Y)$. Consider the isometric $*$-homomorphism $\gamma : I \to C(U)$ whereby $\gamma(f) = f_{|U}$ for every $f \in I$. If $\varepsilon > 0$, then

$$\{t \in U \mid |f(t)| \geq \varepsilon\} = \{t \in U \cup Z \mid |f(t)| \geq \varepsilon\};$$

the set on the right-hand side is compact, since $f \in C_0(X)$. Therefore, $\gamma(f) \in C_0(U)$.

To see that γ is surjective, let $f \in C_0(U)_+$ and define $F : X \to \mathbb{R}^+$ by $F(t) = f(t)$, if $t \in U$, and by $F(t) = 0$, if $t \in Z$. Let $r \in \mathbb{R}^+$. The set $F^{-1}(-\infty, r]$ is $Z \cup f^{-1}(0, r]$, which is closed. Hence, F is lower semicontinuous. The set $F^{-1}[r, \infty) = f^{-1}[r, \infty)$, which is also closed; therefore, F is upper semicontinuous. Thus F is continuous. Moreover, F vanishes at infinity and $f = \gamma(F)$. Since $C_0(U)$ is spanned by functions in $C_0(U)_+$, this proves that γ is surjective. □

11.9 Traces and Factors

The set $M_n(\mathbb{C})$ of $n \times n$ complex matrices is a unital C*-algebra, since $M_n(\mathbb{C}) = B(\mathbb{C}^n)$, the C*-algebra of bounded linear operators on the Hilbert space $\mathbb{C}^n$. Of the many important functions defined on matrices, the trace functional is especially relevant to functional analysis, as it is a positive linear functional that behaves as though $M_n(\mathbb{C})$ were commutative. To be more precise, the function $\mathrm{Tr} : M_n(\mathbb{C}) \to \mathbb{C}$ defined by

$$\mathrm{Tr}\left([\alpha_{ij}]_{i,j=1}^n\right) = \sum_{k=1}^n \alpha_{kk},$$

is linear, has the property that $\mathrm{Tr}(x^*x) \geq 0$ for all $x \in M_n(\mathbb{C})$, and satisfies $\mathrm{Tr}(xy) = \mathrm{Tr}(yx)$, for all $x, y \in M_n(\mathbb{C})$. There is one additional feature: the trace is faithful, which is to say that $\mathrm{Tr}(x^*x) = 0$ only if $x = 0$.

Concerning $M_n(\mathbb{C})$ as an operator algebra, it is of course a von Neumann algebra and its centre, $Z(M_n(\mathbb{C}))$ is trivial in the sense that $Z(M_n(\mathbb{C})) = \{\lambda 1 \mid \lambda \in \mathbb{C}\}$, where, for any ring R, the *centre of R* is the abelian subring $Z(R)$ of R defined by

$$Z(R) = \{x \in R \mid xy = yx, \ \forall y \in R\}.$$

One additional feature of $M_n(\mathbb{C})$ is that its only ideals are $\{0\}$ and $M_n(\mathbb{C})$, making it a *simple algebra*.

The purpose of this section is to introduce and examine operator algebras that exhibit these same algebraic and functional-analytic properties that are present in the matrix algebra $M_n(\mathbb{C})$. The first fundamental concept is that of a trace.

Definition 11.97. A state τ on a C*-algebra A is a *trace* if $\tau(xy) = \tau(yx)$, for all $x, y \in A$. If, in addition, $\tau(x^*x) = 0$ only if $x = 0$, then τ is said to be a *faithful trace*.

Example 11.98. *If G is a countable discrete group, then the group von Neumann algebra $V_\lambda(G)$ has a faithful trace.*

Proof. Let $\xi \in \ell^2(G)$ be the unit vector $\xi = \delta_e$, where e is the identity of G, and define $\tau : V_\lambda(G) \to \mathbb{C}$ by $\tau(x) = \langle x\xi, \xi \rangle$. Thus, τ is a state on $V_\lambda(G)$, and it remains to show that τ is a faithful trace.

If $g, h \in G$, then $\lambda_g \lambda_h[\delta_e] = \delta_{gh}$, and so $\langle \lambda_g \lambda_h[\delta_e], \delta_e \rangle = \langle \delta_{gh}, \delta_e \rangle$, which is nonzero if and only if $gh = e$. But $gh = e$ if and only if $hg = e$; therefore, $\langle \lambda_h \lambda_g[\delta_e], \delta_e \rangle = \langle \lambda_g \lambda_h[\delta_e], \delta_e \rangle$. Hence, if $x, y \in V_\lambda(G)$ are elements of the group algebra $\mathbb{C}[G]$, then $\tau(xy) = \tau(yx)$. Because multiplication in a Banach algebra is norm continuous (Exercise 9.45), we deduce that $\tau(xy) = \tau(yx)$ for all $x, y \in C_\lambda^*(G)$.

To show that $\tau(xy) = \tau(yx)$ for $x, y \in V_\lambda(G)$, it is sufficient to assume that x and y are hermitian and that $\|x\| = \|y\| = 1$ (because the hermitian operators of $V_\lambda(G)$ span $V_\lambda(G)$). By the Density Theorem, there are nets $\{x_\alpha\}_\alpha$ and $\{y_\beta\}_\beta$ of hermitian operators in the unit ball of $C^*_\lambda(G)$ that are SOT-convergent to x and y, respectively. In particular, with the unit vector ξ and making use of the fact that $\langle y_\beta \xi, x_\alpha \xi \rangle = \pi(x_\alpha y_\beta) = \tau(y_\beta x_\alpha) = \langle x_\alpha \xi, y_\beta \xi \rangle$ for all α and β, we see that

$$|\tau(xy) - \tau(yx)| \leq 2 \left(\|x\xi - x_\alpha \xi\| + \|y\xi - y_\beta \xi\| \right),$$

and so $|\tau(xy) - \tau(yx)| = 0$. Thus, τ is a trace on $V_\lambda(G)$.

Suppose now that $x \in V_\lambda(G)$ satisfies $\tau(x^*x) = 0$. Thus, $0 = \langle x^*x\delta_e, \delta_e \rangle = \|x\delta_e\|^2$. Select any other orthonormal basis vector, say δ_g. Thus, $\delta_g = \delta_{eg} = \rho(g)[\delta_e]$, where $\rho : G \to B\left(\ell^2(G)\right)$ is the right regular representation. Because $\rho_g \in V_\lambda(G)'$, we have that $x\delta_g = x\rho_g\delta_e = \rho_g x\delta_e = 0$. Hence, x maps every orthonormal basis vector δ_g to 0, which proves that $x = 0$. □

The proof employed in Example 11.98 shows how a certain type of trace functional on a C*-algebra A extends to a trace on $\overline{A}^{SOT}$. This fact is recorded below for later reference.

Proposition 11.99. *If ξ is a unit cyclic vector for a unital C*-algebra A acting on a Hilbert space H, and if the state $\tau : B(H) \to \mathbb{C}$ defined by $\tau(x) = \langle x\xi, \xi \rangle$ is a trace on A, then τ is also a trace on $M = \overline{A}^{SOT}$.*

Proposition 11.99 is particularly useful in the following form.

Corollary 11.100. *If (H_π, π, ξ) is a GNS-triple for a trace τ on a unital C*-algebra A, then there is a trace τ_M on the von Neumann algebra $M = \overline{\pi(A)}^{SOT}$ such that $\tau = \tau_M \circ \pi$.*

Again motivated by the situation with the matrix algebra $M_n(\mathbb{C})$, von Neumann algebras with trivial centres are of particular interest.

Definition 11.101. A von Neumann algebra M is a *factor* if $Z(M) = \{\lambda 1 \,|\, \lambda \in \mathbb{C}\}$.

The von Neumann algebra $B(H)$ is a factor (Exercise 11.141), and it is known, for infinite discrete groups G, that $V_\lambda(G)$ is a factor if and only if the conjugacy class $\{h^{-1}gh \,|\, h \in G\}$ of $g \in G$ is infinite for every $g \neq e$. (The free group $\mathbb{F}_n$ on $n \geq 2$ generators is one easy example of a group that has infinite conjugacy classes.) Another manner in which factors are obtained is through the tracial state space of a C*-algebra (Proposition 11.103 below).

Definition 11.102. Assume that A is a unital C*-algebra.

1. The set $T(A)$ of all traces on A is called the *tracial state space* of A.
2. If (H_π, π, ξ) is a GNS-triple for a trace $\tau \in T(A)$, and if $\overline{\pi(A)}^{SOT}$ is a factor, then τ is said to be a *factorial trace*.

It may happen that a C*-algebra does not admit a trace (for example, $B(H)$ does not, if H has infinite dimension), or it may admit a continuum of traces.

Proposition 11.103. *If A is a unital C*-algebra, then the tracial state space $T(A)$ is weak*-compact and convex. Moreover, every extreme point of $T(A)$ is a factorial trace.*

Proof. The assertion that $T(A)$ is weak*-compact and convex is left as Exercise 11.142.

Suppose that τ is an extreme point of $T(A)$, and suppose that (H_π, π, ξ) is a GNS-triple for τ, and let $M = \overline{\pi(A)}^{SOT}$. Assume, contrary to what we aim to prove, that τ is not factorial. Thus, the centre $Z(M)$ of M is nontrivial. Because $Z(M)$ is an abelian von Neumann algebra and is nontrivial, there is a projection $p \in Z(M)$ such that $p \neq 0$ and $p \neq 1$. If $\xi_1 = p\xi$, then because $\{\pi(x)\xi \,|\, x \in A\}$ is dense in H_π, and because $\pi(x)\xi_1 = \pi(x)p\xi = p\pi(x)\xi$ for all $x \in A$, it cannot be that $\xi_1 = 0$. Similarly, if $\xi_2 = (1-p)\xi_1$, then $\xi_2 \neq 0$. Thus, $\|\xi_1\|^2 = 1 - \|\xi_2\|^2 < 1$.

For each $j \in \{1,2\}$, let $\eta_j = \|\xi_j\|^{-1}\xi_j$ and $\tau_j(x) = \langle \pi(x)\eta_j, \eta_j \rangle$, for $x \in A$. Thus, τ_j is a state on A. Since $p \in Z(M) \subseteq M = \overline{\pi(A)}^{SOT} = \overline{\pi(A)}^{WOT}$, there is a net $\{x_\alpha\}_\alpha$ of hermitian elements in A such that $\{\pi(x_\alpha)\}_\alpha$ is convergent to p in the weak operator topology of $B(H_\pi)$. Therefore, for fixed $x, y \in A$, we have that

$$\langle \pi(xy)\xi_1, \xi_1 \rangle = \langle p\pi(xy)\xi, \xi \rangle = \lim_\alpha \langle \pi(x_\alpha)\pi(xy)\xi, \xi \rangle = \lim_\alpha \langle \pi(x_\alpha xy)\xi, \xi \rangle$$

$$= \lim_\alpha \tau(x_\alpha xy) = \lim_\alpha \tau(yx_\alpha x) = \lim_\alpha \langle \pi(x_\alpha)\pi(x)\xi, \pi(y)^*\xi \rangle$$

$$= \langle p\pi(x)\xi, \pi(y)^*\xi \rangle = \langle \pi(y)p\pi(x)\xi, \xi \rangle = \langle p\pi(yx)\xi, \xi \rangle$$

$$= \langle \pi(yx)\xi_1, \xi_1 \rangle.$$

Hence, τ_1 is a trace on A. Similarly, $\tau_2 \in T(A)$.

Let $s_j = \|\xi_j\|^2 \in (0,1)$ so that $s_1 + s_2 = 1$ and $\tau = s_1\tau_1 + s_2\tau_2$. Because each $\tau(x_\alpha) = \langle \pi(x_\alpha)\xi, \xi \rangle$, the net $\{\tau(x_\alpha)\}_\alpha$ is convergent to $\langle p\xi, \xi \rangle = \|\xi_1\|^2 < 1$, whereas $\tau_1(x_\alpha) = \|\xi_1\|^{-2}\langle \pi(x_\alpha)\xi_1, \xi_1 \rangle$ implies that net $\{\tau_1(x_\alpha)\}_\alpha$ is convergent to $\|\xi_1\|^{-2}\langle p\xi_1, \xi_1 \rangle = \|\xi_1\|^{-2}\|\xi_1\|^2 = 1$. Therefore, $\tau_1 \neq \tau$, in contradiction to the hypothesis that τ is an extreme point of $T(A)$. Hence, it must be that $Z(M)$ is trivial, which is to say that τ is factorial. □

With the theory and examples developed to this point, we conclude this chapter with the construction of a unital simple C*-algebra with a unique trace.

Let $\tau_n : M_{2^n}(\mathbb{C}) \to \mathbb{C}$ be given by $\tau\left([\alpha_{ij}]_{i,j=1}^n\right) = \frac{1}{2^n} \sum_{j=1}^n \alpha_{jj}$. Thus, τ_n is a trace on $M_{2^n}(\mathbb{C})$ and basic linear algebra shows that it is the only element of the tracial state space $T(M_{2^n}(\mathbb{C}_n))$. Denote the zero and identity matrices in $M_{2^n}(\mathbb{C})$ by 0_n and 1_n, respectively.

For $n \in \mathbb{N}$, let $\vartheta_n : M_{2^n}(\mathbb{C}) \to M_{2^{n+1}}(\mathbb{C})$ be defined by

$$\vartheta_n(x) = x \oplus x,$$

and note that ϑ_n is an injective $*$-homomorphism and is trace preserving in the sense that $\tau_n = \tau_{n+1} \circ \vartheta_n$. Consider now

$$B = \{(x_n)_{n \in \mathbb{N}} \,|\, x_n \in M_{2^n}(\mathbb{C}) \text{ and } \sup_n \|x_n\| < \infty\}$$

and

$$J = \{(x_n)_{n \in \mathbb{N}} \in B \,|\, \lim_n \|x_n\| = 0\}.$$

The set B is clearly a unital C^*-algebra with respect to the algebraic operations and involution induced by each coordinate, and the norm $\|(x_n)_n\| = \sup_n \|x_n\|$, while J is evidently an ideal of B. Let $Q = B/J$ and denote the identity of Q by 1 and the canonical quotient $*$-homomorphism $B \to Q$ by q. For each $n \in \mathbb{N}$, let $\iota_n : M_{2^n}(\mathbb{C}_n) \to B$ and $\pi_n : M_{2^n}(\mathbb{C}_n) \to Q$ be defined by

$$\iota_n(x) = (0_1, \ldots, 0_{n-1}, x, \vartheta_n(x), \vartheta_{n+1} \circ \vartheta_n(x), \vartheta_{n+2} \circ \vartheta_{n+1} \circ \vartheta_n(x), \ldots)$$

and

$$\pi_n = q \circ \iota_n.$$

Both ι_n and π_n are injective $*$-homomorphisms, and $\pi_n(1_n) = 1$ (see Exercise 11.143). Let $A_n = \pi_n(M_{2^n}(\mathbb{C}_n))$, which is a finite-dimensional unital C^*-subalgebra of Q, and observe that $A_n \subseteq A_{n+1}$ for every n (Exercise 11.143). Let

$$A_0 = \bigcup_{n \in \mathbb{N}} A_n \quad \text{and} \quad A = \overline{A_0}.$$

The unital C^*-subalgebra A of Q above is called the *Fermion C^*-algebra*.

Proposition 11.104. *The Fermion C^*-algebra is a simple unital C^*-algebra with trivial centre and a unique trace τ. Furthermore, the trace τ is faithful and factorial.*

Proof. Let $K \subseteq A$ be an ideal of A such that $K \neq \{0\}$. If $K \cap A_n = \{0\}$ for every $n \in \mathbb{N}$, then because $\bigcup_{n \in \mathbb{N}} (A_n \cap K)$ is dense in K, we would have $K = \{0\}$; therefore, it must be that $K \cap A_n \neq \{0\}$, for some n. Because $A_n \cong M_{2^n}(\mathbb{C})$ is simple, we deduce that $K \cap A_n = A_n$. However, as $1 \in A_n$, we also have that $1 \in K$, and so $K = A$, which proves that A is a simple C^*-algebra. The fact that $Z(A) = \{\lambda 1 \,|\, \lambda \in \mathbb{C}\}$ is left as Exercise 11.143.

If $x \in A_n$ for some n, then $x = \pi_n(a)$ for some matrix $a \in M_{2^n}(\mathbb{C})$, and we may define the trace $\tau_0(x)$ of x by $\tau_n(a)$. If $k > n$, then there is a unique $b \in M_{2^k}(\mathbb{C})$ with $x = \pi_k(b)$, namely $b = \vartheta_{k-1} \circ \cdots \circ \vartheta_n(a) \in M_{2^k}(\mathbb{C})$. (The uniqueness of b is on account of the injectivity of π_k.) Moreover, the trace $\tau_k(b)$ of b is the same as the trace $\tau_n(a)$ of a. Hence, the function $\tau_0 : A_0 \to \mathbb{C}$ defined by $\tau_0(x) = \tau_n(a)$, if $x = \pi_n(a)$ for some n and some $a \in M_{2^n}(\mathbb{C})$, is well defined, linear, positive, bounded, and satisfies $\tau_0(xy) = \tau_0(yx)$ for all $x, y \in A_0$. Hence, τ_0 extends to a trace τ on A. The uniqueness of τ follows from the fact that A_0 is dense in A and from the fact that each matrix algebra $M_{2^n}(\mathbb{C})$ has a unique tracial state. Therefore, $T(A) = \{\tau\}$, which implies that τ is an extreme point of $T(A)$. Therefore, by Proposition 11.103, τ is a factorial trace.

To show that τ is faithful, consider the set $D = \{x \in A_+ \,|\, \tau(x) = 0\} \subseteq A_+$. If $x \in D$ and $y \in A_+$, then $|\tau(yx)| = |\tau(xy)| = \tau(x^{1/2}yx^{1/2}) \le \|y\|\tau(x) = 0$. Therefore, if $x \in \mathrm{Span}\,D$ and if $y \in A_+$, then it is also true that $|\tau(yx)| = |\tau(xy)| = 0$. Hence, because A_+ spans A, and using the continuity of τ, we obtain $\tau(xy) = 0$ for every $x \in K = \overline{\mathrm{Span}\,D}$ and every $y \in A$. In other words, K is an ideal of A. Evidently $1 \notin K$; thus, by the fact that A is simple, it must be that $K = \{0\}$. Hence, $D = \{0\}$, which proves that τ is a faithful trace. $\qquad\qquad\square$

Corollary 11.105. *If (H_π, π, ξ) is a GNS-triple for the unique faithful trace τ on the Fermion algebra A, then $R = \overline{\pi(A)}^{SOT}$ is a factor and has a faithful trace τ_R such that $\tau_R(\pi(a)) = \tau(a)$, for every $a \in A$.*

Proof. Proposition 11.100 shows the existence of the trace τ_R with the stated property $\tau_R \circ \pi = \tau$, defined by $\tau_R(x) = \langle x\xi, \xi \rangle$ for $x \in R$, while Proposition 11.104 asserts that τ is factorial. All that remains is to show that the trace τ_R is faithful.

Suppose that $x \in R$ satisfies $\tau_R(x^*x) = 0$. Thus, $0 = \langle x^*x\xi, \xi \rangle = \|x\xi\|^2$, and so $x\xi = 0$. For any $a \in A$,

$$\|x\pi(a)\xi\|^2 = \langle x\pi(a)\xi, x\pi(a)\xi \rangle = \tau_R(\pi(a)^*x^*x\pi(a))$$

$$= \tau_R(\pi(a)\pi(a)^*x^*x) = \langle \pi(a)\pi(a)^*x^*x\xi, \xi \rangle$$

$$= 0,$$

as $x\xi = 0$. Thus, x is zero on the dense linear submanifold $\{\pi(a)\xi \,|\, a \in A\}$, and therefore $x = 0$. $\qquad\qquad\square$

Definition 11.106. The factor R described by Corollary 11.105 is called the *hyperfinite II_1-factor*.

Of course, the definition of the hyperfinite II_1-factor R depends on the choice of GNS triple for the trace τ on the Fermion algebra A; however, this is not really an issue because of the uniqueness assertion in the GNS theorem. The distinguishing features of R are that it is a factor, it possesses a faithful trace, and it has a unital separable strongly dense C*-subalgebra attained from an increasing sequence of finite-dimensional factors (that is, simple matrix algebras).

Problems

11.107. Assume that H is a Hilbert space, and consider the C*-algebra $K(H)$ of compact operators.

1. Show that, with respect to the strong operator topology of $B(H)$, the identity operator $1 \in B(H)$ is the limit of a net of compact operators.
2. Show that $K(H)$ is a von Neumann algebra if and only if H has finite dimension.

11.108. Prove that if (X, Σ, μ) is a σ-finite measure space, then $L^\infty(X, \Sigma, \mu)$ is a von Neumann algebra when considered as a C*-algebra of multiplication operators acting on the Hilbert space $L^2(X, \Sigma, \mu)$.

11.109. Assume that X is a compact Hausdorff space and that μ is a finite regular Borel measure on the σ-algebra of Borel sets of X. Consider the following unital operator algebras C and M acting on $L^2(X, \Sigma, \mu)$ as algebras of multiplication operators:

$$C = \{M_f \,|\, f \in C(X)\} \quad \text{and} \quad M = \{M_\psi \,|\, \psi \in \mathscr{L}^\infty(X, \Sigma, \mu)\}.$$

1. Prove that $C' = M$.
2. Prove that $\overline{C}^{SOT} = M$.

11.110. Prove that a $*$-subalgebra $A \subseteq B(H)$ is irreducible if and only if the only projections $p \in B(H)$ that belong to the commutant A' of A are $p = 0$ and $p = 1$.

11.111. Show that the von Neumann algebra generated by the unilateral shift operator on $\ell^2(\mathbb{N})$ is $B(\ell^2(\mathbb{N}))$.

11.112. Show that the von Neumann algebra $L^\infty([0, 1], \mathfrak{M}, m)$ is not irreducible.

11.113. Let A be a unital C*-subalgebra of $B(H)$ and let $M = \overline{A}^{SOT}$.

1. Prove that the matrix operator algebras $M_2(A)$ and $M_2(M)$ acting on $H \oplus H$ satisfy $M_2(N) = \overline{M_2(A)}^{SOT}$.
2. Prove that if $x \in M$ has norm $\|x\| \le 1$, then the hermitian operator matrix $X = \begin{bmatrix} 0 & x \\ x^* & 0 \end{bmatrix}$ has norm $\|X\| \le 1$.
3. Prove that there exists a net of hermitian, contractive operator matrices $(B_\alpha)_{\alpha \in \Lambda}$ in $M_2(A)$ converging to X in the strong operator topology of $M_2(B(H))$.
4. Prove that there exists a net $(b_\alpha)_{\alpha \in \Lambda}$ of contractive operators $b_\alpha \in A$ converging to x in the strong operator topology of $B(H)$.

11.114. Assume that H is an infinite-dimensional separable Hilbert space. Let $B(H)_1$ denote the closed unit ball of $B(H)$, and let $\mathscr{T}_{SOT}$ denote the strong operator topology on $B(H)_1$.

1. Use the Kaplansky Density Theorem to show that $(B(H)_1, \mathcal{T}_{SOT})$ is a separable topological space.
2. Show that $(B(H)_1, \mathcal{T}_{SOT})$ is a metrisable topological space. (Suggestion: select a countable dense subset $\{\xi_n\}_n$ of the closed unit ball of H and show that d, defined by $d(x,y) = \sum_{n=1}^{\infty} \frac{1}{2^n} \|x\xi_n - y\xi_n\|$, is a metric inducing the topology $\mathcal{T}_{SOT}$.)

11.115. Suppose that a normal operator N on a separable Hilbert space has the property that the von Neumann algebra N generates has a cyclic vector. Prove that there exists a regular Borel probability measure on the Borel sets Σ of $\sigma(N)$ and a unital isometric $*$-isomorphism $\rho : W^*(N) \to L^{\infty}(\sigma(N), \Sigma, \mu)$ such that $\rho(N) = \dot{\psi}$, where $\psi : \sigma(N) \to \mathbb{C}$ is given by $\psi(t) = t$.

11.116. If M is a von Neumann algebra acting on a Hilbert space H, and if $a, b \in M_+$, then prove that the following statements are equivalent for $x \in M$:

1. $\begin{bmatrix} a & x \\ x^* & b \end{bmatrix}$ is a positive operator on $H \oplus H$;
2. $x = a^{1/2} y b^{1/2}$ for some $y \in M$ with $\|y\| \leq 1$.

11.117. Prove that a unital abelian C*-algebra A is a semisimple (Definition 9.32).

11.118. Prove that if A is a nonunital C*-algebra, then $0 \in \sigma(x)$, for all $x \in A$.

11.119. Suppose that A is a C*-algebra with norm $\| \cdot \|$. Prove that if $\| \cdot \|'$ is a norm on A that satisfies all of the axioms of a C*-norm, then $\|x\|' = \|x\|$ for all $x \in A$.

11.120. If A is a unital C*-algebra and if $u \in A$ is unitary, then prove that $\sigma(u) \subseteq \partial\mathbb{D}$, where $\mathbb{D}$ is the open unit disc of the complex plane.

11.121. If A is a unital C*-algebra, then prove that the set of unitary elements of A if the form e^{ih} for some hermitian $h \in A$ is a path-connected set.

11.122. If A is a unital C*-algebra and if $x \in A$ (not necessarily normal), then prove or find a counterexample to each of the following statements.

1. $\sigma(x^*) = \{\overline{\lambda} \mid \lambda \in \sigma(x)\}$.
2. $x^* x$ is invertible if x is invertible.
3. x is invertible if $x^* x$ is invertible.

11.123. Suppose that $X \subset \mathbb{R}$ is a compact set such that $0 \in X$. Prove that if $f \in C(X)$ satisfies $f(0) = 0$ and if $\varepsilon > 0$, then there is a polynomial p such that $p(0) = 0$ and $|f(t) - p(t)| < \varepsilon$ for all $t \in X$.

11.124. In a unital Banach algebra A, an element $x \in A$ is *quasinilpotent* if $\sigma(x) = \{0\}$, and x is *properly quasinilpotent* if $\sigma(xy) = \{0\}$ for all $y \in A$. Prove that if A is a unital C*-algebra, then the only properly quasinilpotent element $x \in A$ is $x = 0$.

11.125. If A is a C*-algebra, and if $\gamma \in \mathbb{R}_+$ and $h, k \in A_+$, then prove the following assertions:

1. $\gamma h \in A_+$;
2. $h + k \in A_+$; and
3. $-h \in A_+$ only if $h = 0$.

11.126. Suppose that $a, b, c \in A_{sa}$. Prove the following assertions:

1. $a \leq a$;
2. if $a \leq b$ and $b \leq a$, then $b = a$; and
3. if $a \leq b$ and $b \leq c$, then $a \leq c$.

11.127. Suppose that A is a C*-algebra with $x \in A$, $h \in A_+$, and $xh = hx$. Prove that $xh^{1/2} = h^{1/2}x$.

11.128. If A is a C*-algebra and if $a, b \in A_+$ satisfy $a \leq b$ and $ab = ba$, then prove that $a^2 \leq b^2$. Show by example that $a \leq b$ does not always imply $a^2 \leq b^2$ if $ab \neq ba$.

11.129. If A is a unital C*-algebra, prove that $1 + x^*x$ is invertible for every $x \in A$.

11.130. Prove that $\sigma(ab) \subset \mathbb{R}_+$ for all positive elements a and b in a C*-algebra A.

11.131. Prove that if A is a unital C*-algebra and if $x \in A$ is invertible, then there is a unitary $u \in A$ such that $x = u|x|$.

11.132. Prove that if $h \in A_{sa}$, where A is a unital C*-algebra, and if $\alpha \geq \|h\|$, then

1. $h \leq \alpha 1$, and
2. $\|\alpha 1 - h\| \leq \alpha$.

11.133. Suppose that J is an ideal of a nonunital C*-algebra A. Consider the inclusion of A in its unitisation A^1. Show that J is an ideal of A^1.

11.134. Suppose that J is a proper ideal of a unital C*-algebra A. Define $J + \mathbb{C}1$ by

$$J + \mathbb{C}1 = \{x + \lambda 1 \mid x \in J, \lambda \in \mathbb{C}\}.$$

1. Prove that $J + \mathbb{C}1$ is a unital C*-subalgebra of A.
2. Prove or find a counterexample to the following statement: the C*-algebras J^1 and $J + \mathbb{C}1$ are isometrically isomorphic.

11.135. Prove that if A and B are unital C*-algebras and if $\rho : A \to B$ is a *-homomorphism, then $\sigma(\rho(x)) \subseteq \sigma(x)$, for all $x \in A$.

11.136. Prove that if A and B are C*-algebras and if $\rho : A \to B$ is a *-homomorphism, then $\text{spr}\,\rho(x^*x) \leq \text{spr}\,(x^*x)$, for all $x \in A$. (A and B are not assumed to be unital.)

11.137. Let $q : B(H) \to B(H)/K(H)$ be the unital *-homomorphism in which $q(x) = \dot{x}$, for every $x \in B(H)$. Compute the spectrum of $q(s)$, where $s \in B(H)$ is the unilateral shift operator on $H = \ell^2(\mathbb{N})$.

11.138. Assume that φ is a state on a unital C*-algebra A and that (H_π, π, ξ) and (H_ρ, ρ, η) are GNS-triples for φ. Prove that there is a surjective isometry $u : H_\pi \to H_\rho$ such that $u\xi = \eta$ and $u\pi(x) = \rho(x)u$, for every $x \in A$.

11.139. If $\{\xi_n\}_n$ is a sequence in a Hilbert space H such that $\sum_{n=1}^{\infty} \|x_n\|^2$ converges, and if $\xi_n \neq 0$ for at least one n, then prove that the positive linear functional φ on $B(H)$ defined by

$$\varphi(x) = \sum_{n=1}^{\infty} \langle x\xi, \xi \rangle,$$

has the property that $\varphi(k) \neq 0$ for at least one compact operator $k \in B(H)$.

11.140. Prove that there exists a unique topology $\mathscr{T}$ on Prim$A$ in which a subset $\mathscr{F}$ of Prim$A$ is closed if and only if

$$\mathscr{F} = \{J \in \mathrm{Prim}A \mid \bigcap_{I \in \mathscr{F}} I \subseteq J\}.$$

11.141. Prove that $B(H)$ is a factor.

11.142. Prove that the tracial state space $T(A)$ of a unital C*-algebra A is weak*-compact and convex.

11.143. Let $J = \{(x_n)_{n \in \mathbb{N}} \in B \mid \lim_n \|x_n\| = 0\}$, where

$$B = \{(x_n)_{n \in \mathbb{N}} \mid x_n \in M_{2^n}(\mathbb{C}) \text{ and } \sup_n \|x_n\| < \infty\},$$

and let $Q = B/J$. Denote the quotient map $B \to Q$ by q and the identity elements of $M_{2^n}(\mathbb{C})$ and Q by 1_n and 1, respectively. Let $\vartheta_n : M_{2^n}(\mathbb{C}) \to M_{2^{n+1}}(\mathbb{C})$ be defined by $\vartheta_n(x) = x \oplus x$.

1. Prove that $\pi_n : M_{2^n}(\mathbb{C}) \to Q$ is an isometric *-homomorphism such that $\pi_n(1_n) = 1$, where $\pi_n = q \circ \iota_n$ and $\iota_n : M_{2^n}(\mathbb{C}) \to B$ is defined by

$$\iota_n(x) = (0_1, \ldots, 0_{n-1}, x, \vartheta_n(x), \vartheta_{n+1} \circ \vartheta_n(x), \ldots).$$

2. If $A_n = \overline{\pi_n(M_{2^n}(\mathbb{C}))}$ for each $n \in \mathbb{N}$, then prove that $A_n \subseteq A_{n+1}$.
3. If $A = \overline{\bigcup_{n \in \mathbb{N}} A_n}$, then prove that the centre $Z(A)$ of A is $Z(A) = \{\lambda 1 \mid \lambda \in \mathbb{C}\}$.

11.144. Prove that the trace τ_R on the hyperfinite II$_1$-factor R is unique.

References

1. Akemann, C., Anderson, J., & Pedersen, G. K. (1982). A triangle inequality in operator algebras. *Linear Multilinear Algebra, 11*, 1679–178.
2. Ando, T. (1963). Note on invariant subspaces of a compact normal operator. *Archiv der Mathematik, 14*, 337–340.
3. Arveson, W. B. (1976). *An invitation to C*-algebra*. New York: Springer-Verlag.
4. Arveson, W. B. (2002). *A short course in spectral theory*. New York: Springer-Verlag.
5. Alexander, H., & Wermer, J. (1998). *Several complex variables and Banach algebras*. New York: Springer-Verlag.
6. Bartle, R. G. (1966). *The elements of integration*. New York: John Wiley & Sons, Inc.
7. Bourdon, P. S. (2016). Spectra of composition operators with symbols in $\mathscr{S}(2)$. *Journal of Operator Theory, 75*, 321–348.
8. Brown, J. W., & Churchill, R. V. (2009). *Complex variables and applications*. New York: McGraw-Hill, Inc.
9. Chaktoura, M. (2013). Personal communication.
10. Cohn, D. L. (2013). *Measure theory* (2nd ed.). New York: Birkhäuser.
11. Conway, J. B. (1972). The compact operators are not complemented in $B(H)$. *Proceedings of the American Mathematical Society, 32*, 549–550.
12. Conway, J. B. (1985). *A course in functional analysis*. New York: Springer-Verlag.
13. Conway, J. B. (2000). *A course in operator theory*. Providence: American Mathematical Society.
14. Davidson, K. R., & Donsig, A. P. (2002). *Real analysis with real applications*. Upper Saddle River, NJ: Prentice Hall, Inc.
15. Davies, E. B. (1976). *Quantum theory of open systems*. New York: Academic Press.
16. Davies, E. B. (2007). *Linear operators and their spectra*. Cambridge: Cambridge University Press.
17. Davis, C. (1963). Notions generalizing convexity for functions defined on spaces of matrices. *Proceedings of Symposia in Pure Mathematics, 7*, 1987–201.
18. DiBenedetto, E. (2002). *Real analysis*. Boston: Birkhäuser.
19. Douglas, R. G. (1964). On extremal measures and subspace density. *Michigan Mathematical Journal, 11*, 243–246.
20. Douglas, R. G. (1998). *Banach algebra techniques in operator theory* (2nd ed.). New York: Springer-Verlag.
21. Dugundji, J. (1966). *Topology*. Boston: Allyn and Bacon, Inc.
22. Dunford, N., & Schwartz, J. T. (1957). *Linear operators, Part I*. New York: John Wiley & Sons, Inc.

© Springer International Publishing Switzerland 2016

D. Farenick, *Fundamentals of Functional Analysis*, Universitext,
DOI 10.1007/978-3-319-45633-1

23. Duren, P. (2012). *Invitation to classical analysis*. Providence: American Mathematical Society.
24. Enflo, P. (1973). A counterexample to the approximation problem in Banach spaces. *Acta Mathematica, 130*, 309–317.
25. Fabian, M., Habala, P., Hájek, P., Montesinos, V., & Zizler, V. (2011). *Banach space theory*. New York: Springer-Verlag.
26. Farenick, D. R. (2000). *Algebras of linear transformations*. New York: Springer-Verlag.
27. Farenick, D., Kavruk, A., & Paulsen, V. I. (2013). C^*-algebras with the weak expectation property and a multivariable analogue of Ando's theorem on the numerical radius. *Journal of Operator Theory, 70*, 573–590.
28. Gillman, L., & Jerison, M. (1960). *Rings of continuous functions*. New York: Van Nostrand.
29. Gohberg, I. C., & Kreĭn, M. G. (1969). *Introduction to the theory of linear nonselfadjoint operators*. Providence: American Mathematical Society.
30. Hadwin, D., & Ma, X. (2008). A note on free products. *Operators and Matrices, 2*, 53–65.
31. Halmos, P. R. (1957). *Introduction to Hilbert space* (2nd ed.). New York: Chelsea.
32. Halmos, P. R. (1982). *A Hilbert space problem book* (2nd ed.). New York: Springer-Verlag.
33. Herstein, I. M. (1964). *Topics in algebra*. New York: Blaisdell.
34. Enflo, P. (1970). Atomic and nonatomic measures. *Proceedings of the American Mathematical Society, 25*, 650–655.
35. Kadison, R. V., & Ringrose, J. R. (1983). *Fundamentals of the theory of operator algebras* (Vol. I). New York: Academic Press.
36. Katznelson, Y. (2004). *An introduction to harmonic analysis* (3rd ed.). Cambridge: Cambridge University Press.
37. Lang, S. (1993). *Real and functional analysis* (3rd ed.). New York: Springer-Verlag.
38. Lindenstrauss, J. (1966). A short proof of Liapounoff's convexity theorem. *Journal of Mathematics and Mechanics, 15*, 971–972.
39. McCann, S. J. (2008). *Equivalent forms of the Connes embedding problem*. M.Sc. Thesis, University of Regina.
40. Michaels, A. J. (1977). Hilden's simple proof of Lomonosov's invariant subspace theorem. *Advances in Mathematics, 25*, 56–58.
41. Munkres, J. R. (2000). *Topology* (2nd ed.). Upper Saddle River, NJ: Prentice Hall, Inc.
42. Paulsen, V. I. (2001). *Completely bounded maps on operator algebras*. Cambridge: Cambridge University Press.
43. Phelps, R. R. (2001). *Lectures on Choquet theory* (2nd ed.). Lecture Notes in Mathematics, Vol. 1757. New York: Springer-Verlag.
44. Radjavi, H., & Rosenthal, P. (1973). *Invariant subspaces*. New York: Springer-Verlag.
45. Raeburn, I., & Williams, D. P. (1998). *Morita equivalence and continuous-trace C^*-algebras*. Mathematical Surveys and Monographs, Vol. 60. Providence: American Mathematical Society.
46. Read, C. J. (1986). A short proof concerning the invariant subspace problem. *Journal of the London Mathematical Society, 34*, 335–348.
47. Rosenholtz, I. (1976). Another proof that any compact metric space is the continuous image of the Cantor set. *American Mathematical Monthly, 83*, 646–647.
48. Rosenthal, P. (1968). Completely reducible operators. *Proceedings of the American Mathematical Society, 19*, 826–830.
49. Rudin, W. (1964). *Principles of mathematical analysis*. New York: McGraw-Hill, Inc.
50. Rudin, W. (1987). *Real and complex analysis* (3rd ed.). New York: McGraw-Hill, Inc.
51. Rudin, W. (1991). *Functional analysis* (2nd ed.). New York: McGraw-Hill, Inc.
52. Stein, E. M., & Shakarchi, R. (2005). *Real analysis*. Princeton: Princeton University Press.
53. Stein, E. M., & Shakarchi, R. (2011). *Functional analysis*. Princeton: Princeton University Press.
54. Takesaki, M. (1979). *Theory of operator algebras, I*. New York: Springer-Verlag.
55. Thompson, R. C. (1976). Convex and concave functions of singular values of matrix sums. *Pacific Journal of Mathematics, 66*, 285–290.

Index

© Springer International Publishing Switzerland 2016
D. Farenick, *Fundamentals of Functional Analysis*, Universitext,
DOI 10.1007/978-3-319-45633-1

Printed in the United States
By Bookmasters